Lecture Notes in Computer Science 7918

Commenced Publication in 1973
Founding and Former Series Editors:
Gerhard Goos, Juris Hartmanis, and Jan van Leeuwen

Amr Youssef   Abderrahmane Nitaj
Aboul Ella Hassanien (Eds.)

Progress in Cryptology – AFRICACRYPT 2013

6th International Conference on Cryptology in Africa
Cairo, Egypt, June 22-24, 2013
Proceedings

 Springer

Volume Editors

Amr Youssef
Concordia University
Concordia Institute for Information Systems Engineering
1515 St. Catherine Street West, Montreal, QC, H3G 2W1, Canada
E-mail: youssef@ciise.concordia.ca

Abderrahmane Nitaj
Université de Caen Basse-Normandie
Laboratoire de Mathématiques Nicolas Oresme
BP 5186, 14032 Caen, France
E-mail: abderrahmane.nitaj@unicaen.fr

Aboul Ella Hassanien
Cairo University, Department of Information Technology
5 Dr. Ahmed Zewail Street, 12613 Cairo, Giza, Egypt
E-mail: aboitcairo@fci-cu.edu.eg

ISSN 0302-9743   e-ISSN 1611-3349
ISBN 978-3-642-38552-0   e-ISBN 978-3-642-38553-7
DOI 10.1007/978-3-642-38553-7
Springer Heidelberg Dordrecht London New York

Library of Congress Control Number: 2013938371

CR Subject Classification (1998): E.3, K.6.5, E.4, K.4.4, G.1, F.1, F.2, C.2.0

LNCS Sublibrary: SL 4 – Security and Cryptology

© Springer-Verlag Berlin Heidelberg 2013

Typesetting: Camera-ready by author, data conversion by Scientific Publishing Services, Chennai, India

Printed on acid-free paper

Springer is part of Springer Science+Business Media (www.springer.com)

Preface

This volume contains the papers accepted for presentation at Africacrypt 2013, the 6th International Conference on the Theory and Application of Cryptographic Techniques in Africa. The aim of this series of conferences is to provide an international forum for practitioners and researchers from industry, academia, and government agencies from all over the world for a wide-ranging discussion of all forms of cryptography and its applications.

The initiative of organizing Africacrypt started in 2008 where it was first held in Morocco. Subsequent yearly events were held in Tunisia, South Africa, Senegal, and Morocco. This year, on the initiative of the organizers from Cairo University, Africacrypt 2013, which is organized in cooperation with the International Association for Cryptologic Research (IACR), was held in the conference center of Cairo University, Egypt, during June 22–24.

We received 77 submissions authored by researchers from 26 different countries. After a reviewing process that involved 36 Technical Program Committee members from 18 countries and 74 external reviewers, the Technical Program Committee went through a significant online discussion phase before deciding to accept 26 papers. We are indebted to the members of the Program Committee and the external reviewers for their diligent work and fruitful discussions. We are also grateful to the authors of all submitted papers for supporting the conference. The authors of accepted papers are thanked again for revising their papers according to the suggestions of the reviewers. The revised versions were not checked again by the Program Committee, so authors bear full responsibility for their content.

Besides the peer-reviewed accepted papers, the technical program included two invited talks by Taher Elgamal and Martin Schläffer.

In Dr. Schläffer's talk, he gave an introduction to the cryptanalysis of hash functions and discussed the main idea of the attacks by Wang et al. He also presented new design ideas submitted to the NIST SHA-3 competition, discussed some simple attacks on weak submissions, and highlighted common pitfalls made. Dr. Martin Schläffer also presented Keccak (SHA-3) and new cryptanalysis results on SHA-2 and SHA-3.

Finally, we would like to thank everyone who contributed to the success of this conference. The local Organizing Committee from Cairo University were always a pleasure to work with. We are deeply thankful that they hosted Africacrypt 2013. We are also thankful to the staff at Springer for their help with producing the proceedings.

April 2013

Amr Youssef
Abderrahmane Nitaj
Aboul Ella Hassanien

Organization

Honorary Chair

Mohamed Fahmy Tolba   Ain Shams University, Egypt

General Chair

Aboul Ella Hassanien   Cairo University, Egypt

Program Chairs

Abderrahmane Nitaj   University of Caen, France
Amr Youssef   Concordia University, Canada

Publicity Chairs

Ali Ismail Awad   Scientific Research Group, Egypt
Ahmad Taher Azar   Scientific Research Group, Egypt
Nashwa El Bendary   Scientific Research Group, Egypt

Local Organizing Committee

Ali Ismail Awad (Chair)   Scientific Research Group, Egypt
Neveen Ghali   Scientific Research Group, Egypt
Nashwa El Bendary   Scientific Research Group, Egypt
Mostafa Salama   Scientific Research Group, Egypt
Mohamed Mostafa   Scientific Research Group, Egypt
Heba Eid   Scientific Research Group, Egypt
Kareem Kamal   Scientific Research Group, Egypt
Mohamed Tahoun   Scientific Research Group, Egypt

Program Committee

Roberto Avanzi   Qualcomm Research, Germany
Abdelhak Azhari   ENS - Casablanca, Hassan II University, Morocco
Hatem M. Bahig   Ain Shams University, Egypt
Hussain Benazza   Ensam-Meknès, Moulay Ismail University, Morocco

Chattopadhyay, Anupam
Chen, Jiageng
Dent, Alexander
Duc, Alexandre
Dunkelman, Orr
El Mrabet, Nadia
Elkadi, Mohamed
Gama, Nicolas
Gangopadhyay, Sugata
Grosso, Vincent
Guillot, Philippe
Heen, Olivier
Henry, Ryan
Hermans, Jens
Herranz, Javier
Jambert, Amandine
Jhanwar, Mahabir P.
Järvinen, Kimmo
Karabudak, Ersin
Kawachi, Akinori
Keller, Marcel
Kircanski, Aleksandar
Kurosawa, Kaoru
Laguillaumie, Fabien
Lange, Tanja
Le Duc, Phong
Le, Duc-Phong
Leander, Gregor
Mahdy, Riham
Maitra, Arpita
Mandal, Kalikinkar
Manulis, Mark

Matsuda, Takahiro
Moradi, Amir
Naya-Plasencia, María
Nojoumian, Mehrdad
Ohkubo, Miyako
Otmani, Ayoub
Pandit, Tapas
Peters, Christiane
Prouff, Emmanuel
Regazzoni, Francesco
Renault, Guenael
Salvail, Louis
Sarkar, Santanu
Schwabe, Peter
Shahandashti, Siamak
Sica, Francesco
Soleimany, Hadi
Takashima, Katsuyuki
Tan, Yin
Toz, Deniz
Upadhyay, Jalaj
van de Pol, Joop
Vaudenay, Serge
Vergnaud, Damien
Veyrat-Charvillon, Nicolas
Wang, Pengwei
Yamaguchi, Teruyoshi
Yanagihara, Shingo
Yang, Guoming
Zhang, Liangfeng
Zhu, Bo

Table of Contents

Adapting Lyubashevsky's Signature Schemes to the Ring Signature Setting

Carlos Aguilar Melchor[1], Slim Bettaieb[1], Xavier Boyen[2], Laurent Fousse[3], and Philippe Gaborit[1]

[1] XLIM-DMI, Université de Limoges, France
{carlos.aguilar,slim.bettaieb,philippe.gaborit}@xlim.fr
[2] Queensland University of Technology, Brisbane, Australia
xb@boyen.org
[3] Laboratoire Jean-Kuntzmann, Université de Grenoble, France
laurent.fousse@imag.fr

Abstract. Basing signature schemes on strong lattice problems has been a long standing open issue. Today, two families of lattice-based signature schemes are known: the ones based on the hash-and-sign construction of Gentry et al.; and Lyubashevsky's schemes, which are based on the Fiat-Shamir framework.

In this paper we show for the first time how to adapt the schemes of Lyubashevsky to the ring signature setting. In particular we transform the scheme of ASIACRYPT 2009 into a ring signature scheme that provides strong properties of security under the random oracle model. Anonymity is ensured in the sense that signatures of different users are within negligible statistical distance even under full key exposure. In fact, the scheme satisfies a notion which is stronger than the classical full key exposure setting as even if the keypair of the signing user is adversarially chosen, the statistical distance between signatures of different users remains negligible.

Considering unforgeability, the best lattice-based ring signature schemes provide either unforgeability against arbitrary chosen subring attacks or insider corruption in log-sized rings. In this paper we present two variants of our scheme. In the basic one, unforgeability is ensured in those two settings. Increasing signature and key sizes by a factor k (typically $80 - 100$), we provide a variant in which unforgeability is ensured against insider corruption attacks for arbitrary rings. The technique used is pretty general and can be adapted to other existing schemes.

Keywords: Ring signatures, lattices.

1 Introduction

In 2001 Rivest, Shamir and Tauman [1] introduced the concept of ring signature. In such a scheme, each user has a keypair; a secret signing key and a public verification key. Any of them can choose a subset of the public keys (including his own), the ring, and sign on behalf of the associated subset of users, without

A. Youssef, A. Nitaj, A.E. Hassanien (Eds.): AFRICACRYPT 2013, LNCS 7918, pp. 1–25, 2013.
© Springer-Verlag Berlin Heidelberg 2013

permission or assistance. This signature can be verified using the ring of public keys. Such a scheme must have the classic unforgeability property (it is not possible to sign on behalf of a ring without knowing one of the associated secret keys) as well as an anonymity property: it is not possible to know which secret key was used, just that it is associated to one of the public keys in the ring.

In 2006 [2], Bender et al. noted that the usual security setting with respect to anonymity and unforgeability did not take into account that some of the public keys in the ring may have been issued adversarially by an attacker. They therefore proposed adapted security definitions for such a situation as well as for the one in which all the secret keys including the one of the signers would be exposed. Of course if all the keys are exposed it is no longer possible to ensure unforgeability but anonymity of previously issued signatures may be preserved.

Such a strong anonymity property is a reassuring guarantee for a user hesitating to leak a secret, specially if the consequences of an identification are dire. It also seems reasonable, specially if anonymity must be preserved for a few decades (e.g. depending on the statute of limitations) not to rely on an estimation of the computational complexity of a given problem and require unconditional anonymity. The definitions of Bender et al. can be easily translated to the unconditional setting and thus cover such a requirement.

A close but different setting, introduced by Chaum and van Heyst in 1991 [3], is the one of group signature. In this setting there is an anonymity revocation mechanism that allows a given group manager to reveal who was the signer of a given message. This property comes however at a cost as it requires a group setup procedure which is not needed in the ring signature setting.

Most of the existing ring signature schemes are based on number theory assumptions: large integer factorization [4,1], discrete logarithm problem [5,6] and bilinear pairing problems [7,8,9].

There are also a few ring signature schemes with security based on standard lattice problems. In [10], Brakerski and Kalai propose a ring signature scheme in the standard model based on the Small Integer Solution (SIS) problem using the hash-and-sign/bonsai-tree [11,12] approach. Using again this approach Wang and Sun propose in [13] two other ring signature schemes, one under the random oracle model and one on the standard model. Both papers provide constructions in the standard model but, on the other hand, use Gentry-Peikert-Vaikuntanathan's (hereafter GPV) [11] strong trapdoors. These trapdoors are known to be very versatile and the ring signature constructions come naturally. However, when using these strong trapdoors a (hidden) structure is added to the underlying lattice which is, from a theoretical point of view, an important price to pay. In practice, it is possible to give public bases of these lattices which are closer to uniform, but some parameters, namely the dimension of the "perp" lattice (for a definition and details see [14]), are increased significantly in the process.

Two other schemes, one by Kawachi et al. [15], and one by Cayrel et al. [16], follow a very different approach. These schemes are based on weak trapdoors in which the underlying lattice is completely uniform except for some public syndromes given (which corresponds to a small vectors in the "perp" lattice). Even if we consider the lattice together with the syndromes, very little structure is added and in practice, it is close to uniform for smaller parameters than with GPV trapdoors. Such "weak trapdoors" can be used in not that many contexts (when compared to the strong GPV trapdoors) and [15,16] prove that they are enough for ring signature. On the other hand, these schemes use pretty straight-forwardly Stern's construction for signature schemes [17] which is somehow a bad property. Of course, it is clearly important to note that the code-based constructions can be translated to the lattice setting. In the case of the scheme by Cayrel et al. this has the added benefit that by adapting the code-based construction of Aguilar et al. [18] they obtain a threshold scheme, which is a pretty hard feature to obtain. However, we believe that finding ring signatures which follow the more recent and promising techniques of lattice-based cryptography, without the cumbersome zero-knowledge proofs used in code-based cryptography, is of independent interest.

Our Contributions. In this paper we present a lattice-based ring signature algorithm. As in the other lattice-based ring signature schemes, each signature and verification key is composed of a linear amount of sub-elements in the ring size.

Our main contribution is that our scheme is the first one to be based on Lyubashevsky's approach to lattice-based signature [19,20,21]. This is interesting from a theoretical point of view for two reasons. First, it is one of the major approaches to build standard signatures and no ring signature was until now based on it. Second, as Lyubashevsky's signatures, our scheme uses a weak trapdoor (a uniformly random lattice with a single syndrome) *without* Stern's zero-knowledge proofs (previous lattice-based schemes used either GPV strong trapdoors or Stern's proofs).

We describe our scheme as a modification of the scheme of ASIACRYPT 2009 [20]. The ideas and proofs can be re-used for the more recent schemes of [21], but the whole presentation gets trickier and the practical benefits are of limited interest (unlike in the standard signature setting).

As a second contribution we present a modification of our scheme, which can be applied to other lattice-based ring signature schemes to provide unforgeability in the insider corruption setting, even if the ring is of polynomial size in the security parameter. To the best of our knowledge this is the first time such a security property is obtained with lattice-based schemes.

Indeed, the schemes of Kawachi et al., Cayrel et al. and Brakerski and Kalai only provide a proof for the fixed ring or chosen subring attack settings. Wang and Sun provide a proof for insider corruption which only works for log-sized rings. More precisely, the advantage of the SIS attacker in their proof is in $O(1/\binom{q_E}{q_E/2})$, q_E being an upper-bound on the ring size (see first line of the Setup step in the proof of Theorem 2 in [13]).

The third contribution we would like to mention is that our ring signature scheme provides unconditional anonymity even if the secret key of the signer and the public parameters have not been generated following the key generation process (in fact, even if they are adversarially chosen). In the light of recent results on the high percentage of unsure RSA keypairs [22], we find such a result a reassuring property for the users of a ring signature scheme.

Finally, we would like to note that the underlying scheme on which we are based seems to be more efficient than the schemes on which the alternative works are based. However, all ring signature lattice-based schemes, including ours, are still pretty unpractical and in any case far behind the best results of number theoretic schemes, such as [9], in size as well as in versatility. We therefore focus on the theoretical contributions and leave aside practical parameter comparisons.

2   Preliminaries

2.1   Notations

Polynomials and Vectors of Polynomials. Let $\mathbb{Z}_p$ denote the quotient ring $\mathbb{Z}/p\mathbb{Z}$. In this work we build our cryptographic construction upon the ring $\mathcal{D} = \mathbb{Z}_p[x]/\langle x^n+1 \rangle$; where x^n+1 is irreducible, n is a power of two, and p is a prime such that $p = 3 \mod 8$. The elements of $\mathcal{D}$ will be represented by polynomials of degree $n-1$ having coefficients in $\{-(p-1)/2, \ldots, (p-1)/2\}$.

We will denote polynomials by roman letters $(a, b, \ldots)$, vectors of polynomials will be denoted by a roman letter with a hat $(\hat{a}, \hat{b}, \ldots)$. Let m some positive integer such that, $a_1, \ldots, a_m$ are polynomials in $\mathcal{D}$, then we can write $\hat{a} = (a_1, \ldots, a_m)$. For any polynomial a, the infinity norm ℓ_∞ is defined by $\|a\|_\infty = max_i|a^{(i)}|$, with $a^{(i)}$ the coefficients of the polynomial, and for a vector of polynomials by $\|\hat{a}\|_\infty = max_i\|a_i\|_\infty$.

Sets. For a positive integer i, $[i]$ denotes the set $\{1, \ldots, i\}$. For a given set S, the notation $x \leftarrow S$ represents a uniformly random sample from the set, and for a given randomized algorithm $x \leftarrow \mathtt{RandomizedAlgorithm}$ represents a sampling from the possible outputs following the distribution given by the algorithm.

In our scheme, forgery attacks become easier as the ring size grows (this is also true for other schemes). We therefore suppose that there is a constant c such that acceptable ring sizes are bounded from above by k^c, k being the security parameter. As signature and verification key sizes have an amount of sub-elements proportional to the ring size the reader can replace c by 1 or 2 which will cover any reasonable use of these signatures. The table below defines different sets we use and the parameters associated to these sets.

Rings and Random Oracles. Each keypair of the ring signature scheme is in any protocol, game or experiment we may present always uniquely defined by an integer (its index). We define a ring R as a set of verification keys. We consider that there is a bijection between users and keypairs and sometimes we will implicitly use this bijection saying that a user belongs to a ring. We also define $\#R$ as the size of the ring (i.e. the amount of verification keys it contains), and as $index(R)$ the set of integers corresponding to the indexes of the verification keys in R (each keypair being uniquely defined by its index).

Table 1. Sets and parameters

n	power of 2 greater than the security parameter k
p	prime of order $\Theta(n^{4+c})$ such that $p \equiv 3 \bmod 8$
m_u	$(3 + 2c/3) \log n$
m	$(3 + 2c/3)n^c \log n$
D_h	$\{g \in \mathcal{D} : \|g\|_\infty \leq mn^{1.5} \log n + \sqrt{n} \log n\}$
D_y	$\{g \in \mathcal{D} : \|g\|_\infty \leq mn^{1.5} \log n\}$
D_z	$\{g \in \mathcal{D} : \|g\|_\infty \leq mn^{1.5} \log n - \sqrt{n} \log n\}$
$D_{s,c}$	$\{g \in \mathcal{D} : \|g\|_\infty \leq 1\}$

$H : \{0,1\}^* \to D_{s,c}$ denotes a random oracle. We describe a ring $R = \{pk_{i_1}, \ldots, pk_{i_{\#R}}\}$, for example when using it as input to the random oracle, by $desc(pk_{i_1})\| \ldots \|desc(pk_{i_{\#R}})$ where $desc(pk)$ is a binary description of a public key, $\|$ is the concatenation operator and $i_1 < \cdots < i_{\#R}$ (the representation is thus unique). When possible we will just skip $desc()$ in such notations, $f_1\|f_2$ meaning the concatenation of the description of functions f_1 and f_2.

2.2 Collision-Resistant Hash Functions

In [23] Lyubashevsky and Micciancio introduced a family $\mathcal{H}$ of collision-resistant hash functions with security based on the worst-case hardness of standard lattice problems over ideal lattices.[1]

Definition 1. *For any integer m_u and $D_\times \subseteq \mathcal{D}$, let $\mathcal{H}(\mathcal{D}, D_\times, m_u) = \{h_{\hat{a}} : \hat{a} \in \mathcal{D}^{m_u}\}$ be the function family such that for any $\hat{z} \in D_\times^{m_u}$, $h_{\hat{a}}(\hat{z}) = \hat{a} \cdot \hat{z} = \sum a_i z_i$, where $\hat{a} = (a_1, ..., a_{m_u})$ and $\hat{z} = (z_1, ..., z_{m_u})$ and all the operations $a_i z_i$ are performed in the ring $\mathcal{D}$.*

Note that hash functions in $\mathcal{H}(\mathcal{D}, D_\times, m_u)$ satisfy the following two properties for any $\hat{y}, \hat{z} \in \mathcal{D}^{m_u}$ and $c \in \mathcal{D}$:

$$h(\hat{y} + \hat{z}) = h(\hat{y}) + h(\hat{z}) \tag{1}$$

$$h(\hat{y}c) = h(\hat{y})c \tag{2}$$

Moreover, when the input domain is restricted to a strategically chosen set $D_\times^{m_u} \subset \mathcal{D}^{m_u}$, the function family is collision resistant. We first introduce the collision finding problem and then present the security reduction result for a well-chosen $D_\times$.

---

[1] In this work, by ideal lattices, we make reference to the discrete subgroups of $\mathbb{Z}_p^n$ that can be mapped from ideals in rings of the form $\mathbb{Z}_p[x]/\langle f \rangle$ for some irreducible polynomial of degree n. The mapping between ideals and ideal lattices is trivially derived from the canonical isomorphism between polynomials $v^{(0)} + v^{(1)}x + \ldots + v^{(n-1)}x^{n-1}$ in $\mathbb{Z}_p[x]/\langle f \rangle$ and vectors $v = (v^{(0)}, \ldots, v^{(n-1)})$ in $\mathbb{Z}_p^n$.

Definition 2 (Collision Problem). *Given an element $h \in \mathcal{H}(\mathcal{D}, D_\times, m)$, the collision problem $Col(h, D_\times)$ (where $D_\times \subset \mathcal{D}$) asks to find distinct elements $\hat{z}_1, \hat{z}_2 \in D_\times$ such that $h(\hat{z}_1) = h(\hat{z}_2)$.*

It was shown in [23] that, when $D_\times$ is restricted to a set of small norm polynomials, solving $Col(h, D_\times)$ is as hard as solving $SVP_\gamma(\mathcal{L})$ in the worst case over lattices that correspond to ideals in $\mathcal{D}$.

Theorem 1 (Theorem 1 in [20], applied to our setting). *Let $\mathcal{D}$ be the ring $\mathbb{Z}_p[x] / \langle x^n + 1 \rangle$ for n a power of two. Define the set $D_\times = \{y \in \mathcal{D} \mid \|y\|_\infty \leq d\}$ for some integer d. Let $\mathcal{H}(\mathcal{D}, D_\times, m)$ be a hash function family as in Definition 1 such that $m > \frac{\log p}{\log 2d}$ and $p \geq 4dmn^{1.5} \log n$. If there is a polynomial-time algorithm that solves $Col(h, D_\times)$ for random $h \in \mathcal{H}(\mathcal{D}, D_\times, m)$ with some non-negligible probability, then there is a polynomial-time algorithm that can solve $SVP_\gamma(\mathcal{L})$ for every lattice corresponding to an ideal in $\mathcal{D}$, where $\gamma = 16dmn \log^2 n$.*

In this paper we will set $d = mn^{1.5} \log n + \sqrt{n} \log n$, $D_\times = D_h$, and n, p, m as suggested in Table 1. This ensures that the conditions required by the above theorem are verified and that finding collisions for $\mathcal{H}(\mathcal{D}, D_h, m)$ implies an algorithm for breaking SVP in the worst-case over ideal lattices for polynomial gaps (in n and therefore in k).

In the ring signature scheme we present, the manipulated hash functions will always belong to sets $\mathcal{H}(\mathcal{D}, D_h, m')$ with $m' \leq m$. It is important to note that if an attacker is able to solve the above problem for $m' \leq m$ he can also solve it for m. Indeed, when given a challenge $h \in \mathcal{H}(\mathcal{D}, D_h, m)$ the attacker can puncture the tuple of polynomials describing h, to obtain a tuple of m' polynomials, solve the collision problem for m' and the pad the obtained solution with zeros on the punctured coordinates to obtain a solution to the problem for h.

2.3 Statistical Distance

The statistical distance measures how different are two probability distributions. In this paper we will use this tool to prove that the ring signature scheme presented is anonymous.

Definition 3 (Statistical Distance). *Let X and X' be two random variables over a countable set S. We define by:*

$$\Delta(X, X') = \frac{1}{2} \sum_{x \in S} |Pr[X = x] - Pr[X' = x]|$$

the statistical distance between X and X'.

One important property of the statistical distance is that it cannot be increased by a randomized algorithm which is formalized by the following proposition.

Proposition 1 (Proposition 8.10 in [24]). *Let X, X' be two random variables over a common set A. For any (possibly randomized) function f with domain A, the statistical distance between $f(X)$ and $f(X')$ is at most*

$$\Delta(f(X), f(X')) \leq \Delta(X, X').$$

This proposition implies that if the statistical distance of two families of random variables (X_k) and (X'_k) is negligible,[2] an attacker given a sample will only obtain a negligible advantage over a wild guess when trying to distinguish between the distributions of (X_k) and the ones of (X'_k). Note that the proposition above does not make any assumption on the computational complexity of f and thus this is true whether the attacker is computationally bounded or unbounded.

Note that the statistical distance may grow if we consider multiple variables. It is easy to verify from definition 3 that if X, Y follow a distribution ϕ and X', Y' a distribution ϕ', we have

$$2\Delta(X, X') \geq \Delta((X, Y), (X', Y')) \geq \Delta(X, X'). \tag{3}$$

Thus, if an attacker is given many samples of the same distribution he may be able to distinguish better than with just one sample. More specifically, using (3) iteratively, if the attacker is given $\#s$ samples of the same distribution and the families of random variables have an upper-bound $\epsilon(k)$ on the statistical distance, the advantage over a wild guess for such an attacker will be bounded from above by $\#s * \epsilon(k)$.

In Section 4.1 we prove that, for our scheme, the signatures of two different users have a statistical distance which is exponentially small in k and thus, even a computationally unbounded attacker given an exponential amount of signatures of the same user will have a negligible advantage over a wild guess when trying to break anonymity.[3]

Attacker with Additional Information. An attacker trying to distinguish between the distributions of two random variables X_k, X'_k may have some extra information which we model as a third random variable Z_k. This information may for example be obtained during an indistinguishability game prior to obtaining a sample (e.g. two public keys). If X_k and X'_k are not dependent on Z_k, this extra information is of no use to the attacker as then, using proposition 8.8 from [24], $\Delta((X_k, Z_k), (X'_k, Z_k)) = \Delta(X_k, X'_k)$ and therefore we still have $\Delta(f(X_k, Z_k), f(X'_k, Z_k)) \leq \Delta(X_k, X'_k)$.

If X_k or X'_k depend on Z_k (as signatures depend on the public keys in our case) we cannot use the same argument, as proposition 8.8 from [24] only applies to independent variables. However, noting $X_{k,z}$ and $X'_{k,z}$ the random variables conditioned on $Z = z$, if we have an upper-bound

$$\Delta(X_{k,z}, X'_{k,z}) < \epsilon(k)$$

---

[2] I.e. asymptotically bounded from above by k^{-c} for any c, k being the security parameter.

[3] For example for $2^{k/2}$ samples, and $\epsilon(k) = 2^{-k}$ we have $\#s * \epsilon(k) = 2^{-k/2}$.

which is independent of z it is also an upper-bound for $\Delta(f(X_k, Z_k), f(X'_k, Z'_k))$. Indeed, we have

$$
\begin{aligned}
\Delta(f(X_k, Z_k), f(X'_k, Z_k)) &\leq \Delta((X_k, Z_k), (X'_k, Z_k)) \\
&= \frac{1}{2} \sum_{x,z} |Pr[(X_k, Z_k) = (x, z)] - Pr[(X'_k, Z_k) = (x, z)]| \\
&= \frac{1}{2} \sum_z Pr[Z_k = z] \sum_x |Pr[X_k = x \mid Z_k = z] - Pr[X'_k = x \mid Z_k = z]| \\
&= \sum_z Pr[Z_k = z] \Delta(X_{k,z}, X'_{k,z})
\end{aligned}
$$

$$\Delta(f(X_k, Z_k), f(X'_k, Z_k)) \leq \epsilon(k).$$

Note that the upper bound is valid independently of the distribution followed by Z_k, and thus we can include in this random variable parameters adversarially chosen by the attacker.

2.4   Ring Signature Schemes: Definitions and Properties

A ring signature schemes gives means to individual users to define an arbitrary set of public keys R (the ring), and issue a signature using a secret key associated to one of the public keys in the ring. Using the set R and a verification algorithm it is possible to verify that a signature has been issued by a member of the ring (i.e. by a user who knows a secret key associated to one of the public keys in R), but it is not possible to learn whom.

Ring Signature Scheme. We will describe a ring signature scheme by triple of algorithms (Ring−gen, Ring−sign, Ring−verify):

- Ring−gen(1^k): A probabilistic polynomial time algorithm that takes as input a security parameter k, outputs a a public key pk and a secret key sk. For many schemes, the users in a ring must share in common some public information derived from k. We thus suppose that Ring−gen(1^k) has two sub-algorithms: Ring−gen−params(1^k) which generates a set of public parameters $\mathcal{P}$ which are used in all the algorithms; and Ring−gen−keys($\mathcal{P}$) which generates keypairs based on the public parameters. We suppose that the constant c is defined in the description of the scheme (we could also define it as an input parameter of Ring−gen−params).
- Ring−sign($\mathcal{P}, sk, \mu, R$): A probabilistic polynomial time algorithm that takes as input a set of parameters $\mathcal{P}$, a signing key sk, a message $\mu \in \mathcal{M}$ ($\mathcal{M}$ being the message space of the scheme) and a set of public keys R (the ring). It returns a signature σ for μ under sk, or failed.
- Ring−verify($\mathcal{P}, \sigma, \mu, R$): A deterministic algorithm that takes as input a set of parameters $\mathcal{P}$, a ring signature σ on a message μ and a set of public keys R, and outputs 1 or 0 for accept or reject respectively.

We require the following correctness condition: for any k, any ℓ (bounded by a polynomial in k), any $\mathcal{P} \in$ Ring−gen−params(1^k), any $\{(pk_i, sk_i)\}_{i\in[\ell]} \subset$ Ring−gen−keys($\mathcal{P}$), any $i_0 \in [\ell]$, any message μ, and any $\sigma \in$ Ring−sign

$(\mathcal{P}, sk_{i_0}, \mu, \{pk_i\}_{i\in[\ell]})$, $\sigma \neq$ failed, we have Ring$-$verify$(\mathcal{P}, \mu, \sigma, \{pk_i\}_{i\in[\ell]}) = 1$. Moreover, in order to be considered secure, a ring signature scheme must satisfy some anonymity and unforgeability properties. Our goal will be to obtain the following two properties:

Anonymity. In [2], Bender et al. define various levels of anonymity for a ring signature scheme. Among them the highest is *anonymity against full key exposure*. The authors of [2] use the same definition for two levels of security (attribution attacks and full-key exposure) which results in a definition which is slightly too complex for our needs. We use here a definition which is stronger and as a consequence, much simpler. Indeed, we do not need to define a signing or corruption oracle as the attacker knows all the secrets and can thus simulate both oracles effectively, and we don't have a first step in which the challenger generates the parameters and keys as these can be adversarially chosen. The definition is given by the following game:

Unconditional Anonymity Against Chosen Setting Attacks

---

1. $\mathcal{A}$ outputs a set of public parameters $\mathcal{P} = (k, n, m_u, p, S)$, a ring $R = \{pk_1, \ldots, pk_\ell\}$ for ℓ in $[k^c]$, two distinct indices $i_0, i_1 \in [k^c]$, two secret keys sk_{i_0}, sk_{i_1}, and a message μ.
2. Two signatures $\sigma_0 \leftarrow$ Ring$-$sign$(\mathcal{P}, sk_{i_0}, \mu, R)$, $\sigma_1 \leftarrow$ Ring$-$sign$(\mathcal{P}, sk_{i_1}, \mu, R)$ are generated and a random bit b is chosen. If $\sigma_0 \neq$ failed and $\sigma_1 \neq$ failed, $\mathcal{A}$ is given σ_b, else the game is restarted.
3. $\mathcal{A}$ outputs a bit b' and succeeds if $b' = b$.

The ring signature scheme achieves unconditional anonymity against chosen setting attacks if any adversary $\mathcal{A}$ has an advantage with respect to a wild guess which is negligible in the security parameter k.

This definition can be easily generalized to the case in which the adversary is given sets of samples instead of a single sample. In order to simplify the definition we use just one sample, but the proofs of Section 4.1 show that the anonymity is ensured even if the attacker is given an exponential amount of samples.

Unforgeability. In [2] different notions of unforgeability are introduced. We present here all of them as going from the simplest to the most complex definition helps in the presentation.

For a ring signature scheme with ℓ members, the unforgeability against fixed-ring attacks is defined using the following experiment. The challenger firstly runs the algorithm Ring$-$gen to obtain compatible keypairs $(pk_1, sk_1), \ldots, (pk_\ell, sk_\ell)$ for the signature scheme and sends $R = \{pk_i\}_{i\in[\ell]}$ to the forger. The forger can then make polynomially many signing queries. A ring signing query is of the form $(i_{\text{signer}}, \mu, R)$ for varying $\mu \in \mathcal{M}$, $i_{\text{signer}} \in index(R)$. The challenger replies with $\sigma \leftarrow$ Ring$-$sign$(\mathcal{P}, sk_{i_{\text{signer}}}, \mu, R)$. Finally the forger outputs $(\sigma^\star, \mu^\star, R)$ and it wins if $\mathsf{Ring{-}verify}(\mathcal{P}, \sigma^\star, \mu^\star, R)$ outputs accept and $\mu^\star$ is not one of the messages for which a signature was queried.

An intermediate, stronger, definition is unforgeability against chosen subring attacks in which signing queries and the final forgery can be done with respect to any subring $S \subset R$. The strongest definition proposed is unforgeability with respect to insider corruption. In this setting, signing queries can be done with respect to *any* ring (i.e. the attacker can add to these rings adversarially generated public keys). The attacker is also given a corruption oracle that for any $i \in index(R)$ returns sk_i. Forgeries are valid if they do not correspond to a signing query, the subring of the forgery is a subset of R, and none of the secret keys of the subring have been obtained through the corruption oracle.

In this paper we will first show that the proof of [19] can be adapted to our scheme in order to prove unforgeability with respect to subring attacks. We will then prove that for an alternative version of this scheme it is also possible to ensure unforgeability with respect to insider corruption.

3   Our Scheme

3.1   Informal Description

In this section we first provide a high-level description of the tree-less signature scheme in [19], and then we show how to transform it into a ring signature scheme. In [19], the signer has as (secret) signing key $\hat{s}$ and a (public) verification key (h, S) such that $h(\hat{s}) = S$. The domains to which these keys belong will be made explicit later.

To make a signature of some message μ, the signer will prove that he can:

– Choose a random vector of polynomials $\hat{y}$ (which he does not reveal)
– Output a vector of polynomials whose difference with $\hat{y}$ is $\hat{s}$ (his secret key) times a small polynomial e that is not of his choice

In order to do this the signer will select some random $\hat{y}$, compute $e = H(h(\hat{y}), \mu)$ and output $(\hat{z}, e)$ with $\hat{z} = \hat{s}e + \hat{y}$. The verifier tests that $e = H(h(\hat{z}) - Se, \mu)$. This is true for a correct signature thanks to the linearity of h as $h(\hat{z}) - Se = h(\hat{s}e + \hat{y}) - Se = h(\hat{y}) + Se - Se = h(\hat{y})$.

To obtain a ring signature from this scheme we make two major modifications. The first one is to ensure that each user has in his public key a function h_i that satisfies $h_i(\hat{s}_i) = S$ where $\hat{s}_i$ is the secret key and S is a fixed standard polynomial (not null). Consider a ring $R = \{h_i\}_{i \in [\ell]}$. The second modification keeps the real signer anonymous when he signs a message. We do this by simply adding $\ell - 1$ random variables that will corresponds to the ring members except the real signer. For example, suppose that the real signer is indexed by $j \in [\ell]$, the signer sends the signature $(\hat{z}_i; i \in [\ell], e)$ with $e = H(\sum_{i \in [\ell]} h_i(\hat{y}_i), \mu)$, $\hat{z}_j = \hat{s}_j e + \hat{y}_j$ and $\hat{z}_i = \hat{y}_i$ for $i \in [\ell] \setminus \{j\}$. Therefore, the final signature will contain $[\ell]$ elements one for each member in the ring. Now we go gack to the first modification and we will show its utility in the correctness of the scheme. In the verification step, the verifier checks if the hash value of $(\sum_{i \in [\ell]} h_i(\hat{z}_i) - Se, \mu)$ is equal to e. Note that this will be true only if $\sum_{i \in [\ell]} h_i(\hat{z}_i) - Se$ is equal to $\sum_{i \in [\ell]} h_i(\hat{y}_i)$. In fact using the linearity of h_j we have $h_j(\hat{z}_j) = h_j(\hat{s}_j e + \hat{y}_j) = h_j(\hat{y}_j) + Se$. Since all the ring members have key pairs $(h_i, \hat{s}_i)$ such that $h_i(\hat{s}_i) = S$, the verifier will always accept when one of them produces a ring signature.

In order to resist to chosen subring attacks against unforgeability we must modify the scheme to include in the random oracle call a description of the ring for which the signature is valid (if not, it is easy to reuse a signature to generate a forged signature on a larger ring). In order to resist to attacks on adversarially chosen parameters, the ring signature algorithm starts with an initial step in which the inputs are required to pass simple tests (bounds on the amount of coordinates, scalar sizes, etc.). The security proofs proposed by Lyubashevsky must also be modified to take into account the existence of multiple hash functions and signature elements. Moreover, using modified hash functions also requires that we introduce a few new lemmas and propositions to complete the proofs.

3.2   A More Formal Description

Ring Signature Scheme

Ring$-$gen$-$params(1^k):
Given an integer k define the common public parameters.

1. Set n as a power of two larger than k
2. Set $m_u = 3 \log n$, and p as a prime larger than n^4 such that $p = 3$ mod 8
— Note: these parameters define the sets $\mathcal{D}, D_h, D_z, D_y, D_{s,c}$ and the family $\mathcal{H}$.
3. Set $S \leftarrow \mathcal{D}, S \neq 0$
4. Output $\mathcal{P} = (k, n, m_u, p, S)$

Ring$-$gen$-$keys($\mathcal{P}$):
Generate a keypair.

1. Set $\hat{s} = (s_1, s_2, \ldots, s_{m_u}) \leftarrow D_{s,c}^{m_u}$
2. If none of the s_i is invertible, go to 1.
3. Let $i_0 \in \{1, \ldots, m\}$ such that s_{i_0} is invertible.
4. $(a_1, a_2, \ldots, a_{i_0-1}, a_{i_0+1}, \ldots, a_{m_u}) \leftarrow \mathcal{D}^{m_u-1}$.
5. Let $a_{i_0} = s_{i_0}^{-1}(S - \sum_{i \neq i_0} a_i s_i)$ and note $\hat{a} = (a_1, \ldots, a_{m_u})$
6. Output $(pk, sk) = (h, \hat{s})$, h being the hash function in $\mathcal{H}$ defined by $\hat{a}$

Ring−sign($\mathcal{P}, sk, \mu, R$):

Given a message $\mu \in \mathcal{M}$, a ring of ℓ members with public keys $R = \{h_i\}_{i \in [\ell]} \subset \mathcal{H}(\mathcal{D}, D_h, m_u)$, and a private key $sk = \hat{s}_j$ associated to one of the public keys h_j in R, generate a ring signature for the message.

0. Verify that: the public parameters respect the constraints of steps $1 - 3$ in Ring−gen−params ; sk is in $D_{s,c}^{m_u}$; R is of size bounded by k^c ; one of the public keys in R is associated to sk. If the verification fails output failed.
1. For all $i \in [\ell]$; $i \neq j$;$\hat{y}_i \leftarrow D_z^{m_u}$
2. For $i = j$; $\hat{y}_j \leftarrow D_y^{m_u}$
3. Set $e \leftarrow H(\sum_{i \in [\ell]} h_i(\hat{y}_i), R, \mu)$ (e is therefore in $D_{s,c}$)
4. For $i = j$, $\hat{z}_j \leftarrow \hat{s}_j \cdot e + \hat{y}_j$
5. If $\hat{z}_j \notin D_z^{m_u}$ then go to Step 2
6. For $i \neq j$, $\hat{z}_i = \hat{y}_i$
7. Output $\sigma = (\hat{z}_i; i \in [\ell], e)$

Ring−verify($\mathcal{P}, \mu, R, \sigma$):

Given a message μ, a ring $R = \{h_i\}_{i \in [\ell]}$ and a ring signature $\sigma = (\hat{z}_i; i \in [\ell], e)$, the verifier accepts the signature only if both of the following conditions satisfied:

1. $\hat{z}_i \in D_z^{m_u}$ for all $i \in [\ell]$
2. $e = H(\sum_{i \in \{1,\ldots,\ell\}} h_i(\hat{z}_i) - S \cdot e, R, \mu)$

Otherwise, the verifier rejects.

3.3 Correctness and Convergence of the Algorithms

The correctness of the signing algorithm is pretty straightforward. Indeed, let $\sigma = (\hat{z}_i; i \in [\ell], \{h_i\}_{i \in [\ell]}, e) \leftarrow$ Ring−sign($\mathcal{P}, \hat{s}_j, \mu, \{h_i\}_{i \in [\ell]}$) be a signature with $j \in [\ell]$ and $(h_j, \hat{s}_j)$ a given keypair. The first test in Ring−verify is always passed by a valid signature as steps 2 and 5 of Ring−sign ensure that signatures only contain elements in $D_z^{m_u}$. With respect to the second test we have:

$$\sum_{i \in [\ell]} h_i(\hat{z}_i) - S \cdot e = h_j(\hat{z}_j) - S \cdot e + \sum_{i \in [\ell] \setminus \{j\}} h_i(\hat{z}_i)$$

$$= h_j(\hat{s}_j e + \hat{y}_j) - S \cdot e + \sum_{i \in [\ell] \setminus \{j\}} h_i(\hat{y}_i)$$

by replacing $\hat{z}_j$ by $\hat{s}_j \cdot e + \hat{y}_j$ and $\hat{z}_i$ by $\hat{y}_i$,

$$= h_j(\hat{s}_j) \cdot e + h_j(\hat{y}_j) - S \cdot e + \sum_{i \in [\ell] \setminus \{j\}} h_i(\hat{y}_i)$$

using the homomorphic properties of $h_j \in \mathcal{H}$,

$$= \sum_{i \in [\ell]} h_i(\hat{y}_i) \qquad \text{as } h_j(\hat{s}_j) = S.$$

As $e = H(\sum_{i \in [\ell]} h_i(\hat{y}_i), \{h_i\}_{i \in [\ell]}, m)$ the second test of Ring$-$verify is therefore always satisfied by a valid signature.

A correctly issued signature is therefore always verified. Let's consider now the expected running time of the different algorithms.

Proposition 2. *The expected running times of* Ring$-$gen$-$params, Ring$-$gen$-$keys, Ring$-$sign *and* Ring$-$verify *are polynomial in the security parameter.*

Proof. All the operations in the different algorithms can be executed in polynomial time in the security parameter. Thus, Ring$-$gen$-$params and Ring$-$verify run in polynomial time and the only possible issue would be the amount of iterations in the loops of Ring$-$gen$-$keys and Ring$-$sign.

Lemma 3, proved in the appendix, states that each of the polynomials chosen in step 1 of algorithm Ring$-$gen$-$keys is invertible with probability exponentially close to one and thus the expected amount of iterations in the associated loop is roughly one. Thus the expected running time of Ring$-$gen$-$keys is polynomial.

In Ring$-$sign the outer loop has a polynomial amount of iterations, on the other hand inner loop between steps 2 and 5 which will continue as long as $\hat{z}_j = \hat{s}_j \cdot e + \hat{y}_j \notin D_z^{m_u}$. Corollary 6.2 from [19] states that for any $\hat{s} \in D_s^{m_u}$,

$$Pr_{c \leftarrow D_{s,c}, \hat{y} \leftarrow D_y^{m_u}} [\hat{s}c + \hat{y} \in D_z^{m_u}] = \frac{1}{e} - o(1).$$

As e and $\hat{y}_j$ are drawn uniformly from $D_{s,c}$ and $D_y^{m_u}$ we can use this result. This implies that the expected amount of iterations in the inner loop of Ring$-$sign is less than 3, and therefore that Ring$-$sign also runs in expected polynomial time. □

4 Security of the Proposed Scheme

4.1 Anonymity

In the anonymity against chosen setting attacks game, the adversary receives a signature which depends on a random bit b as well as on a set of public parameters $\mathcal{P} = (k, n, m_u, p, S)$, two secret keys sk_{i_0}, sk_{i_1}, a message μ, and a ring of public keys R. All of these parameters have been chosen adversarially except the random bit b.

Let $X_{b, \mathcal{P}, sk_{i_b}, \mu, R}$ be the random variable that represents the signature received by the adversary for a given set of parameters. The following theorem states that for any choice of $\mathcal{P}, sk_{i_0}, sk_{i_1}, \mu, R$ which does not result in a game restart, the statistical distance between $X_{0, \mathcal{P}, R, sk_{i_0}, \mu}$ and $X_{1, \mathcal{P}, R, sk_{i_1}, \mu}$ is negligible in k.

Theorem 2 (Anonymity). *For* $b \in \{0, 1\}$, *let* $X_{b, \mathcal{P}, sk_{i_b}, \mu, R}$ *be the random variable describing the output of* Ring$-$sign$(\mathcal{P}, sk_{i_b}, \mu, R)$ *with* $\mathcal{P} = (k, n, m_u, p, S)$, sk_{i_b}, μ, R *a set of arbitrary inputs to the algorithm. If the domains of these variables are both different from* {failed} *we have*

$$\Delta(X_{0, \mathcal{P}, sk_{i_0}, \mu, R}, X_{1, \mathcal{P}, sk_{i_1}, \mu, R}) = n^{-\omega(1)}.$$

The proof of this theorem is available on the appendix. Using the properties of statistical distance presented in Section 3 this implies that our scheme ensures unconditional anonymity against chosen setting attacks.

4.2 Unforgeability Against Chosen Subring Attacks

In this section we will show that an adversary able to break the unforgeability in the chosen subring setting for our scheme is also able to break the unforgeability of Lyubashevsky's scheme. Given the results of [19], this implies an efficient algorithm to solve SVP_γ on ideal lattices (for γ defined as in Theorem 1). It is important to note that in Table 1 we change the parameters given by Lyubashevsky by increasing m and p, but the reader can easily check (using the sketch proof of the corollary below) that the proofs given in [19] are still valid and that the inequalities in Theorem 1 are verified.

Corollary 1 (of Theorems 6.5 and 6.6 in [19]). *If there is a polynomial time algorithm that can break the unforgeability of the signature scheme proposed in [19] for the parameters presented in Table 1, and more precisely for hash functions with $m = n^c * m_u$ columns, there is a polynomial time algorithm that can solve SVP_γ for $\gamma = O(n^{2.5+2c})$ for every lattice L corresponding to an ideal in D.*

Proof (Sketch.). The unforgeability proof of [19] is split in two. First it reduces signature unforgeability from collision finding and then collision finding from SVP (for a given gap). The first part of the reduction is given by Theorem 6.6 which is almost unchanged for our parameters. As we have a major modification on m, the amount of columns of the hash functions, the only point which could raise an issue is when Lemma 5.2 and Theorem 6.5 are used. These prove that for the parameters in [19] there is a second pre-image for a given output of the hash function with high probability and that signatures using two such pre-images are indistinguishable. As we increase the amount of columns in the hash functions used, the probability of a second pre-image and the uniformity of the output are increased and thus we can still use these results.

Therefore using Theorem 6.6 of [19] we can deduce that breaking unforgeability for our parameters implies finding collisions for the corresponding hash functions. As our hash functions have more columns than in [19] it is easier to find such collisions than with the original parameters. Using Theorem 1 on our parameters shows that collision finding can be reduced from SVP_γ for $\gamma = O(n^{2.5+2c})$ (instead of $O(n^{2.5})$ for the original scheme).   □

In our ring signature scheme, an unforgeability challenge is a set of verification hash functions defined by a set of ℓ tuples of m_u polynomials. In Lyubashevsky's signature scheme an unforgeability challenge is a single tuple of m' polynomials. The idea is thus, considering an attacker to our scheme, to set $m' = \ell \times m_u$, transform this challenge into a set of ℓ tuples and show that if we give this as a ring signature challenge to the attacker, we will obtain with non-negligible probability a valid forgery for Lyubashevsky's scheme.

The are two key issues. First, the polynomials in the m'-tuple are uniformly random but the polynomials in the m_u-tuples that the attacker expects to receive are not. We address this issue by proving, in Corollary 2, that the statistical distance between public keys in our ring signature scheme, and hash functions chosen uniformly at random is negligible. This implies that an attacker able to forge signatures for hash functions associated to the public keys is also able to forge them for hash functions chosen uniformly at random. Indeed, if we model the attacker by a randomized function, it cannot increase the statistical distance between the two families of functions and thus if it succeeds with non-negligible probability for a family he also will for the other. The second issue is that we do not know the private keys associated to the challenge given to the attacker but we must be able to answer to his signing queries. We solve this issue simply by programming the random oracle that the attacker uses.

In order to prove Corollary 2, we will use the following theorem.

Theorem 3 (Adapted from [25], Theorem 3.2). *For $g_1, \ldots, g_{m_u-1} \in \mathcal{D}$, we denote by $F(g_1, \ldots, g_{m_u-1})$ the random variable $\sum_{i \in [m_u-1]} s_i g_i \in D$ where $s_1, \ldots, s_{m_u-1}$ are chosen uniformly at random in $D_{s,c}$. Noting $U_1, \ldots, U_{m_u}$ independent uniform random variables in D, we have*

$$\Delta((U_1, \ldots, U_{m_u-1}, F(U_1, \ldots, U_{m_u-1})), (U_1, \ldots, U_{m_u})) \le \frac{1}{2}\sqrt{\left(1 + \left(\frac{p}{3^{m_u-1}}\right)^{n/2}\right) - 1}$$

Proof. Just apply Theorem 3.2 from [25] to our setting, using the fact that by Lemma 2.3 from [25] our choice of parameters ensures that $x^n + 1 = f_1 f_2 \mod p$ where each f_i is irreducible in $\mathbb{Z}_p[x]$ and can be written $f_i(x) = x^{n/2} + t_i x^{n/4} - 1$ with $t_i \in \mathbb{Z}_p$. $\square$

Corollary 2 (Uniformity of the public key). *Let $X_{\mathcal{P}}$ be a random variable describing the distribution of the hash functions resulting from the key generation algorithm* Ring$-$gen$-$keys$(\mathcal{P})$ *and $U_1, \ldots, U_{m_u}$ denote independent uniform random variables in D. Then*

$$\Delta(X_{\mathcal{P}}, (U_1, \ldots, U_{m_u})) \le n^{-\omega(1)}.$$

Proof. We describe a hash function $h_{\hat{a}}$ by the set of polynomials $\hat{a} = (a_1, \ldots, a_{m_u})$. We suppose, w.l.o.g. that $(a_1, \ldots, a_{m_u}) = (a_1, \ldots, a_{m_u-1}, s_{m_u}^{-1}(S - \sum_{i \in [m_u-1]} a_i s_i))$, where $\hat{s} = (s_1, \ldots, s_{m_u})$ is the secret key corresponding to $h_{\hat{a}}$.

We first note that the function on the right side of the inequality in Theorem 3 is negligible for our parameters as $p = \Theta(n^{4+c})$ and $3^{m_u-1} = 3^{(3+2c/3)\log_2 n}/3 = n^{(3+2c/3)\log_2 3}/3 > n^{4.5+c}/3$. Thus, using Theorem 3, we have

$$\Delta((U_1, \ldots, U_{m_u-1}, F(U_1, \ldots, U_{m_u-1})), (U_1, \ldots, U_{m_u})) \le n^{-\omega(1)}.$$

Proposition 1 states that a function cannot increase the statistical distance. Noting $f(g_1, \ldots, g_{m_u})$ the function that leaves unchanged the $m_u - 1$ first coordinates and replaces g_{m_u} by $s_{m_u}^{-1}(S - g_{m_u})$ we get

$$\Delta(f(U_1, \ldots, U_{m_u-1}, F(U_1, \ldots, U_{m_u-1})), f(U_1, \ldots, U_{m_u})) \le n^{-\omega(1)}.$$

To prove the corollary we just need to note that $f(U_1, \ldots, U_{m_u-1}, F(U_1, \ldots, U_{m_u-1}))$ has exactly the same distribution as $X_{\mathcal{P}}$, and $f(U_1, \ldots, U_{m_u})$ the same as $(U_1, \ldots, U_{m_u})$. The first assertion is trivially true given the Ring−gen−keys algorithm. The second one comes from the fact that adding an element (in our case S) or multiplying by an invertible element (in our case $s_{m_u}^{-1}$) in D just permutes the elements of the ring and thus the uniform distribution remains unchanged. We therefore have

$$\Delta(f(U_1, \ldots, U_{m_u-1}, F(U_1, \ldots, U_{m_u-1})), f(U_1, \ldots, U_{m_u})) = \Delta(X_{\mathcal{P}}, (U_1, \ldots, U_{m_u}))$$

and thus, $\Delta(X_{\mathcal{P}}, (U_1, \ldots, U_{m_u})) \leq n^{-\omega(1)}$.   □

We are now ready to prove our theorem on unforgeability against chosen subring attacks.

Theorem 4. *If there is a polynomial time algorithm that can break the unforgeability under chosen subring attacks of the ring signature scheme described in Section 3.2 for the parameters presented in Table 1, there is a polynomial time algorithm that can solve $SVP_\gamma(\mathcal{L})$ for $\gamma = \tilde{O}(n^{2.5+2c})$ for every lattice $\mathcal{L}$ corresponding to an ideal in $\mathcal{D}$.*

Proof (Sketch.).

   Suppose that we have an adversary $\mathcal{A}$ that can output a forgery for the ring signature scheme with non-negligible advantage in the chosen subring setting. Given Corollary 1, it is enough to prove that using $\mathcal{A}$, we can construct a polynomial time adversary $\mathcal{B}$ that outputs a forgery for Lyubashevsky's scheme with non-negligible advantage for the parameters given in Table 1.

Setup: $\mathcal{B}$ is given as a challenge a description of a hash function (namely a tuple of $m' = \ell \times m_u$ polynomials $(a_1, \ldots, a_{\ell \times m_u})$ in D), an element S of $\mathcal{D}$, and access to the random oracle H_L of the signing algorithm. $\mathcal{B}$ splits the set of polynomials in ℓ sets of m_u polynomials $(a_{i,1}, \ldots, a_{i,m_u})$ for $i \in [\ell]$. Finally, $\mathcal{B}$ initializes $\mathcal{A}$ by giving it the set of tuples generated, the public parameters associated (among which S) and access to the *ring signature* random oracle H which it controls.

Query Phase: $\mathcal{B}$ answers the random oracle and signing queries of $\mathcal{A}$ as follows. For each random oracle query (x_y, x_h, x_m) it will test whether it has already replied to such a query. If so, it will reply consistently. If not, it will reply with $H_L(x_y, x_h \| x_m)$, $\|$ being the concatenator operator and store the result. For each signing query $(\{h_i\}_{i \in T}, i_0, \mu)$ for $i_0 \in T \subseteq [\ell]$, $\mathcal{B}$ programs H to produce a signature. In other words:

1. It follows the steps of the signature by generating a set of $\hat{y}_i \leftarrow D_y^{m_u}$ for $i \in T$
2. It generates at random $r \leftarrow D_{s,c}$
3. It checks whether H_L has been called with parameters $(\sum_i h_i(\hat{y}_i) - S \cdot r, \{h_i\}_{i \in T} \| \mu)$ (if so it aborts)

4. It programs the random oracle H so that $H(\sum_i h_i(\hat{y}_i) - S \cdot r, \{h_i\}_{i \in T}, \mu) = r$, and stores the result

5. Finally, it outputs $(\hat{y}_i; i \in T, r)$.

Forgery Phase: At a given point, $\mathcal{A}$ finishes running and outputs a forgery $((\hat{z}_i; i \in T, e), \mu, \{h_i\}_{i \in T})$, for $T \subseteq [\ell]$ with non-negligible probability. $\mathcal{B}$ just pads the remaining coordinates of the signature with null polynomials $\hat{z}_i = 0$ for $i \in [\ell] \backslash T$ and outputs $((\hat{z}_1 \| \cdots \| \hat{z}_\ell, e), \mu)$, $\hat{z}_1 \| \cdots \| \hat{z}_\ell$ being the vector of polynomials resulting from the concatenation of each of the vectors of polynomials $\hat{z}_i$.

Analysis: In this analysis we will detail completely the reasoning but just sketch the statistical arguments as they are pretty standard and easy to verify.

First of all, note that in step 3 of the protocol above there is the possibility of an abort. The probability of this event is negligible. Indeed, given the sizes of D_y and p^n, m_u is large enough to ensure by leftover hash lemma [26] that $\sum_i h_i(\hat{y}_i) - S \cdot r$ is within negligible distance from an uniformly random distribution. Thus, the probability that $\mathcal{A}$ has generated beforehand a query colliding with the one of the protocol is negligible.

In order to prove that $\mathcal{A}$ will output a ring signature forgery with non-negligible probability we must show that all the inputs it receives are within negligible statistical distance to the ones it would have received in a ring signature challenge. Once this is proved, we must show that the final forgery that $\mathcal{B}$ outputs is valid in the basic signature scheme.

First note that by Corollary 2 and using equation (3) of Section 3, we have that the challenge given to $\mathcal{A}$ is within negligible statistical distance to the one it has in a ring signature challenge. Another family of inputs to consider are the signatures generated by $\mathcal{B}$. These signatures are trivially valid for the given random oracle. Following the same ideas as those used in Theorem 2 we have that these signatures are withing negligible statistical distance from a signature generated using the signing algorithm (omitted). The last inputs to consider are the ones coming from the random oracle H which are pairs $(preimage, image)$ of the graph of H. All of the images of H are chosen uniformly at random over $D_{s,c}$. The issue is that we have programmed H and thus, in our protocol the first coordinate of the pre-image and the image are related. However, using the leftover hash lemma as in the first paragraph of this analysis (when considering aborts) we have that we add to this coordinate $\sum_i h_i(\hat{y}_i)$ which is close to uniform and independent of the image. This implies that (omitting the coordinates set by $\mathcal{A}$ with his query), the pairs $(preimage, image)$ are statistically close to uniform.

We therefore can ensure that $\mathcal{A}$ outputs a forgery for the ring signature scheme with non-negligible probability and the only issue left to prove is the validity of the forgery given by $\mathcal{B}$. The basic idea we will use in the rest of this proof is that in the forgery $((\hat{z}_i; i \in T, e), \mu, \{h_i\}_{i \in T})$ that $\mathcal{A}$ outputs, if e has been obtained during a direct call to the random oracle we can prove that the forgery is valid in the basic signature scheme. If not, we use the same ideas than in Theorem 6.6 of [19].

A wild guess of e can only happen with negligible probability. Indeed, as $D_{s,c}$ is exponentially large, if H has not been queried the probability that the output of $\mathcal{A}$ will give e when using H in the verification step is exponentially small. If we therefore suppose that e has been obtained through a call to H there are two possibilities: either e has been generated during a signing query, or it has been generated during a direct random oracle query. We leave the latter option for the end of the proof.

Suppose e has been generated during a signing query which resulted in an output $(\hat{z}'_i; i \in T', e)$ for a ring $\{h_i\}_{i \in T'}$ and a message μ'. In order to be valid, the forgery must be different from $((\hat{z}'_i; i \in T', e), \mu', \{h_i\}_{i \in T'})$ and thus we must have either $\{h_i\}_{i \in T'} \| \mu' \neq \{h_i\}_{i \in T} \| \mu$ or $(\hat{z}'_i; i \in T') \neq (\hat{z}_i; i \in T)$. The former case can only happen with negligible probability as it implies a collision for H. Indeed, $\{h_i\}_{i \in T'} \| \mu' \neq \{h_i\}_{i \in T} \| \mu$ implies $\{h_i\}_{i \in T'} \neq \{h_i\}_{i \in T}$ or $\mu' \neq \mu$ and in both cases we have a collision as $H(\sum_{i \in T'} h_i(\hat{z}'_i) - S \cdot e, \{h_i\}_{i \in T'}, \mu') = H(\sum_{i \in T} h_i(\hat{z}_i) - S \cdot e, \{h_i\}_{i \in T}, \mu) = r$ (we have twice the same image for two pre-images that are different either in the second or the third coordinate). In the latter case, using the same reasoning, the first coordinate of H is the same in the two cases with overwhelming probability and thus we have $\sum_{i \in T'} h_i(\hat{z}'_i) - S \cdot e = \sum_{i \in T} h_i(\hat{z}_i) - S \cdot e$ for $(\hat{z}'_i; i \in T') \neq (\hat{z}_i; i \in T)$. Setting $\hat{z}_i = 0$ for $i \in [\ell] \backslash T$ and $\hat{z}'_i = 0$ for $i \in [\ell] \backslash T'$ we obtain a collision for a random hash function of $\mathcal{H}(\mathcal{D}, D_h, \ell)$. If this event happens with non-negligible probability, using Theorem 1 we deduce that we can solve SVP_γ.

We conclude this proof by working on the case in which e has been generated during a direct random oracle query. We have the forgery $((\hat{z}_i; i \in T, e), \mu, \{h_i\}_{i \in T})$ and as the ring signature must be valid we have $H(\sum_{i \in T} h_i(\hat{z}_i) - S \cdot e, \{h_i\}_{i \in T}, \mu) = e$. Noting $x_y = \sum_{i \in T} h_i(\hat{z}_i) - S \cdot e$ and given the algorithm followed by $\mathcal{B}$, we also have $H_L(x_y, \{h_i\}_{i \in T} \| \mu) = e$. If we pad the $\hat{z}_i$ with $\hat{z}_i = 0$ for $i \in [\ell] \backslash T$ we still have $\sum_{i \in [\ell]} h_i(\hat{z}_i) - S \cdot e = x_y$ and thus $H_L(\sum_{i \in [\ell]} h_i(\hat{z}_i) - S \cdot e, \{h_i\}_{i \in T} \| \mu) = e$. We therefore have that $\sigma' = (\hat{z}_1 \| \cdots \| \hat{z}_\ell, e)$ is a signature of $\mu' = \{h_i\}_{i \in T} \| \mu$ and, as we have not done any signing query in the basic signature scheme, $(\sigma', \mu', [\ell])$ is a valid forgery.

4.3   Unforgeability Against Insider Corruption Attacks

Suppose that we have an algorithm $\mathcal{A}$ able to break the unforgeability of our scheme with insider corruption attacks. We would like to prove again that there exists an algorithm $\mathcal{B}$ which using $\mathcal{A}$ can break the unforgeability of the underlying basic signature scheme. The main issue is that when $\mathcal{B}$ gets the challenge from the basic signature scheme he can split the tuple of polynomials but he does not know the signing keys associated to those split tuples and thus he cannot answer to the corresponding corruption queries.

One idea would be to pass to $\mathcal{A}$ modified versions of these tuples so that $\mathcal{B}$ knows the signing keys associated to them, but we have been unable to find a way to use the final forgery with such a strategy. A more standard approach to solve this issue is the one used by Wang and Sun in [13], which consists in giving to $\mathcal{A}$ more tuples than the ones obtained from splitting the initial challenge. These additional tuples are generated using the key generation algorithm and $\mathcal{B}$ knows the associated signing keys. If there are enough of them it is feasible to obtain a run from $\mathcal{A}$ in which all the corruption queries correspond to tuples for which $\mathcal{B}$ has the signing key.

Unfortunately, this creates a new issue in [13] as well as in our case. If we have a lot of tuples which do not correspond to the challenge then, for some strategies of $\mathcal{A}$ which are plausible, the final forgery is for a ring which contains tuples which were not in the challenge with overwhelming probability. In some number theory schemes this is not an issue, but in our case such forgeries are useless as $\mathcal{B}$ could have generated them (as the statistical distance between ring signatures by different members of the ring is negligible).

In the end, if the ring size for which $\mathcal{A}$ works is polynomial in the security parameter, for some strategies of $\mathcal{A}$ (in particular if he corrupts half of the ring and gives a ring signature for the other half), all the trade-offs fail. If we have a super logarithmic amount of new tuples there is an overwhelming probability to have a forgery for an inappropriate ring, and if not there is an overwhelming probability that $\mathcal{B}$ will be unable to answer to some of the corruption queries.

Our Approach. In order to solve these issues we modify the key generation process. Each user generates a set of k verification keys, k being the security parameter. Among these keys $k/2$ are generated through the original key generation algorithm Ring$-$gen$-$keys, and the user stores the associated signing keys. The other $k/2$ verification keys are just uniformly chosen hash functions. The k verification keys are numbered, the order being chosen at random (mixing both types of keys).

Ring$-$gen$-$keys$-$ic$(\mathcal{P})$:

1. Choose randomly a subset $T \subset [k]$ of size $k/2$.
2. For $i \in T$, set $pk_i \leftarrow \mathcal{H}(\mathcal{D}, D_h, m)$ and $sk_i = 0$.
3. For $i \in [k] \backslash T$, set $(pk_i, sk_i) \leftarrow$ Ring$-$gen$-$keys$(\mathcal{P})$.
4. Output $(pk, sk) = (\{pk_i\}_{i \in [k]}, \{sk_i\}_{i \in [k]})$.

For a set of users $S \subset \mathbb{Z}$ (we associate users to integers), we define fulldesc(S) as a description of the full set of verification keys of the users of S.

When signing a message μ for a set of users S, the user calls a first random oracle with input $(\mu, \text{fulldesc}(S))$. The output of this random oracle is $\{T_{\sigma,i}\}_{i \in S}$, a set of subsets of $[k]$, each of them of size $k/2$. In order to sign, the user will create a ring of verification keys T which includes for each user i the subset of verification keys indexed by $T_{\sigma,i}$.

Ring$-$sign$-$ic$(\mathcal{P}, sk, \mu, R)$:

0. Verify that: the public parameters respect the constraints of steps $1-3$ in Ring$-$gen$-$params ; each $sk_i \in sk$ is in $D_{s,c}^{m_u}$; R is of size bounded by k^c ; one of the public keys in R is associated to sk. If the verification fails output `failed`.
1. Set $\{T_{\sigma,i}\}_{i \in R} \leftarrow H_\sigma(\mu, \mathrm{keys}(R))$
2. Define $\mathrm{keys}(T) = (pk_{i,j})_{i \in R, j \in T_{\sigma,i}}$
3. Let i_0 denote the index of the signer in R. Choose randomly $sk_i \in sk$ with $i \in T_{\sigma,i_0}$ such that $sk_i \neq 0$. If none exists, abort.
4. Output Ring$-$sign$(\mathcal{P}, sk_i, \mu, T)$.

Note that since the random oracle chooses $k/2$ verification keys of the signer at random, the probability that they all are uniformly chosen random hash functions is exponentially small and thus the probability of an abort in Step 3 is negligible.

The adaptation of the verification algorithm to this new setting is straightforward.

Sketch of Proof in the Insider Corruption Setting. We just outline the ideas of the proof which can be obtained by small modifications in the proof of Theorem 4.

We use an insider corruption attacker to break a challenge of Lyubashevsky's signature. In order to do this we get as a challenge a tuple of $m' = k/2 \times \ell \times m_u$ polynomials. We use them in our ring signature algorithm instead of the uniformly chosen hash function obtained in the usual key generation process. The rest of the keys are generated according to Ring$-$gen$-$keys$-$ic (i.e. through the original algorithm Ring$-$gen$-$keys). Note that as half of the verification keys of each user are generated using this algorithm, we can answer to the corruption queries of the attacker.

In order to answer to the challenge we want the attacker to output a signature that only uses as verification keys the ones corresponding to the challenge and none of the ones generated by Ring$-$gen$-$keys. The main idea is to call the attacker with a controlled random oracle H_σ, pre-generate a polynomial number of the outputs of this oracle sampling uniformly its range and guess which one of these outputs will be used in the forgery. There is a polynomial loss in the reduction but using this approach we can then order the keys of the different users so that if the attacker uses the random oracle reply we expect, he will only use keys for which we do not have the associated signing key. When this happens (after a polynomial number of calls to the attacker), we obtain a forgery which can be used to answer the initial challenge.

References

1. Rivest, R.L., Shamir, A., Tauman, Y.: How to leak a secret. In: Boyd, C. (ed.) ASIACRYPT 2001. LNCS, vol. 2248, pp. 552–565. Springer, Heidelberg (2001)
2. Bender, A., Katz, J., Morselli, R.: Ring signatures: Stronger definitions, and constructions without random oracles. In: Halevi, S., Rabin, T. (eds.) TCC 2006. LNCS, vol. 3876, pp. 60–79. Springer, Heidelberg (2006)
3. Chaum, D., van Heyst, E.: Group signatures. In: Davies, D.W. (ed.) EUROCRYPT 1991. LNCS, vol. 547, pp. 257–265. Springer, Heidelberg (1991)
4. Dodis, Y., Kiayias, A., Nicolosi, A., Shoup, V.: Anonymous identification in *ad hoc* groups. In: Cachin, C., Camenisch, J.L. (eds.) EUROCRYPT 2004. LNCS, vol. 3027, pp. 609–626. Springer, Heidelberg (2004)
5. Abe, M., Ohkubo, M., Suzuki, K.: 1-out-of-n signatures from a variety of keys. In: Zheng, Y. (ed.) ASIACRYPT 2002. LNCS, vol. 2501, pp. 415–432. Springer, Heidelberg (2002)
6. Herranz, J., Sáez, G.: Forking lemmas for ring signature schemes. In: Johansson, T., Maitra, S. (eds.) INDOCRYPT 2003. LNCS, vol. 2904, pp. 266–279. Springer, Heidelberg (2003)
7. Shacham, H., Waters, B.: Efficient ring signatures without random oracles. In: Okamoto, T., Wang, X. (eds.) PKC 2007. LNCS, vol. 4450, pp. 166–180. Springer, Heidelberg (2007)
8. Zhang, F., Safavi-Naini, R., Susilo, W.: An efficient signature scheme from bilinear pairings and its applications. In: Bao, F., Deng, R., Zhou, J. (eds.) PKC 2004. LNCS, vol. 2947, pp. 277–290. Springer, Heidelberg (2004)
9. Boyen, X.: Mesh signatures. In: Naor, M. (ed.) EUROCRYPT 2007. LNCS, vol. 4515, pp. 210–227. Springer, Heidelberg (2007)
10. Brakerski, Z., Kalai, Y.T.: A framework for efficient signatures, ring signatures and identity based encryption in the standard model. Technical report, Cryptology ePrint Archive, Report 2010086 (2010)
11. Gentry, C., Peikert, C., Vaikuntanathan, V.: Trapdoors for hard lattices and new cryptographic constructions. In: Proceedings of the 40th Annual ACM Symposium on Theory of Computing, pp. 197–206. ACM (2008)
12. Cash, D., Hofheinz, D., Kiltz, E., Peikert, C.: Bonsai trees, or how to delegate a lattice basis. In: Gilbert, H. (ed.) EUROCRYPT 2010. LNCS, vol. 6110, pp. 523–552. Springer, Heidelberg (2010)
13. Wang, J., Sun, B.: Ring signature schemes from lattice basis delegation. In: Qing, S., Susilo, W., Wang, G., Liu, D. (eds.) ICICS 2011. LNCS, vol. 7043, pp. 15–28. Springer, Heidelberg (2011)
14. Micciancio, D., Peikert, C.: Trapdoors for lattices: Simpler, tighter, faster, smaller. In: Pointcheval, D., Johansson, T. (eds.) EUROCRYPT 2012. LNCS, vol. 7237, pp. 700–718. Springer, Heidelberg (2012)
15. Kawachi, A., Tanaka, K., Xagawa, K.: Concurrently secure identification schemes based on the worst-case hardness of lattice problems. In: Pieprzyk, J. (ed.) ASIACRYPT 2008. LNCS, vol. 5350, pp. 372–389. Springer, Heidelberg (2008)
16. Cayrel, P.-L., Lindner, R., Rückert, M., Silva, R.: A lattice-based threshold ring signature scheme. In: Abdalla, M., Barreto, P.S.L.M. (eds.) LATINCRYPT 2010. LNCS, vol. 6212, pp. 255–272. Springer, Heidelberg (2010)
17. Stern, J.: A new identification scheme based on syndrome decoding. In: Stinson, D.R. (ed.) CRYPTO 1993. LNCS, vol. 773, pp. 13–21. Springer, Heidelberg (1994)

18. Aguilar Melchor, C., Cayrel, P.L., Gaborit, P., Laguillaumie, F.: A new efficient threshold ring signature scheme based on coding theory. IEEE Transactions on Information Theory 57(7), 4833–4842 (2010)
19. Lyubashevsky, V.: Towards practical lattice-based cryptography. PhD thesis, University of California, San Diego (2008)
20. Lyubashevsky, V.: Fiat-shamir with aborts: Applications to lattice and factoring-based signatures. In: Matsui, M. (ed.) ASIACRYPT 2009. LNCS, vol. 5912, pp. 598–616. Springer, Heidelberg (2009)
21. Lyubashevsky, V.: Lattice signatures without trapdoors. In: Pointcheval, D., Johansson, T. (eds.) EUROCRYPT 2012. LNCS, vol. 7237, pp. 738–755. Springer, Heidelberg (2012)
22. Lenstra, A.K., Hughes, J.P., Augier, M., Bos, J.W., Kleinjung, T., Wachter, C.: Ron was wrong, whit is right. Cryptology ePrint Archive, Report 2012/064 (2012), http://eprint.iacr.org/
23. Lyubashevsky, V., Micciancio, D.: Generalized compact knapsacks are collision resistant. In: Bugliesi, M., Preneel, B., Sassone, V., Wegener, I. (eds.) ICALP 2006. LNCS, vol. 4052, pp. 144–155. Springer, Heidelberg (2006)
24. Micciancio, D., Goldwasser, S.: Complexity of Lattice Problems: a cryptographic perspective. The Kluwer International Series in Engineering and Computer Science, vol. 671. Kluwer Academic Publishers, Boston (2002)
25. Stehlé, D., Steinfeld, R., Tanaka, K., Xagawa, K.: Efficient public key encryption based on ideal lattices. In: Matsui, M. (ed.) ASIACRYPT 2009. LNCS, vol. 5912, pp. 617–635. Springer, Heidelberg (2009)
26. Bennett, C.H., Brassard, G., Robert, J.M.: Privacy amplification by public discussion. SIAM J. Comput. 17(2), 210–229 (1988)

A   Proof of Theorem 2

The contents of this section are closely related to Theorem 6.5 in [19], and the associated proof, which deals with statistical distance between signatures with two related secret keys in the signature scheme we are based on. The first major difference when dealing with statistical distance comes from the fact that we do not have a single random polynomial $y \leftarrow D_y^{m_u}$ (resp. a random hash function) but a set of ℓ polynomials (resp. a set of hash functions) on each signature. The second major difference is that we have to verify that the adversarial nature of some parameters does not result in an explosion of the statistical distance. In order to make that clear we first introduce a lemma that we will need in the main proof.

In order to get lighter notations we will denote the random variables of the theorem X_0 and X_1 dropping the other parameters. As the output of the ring signature algorithm is a vector of $\ell+1$ coordinates we will note $X_b^{(i)}$ for $i \in [\ell+1]$ and $b \in \{0,1\}$ the random variable associated to the i-th coordinate of X_b. Suppose that none of these variables has {failed} as domain. We can then guarantee that the parameters given in the theorem verify the properties tested on step 0 of the algorithm Ring−sign. We will say that these parameters have passed the *sanity check*.

As in [19], we therefore start by noting that the set

$$D_{s,c}(sk_{i_0}, sk_{i_1}) = \{c \in D_{s,c} : \|sk_{i_0} c\|_\infty, \|sk_{i_1} c\|_\infty \leq \sqrt{n} \log n\}$$

has a cardinality negligibly close (in a relative sense) to the one of $D_{s,c}$. As the secret keys have passed the sanity check, they belong to $D_{s,c}$ and therefore, even if they are chosen adversarially, Lemma 1 guarantees that

$$\frac{|D_{s,c}(sk_{i_0}, sk_{i_1})|}{|D_{s,c}|} = 1 - n^{-\omega(1)}. \tag{4}$$

Note that n is also an adversarially chosen parameter but, again, the sanity check ensures that $n \geq k$ and thus that $n^{-\omega(1)}$ is a negligible function. Splitting the statistical distance in two we get

$$\Delta(X_0, X_1) = \frac{1}{2} \sum_{\hat{\alpha}_i \in D_z^{mu}; i \in [\ell], \beta \notin D_{s,c}(sk_{i_0}, sk_{i_1})}$$
$$|Pr[X_0 = (\hat{\alpha}_i; i \in [\ell], \beta)] - Pr[X_1 = (\hat{\alpha}_i; i \in [\ell], \beta)]| \tag{5}$$
$$+ \frac{1}{2} \sum_{\hat{\alpha}_i \in D_z^{mu}; i \in [\ell], \beta \in D_{s,c}(sk_{i_0}, sk_{i_1})}$$
$$|Pr[X_0 = (\hat{\alpha}_i; i \in [\ell], \beta)] - Pr[X_1 = (\hat{\alpha}_i; i \in [\ell], \beta)]|. \tag{6}$$

In order to prove that the statistical distance is negligible, we will first prove that (5) is negligible and then that (6) is equal to zero. The first assertion is almost trivial, as the last coordinate in the signature comes from a random oracle and thus the probability it does not belong to $D_{s,c}(sk_{i_0}, sk_{i_1})$ is negligible. Noting that this is true for X_0 as for X_1 and that $\sum |Pr[X_0 = (\hat{\alpha}_i; i \in [\ell], \beta)] - Pr[X_1 = (\hat{\alpha}_i; i \in [\ell], \beta)]| \leq \sum Pr[X_0 = (\hat{\alpha}_i; i \in [\ell], \beta)] + \sum Pr[X_0 = (\hat{\alpha}_i; i \in [\ell], \beta)]$ we can prove that

$$(5) \leq 1 - \frac{|D_{s,c}(sk_{i_0}, sk_{i_1})|}{|D_{s,c}|} = n^{-\omega(1)}. \tag{7}$$

More formally, we have

$$(5) \leq \frac{1}{2} \sum_{\hat{\alpha}_i \in D_z^{mu}; i \in [\ell], \beta \notin D_{s,c}(sk_{i_0}, sk_{i_1})} Pr[X_0 = (\hat{\alpha}_i; i \in [\ell], \beta)] + Pr[X_1 = (\hat{\alpha}_i; i \in [\ell], \beta)]$$

and for any $b \in \{0, 1\}$, noting that $\sum_{\forall A} Pr[A \wedge B] = \sum_{\forall A} Pr[A|B]Pr[B] = Pr[B]$, we have

$$\sum_{\hat{\alpha}_i \in D_z^{mu}; i \in [\ell], \beta \notin D_{s,c}(sk_{i_0}, sk_{i_1})} Pr[X_b = (\hat{\alpha}_i; i \in [\ell], \beta)] = \sum_{\beta \notin D_{s,c}(sk_{i_0}, sk_{i_1})} Pr[X_b^{(\ell+1)} = \beta].$$

Finally, noting that $X_b^{(\ell+1)}$ is obtained through a call to a random oracle $H(\sum_{i \in [\ell]} h_i(y_i), R, \mu)$, the probability it is equal to a given β is $1/|D_{s,c}|$. It is important to note that this is true, independently of the distribution of the input. Thus, even if the h_i in $H(\sum h_i(y_i), R, r\mu)$ are adversarially chosen (say, $h_i(x) = 0$ for all x and i), the probability given is still valid. Using this probability for all $\beta \notin D_{s,c}(sk_{i_0}, sk_{i_1})$, equation (7) follows immediately.

In order to prove that (6) is equal to zero we will show that each term in the sum is null. As the last coordinate of a signature is generated through a call to a random oracle we have for both random variables the same probability to obtain β. Therefore, we must prove that for each term in (6):

$$Pr\left[(X_0^{(i)}; i \in [\ell]) = (\hat{\alpha}_i; i \in [\ell]) | X_0^{(\ell+1)} = \beta\right] = Pr\left[(X_1^{(i)}; i \in [\ell]) = (\hat{\alpha}_i; i \in [\ell]) | X_1^{(\ell+1)} = \beta\right] \quad (8)$$

We will prove that for all $b \in \{0,1\}$, $Pr[X_b^{(i)} = \hat{\alpha}_i | X_b^{(\ell+1)} = \beta]$ is equal to $1/|D_z^{m_u}|$ if $i \in [\ell] \setminus i_b$ and to $1/|D_y^{m_u}|$ if $i = i_b$. This is enough to prove (8) as the first ℓ coordinates of the random variables are independently chosen in the signature algorithm.

We note $\hat{y}_{b,i}$ the variable $\hat{y}_i$ corresponding to an execution of the signature algorithm in which sk_{i_b} is used. For $i \neq i_b$, we have $X_b^{(i)} = \hat{\alpha}_i$ if $\hat{y}_{b,i} = \hat{\alpha}_i$. As $\hat{y}_{b,i}$ is drawn uniformly at random from $D_z^{m_u}$, and $\hat{\alpha}_i \in D_z^{m_u}$, the probability that both values are equal is $1/|D_z^{m_u}|$. For $i = i_b$, we have $X_b^{(i_b)} = \hat{\alpha}_{i_b}$ if $\hat{y}_{b,i_b} = \hat{\alpha}_{i_b} - sk_{i_b}\beta$. As $\hat{y}_{b,i}$ is drawn uniformly at random from $D_y^{m_u}$, the probability that it is equal to a given value is $1/|D_y^{m_u}|$ if this value is in $D_y^{m_u}$, and 0 if not. By the definition of $D_{s,c}(sk_{i_0}, sk_{i_1})$, to which β belongs, we have $sk_{i_b}\beta \leq \sqrt{n}\log n$ and thus $\hat{\alpha}_{i_b} - sk_{i_b}\beta$ belongs to $D_y^{m_u}$ which completes the proof. $\qquad\square$

B   Lemmas

Lemma 1 (Lemma 2.11 in [19] restricted to our setting). *Let a be any polynomial in $D_{s,c}$ and b a polynomial uniformly chosen in $D_{s,c}$. Then*

$$Pr[\|ab\|_\infty \geq \sqrt{n}\log n] \leq 4ne^{-\frac{\log^2 n}{8}}.$$

This lemma will be used to show that for any two secret keys, an overwhelming fraction of the polynomials in $D_{s,c}$ result in a small polynomial when multiplied by any of the two keys. Note that the lemma is valid for *any* polynomial a in $D_{s,c}$ (i.e. even if it is adversarially chosen).

Before proving Lemma 3, which is used in Proposition 2, we recall the following lemma which is adapted from Lemma 3 in [25].

Lemma 2. *Let $f = x^n + 1$, $r \geq 2$ and $n = 2^r$ and p is a prime with $p \equiv 3 \mod 8$, then there exist f_1, f_2 such that $f = f_1 f_2 \mod p$ where each f_i is irreducible in $\mathbb{Z}_p[x]$ and can be written $f_i(x) = x^{n/2} + t_i x^{n/4} - 1$ with $t_i \in \mathbb{Z}_p$*

Lemma 3. *Let $D_{s,c}^\times$ denote the set of non-invertible polynomials of $D_{s,c}$. We have*

$$Pr_{f \leftarrow D_{s,c}}\left[f \in D_{s,c}^\times\right] \leq \frac{2}{3^{n/2}}.$$

Proof (Sketch.). We have $\mathcal{D} = \mathbb{Z}_p[x]/\langle x^n + 1\rangle$ and by Lemma 2 and Table 1 we have $x^n + 1 = f_1 f_2 \mod p$, both factors being of degree $n/2$ and irreducible over $\mathbb{Z}_p[x]$. As these factors are irreducible, we have that the non-invertible polynomials of $\mathcal{D}$ are such that $f = 0 \mod f_1$ or $f = 0 \mod f_2$.

As $D_{s,c} = \{g \in \mathbb{Z}_p[x]/\langle x^n + 1\rangle : \|g\|_\infty \leq 1\}$ we have

$$D_{s,c}^\times = \{f \in \mathbb{Z}_p[x]/\langle x^n + 1\rangle : \|f\|_\infty \leq 1 \text{ and } (f = 0 \mod f_1 \text{ or } f = 0 \mod f_2)\}.$$

By the union bound, we have

$$Pr_{f \leftarrow D_{s,c}}\left[f \in D_{s,c}^\times\right] \leq Pr_{f \leftarrow D_{s,c}}\left[f = 0 \mod f_1\right] + Pr_{f \leftarrow D_{s,c}}\left[f = 0 \mod f_1\right].$$

It is easy to see that for each choice of $n/2$ higher order terms of f there is as most one choice on the $n/2$ lower order terms that satisfies $f = 0 \mod f_1$. The same can be said for $f = 0 \mod f_2$. Thus the probability of each of the right-hand terms is bounded from above by $1/3^{n/2}$ which immediately proves the result.

GPU-Based Implementation of 128-Bit Secure Eta Pairing over a Binary Field

Utsab Bose, Anup Kumar Bhattacharya, and Abhijit Das

Department of Computer Science and Engineering
Indian Institute of Technology Kharagpur, West Bengal, India
utsab.bose@yahoo.co.in, bhattacharya.anup@gmail.com,
abhij@cse.iitkgp.ernet.in

Abstract. Eta pairing on a supersingular elliptic curve over the binary field $F_{2^{1223}}$ used to offer 128-bit security, and has been studied extensively for efficient implementations. In this paper, we report our GPU-based implementations of this algorithm on an NVIDIA Tesla C2050 platform. We propose efficient parallel implementation strategies for multiplication, square, square root and inverse in the underlying field. Our implementations achieve the best performance when López-Dahab multiplication with four-bit precomputations is used in conjunction with one-level Karatsuba multiplication. We have been able to compute up to 566 eta pairings per second. To the best of our knowledge, ours is the fastest GPU-based implementation of eta pairing. It is about twice as fast as the only reported GPU implementation, and about five times as fast as the fastest reported single-core SIMD implementation. We estimate that the NVIDIA GTX 480 platform is capable of producing the fastest known software implementation of eta pairing.

Keywords: Supersingular elliptic curve, eta pairing, binary field, parallel implementation, GPU.

1   Introduction

Recently GPUs have emerged as a modern parallel computing platform for general-purpose programming. Many cryptographic algorithms have been implemented efficiently using GPU-based parallelization. Pairing (assumed symmetric) is a bilinear mapping of two elements in a group to an element in another group. Elliptic curves are widely used to realize various forms of pairing, like Weil pairing, Tate pairing, and eta pairing. Eta pairing, being one of the most efficient pairing algorithms, has extensive applications in identity-based and attribute-based encryption, multi-party communication, identity-based and short signatures, and autonomous authentication [6]. Investigating the extent of parallelizing eta pairing on GPU platforms is an important area of current research. Although several implementations of eta pairing have already been published in the literature [3,5,11], most of them are CPU-based, and aim at improving the performance of single eta-pairing computations. However,

A. Youssef, A. Nitaj, A.E. Hassanien (Eds.): AFRICACRYPT 2013, LNCS 7918, pp. 26–42, 2013.

many applications (like authentication in vehicular ad hoc networks) require computing large numbers of eta pairings in short intervals.

In this paper, we report efficient implementations of eta pairing on a supersingular elliptic curve defined over a field of characteristic two. This is a standard curve studied in the literature. At the time this work was done, this curve was believed to provide 128-bit security in cryptographic applications. Recent developments [13] tend to indicate that this security guarantee may be somewhat less.[1] To the best of our knowledge, there are no previous GPU-based implementations for this curve. The only GPU-based implementation of eta pairing reported earlier [15] is on a supersingular elliptic curve defined over a field of characteristic three, which also used to provide 128-bit security (but no longer now; see [19] and also [13]). Our implementation is about twice as efficient as this only known GPU implementation. We attempt to parallelize each pairing computation alongside multiple pairing computations, so as to exploit the GPU hardware as effectively as possible. In other words, we use both intra- and inter-pairing parallelization.

We start with parallel implementations of the binary-field arithmetic. We use López-Dahab multiplication [16] with four-bit windows in tandem with one-level Karatsuba multiplication [14] to obtain the best performance. We report the variation in performance of multiplication with the number of threads. We also report our GPU implementations of the square, square-root, inverse and reduction operations in the field. Finally, we use these parallel field-arithmetic routines for implementing eta pairing. Our best eta-pairing implementation is capable of producing up to 566 eta pairings per second on an NVIDIA TESLA C2050 platform. This indicates an average time of 1.76 ms per eta-pairing computation, which is comparable with the fastest known (1.51 ms) software implementation [3] (which is a multi-core SIMD-based implementation). The fastest reported single-core SIMD implementation [3] of eta pairing on this curve takes about 8 ms for each eta pairing. The only reported GPU-based implementation [15] of eta pairing is capable of producing only 254 eta pairings per second. We estimate, based upon previously reported results, that our implementation when ported to an NVIDIA GTX 480 platform is expected to give a 30% improvement in throughput, thereby producing 736 eta pairings per second, that is, 1.36 ms for computing one eta pairing.

It is worthwhile to note here that our work deals with only software implementations. Hardware implementations, significantly faster than ours, are available in the literature. For example, some recent FPGA implementations are described in [1,9]. The paper [1] also reports ASIC implementations. Both these papers use the same curve as we study here. Other types of pairing functions (like Weil, Tate, ate, and R-ate pairings) are also widely studied from implementation perspectives. The hardware and software implementations reported in [2,7,11] (to name a recent few) use other types of pairing.

---

[1] We are not changing the title of this paper for historical reasons, and also because of that the analysis presented in [13] is currently only heuristic.

The rest of the paper is organized as follows. In Section 2, a short introduction to GPUs and GPU programming is provided. Section 3 sketches the eta-pairing algorithm which we implement. Our implementation strategies for the binary-field arithmetic are detailed in Section 4. This is followed in Section 5 by remarks about our implementation of eta pairing. In Section 6, our experimental results are supplied and compared with other reported results. Section 7 concludes this paper after highlighting some future scopes of work.

2 NVIDIA Graphics Processing Units

In order to program in GPU, it is important to know the architectural details and how tasks are divided among threads at software level. Here, we briefly describe the Fermi architecture of CUDA and the programming model. Some comments on the GPU memory model are also in order.

2.1 GPU Architecture

The next-generation CUDA architecture, code-named *Fermi* [10], is an advanced and yet commonly available GPU computing architecture. With over three billion transistors and featuring up to 512 CUDA cores, it is one of the fastest GPU platforms provided by CUDA. Our implementations are made on one such Fermi-based GPU called TESLA C2050.

In TESLA C2050, there are 14 streaming multiprocessors (SMs) with a total of 448 CUDA cores. Each SM contains 32 CUDA cores along with 4 special function units and 16 load/store units. The detailed specification can be found in [17]. The 32 CUDA cores are arranged in two columns of 16 cores each. A program is actually executed in groups of 32 threads called *warps*, and a Fermi multiprocessor allocates a group of 16 cores (half warp) to execute one instruction from each of the two warps in two clock cycles. This allocation is done by two warp schedulers which can schedule instructions for each half warp in parallel. Each SM has a high-speed shared memory to be used by all threads in it.

2.2 GPU Programing Model

In order to get a good performance by parallel execution in GPUs, it is important to know how the threads can be organized at software level, and how these threads map to the hardware. A CUDA program is written for each thread. There is a programmer- or compiler-level organization of threads which directly project on to the hardware organization of threads. At user level, threads are grouped into blocks, and each block consists of several threads. A *thread block* (also called a *work group*) is a set of concurrently executing threads that can cooperate among themselves through barrier synchronization and shared memory. All the threads of a block must reside in the same SM. As the number of threads in a SM is limited because of limited number of registers and shared memory, there is a bound on the maximum number of threads that a single work group can have.

But we can have as many work groups as we want. In Fermi, the maximum work-group size is 1024. However, the threads within a work group can be arranged in one, two or three dimensions, and the work groups themselves can again be organized into one of the three dimensions. Each thread within a work group has a local ID, and each work group has a block ID. The block ID and the local ID together define the unique global ID of a thread.

In Fermi, there can be 48 active warps (1536 threads) and a maximum of eight active work groups per SM, that can run concurrently. So it is preferred to have a work group size of 192 in order to perfectly utilize all the active threads, provided that there is scope for hiding memory latency. The number of resident work groups in a SM is also bounded by the amount of shared memory consumed by each work group and by the number of registers consumed by each thread within each group.

2.3 GPU Memory Architecture

Each GPU is supplied with 3GB of device memory (also known as the global memory), which can be accessed by all the threads from all the multiprocessors of the GPU. One of the major disadvantages of this memory is its low band-width. When a thread in a warp (a group of 32 threads) issues a device memory operation, that instruction may eat up even hundreds of clock cycles. This per-formance bottleneck can be overcome to some extent by memory coalescing, where multiple memory requests from several threads in a warp are coalesced into a single request, making all the threads request from the same memory seg-ment. There is a small software-managed data cache (also known as the shared memory) associated with each multiprocessor. This memory is shared by all the threads executing on a multiprocessor. This low-latency high-bandwidth index-able memory running essentially at the register speed is configurable between 16KB and 48KB in Fermi architectures. In TESLA C2050, we have 48KB shared memory. We additionally have 16KB of hardware cache meant for high-latency global memory data. The hardware cache is managed by the hardware. Software programs do not have any control over the data residing in the hardware cache.

3 Eta Pairing in a Field of Characteristic Two

Here, we present the algorithm for eta (η_T) pairing [4] over the supersingular curve $y^2 + y = x^3 + x$ (embedding degree four) defined over the binary field $F_{2^{1223}}$ represented as an extension of F_2 by the irreducible polynomial $x^{1223} + x^{255} + 1$. As the embedding degree of the supersingular curve is four, we need to work in the field $F_{(2^{1223})^4}$. This field is represented as a tower of two quadratic extensions over $F_{2^{1223}}$, where the basis for the extension is given by $(1, u, v, uv)$ with $g(u) = u^2 + u + 1$ being the irreducible polynomial for the first extension, and with $h(v) = v^2 + v + u$ defining the second extension. The distortion map is given by $\phi(x, y) = (x + u^2, y + xu + v)$.

All the binary-field operations (addition, multiplication, square, square-root, reduction and inverse) are of basic importance in the eta-pairing algorithm, and are discussed in detail in Section 4. Now, we show the eta-pairing algorithm which takes two points P and Q on the supersingular curve $y^2 + y = x^3 + x$ as input with P having a prime order r, and which computes an element of μ_r as output, where μ_r is the group (contained in $F^*_{(2^{1223})^4}$) of the r-th roots of unity.

Algorithm 1. Eta-pairing algorithm for a field of characteristic two

---

Input: $P = (x_1, y_1), Q = (x_2, y_2) \in E(F_{2^{1223}})[r]$
Output: $\eta_T(P, Q) \in \mu_r$

1 **begin**
2 | $T \leftarrow x_1 + 1$
3 | $f \leftarrow T \cdot (x_1 + x_2 + 1) + y_1 + y_2 + (T + x_2)u + v$
4 | **for** $i = 1$ **to** 612 **do**
5 | | $T \leftarrow x_1$
6 | | $x_1 \leftarrow \sqrt{x_1},\ y_1 \leftarrow \sqrt{y_1}$
7 | | $g \leftarrow T \cdot (x_1 + x_2) + y_1 + y_2 + x_1 + 1 + (T + x_2)u + v$
8 | | $f \leftarrow f \cdot g$
9 | | $x_2 \leftarrow x_2^2,\ y_2 \leftarrow y_2^2$
10 | **end**
11 | **return** $f^{(q^2-1)(q-2\sqrt{q}+1)}$, *where* $q = 2^{1223}$
12 **end**

In Algorithm 1, $f \leftarrow f \cdot g$ is a multiplication in $F_{(2^{1223})^4}$ requiring eight multiplications in $F_{2^{1223}}$. This number can be reduced to six with some added cost of linear operations, as explained in [11]. Thus, the entire for loop (called the *Miller loop*) executes 1224 square-roots, 1224 squares, and 4284 multiplications. Multiplication being the most frequently used operation, its efficiency has a direct consequence on the efficiency of Algorithm 1.

4 Arithmetic of the Binary Field

An element of $F_{2^{1223}}$ is represented by 1223 bits packed in an array of twenty 64-bit words. All binary-field operations discussed below operate on these arrays.

4.1 Addition

Addition in $F_{2^{1223}}$ is word-level bit-wise XOR of the operands, and can be handled by 20 threads in parallel. In binary fields, subtraction is same as addition.

4.2 Multiplication

The multiplication operation is associated with some precomputation, where the results of multiplying the multiplicand with all four-bit patterns are stored in a two-dimensional array, called the precomputation matrix P. The quadratic multiplication loop involves processing four bits together from the multiplier. See [16] for the details. After the multiplication, the 40-word intermediate product is reduced back to an element of $F_{2^{1223}}$ using the irreducible polynomial $x^{1223} + x^{255} + 1$. To sum up, the multiplication consists of three stages: precomputation, computation of the 40-word intermediate product, and polynomial reduction. These three stages are individually carried out in parallel, as explained below.

Precomputation. The precomputation matrix P has 320 entries (words). We can use 320 threads, where each thread computes one $P(i, j)$. It is also possible to involve only 160 or 80 threads with each thread computing two or four matrix entries. The exact number of threads to be used in this precomputation stage is adjusted to tally with the number of threads used in the second stage (generation of the intermediate product).

Let us use 320 threads in this stage. Each thread uses its thread ID to determine which entry in P it should compute. Let $\Theta_{i,j}$ denote the thread responsible for computing $P(i, j)$. Let us also denote the i-th word of the multiplicand A by $A_i = (a_{64i+63}a_{64i+62} \ldots a_{64i+1}a_{64i})$, where each a_k is a bit, and the most significant bit in the word A_i is written first. Likewise, w_j is represented by the bit pattern $(b_3b_2b_1b_0)$. The thread $\Theta_{i,j}$ performs the following computations:

> Initialize $P(i, j)$ to $(000 \ldots 0)$.
> If $b_0 = 1$, XOR $P(i, j)$ with $(a_{64i+63}a_{64i+62} \cdots a_{64i+1}a_{64i})$.
> If $b_1 = 1$, XOR $P(i, j)$ with $(a_{64i+62}a_{64i+61} \cdots a_{64i}a_{64i-1})$.
> If $b_2 = 1$, XOR $P(i, j)$ with $(a_{64i+61}a_{64i+60} \cdots a_{64i-1}a_{64i-2})$.
> If $b_3 = 1$, XOR $P(i, j)$ with $(a_{64i+60}a_{64i+59} \cdots a_{64i-2}a_{64i-3})$.

In addition to the word A_i, the thread $\Theta_{i,j}$ needs to read the three most significant bits $a_{64i-1}a_{64i-2}a_{64i-3}$ from A_{i-1}. This is likely to incur conflicts during memory access from L1 cache, since the thread $\Theta_{i-1,j}$ also accesses the word A_{i-1}. In order to avoid this, three most significant bits of the words of A are precomputed in an array M of size 20. As a result, $\Theta_{i,j}$ reads only from A_i and M_i, whereas $\Theta_{i-1,j}$ reads from the different locations A_{i-1} and M_{i-1}. Since M depends only on A (not on w_j), only 20 threads can prepare the array M, and the resulting overhead is negligible compared to the performance degradation that was associated with cache conflicts.

Intermediate Product Computation. This stage proceeds like school-book multiplication. Instead of doing the multiplication bit by bit, we do it by chunks of four bits. Each word of the multiplier B contains sixteen such four-bit chunks. Figure 1 shows the distribution of the work among several threads. The threads use a temporary matrix R for storing their individual contributions. The use of R is necessitated by that the different threads can write in mutually exclusive cells of R. In practice, R is implemented as a one-dimensional array. However, Figure 1 shows it as a two-dimensional array for conceptual clarity. After all the entries in R are computed by all the threads, the 40-word intermediate product is obtained by adding the elements of R column-wise.

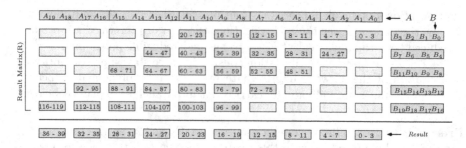

Fig. 1. Result Matrix Computation with 120 Threads

In Figure 1, we show how the result matrix R is computed by 120 threads. Here, R is of size 200 words, and is maintained as a 40×5 matrix. The figure, however, does not show these individual words (for lack of space). Instead, each box of R represents a group of four consecutive words. The range of four consecutive integers written within a box of R represents the IDs of the four threads that compute the four words in that box. In order to prevent synchronization overheads arising out of the race condition, different threads compute pair-wise different cells in R.

The first row of R corresponds to the multiplication of A by the least significant four words of B (that is, B_0, B_1, B_2, B_3). This partial product occupies 24 words which are computed by the threads with IDs 0–23, and stored in the columns 0–23 (column numbering begins from right). The second row of R is meant for storing the product of A with the next more significant four words of B (that is, B_4, B_5, B_6, B_7). This 24-word partial product is computed by a new set of 24 threads (with IDs 24–47), and stored in the second row of R with a shift of four words (one box). Likewise, the third, fourth and fifth rows of R are computed by 72 other threads. Notice that each row of R contains sixteen unused words, and need to be initialized to zero. After the computation of R, 40 threads add elements of R column-wise to obtain the intermediate product of A and B. Algorithm 2 elaborates this intermediate product generation stage. The incorporation of the precomputation table P is explicitly mentioned there. We use the symbols $\oplus$ (XOR), AND, OR, LEFTSHIFT and RIGHTSHIFT to stand for standard bit-wise operations.

In Algorithm 2, the values of $c - 4r - ID$ and $c - 4r - ID - 1$ at Lines 12 and 13 may become negative. In order to avoid the conditional check for negative values (which degrades performance due to warp divergence [8]), we maintain the precomputation matrix P as a 28×16 matrix instead of a 20×16 matrix. The first four and the last four columns of each row are initialized with zero entries. The actual values are stored in the middle 20 columns in each row.

---

[1] By a barrier, we mean that any thread must start executing instructions following MEM FENCE, only after all the currently running threads in a work group have completed execution up to the barrier. This ensures synchronization among the running threads.

Algorithm 2. Code for the i-th thread during intermediate product computation

---

Input: $A, B \in F_{2^{1223}}$ with precomputations on A stored in P
Output: The intermediate product $C = A \times B$ (a 40-word polynomial)

```
1  begin
2  │   r ← i/24
3  │   c ← i mod 24
4  │   r ← r + 4c
5  │   t₁ ← 0, t₂ ← 0
6  │   r ← r + 4
7  │   for ID = 0 to 4 do
8  │   │   w₂ ← B₄ᵣ₊ᵢᴅ
9  │   │   for j = 0 to 16 do
10 │   │   │   bit ← RIGHTSHIFT(w₂, 4j) AND 0x0F
11 │   │   │   w₁ ← P(bit, c − 4r − ID)
12 │   │   │   w₀ ← P(bit, c − 4r − ID − 1)
13 │   │   │   t₁ ← LEFTSHIFT(w₁, 4j) ⊕ RIGHTSHIFT(w₀, 64 − 4j)
14 │   │   │   t₂ ← t₂ ⊕ t₁
15 │   │   end
16 │   end
17 │   r ← r − 4
18 │   Rᵣ,ᵪ ← t₂
19 │   barrier(MEM FENCE)²
20 │   if i < 40 then
21 │   │   Resultᵢ ← R₀,ᵢ ⊕ R₁,ᵢ ⊕ R₂,ᵢ ⊕ R₃,ᵢ ⊕ R₄,ᵢ
22 │   end
23 │   barrier(MEM FENCE)
24 end
```

In the above implementation, each thread handles four words of the multiplier B. This can, however, be improved. If each thread handles only three consecutive words of B, we need $\lceil 20/3 \rceil = 7$ rows in R and $20 + 3 = 23$ used columns in each row. This calls for $7 \times 23 = 161$ threads. Since the 40-th word of the result is necessarily zero (the product of two polynomials of degrees < 1223 is at most $2444 \leq 2496 = 39 \times 64$), we can ignore the 161-st thread, that is, we can manage with 160 threads only.

Each thread can similarly be allowed to handle 1, 2, 5, 7, 10, or 20 words of B. The number of threads required in these cases are respectively 420, 220, 100, 60, 80 and 40. Increasing the number of threads to a high value may increase the extent of parallelism, but at the same time it will restrict the number of work groups that can be simultaneously active for concurrent execution in each multiprocessor. On the other hand, decreasing the number of threads to a small value increases the extent of serial execution within each thread. Thus, we need to reach a tradeoff between the number of threads and the amount of computation by each thread. Among all the above choices, we obtained the best performance with 160 threads, each handling three words of B.

One-level Karatsuba Multiplication. Using Karatsuba multiplication in conjunction with López-Dahab multiplication can speed up eta-pairing computations further, because Karatsuba multiplication reduces the number of F_2 multiplications at the cost of some linear operations. If we split each element A of $F_{2^{1223}}$ in two parts A_{hi} and A_{lo} of ten words each, López-Dahab multiplication of A and B requires four multiplications of ten-word operands ($A_{hi}B_{hi}$, $A_{lo}B_{lo}$, $A_{lo}B_{hi}$, and $A_{hi}B_{lo}$), as shown below (where $n = 640$):

$$(A_{hi}x^n + A_{lo})(B_{hi}x^n + B_{lo}) = A_{hi}B_{hi}x^{2n} + (A_{hi}B_{lo} + A_{lo}B_{hi})x^n + A_{lo}B_{lo}.$$

Karatsuba multiplication computes only the three products $A_{hi}B_{hi}$, $A_{lo}B_{lo}$, and $(A_{hi} + A_{lo})(B_{hi} + B_{lo})$, and obtains

$$A_{hi}B_{lo} + A_{lo}B_{hi} = (A_{hi} + A_{lo})(B_{hi} + B_{lo}) + A_{hi}B_{hi} + A_{lo}B_{lo}$$

using two addition operations. Each of the three ten-word multiplications is done by López-Dahab strategy. Each thread handles two words of the multiplier, and the temporary result matrix R contains five rows and twenty columns. Twelve threads write in each row, so the total number of threads needed for each ten-word multiplication is 60. All the three ten-word multiplications can run concurrently, thereby using a total of 180 threads.

Two levels of Karatsuba multiplication require only nine five-word multiplications instead of 16. Although this seems to yield more speedup, this is not the case in practice. First, the number of linear operations (additions and shifts that cannot be done concurrently) increases. Second, precomputation overheads associated with López-Dahab multiplication also increases. Using only one level of Karatsuba multiplication turns out to be the optimal strategy.

Reduction. The defining polynomial $x^{1223} + x^{255} + 1$ is used to reduce the terms of degrees ≥ 1223 in the intermediate product. More precisely, for $n \geq 0$, the non-zero term x^{1223+n} is replaced by $x^{255+n} + x^n$. This is carried out in parallel by 20 threads, where the i-th thread reduces the $(20 + i)$-th word. All the threads first handle the adjustments of x^{255+n} concurrently. Reduction of the entire $(20 + i)$-th word affects both the $(5 + i)$-th and the $(4 + i)$-th words. In order to avoid race condition, all the threads first handle the $(5 + i)$-th words concurrently. Subsequently, after a synchronization, the $(4 + i)$-th words are handled concurrently again. The adjustments of x^n in a word level are carried out similarly. Note that for large values of n, we may have $255 + n \geq 1223$. This calls for some more synchronization of the threads.

4.3   Square

We precompute the squares of all 8-bit patterns, and store them in an array Q of 256 words [12]. The parallel implementation of squaring in $F_{2^{1223}}$ is done by a total of 40 threads. Each thread handles a half word (that is, 32 bits) of the operand A, consults the precomputation table Q four times, and stores the partial result in an array R. More precisely, the threads $2i$ and $2i + 1$ read the word A_i. The $2i$-th thread reads the least significant 32 bits of A_i, and writes the corresponding square value in R_{2i}, whereas the $(2i + 1)$-st thread writes the square of the most significant 32 bits of A_i in R_{2i+1}. The threads write in pairwise distinct words of R, so they can run concurrently without synchronization.

Algorithm 3. Code for the i-th thread during squaring

Input: An element of $A \in F_{2^{1223}}$ and the precomputed table Q
Output: The intermediate 40-word square $R = A^2$

```
1  begin
2  │   T ← A_{i/2}
3  │   if  i is odd then
4  │   │   T ← RIGHTSHIFT(T, 32)
5  │   end
6  │   RT ← 0
7  │   for j = 0 to 3 do
8  │   │   byte ← T AND 0xFF
9  │   │   T ← RIGHTSHIFT(T, 8)
10 │   │   RT ← RT ⊕ RIGHTSHIFT(Q[byte], 16j)
11 │   end
12 │   Result_i ← RT
13 │   barrier(MEM FENCE)
14 end
```

4.4   Square-Root

Write an element A of $F_{2^{1223}}$ as $A = A_{even}(x) + xA_{odd}(x)$, where

$$A_{even}(x) = a_{1222}x^{1222} + a_{1220}x^{1220} + \cdots + a_2x^2 + a_0,$$
$$A_{odd}(x) = a_{1221}x^{1220} + a_{1219}x^{1218} + \cdots + a_3x^2 + a_1.$$

Then,

$$\sqrt{A} = A_{even}(\sqrt{x}) + \sqrt{x}A_{odd}(\sqrt{x})$$
$$= (a_{1222}x^{611} + a_{1220}x^{610} + \cdots + a_2x + a_0) +$$
$$\sqrt{x}(a_{1221}x^{610} + a_{1219}x^{609} + \cdots + a_3x + a_1).$$

Moreover, since $x^{1223} + x^{255} + 1 = 0$, we have $\sqrt{x} = x^{612} + x^{128}$.

We use 40 threads for this computation. Twenty threads $\Theta_{even,i}$ compute $A_{even}(\sqrt{x})$, and the remaining twenty threads $\Theta_{odd,i}$ compute $A_{odd}(\sqrt{x})$. For $j = 0, 1, 2, \ldots, 9$, the thread $\Theta_{even,2j}$ reads only the even bits of A_{2j}, that is, $(a_{128j+62} \cdots a_{128j+2}a_{128j})$, and stores them in the least significant 32 bits of an array $T_{even}[0][j]$. On the other hand, for $j = 0, 1, 2, \ldots, 9$, the thread $\Theta_{even,2j+1}$

reads only the even bits of A_{2j+1}, that is, $(a_{128j+126}\ldots a_{128j+66}a_{128j+64})$ and stores them in the most significant 32 bits of $T_{even}[1][j]$. Likewise, for $j = 0, 1, 2, \ldots, 9$, $\Theta_{odd,2j}$ writes $(a_{128j+63}\ldots a_{128j+3}a_{128j+1})$ in the least significant 32 bits of $T_{odd}[0][j]$, and $\Theta_{odd,2j+1}$ writes $(a_{128j+127}\ldots a_{128j+67}a_{128j+65})$ in the most significant 32 bits of $T_{odd}[1][j]$. After all these threads finish, ten threads add T_{even} column-wise, and ten other threads add T_{odd} column-wise. The column-wise sum T_{odd} is shifted by 612 and 128 bits, and the shifted arrays are added to the column-wise sum of T_{even}. The details are shown as Algorithm 4, where we have used flattened representations of various two-dimensional arrays.

Algorithm 4. Code for the i-th thread for square-root computation

Input: An element $A \in F_{2^{1223}}$
Output: $R = \sqrt{A}$

```
 1 begin
 2 │   d ← i/20, bit ← i/20
 3 │   ID ← i mod 20
 4 │   word ← A_i
 5 │   for j = 0 to 31 do
 6 │   │   W ← RIGHTSHIFT(word, bit) AND 1
 7 │   │   T ← T ⊕ LEFTSHIFT(W, bit/2)
 8 │   │   bit ← bit + 2
 9 │   end
10 │   if i is odd then
11 │   │   T ← LEFTSHIFT(T, 32)
12 │   end
13 │   R_{20d+10×(ID AND 1)+id/2} ← T
14 │   EvenOdd_{20d+ID} ← R_{20d+i} ⊕ R_{20d+10+i}
15 │   odd1_i ← ShiftBy612Bits(EvenOdd_{20+i}) // multiply by x^612
16 │   odd2_i ← ShiftBy128Bits(EvenOdd_{20+i}) // multiply by x^128
17 │   if i < 20 then
18 │   │   R_i ← odd1_i ⊕ odd2_i ⊕ EvenOdd_i
19 │   end
20 end
```

4.5 Inverse

Inverse is used only once during final exponentiation [18]. This is computed in parallel by the extended Euclidean gcd algorithm for polynomials [12]. In addition to finding the gcd γ of two polynomials A and B, it also finds polynomials g and h satisfying the Bézout relation $gA + hB = \gamma$. In the inverse computation, $B = f(x) = x^{1223} + x^{255} + 1$ is irreducible, and $A \neq 0$ is of degree < 1223, so $\gamma = 1$. Thus, the polynomial g computed by the algorithm is the inverse of A (modulo f). It is not necessary to compute the other polynomial h.

Algorithm 5. Code for the i-th thread for computing inverse

Input: A non-zero binary polynomial A, and the irreducible polynomial
$\quad f(x) = x^{1223} + x^{255} + 1$, each occupying 20 words

Output: $A^{-1} \bmod f$

```
 1  begin
 2  │   Uᵢ ← fᵢ, Vᵢ ← Aᵢ, g1ᵢ ← 0, g2ᵢ ← 0
 3  │   SVᵢ ← 0, SVᵢ₊₂₀ ← 0, SGᵢ ← 0, SGᵢ₊₂₀ ← 0
 4  │   if i = 0 then
 5  │   │   g2ᵢ ← 1
 6  │   end
 7  │   while TRUE do
 8  │   │   DEGᵢ ← WordDegree(Uᵢ)   // highest position of a 1-bit in Uᵢ
 9  │   │   DEGᵢ₊₂₀ ← WordDegree(Vᵢ) // highest position of a 1-bit in Vᵢ
10  │   │   barrier(MEM FENCE)
11  │   │   if i = 0 then
12  │   │   │   shared terminate ← 0
13  │   │   │   degU ← GetDegree(DEG, 20) // Find deg(U) from DEG[0–19]
14  │   │   │   degV ← GetDegree(DEG, 40) // Find deg(U) from DEG[20–39]
15  │   │   │   shared diff ← degU − degV
16  │   │   end
17  │   │   barrier(MEM FENCE)
18  │   │   d ← diff
19  │   │   if d ≤ 1 then
20  │   │   │   Uᵢ ↔ Vᵢ, g1ᵢ ↔ g2ᵢ, d ← −d
21  │   │   end
22  │   │   barrier(MEM FENCE)
23  │   │   k ← d/64, w ← d mod 64
24  │   │   SVᵢ₊ₖ ← Vᵢ, SGᵢ₊ₖ ← g2ᵢ
25  │   │   vw ← GetRightMostBits(Vᵢ₋₁, w)
26  │   │   gw ← GetRightMostBits(g2ᵢ₋₁, w)
27  │   │   SVᵢ₊ₖ ← LEFTSHIFT(SVᵢ₊ₖ, w) ⊕ vw
28  │   │   SGᵢ₊ₖ ← LEFTSHIFT(SGᵢ₊ₖ, w) ⊕ gw
29  │   │   barrier(MEM FENCE)
30  │   │   Uᵢ ← Uᵢ ⊕ SVᵢ, g1ᵢ ← g1ᵢ ⊕ SGᵢ
31  │   │   if i ≠ 0 and Uᵢ ≠ 0 then
32  │   │   │   terminate ← 1
33  │   │   end
34  │   │   barrier(MEM FENCE)
35  │   │   if (terminate = 0 and U₀ = 1) then
36  │   │   │   Terminate the loop
37  │   │   end
38  │   end
39  end
```

Two polynomials U and V are initialized to f and A. Moreover, g_1 is initialized to 0, and g_2 to 1. In each iteration, 20 threads compute the word degrees $\deg(U_i)$ and $\deg(V_i)$. Then, one thread computes $d = \deg(U) - \deg(V)$. If d is negative, the 20 threads swap U and V, and also g_1 and g_2. Finally, the threads subtract (add) $x^d V$ from U and also $x^d g_2$ from g_1. This is repeated until U reduces to 1. The detailed code is supplied as Algorithm 5.

In our implementation, this algorithm is further parallelized by using 40 threads (Algorithm 5 uses only 20 threads, for simplicity). The degree calculations of U and V can proceed in parallel. The swapping of (U, V) and (g_1, g_2) can also be parallelized. Finally, the two shifts $x^d V$ and $x^d g_2$ can proceed concurrently, and so also can the two additions $U + x^d V$ and $g_1 + x^d g_2$.

5   Parallel Implementations of Eta Pairing

Suppose that we want to compute n eta pairings in parallel. In our implementation, only two kernels are launched for this task. The first kernel runs the Miller loop for all these n eta-pairing computations. The output of the Miller loop is fed to the second kernel which computes the final exponentiation. Each kernel launches n work groups, each with 180 threads. Threads are launched as warps, so even if we use only 180 threads, the GPU actually launches six warps (that is, 192 threads) per work group.

At the end of each iteration of the Miller loop, the threads of each work group are synchronized. Out of the 180 threads in a work group, 80 threads are used to compute the two squares x_2^2 and y_2^2 in parallel, and also the two square-roots $\sqrt{x_1}$ and $\sqrt{y_1}$ in parallel. For these operations, only 44.45% of the threads are utilized. For most part of the six multiplications in an iteration, all the threads are utilized (we have used Karatsuba and López-Dahab multiplications together). The linear operations (assignments and additions) in each iteration are usually done in parallel using 20 threads. In some cases, multiple linear operations can proceed in parallel, thereby utilizing more threads. Clearly, the square, square-root, and linear operations are unable to exploit available hardware resources (threads) effectively. Nevertheless, since multiplication is the most critical field operation in the eta-pairing algorithm, our implementation seems to exploit parallelism to the best extent possible.

Each multiprocessor can have up to eight active work groups capable of running concurrently. Moreover, there are 14 multiprocessors in our GPU platform. Therefore, a total of 112 work groups can be simultaneously active. As a result, at most 112 eta-pairing computations can run truly in parallel, at least in theory. Our implementations corroborate this expected behavior.

6   Experimental Results

Our implementations are done both in CUDA and in OpenCL. Here, we report our OpenCL results only, since OpenCL gives us slightly better results, potentially because of the following reasons.

- Kernel initialization time is less in OpenCL than in CUDA. This may account for better performance of OpenCL over CUDA for small data sets. For large data sets, both OpenCL and CUDA are found to perform equally well.
- OpenCL can be ported to many other devices (like Intel and AMD graphics cards) with minimal effort.
- CUDA's synchronization features are not as flexible as those of OpenCL. In OpenCL, any queued operation (like memory transfer and kernel execution) can be forced to wait on any other set of queued operations. CUDA's instruction streams at the time of implementation are comparatively more restrictive. In other words, the in-line synchronization features of OpenCL have been useful in our implementation.

We have used the `CudaEvent()` API call for measuring time in CUDA, and the `clGetEventProfilingInfo()` API call in OpenCL. Table 1 shows the number of field multiplications (in millions/sec) performed for different numbers of threads. In the table, ω represents the number of words in the multiplier, that each thread handles. This determines the number of threads to be used, as explained in Section 4.2. Only the entry with 180 threads ($\omega = 3$) uses Karatsuba multiplication in tandem with López-Dahab multiplication.

Table 1. Number of $F_{2^{1223}}$ multiplications with different thread-block sizes

ω	Number of threads	Number of multiplications (millions/sec)
20	40	1.3
10	60	1.7
7	80	1.9
5	100	2.3
4	120	2.8
3	160	3.3
2	180 *	3.5
2	220	3.1
1	420	2.3

* With Karatsuba multiplication

Table 1 shows that the performance gradually improves with the increase in the number of threads, reaches the best for 180 threads, and then decreases beyond this point. This is because there can be at most 48 concurrent warps in a multiprocessor, and the number of work groups that can reside in each multiprocessor is 8. Since each warp has 32 threads, a work-group size of $(48/8) \times 32 = 192$ allows concurrent execution of all resident threads, thereby minimizing memory latency. With more than 192 threads, the extent of concurrency is restricted, and the performance degrades.

Table 2 shows the numbers of all field operations computed per second, along with the number of threads participating in each operation. The multiplication and square operations include reduction (by the polynomial $x^{1223} + x^{255} + 1$).

Table 2. Performance of binary-field arithmetic

Field operation	Number of threads	Number of operations (millions/sec)
Addition	20	100.09
Reduction	20	62.5
Square *	40	15.8
Square root	40	6.4
Multiplication *	180	3.5
Inverse	40	0.022

* Including reduction

Table 3 presents a comparative study of the performances of some eta-pairing implementations. Currently, the fastest software implementation of 128-bit secure eta pairing over fields of small characteristics is the eight-core CPU implementation of Aranha et al. [3]. Our implementation on Tesla C2050 is slightly slower than this. Our codes are readily portable to the GTX 480 platform, but unavailability of such a machine prevents us from carrying out the actual experiment. We, however, estimate that our implementation ported to a GTX 480 platform can be the fastest software implementation of 128-bit secure eta pairing.

Table 3. Comparison with other software implementations of η_T pairing

Implementation	Field	Platform	# cores	Clock freq (GHz)	Time per eta pairing (ms)
Hankerson et al. [11]	$F_{2^{1223}}$	CPU, Intel Core2	1	2.4	16.25
	$F_{3^{509}}$	CPU, Intel Core2	1	2.4	13.75
	$F_{p_{256}}$	CPU, Intel Core2	1	2.4	6.25
Beuchat et al. [5]	$F_{3^{509}}$	CPU, Intel Core2	1	2.0	11.51
	$F_{3^{509}}$	CPU, Intel Core2	2	2.0	6.57
	$F_{3^{509}}$	CPU, Intel Core2	4	2.0	4.54
	$F_{3^{509}}$	CPU, Intel Core2	8	2.0	4.46
Aranha et al. [3]	$F_{2^{1223}}$	CPU, Intel Core2	1	2.0	8.70
	$F_{2^{1223}}$	CPU, Intel Core2	2	2.0	4.67
	$F_{2^{1223}}$	CPU, Intel Core2	4	2.0	2.54
	$F_{2^{1223}}$	CPU, Intel Core2	8	2.0	1.51
Katoh et al. [15]	$F_{3^{509}}$	GPU, Tesla C2050	448	1.1	3.93
	$F_{3^{509}}$	GPU, GTX 480	480	1.4	3.01
This Work	$F_{2^{1223}}$	GPU, Tesla C2050	448	1.1	1.76
This Work	$F_{2^{1223}}$	GPU, GTX 480	480	1.4	1.36 *

* Estimated running time

7   Conclusion

In this paper, we report our GPU-based implementations of eta pairing on a supersingular curve over the binary field $F_{2^{1223}}$. Cryptographic protocols based on this curve offer 128-bit security, so efficient implementation of this pairing is an important issue to cryptographers. Throughout the paper, we report our optimization strategies for maximizing efficiency on a popular GPU architecture. Our implementations can be directly ported to similar GPU architectures, and have the potential of producing the fastest possible implementation of 128-bit secure eta pairing. Our implementations are most suited to applications where a large number of eta pairings need to be computed.

We end this paper after highlighting some possible extensions of our work.

- Our implementations can be applied *mutatis mutandis* to compute eta pairing on another popular supersingular curve defined over the field $F_{3^{509}}$.
- Other types of pairing (like Weil and Tate) and other types of curves (like ordinary) may also be studied for GPU-based implementations.
- Large prime fields involve working with large integers. Since integer arithmetic demands frequent carry manipulation, an efficient GPU-based implementation of prime-field arithmetic is a very challenging exercise, differing substantially in complexity from implementations of polynomial-based arithmetic of extension fields like $F_{2^{1223}}$ and $F_{3^{509}}$.

Acknowledgement. The authors wish to gratefully acknowledge many useful improvements suggested by the anonymous referees.

References

1. Adikari, J., Hasan, M.A., Negre, C.: Towards faster and greener cryptoprocessor for eta pairing on supersingular elliptic curve over $\mathbb{F}_{2^{1223}}$. In: Knudsen, L.R., Wu, H. (eds.) SAC 2012. LNCS, vol. 7707, pp. 166–183. Springer, Heidelberg (2013)
2. Aranha, D.F., Beuchat, J.-L., Detrey, J., Estibals, N.: Optimal eta pairing on supersingular genus-2 binary hyperelliptic curves. In: Dunkelman, O. (ed.) CT-RSA 2012. LNCS, vol. 7178, pp. 98–115. Springer, Heidelberg (2012)
3. Aranha, D.F., López, J., Hankerson, D.: High-speed parallel software implementation of the η_T pairing. In: Pieprzyk, J. (ed.) CT-RSA 2010. LNCS, vol. 5985, pp. 89–105. Springer, Heidelberg (2010)
4. Barreto, P.S.L.M., Galbraith, S., hÉigeartaigh, C.O., Scott, M.: Efficient pairing computation on supersingular Abelian varieties. Designs, Codes and Cryptography, 239–271 (2004)
5. Beuchat, J.-L., López-Trejo, E., Martínez-Ramos, L., Mitsunari, S., Rodríguez-Henríquez, F.: Multi-core implementation of the tate pairing over supersingular elliptic curves. In: Garay, J.A., Miyaji, A., Otsuka, A. (eds.) CANS 2009. LNCS, vol. 5888, pp. 413–432. Springer, Heidelberg (2009)
6. Blake, I., Seroussi, G., Smart, N.: Elliptic curves in cryptography. London Mathematical Society, vol. 265. Cambridge University Press (1999)
7. Fan, J., Vercauteren, F., Verbauwhede, I.: Efficient hardware implementation of $\mathbb{F}_p$-arithmetic for pairing-friendly curves. IEEE Transactions on Computers 61(5), 676–685 (2012)
8. Fung, W.W.L., Sham, I., Yuan, G., Aamodt, T.M.: Dynamic warp formation and scheduling for efficient GPU control flow. Micro 40, 407–420 (2007)

9. Ghosh, S., Roychowdhury, D., Das, A.: High speed cryptoprocessor for η_T pairing on 128-bit secure supersingular elliptic curves over characteristic two fields. In: Preneel, B., Takagi, T. (eds.) CHES 2011. LNCS, vol. 6917, pp. 442–458. Springer, Heidelberg (2011)
10. Glaskowsky, P.: NVIDIA's Fermi: The first complete GPU computing architecture. White paper, NVIDIA Corporation (2009)
11. Hankerson, D., Menezes, A., Scott, M.: Software implementation of pairings. In: Cryptology and Information Security Series, vol. 2, pp. 188–206. IOS Press (2009)
12. Hankerson, D., Menezes, A., Vanstone, S.: Guide to elliptic curve cryptography. Springer (2004)
13. Joux, A.: A new index calculus algorithm with complexity $L(1/4 + o(1))$ in very small characteristic. Cryptology ePrint Archive, Report 2013/095 (2013), http://eprint.iacr.org/2013/095
14. Karatsuba, A., Ofman, Y.: Multiplication of many-digital numbers by automatic computers. Doklady Akad. Nauk. SSSR 145, 293–294 (1962)
15. Katoh, Y., Huang, Y.-J., Cheng, C.-M., Takagi, T.: Efficient implementation of the η_T pairing on GPU. Cryptology ePrint Archive, Report 2011/540 (2011), http://eprint.iacr.org/2011/540
16. López, J., Dahab, R.: High-speed software multiplication in f2m. In: Roy, B., Okamoto, E. (eds.) INDOCRYPT 2000. LNCS, vol. 1977, pp. 203–212. Springer, Heidelberg (2000)
17. NVIDIA Corporation, CUDA: Compute unified device architecture programming guide. Technical Report, NVIDIA (2007)
18. Scott, M., Benger, N., Charlemagne, M., Dominguez Perez, L.J., Kachisa, E.J.: On the final exponentiation for calculating pairings on ordinary elliptic curves. In: Shacham, H., Waters, B. (eds.) Pairing 2009. LNCS, vol. 5671, pp. 78–88. Springer, Heidelberg (2009)
19. Shinohara, N., Shimoyama, T., Hayashi, T., Takagi, T.: Key length estimation of pairing-based cryptosystems using η_T pairing. Cryptology ePrint Archive, Report 2012/042 (2012), http://eprint.iacr.org/2012/042

On Constructions of Involutory MDS Matrices

Kishan Chand Gupta and Indranil Ghosh Ray

Applied Statistics Unit, Indian Statistical Institute,
203, B.T. Road, Kolkata 700108, India
{kishan,indranil_r}@isical.ac.in

Abstract. Maximum distance separable (MDS) matrices have applications not only in coding theory but also are of great importance in the design of block ciphers and hash functions. It is highly nontrivial to find MDS matrices which is involutory and efficient. In a paper in 1997, Youssef et. al. proposed an involutory MDS matrix construction using Cauchy matrix. In this paper we study properties of Cauchy matrices and propose generic constructions of low implementation cost MDS matrices based on Cauchy matrices. In a 2009 paper, Nakahara and Abrahao proposed a 16×16 involutory MDS matrix over $\mathbb{F}_{2^8}$ by using a Cauchy matrix which was used in MDS-AES design. Authors claimed that their construction by itself guarantees that the resulting matrix is MDS and involutory. But the authors didn't justify their claim. In this paper we study and prove that this proposed matrix is not an MDS matrix. Note that this matrix has been designed to be used in the block cipher MDS-AES, which may now have severe weaknesses. We provide an algorithm to construct involutory MDS matrices with low Hamming weight elements to minimize primitive operations such as exclusive-or, table look-ups and xtime operations. In a 2012 paper, Sajadieh et. al. provably constructed involutory MDS matrices which were also Hadamard in a finite field by using two Vandermonde matrices. We show that the same matrices can be constructed by using Cauchy matrices and provide a much simpler proof of their construction.

Keywords: Cauchy matrix, Diffusion, Involutory matrix, MDS matrix, MixColumn operation, Vector space, Subspace, Vandermonde matrix.

1 Introduction

Claude Shannon, in his paper "Communication Theory of Secrecy Systems" [24], defined *confusion* and *diffusion* as two properties, required in the design of block ciphers. One possibility of formalizing the notion of perfect diffusion is the concept of *multipermutation*, which was introduced in [23, 26]. Another way to define it is using MDS matrices. *Maximum Distance Separable (MDS) matrices* offer diffusion properties and is one of the vital constituents of modern age ciphers like Advanced Encryption Standard (AES) [6], Twofish [21, 22], SHARK [18], Square [5], Khazad [1], Clefia [25] and MDS-AES [10]. The stream cipher MUGI [27] uses MDS matrix in its linear transformations. MDS matrices

A. Youssef, A. Nitaj, A.E. Hassanien (Eds.): AFRICACRYPT 2013, LNCS 7918, pp. 43–60, 2013.

are also used in the design of hash functions. Hash functions like Maelstrom [7], Grøstl [8] and PHOTON family of light weight hash functions [9] use MDS matrices as main part of their diffusion layers.

Nearly all the ciphers use predefined MDS matrices for incorporating the diffusion property. Although in some ciphers the possibility of random selection of MDS matrices with some constraints is provided [30]. In this context we would like to mention that in papers [9, 11, 14, 19, 30], different constructions of MDS matrices are provided. In [9], authors constructed lightweight MDS matrices from *companion matrices* by exhaustive search. In [11], authors constructed efficient 4×4 and 8×8 matrices to be used in block ciphers. In [14, 19], authors constructed *involutory* MDS matrices using *Vandermonde matrices*. In [30], authors constructed new involutory MDS matrices using properties of *Cauchy matrices*.

There are two very popular approaches for the design of large MDS matrices. One involves Cauchy matrices [30] and the other uses Vandermonde matrices [14, 19]. In some recent works [9, 20, 29], MDS matrices have been constructed recursively from some suitable companion matrices for lightweight applications.

In [28], authors proposed a special class of *substitution permutation networks (SPNs)* that uses same network for both the encryption and decryption operations. The idea was to use *involutory* MDS matrix for incorporating diffusion. It may be noted that for ciphers like FOX [12] and WIDEA-n [13] that follow the Lai-Massey scheme, there is no need of involutory matrices.

In this paper we revisit and systematize the MDS matrix constructions using Cauchy matrices [30] and generalize it. We also study involutory MDS matrices where the entries are preferably of low *Hamming weight*.

Lacan and Fimes [14] constructed MDS matrices from two Vandermonde matrices. Sajadieh et. al. [19] constructed MDS matrices which were also involutory. They [19] also constructed involutory *Hadamard* MDS matrices in a finite field. In this paper we propose a Cauchy based MDS matrix construction and prove that this is Hadamard in the finite field. We further provide an interesting equivalence of our Cauchy based construction and the Vandermonde based "Hadamard involutory MDS matrix" construction of [19]. By this equivalence we have a much simpler proof of generalization of Corollary 2 of [19]. We also show that our method is faster than the Hadamard involutory MDS matrix construction of [19] in terms of time complexity.

In [10], authors proposed a new diffusion layer for their AES cipher that may replace the original *ShiftRow* and *MixColumn* layers. They proposed a new 16×16 matrix $M_{16 \times 16}$ for designing MDS-AES block cipher, which was claimed to be involutory and MDS. But the authors did not justify their claims. In this paper we prove that their claim is not correct and the constructed $M_{16 \times 16}$ matrix is not an MDS matrix. Our construction (Algorithm 2) may be used to generate 16×16 involutory MDS matrices which may be used in MDS-AES block cipher.

MDS matrices of low Hamming weight are desirable for efficient implementation. In this context it may be noted that multiplication by 1, which is the unit element of $\mathbb{F}_{2^n}$, is trivial. When α is the root of the constructing polynomial of $\mathbb{F}_{2^n}$, the multiplication by α can be implemented by a shift by one bit to the left and a conditional XOR with a constant when a carry bit is set (multiplication by α is often denoted as xtime). Multiplication by $\alpha + 1$ is done by a multiplication by α and one XOR operation. Multiplication by α^2 is done by two successive multiplications by α.

The organization of the paper is as follows: In Section 2 we provide definitions and preliminaries. In Section 3, we construct MDS matrices using Cauchy matrices. In Section 4 we study Cauchy and Vandermonde constructions for FFHadamard involutory MDS matrices. In Section 5 we show that the 16×16 matrix $M_{16 \times 16}$ as proposed in [10] is not MDS. We conclude the paper in Section 6.

2   Definition and Preliminaries

Let $\mathbb{F}_2 = \{0, 1\}$ be the finite field of two elements and $\mathbb{F}_{2^n}$ be the finite field of 2^n elements. Elements of $\mathbb{F}_{2^n}$ can be represented as polynomials of degree less than n over $\mathbb{F}_2$. For example, let $\beta \in \mathbb{F}_{2^n}$, then β can be represented as $\sum_{i=0}^{n-1} b_i \alpha^i$, where $b_i \in \mathbb{F}_2$ and α is the root of generating polynomial of $\mathbb{F}_{2^n}$. Another compact representation uses hexadecimal digits. Here the hexadecimal digits are used to express the coefficients of corresponding polynomial representation. For example $\alpha^7 + \alpha^4 + \alpha^2 + 1 = 1.\alpha^7 + 0.\alpha^6 + 0.\alpha^5 + 1.\alpha^4 + 0.\alpha^3 + 1.\alpha^2 + 0.\alpha + 1 = (10010101)_2 = 95_x \in \mathbb{F}_{2^8}$. We will often denote a matrix by $((a_{i,j}))$, where $a_{i,j}$ is the (i, j)-th element of the matrix.

The Hamming weight of an integer i is the number of nonzero coefficients in the binary representation of i and is denoted by $H(i)$. For example $H(5) = 2$, $H(8) = 1$.

$\mathbb{F}_{2^n}$ and $\mathbb{F}_2^n$ are isomorphic when both of them are regarded as *vector space* over $\mathbb{F}_2$. The isomorphism is given by $x = (x_1 \alpha_1 + x_2 \alpha_2 + \cdots + x_n \alpha_n) \mapsto (x_1, x_2 \cdots, x_n)$, where $\{\alpha_1, \alpha_2, \ldots, \alpha_n\}$ is a basis of $\mathbb{F}_{2^n}$.

Let $(H, +)$ be a *group* and G is a *subgroup* of $(H, +)$ and $r \in H$. Then $r + G = \{r + g : g \in G\}$ is *left coset* of G in H and $G + r = \{g + r : g \in G\}$ is *right coset* of G in H. If the operation $+$ in H is commutative, $r + G = G + r$, i.e. left coset is same as right coset, and $r + G$ is simply called coset of G in H. It follows that any two left cosets (or right cosets) of G in H are either identical or disjoint.

Definition 1. *Let $\mathbb{F}$ be a finite field and p and q be two integers. Let $x \to M \times x$ be a mapping from $\mathbb{F}^p$ to $\mathbb{F}^q$ defined by the $q \times p$ matrix M. We say that it is an MDS matrix if the set of all pairs $(x, M \times x)$ is an MDS code, i.e. a linear code of dimension p, length $p + q$ and minimal distance $q + 1$.*

An MDS matrix provides diffusion properties that have useful applications in cryptography. The idea comes from coding theory, in particular from maximum distance separable code (MDS). In this context we state two important theorems from coding theory.

Theorem 1. *[16, page 33] If C is an $[n, k, d]$ code, then $n - k \geq d - 1$.*

Codes with $n - k = d - 1$ are called maximum distance separable code, or MDS code for short.

Theorem 2. *[16, page 321] An $[n, k, d]$ code C with generator matrix $G = [I|A]$, where A is a $k \times (n-k)$ matrix, is MDS if and only if every square submatrix (formed from any i rows and any i columns, for any $i = 1, 2, \ldots, min\{k, n-k\}$) of A is nonsingular.*

The following fact is another way to characterize an MDS matrix.

Fact: 1 *A square matrix A is an MDS matrix if and only if every square submatrices of A are nonsingular.*

The following fact is immediate from the definition.

Fact: 2 *All square submatrices of an MDS matrix are MDS.*

One of the elementary row operations on matrices is multiplying a row of a matrix by a scalar except zero. MDS property remains invariant under such operations. So we have the following fact.

Fact: 3 *If A is an MDS matrix over $\mathbb{F}_{2^n}$, then A', obtained by multiplying a row (or column) of A by any $c \in \mathbb{F}_{2^n}^*$ is MDS.*

Fact: 4 *If A is an MDS matrix over $\mathbb{F}_{2^n}$, then $c.A$ is MDS for any $c \in \mathbb{F}_{2^n}^*$.*

Recall that many modern block ciphers use MDS matrices as a vital constituent to incorporate diffusion property. In general two different modules are needed for encryption and decryption operations. In [28], authors proposed a special class of SPNs that uses same network for both the encryption and decryption operation. The idea was to use involutory MDS matrices for incorporating diffusion.

Definition 2. *A matrix A is called involutory matrix if it satisfies the condition $A^2 = I$, i.e. $A = A^{-1}$.*

Several design techniques have been used in past for constructing MDS matrices including exhaustive search for small matrices. For large MDS matrices, the designers prefer the following two methods: One method involves Cauchy matrices [30] and the other method uses Vandermonde matrices [14, 19]. In this paper we study construction of involutory MDS matrices using Cauchy matrices. Before going into the construction, we discuss Cauchy matrix and its properties which are of special importance in our constructions.

Definition 3. *Given $x_0, x_1 \ldots, x_{d-1} \in \mathbb{F}_{2^n}$ and $y_0, y_1 \ldots, y_{d-1} \in \mathbb{F}_{2^n}$, such that $x_i + y_j \neq 0$ for all $0 \leq i, j \leq d - 1$, then the matrix $A = ((a_{i,j})), 0 \leq i, j \leq d-1$ where $a_{i,j} = \frac{1}{x_i + y_j}$ is called a Cauchy matrix [16, 30].*

It is known that

$$det(A) = \frac{\prod_{0 \leq i < j \leq d-1}(x_j - x_i)(y_j - y_i)}{\prod_{0 \leq i, j \leq d-1}(x_i + y_j)}.$$

So provided x_i's are distinct and y_j's are distinct and $x_i + y_j \neq 0$ for all $0 \leq i, j \leq d - 1$, $det(A) \neq 0$, i.e. A is nonsingular. So we have the following result.

Fact: 5 *For distinct $x_0, x_1 \ldots, x_{d-1} \in \mathbb{F}_{2^n}$ and $y_0, y_1 \ldots, y_{d-1} \in \mathbb{F}_{2^n}$, such that $x_i + y_j \neq 0$ for all $0 \leq i, j \leq d-1$, the Cauchy matrix $A = ((a_{i,j})), 0 \leq i, j \leq d-1$ where $a_{i,j} = \frac{1}{x_i + y_j}$, is nonsingular.*

From the definition of a Cauchy matrix we have the following fact.

Fact: 6 *Any square submatrix of a Cauchy matrix is a Cauchy matrix.*

From Fact 5 and Fact 6; and for distinct x_i's and y_j's, such that $x_i + y_j \neq 0$, all square submatrices of a Cauchy matrix are nonsingular. This leads to an MDS matrix construction [30]. Towards this we have the following Lemma, which we call a *Cauchy construction*.

Lemma 1. *For distinct $x_0, x_1 \ldots, x_{d-1}$ and $y_0, y_1 \ldots, y_{d-1}$, such that $x_i + y_j \neq 0$ for all $0 \leq i, j \leq d-1$, the matrix $A = ((a_{i,j}))$, where $a_{i,j} = \frac{1}{x_i + y_j}$ is an MDS matrix.*

Proof. It is to be noted that the matrix A is a Cauchy matrix. Also from Fact 6, all of its submatrices are Cauchy matrices. Since $x_0, x_1 \ldots, x_{d-1}$ and $y_0, y_1 \ldots, y_{d-1}$ are distinct and $x_i + y_j \neq 0$ for all $0 \leq i, j \leq d-1$, so from Fact 5, all square submatrices of A are nonsingular. So A is an MDS matrix. $\square$

Lemma 2. *Each row(or each column) of the $d \times d$ MDS matrix A, formed using construction of Lemma 1 has d distinct elements.*

Proof. The elements of i'th row of A are $\frac{1}{x_i + y_j}$ for $j = 0, \ldots, d-1$. Now $\frac{1}{x_i + y_{j_1}} = \frac{1}{x_i + y_{j_2}}$ for any two $j_1, j_2 \in \{0, \ldots, d-1\}$ such that $j_1 \neq j_2$ implies $y_{j_1} = y_{j_2}$, which is a contradiction to the fact that y_j's are distinct. Since i is arbitrary, the result holds for all rows of A. The proof for columns are similar. $\square$

Corollary 1. *The $d \times d$ MDS matrix A, formed using construction of Lemma 1 has at least d distinct elements.*

Definition 4. *[16, 19] The matrix*

$$V = van(v_0, \ldots, v_{d-1}) = \begin{pmatrix} 1 & v_0 & v_0^2 & v_0^3 & \cdots & v_0^{d-1} \\ 1 & v_1 & v_1^2 & v_1^3 & \cdots & v_1^{d-1} \\ \vdots & \vdots & \vdots & \vdots & & \vdots \\ 1 & v_j & v_j^2 & v_j^3 & \cdots & v_j^{d-1} \\ \vdots & \vdots & \vdots & \vdots & & \vdots \\ 1 & v_{d-1} & v_{d-1}^2 & v_{d-1}^3 & \cdots & v_{d-1}^{d-1} \end{pmatrix}$$

is called a Vandermonde matrix, where v_i's are from any finite or infinite field.

Fact: 7 *$det(V) = \prod_{i<j}(v_i - v_j)$, which is non zero if and only if the v_i's are distinct.*

In [14], authors proposed MDS matrix construction from Vandermonde matrices, which we call a *Vandermonde construction*. We record this important result in the following lemma.

Lemma 3. *[14, 19] For distinct $x_0, x_1, \ldots, x_{d-1}$ and $y_0, y_1, \ldots, y_{d-1}$, such that $x_i + y_j \neq 0$, the matrix AB^{-1} is an MDS matrix, where $A = van(x_0, \ldots, x_{d-1})$ and $B = van(y_0, \ldots, y_{d-1})$.*

Authors of [19] proposed techniques to produce involutory MDS matrices. We record this in the following lemma with slightly different notations.

Lemma 4. *[19] Let $A = van(x_0, \ldots, x_{d-1})$ and $B = van(y_0, \ldots, y_{d-1})$ are $d \times d$ invertible Vandermonde matrices in $\mathbb{F}_{2^n}$ satisfying $x_i = y_i + r$ and $x_i \neq y_j, i, j \in \{0, \ldots, d-1\}$, $r \in \mathbb{F}_{2^n}^*$, then AB^{-1} is an involutory MDS matrix.*

In [19], authors constructed a special form of MDS matrices called *Finite Field Hadamard* matrices, which is defined as follows:

Definition 5. *[2, 19] A $2^m \times 2^m$ matrix H is Finite Field Hadamard matrix (FFHadamard) in $\mathbb{F}_{2^n}$ if it can be represented as follows:*

$$H = \begin{pmatrix} U & V \\ V & U \end{pmatrix},$$

where the two submatrices U and V are also FFHadamard.

Fact: 8 *[19] Let $H = ((h_{i,j}))$ be a $2^m \times 2^m$ matrix whose first row is $(x_0 \, x_1 \ldots x_{2^m-1})$ and $h_{i,j} = x_{i \oplus j}$, then H is FFHadamard and is denoted by $H = had(x_0, \ldots, x_{2^m-1})$.*

Let $H = ((h_{i,j})) = had(x_0, \ldots, x_{2^m-1})$, where $x_i \in \mathbb{F}_{2^n}$ for $i \in \{0, \ldots, 2^m - 1\}$. Then clearly $H' = ((h'_{i,j}))$ is FFHadamard, where $h'_{i,j} = r + h_{i,j}$, $r \in \mathbb{F}_{2^n}$. Also if $r + x_i \neq 0$ for $i \in \{0, \ldots, 2^m - 1\}$, then it is easy to check that the matrix $H'' = ((h''_{i,j}))$, where $h''_{i,j} = \frac{1}{h'_{i,j}}$ is also FFHadamard. We now provide Fact 9 which will be used in Theorem 4.

Fact: 9 *[19] Let $G = \{x_0, \ldots, x_{2^m-1}\}$ be an additive subgroup of $\mathbb{F}_{2^n}$, where $x_0 = 0$ and $x_i + x_j = x_{i \oplus j}$. Let $H = ((h_{i,j}))$ be a $2^m \times 2^m$ matrix over $\mathbb{F}_{2^n}$, where $h_{i,j} = \frac{1}{r + x_{i \oplus j}}$, $r \in \mathbb{F}_{2^n} \setminus G$, then H is FFHadamard.*

In [19], authors defined *Special Vandermonde matrix (SV matrix)*, which we restate differently and equivalently.

Definition 6. *Let G be an additive subgroup of $\mathbb{F}_{2^n}$ of order 2^m, which is a linear span of m linearly independent elements $\{x_1, x_2, x_{2^2}, \ldots, x_{2^{m-1}}\}$ such that $x_i = \sum_{k=0}^{m-1} b_k x_{2^k}$, where $(b_0, b_1, \ldots, b_{m-1})$ is the binary representation of i. A Vandermonde matrix $van(y_0, \ldots, y_{2^m-1})$ is called a Special Vandermonde matrix (SV matrix) if $y_i = r + x_i$, where $r \in \mathbb{F}_{2^n}$.*

We restate the generalization of Corollary 2 of [19] in the following lemma.

Lemma 5. *[19] Let $A = van(x_0, \ldots, x_{2^m-1})$ and $B = van(y_0, \ldots, y_{2^m-1})$ are Special Vandermonde matrices in $\mathbb{F}_{2^n}$, where $y_i = x_0 + y_0 + x_i$ and $y_0 \notin \{x_0, \ldots, x_{2^m-1}\}$, then AB^{-1} is an FFHadamard involutory MDS matrix.*

The proof of Corollary 2 of [19] is several pages long. In Section 4 Theorem 5, we propose an alternative and a much simpler proof.

3 Construction of MDS and Involutory MDS Matrices

From Corollary 1, a $d \times d$ matrix constructed using Lemma 1 has at least d distinct elements. In this paper we construct $d \times d$ MDS matrices with exactly d distinct elements. It has two-fold advantage. Firstly, we have to find only d elements of our liking (say of low implementation cost) to form the MDS matrix using Cauchy construction. Secondly, for construction of efficient MDS matrices, it may be desirable to have minimum number of distinct entries to minimize the implementation overheads(See [11]).

Lemma 6. *Let $G = (x_0, x_1, \ldots, x_{d-1})$ be an additive subgroup of $\mathbb{F}_{2^n}$. Let us consider the coset $r + G$, $r \notin G$ of G having elements $y_j = r + x_j$, $j = 0, \ldots, d-1$. Then the $d \times d$ matrix $A = ((a_{i,j}))$, where $a_{i,j} = \frac{1}{x_i + y_j}$, for all $0 \leq i, j \leq d - 1$ is an MDS matrix.*

Proof. We first prove that $x_i + y_j \neq 0$ for all $0 \leq i, j \leq d - 1$. Now, $x_i + y_j = x_i + r + x_j = r + x_i + x_j \in r + G$. But $0 \notin r + G$ (as $r \notin G$ and $0 \in G$). So $x_i + y_j \neq 0$ for all $0 \leq i, j \leq d - 1$. Also all x_i's are distinct elements of the group G and y_j's are distinct elements of the coset $r + G$. Thus from Lemma 1, A is an MDS matrix. $\square$

Remark 1. Lemma 6 gives MDS matrix of order d, where d is a power of 2. When d is not a power of 2, the construction of $d \times d$ MDS matrices over $\mathbb{F}_{2^n}$ ($d < 2^{n-1}$) is done in two steps. Firstly we construct $2^m \times 2^m$ MDS matrix A' over $\mathbb{F}_{2^n}$, where $2^{m-1} < d < 2^m$, using Lemma 6. In the next step, we select $d \times d$ submatrix A of A' of our liking (select d rows and d columns).

Fact: 10 *Lemma 3 of [30] is a particular case of Lemma 6 of this paper.*

Corollary 2. *The matrix A of Lemma 6 is symmetric.*

Proof. From definition, $a_{i,j} = a_{j,i} = \frac{1}{r + x_i + x_j}$ for all $0 \leq i, j \leq d - 1$. Thus A is symmetric matrix. $\square$

Lemma 7. *The $d \times d$ matrix A of Lemma 6 has exactly d distinct entries.*

Proof. In the ith row the elements are $a_{i,j} = \frac{1}{r + x_i + x_j}$ for $j = 0, 1, \ldots, d - 1$. Since x_j's form the additive group G, $x_i + x_j$ for $j = 0, 1, \ldots, d - 1$ gives all d distinct elements of G for a fixed i. Thus $r + x_i + x_j$ for $j = 0, 1, \ldots, d - 1$ gives all d distinct elements of $r + G$. Since i is arbitrary, therefore in each row of A, there are d distinct elements. Since these elements are nothing but the multiplicative inverse of elements of $r + G$ in $\mathbb{F}_{2^n}$, the matrix A has exactly d different elements. $\square$

Corollary 3. *By Lemma 2 and Lemma 7, it is evident that all rows of matrix A constructed by Lemma 6 are the permutations of the first row of A.*

Lemma 6 provides construction of MDS matrices. These matrices may not be involutory. In general, in substitution permutation networks (SPN) decryption needs inverse of A. If A is a low implementation-cost MDS matrix, then it is desirable that $A = A^{-1}$, otherwise implementation of A^{-1} may not be efficient. So we may like to make our MDS matrix to be involutory. Towards this we study the following Lemma which is also given in [30], but in a slightly different setting.

Lemma 8. *Let $A = ((a_{i,j}))$ be the $d \times d$ matrix formed by Lemma 6. Then $A^2 = c^2 I$, where $c = \sum_{k=0}^{d-1} \frac{1}{r+x_k}$.*

Proof. Let $A^2 = H = ((h_{i,j}))$. From Corollary 2, A is symmetric matrix. Therefore $h_{i,j}$ is the inner product of i'th row and j'th row of A. Therefore $h_{i,i} = \sum_{l=0}^{d-1} \frac{1}{(r+x_i+x_l)^2} = \sum_{k=0}^{d-1} \frac{1}{(r+x_k)^2} = c^2$ as x_i's and x_l's are elements of a group which is a subgroup of $\mathbb{F}_{2^n}$ of characteristic 2. Similarly for $i \neq j$, $h_{i,j} = \sum_{k=0}^{d-1} \frac{1}{(r+x_i+x_k)(r+x_j+x_k)} = \frac{1}{x_i+x_j} \sum_{k=0}^{d-1} \frac{1}{(r+x_i+x_k)} + \frac{1}{(r+x_j+x_k)} =$
$\frac{1}{x_i+x_j} \left(\sum_{k=0}^{d-1} \frac{1}{(r+x_i+x_k)} + \sum_{k=0}^{d-1} \frac{1}{(r+x_j+x_k)} \right) =$
$\frac{1}{x_i+x_j} \left(\sum_{l=0}^{d-1} \frac{1}{r+x_l} + \sum_{l'=0}^{d-1} \frac{1}{r+x_{l'}} \right)$. Since $\{r + x_l : l = 0, \ldots, d-1\} = r + G$ and we are working on a field $\mathbb{F}_{2^n}$ of characteristic 2, therefore $\left(\sum_{l=0}^{d-1} \frac{1}{r+x_l} + \sum_{l'=0}^{d-1} \frac{1}{r+x_{l'}} \right) = 0$. So $h_{i,j} = 0$. Thus $A^2 = c^2 I$. $\square$

Corollary 4. *The matrix A of Lemma 6 is involutory if the sum of the elements of any row is 1.*

Proof. The sum of elements of any row of A is equal to $\sum_{i=0}^{d-1} \frac{1}{r+x_i} = c$, where c is as defined in Lemma 8. So if $c = 1$, $c^2 = 1$ and hence $A^2 = I$ (See Lemma 8). $\square$

Corollary 5. *If $d \times d$ MDS matrix A is constructed using Lemma 6, then $\frac{1}{c}A$ is an involutory MDS matrix, where $c = \sum_{k=0}^{d-1} \frac{1}{r+x_k}$.*

Proof. From Lemma 8, $(\frac{1}{c}A)^2 = I$ and from Fact 4, $\frac{1}{c}A$ is MDS. $\square$

Remark 2. Multiplication in $\mathbb{F}_{2^n}$ by 1 is trivial. So for implementation friendly design, it is desirable to have maximum number of 1's in MDS matrices to be used in block ciphers and hash functions. We know that each element in a $d \times d$ matrix A constructed by Lemma 6, occurs exactly d times (See Lemma 7). So in the construction of $d \times d$ matrix A by Lemma 6, maximum d number of 1's can occur in A. It is to be noted that A can be converted to have maximum number of 1's (i.e. d number of 1's) without disturbing the MDS property just by multiplying A by inverse of one of its entries (See Fact 4). Although this will guarantee occurrence of 1's in every row, but with this technique we may not control Hamming weights of other $d-1$ elements. Also if A is an involutory MDS matrix, such conversion will disturbe the involutory property.

Remark 3. In [11], authors introduced the idea of efficient MDS matrices by maximizing the number of 1's and minimizing the number of occurrences of other distinct elements from $\mathbb{F}_{2^n}^*$. It is to be noted that multiplication of each row of $d \times d$ MDS matrix A by inverse of the first elements of the respective rows will lead to an MDS matrix A' having all 1's in first column (See Fact 3). Again by multiplying each columns of $d \times d$ MDS matrix A' (starting from the second column) by inverse of the first elements of the respective columns will lead to an MDS matrix A'' having all 1's in first row and first column. Thus the number of 1's in this matrix is $2d - 1$. Although A'' contains maximum number of 1's that can be achieved starting from the MDS matrix A, but the number of other distinct terms in this case may be greater than $d - 1$. Also A'' will never be involutory.

3.1 Construction of Some Additive Subgroup G of $\mathbb{F}_{2^n}$

Recall that $\mathbb{F}_{2^n}$ and $\mathbb{F}_2^n$ are isomorphic when both of them are regarded as n dimentional vector space over $\mathbb{F}_2$. Any *subspace* of $\mathbb{F}_2^n$ is by definition an additive subgroup of $\mathbb{F}_{2^n}$. Let $B = \{x_0, \ldots, x_{m-1}\}$ be m linearly independent elements of $\mathbb{F}_{2^n}$. Then the linear span of B, denoted by G, is a subspace of $\mathbb{F}_2^n$ of dimension m and is an additive subgroup of $\mathbb{F}_{2^n}$. So G can be used to construct MDS matrix using Lemma 6. Also note that r in Lemma 6 can be any element of $\mathbb{F}_{2^n} \setminus G$. Our aim is to construct efficient MDS matrices. Hamming weights of the elements in the MDS matrix may decide the number of table lookups, xor and xtime operations. The higher order bits of each entries in the matrix affects the number of calls to xtime. In the construction of MDS matrices by Lemma 6, the elements of the matrices are inverses of the elements of $r + G$ (See Lemma 6). So it is desirable that multiplicative inverses of elements of $r + G$ in $\mathbb{F}_{2^n}$ must be of low Hamming weights and also all the 1's should be towards the lower order bits.

3.2 An Algorithm to Construct MDS Matrix

Based on Lemma 6, we now provide Algorithm 1 to construct $2^m \times 2^m$ MDS matrix over $\mathbb{F}_{2^n}$, where $m < n$. Algorithm 1 gives MDS matrix and when the input parameter $b_{Involutory}$ is set *true*, the Algorithm 1 gives involutory MDS matrix of order $d \times d$, where d is power of 2. When d is not a power of 2, the construction of $d \times d$ MDS matrices over $\mathbb{F}_{2^n}$ ($d < 2^{n-1}$) is done in two steps (see Remark 1). Firstly we construct $2^m \times 2^m$ MDS matrix A over $\mathbb{F}_{2^n}$ using Algorithm 1 and keeping input parameter $b_{Involutory} = false$, where $2^{m-1} < d < 2^m$. In the next step, we just select some suitable $d \times d$ submatrix A' of A of our liking (select d rows and d columns of our liking). Note that A'^2 may not be equal to $c^2 I$, where $c \in \mathbb{F}_{2^n}^*$. Although the matrix A' is MDS, it is not involutory (See Example 1).

Remark 4. The additive subgroup $G = \{x_0, \ldots, x_{2^m-1}\}$ in Algorithm 1 is constructed by the linear combination of m linearly independent elements labeled $x_1, x_2, x_{2^2} \ldots, x_{2^{m-1}}$ in Step 1. Note that for such group G, $x_i + x_j = x_{i \oplus j}$, $x_i, x_j \in G$. For such G, the constructed matrix A in Algorithm 1 is FFHadamard

Algorithm 1. Construction of $2^m \times 2^m$ MDS matrix or Involutory MDS matrix over $\mathbb{F}_{2^n}$

Input $n > 1$, the generating polynomial $\pi(x)$ of $\mathbb{F}_{2^n}$, $m < n$ and $b_{Involutory}$.
Output Outputs a $2^m \times 2^m$ MDS matrix A.

1: Select m linearly independent elements, labeled $x_1, x_2, x_{2^2} \ldots, x_{2^{m-1}}$ from $\mathbb{F}_{2^n}$;
2: Construct G, the set of 2^m elements $x_0, x_1, x_2, x_3, \ldots, x_{2^m-1}$, where $x_i = \sum_{k=0}^{m-1} b_k x_{2^k}$, for all $0 \leq i \leq 2^m - 1$, $(b_{m-1}, b_{m-1}, \ldots, b_1, b_0)$ being the binary representation if i;
3: Select some $r \in \mathbb{F}_{2^n} \setminus G$;
4: Construct $r + G$, the set of 2^m elements $y_0, y_1, y_2, y_3, \ldots, y_{2^m-1}$, where $y_i = r + x_i$ for all $0 \leq i \leq 2^m - 1$;
5: **if** $(b_{Involutory} == false)$: Construct $\frac{1}{y_i}$; **else** construct $\frac{1}{cy_i}$ for $i = 0, \ldots, d-1$ in the array ary_s, where $c = \sum_{k=0}^{d-1} \frac{1}{r+x_k}$.
6: Construct the $2^m \times 2^m$ matrix $A = ((a_{i,j}))$, where $a_{i,j} = ary\_s[k]$, where $i \oplus j = k$;

7: Set A as output;

(see Theorem 4). If the ordering is disturbed in Step 1 by labeling the elements differently, so that $x_i + x_j \neq x_{i \oplus j}$, the matrix A may not be FFHadamard, although it will be MDS. We maintain the same ordering while constructing additive subgroup in Algorithm 2. So Algorithm 2 also produces FFHadamard matrices.

Theorem 3. *Algorithm 1 generates $d \times d$ MDS or Involutory MDS matrices over $\mathbb{F}_{2^n}$ where $d = 2^m$, and the complexity is $O(d^2)$ operations in $\mathbb{F}_{2^n}$.*

Proof. The correctness of Algorithm 1 is immediate from Lemma 6, Lemma 8 and Corollary 4. In Algorithm 1, Step 1-Step 5 takes $O(d)$ operations. Step 6 takes $O(d^2)$ operations. Thus the time complexity of Algorithm 1 is $O(d^2)$. □

Example 1: Let $n = 8$, $d = 4$, $\pi(x) = x^8 + x^4 + x^3 + x + 1$ and $b_{Involutory} = false$. Set $r = 1$. Select $x_1 = \alpha^7 + \alpha^3 + \alpha^2$ and $x_2 = \alpha^7 + \alpha^6 + \alpha^5 + \alpha^4 + \alpha^2 + \alpha + 1$. Thus construct $x_0 = 0.x_1 + 0.x_2 = 0$ and $x_3 = 1.x_1 + 1.x_2 = \alpha^6 + \alpha^5 + \alpha^4 + \alpha^3 + \alpha + 1$. So $y_0 = 1$, $y_1 = \alpha^7 + \alpha^3 + \alpha^2 + 1$, $y_2 = \alpha^7 + \alpha^6 + \alpha^5 + \alpha^4 + \alpha^2 + \alpha$ and $y_3 = \alpha^6 + \alpha^5 + \alpha^4 + \alpha^3 + \alpha$. So we have from Lemma 6 (as implemented in Algorithm 1)

$$A = \begin{pmatrix} 01_x & 02_x & 03_x & d0_x \\ 02_x & 01_x & d0_x & 03_x \\ 03_x & d0_x & 01_x & 02_x \\ d0_x & 03_x & 02_x & 01_x \end{pmatrix}, \quad \frac{1}{c}A = \begin{pmatrix} 7a_x & f4_x & 8e_x & 01_x \\ f4_x & 7a_x & 01_x & 8e_x \\ 8e_x & 01_x & 7a_x & f4_x \\ 01_x & 8e_x & f4_x & 7a_x \end{pmatrix},$$

where $01_x = 1$, $02_x = \alpha$ $03_x = \alpha + 1$, $d0_x = \alpha^7 + \alpha^6 + \alpha^4$, $7a_x = \alpha^6 + \alpha^5 + \alpha^4 + \alpha^3 + \alpha$, $f4_x = \alpha^7 + \alpha^6 + \alpha^5 + \alpha^4 + \alpha^2$, $8e_x = \alpha^7 + \alpha^3 + \alpha^2 + \alpha$. Here $c = d0_x$. Note that the matrix A is MDS but not involutory and the matrix $\frac{1}{c}A$ is involutory MDS. To form a 3×3 MDS matrix, we may take a submatrix A' from A or $\frac{1}{c}A$. Let us consider the 3×3 submatrix A' of the involutory MDS matrix $\frac{1}{c}A$ of order 3. Here we take first three rows and columns of $\frac{1}{c}A$ for constructing A'. Thus we have

$$A' = \begin{pmatrix} 7a_x & f4_x & 8e_x \\ f4_x & 7a_x & 01_x \\ 8e_x & 01_x & 7a_x \end{pmatrix}.$$

Note that 2nd and 3rd row of A' are not permutations of its first row.

Remark 5. For an illustration purpose, we count the number of xtime and xor operations for the matrix A and A' of Example 1 without considering any optimization technique. The matrix A requires 9 xtimes and 3 xors each row, i.e. 36 xtimes and 12 xors for one matrix computation. Similarly the matrix A' requires 20 xtimes and 11 xors each row, i.e. 80 xtimes and 44 xors for one matrix computation.

3.3   An Algorithm to Construct Low Hamming Weight Involutory MDS Matrix

Here we present an algorithm (Algorithm 2) to construct efficient $d \times d$ involutory MDS matrices, where d is power of 2. By efficient matrix, we mean a matrix having maximum number of 1's and minimum number of other distinct elements of low Hamming weight (see Remark 3). In the construction using Lemma 6, a $d \times d$ MDS matrix A can have maximum d number of 1's and $d-1$ other distinct elements (See Lemma 7). In the iteration of Algorithm 2, we fix $r = 1$, which ensures that all diagonal elements are 1. Thus we have d number of 1's. For $d = 2^m$, we initially select m distinct elements of first row $a_{0,1}, a_{0,2}, a_{0,2^2} \ldots, a_{0,2^{m-1}}$ which are of low Hamming weight and compute $x_1, x_2, x_{2^2}, \ldots, x_{2^{m-1}}$, where $x_{2^i} = \frac{1}{a_{0,2^i}} + r$, $i = 0, \ldots, m - 1$. We repeat this process by selecting different elements of next lowest possible Hamming weights unless we get m linearly independent elements $x_0, x_1, x_2, x_3, \ldots, x_{2^m-1}$. We next form G and $r + G$ and finally the matrix A using Lemma 6. If the matrix is not involutory, we repeat the process unless we get an involutory MDS matrix A.

Remark 6. Note that we can choose $m + 1$ out of 2^m elements of our liking to have low Hamming weight while constructing involutory MDS matrix using Algorithm 2. But we have no control upon the other $2^m - (m + 1)$ elements of the matrix.

Remark 7. Note that Algorithm 2 is similar to Algorithm 1 and is based on Lemma 6. The Algorithm 2 may not terminate for some conditions in Step 2. If we relax the conditions of low Hamming weight in Step 2, Algorithm 2 will eventually terminate but the time complexity is not clear and may depend upon many conditions.

Remark 8. Algorithm 2 generates $d \times d$ involutory MDS matrix over $\mathbb{F}_{2^n}$ where d is power of 2. The correctness of Algorithm 2 follows from Lemma 6, Lemma 8 and Corollary 4.

Example 2: Let $n = 8$, $d = 4$, $\pi(x) = x^8 + x^4 + x^3 + x + 1$. Set $r = 1$. Also let α be the root of $\pi(x)$. We will select $a_{0,1} = 02_x = \alpha$ and search for the element with next lowest possible Hamming weight for $a_{0,2}$ so that the corresponding values $\frac{1}{a_{0,1}} + 1 = x_1$ and $\frac{1}{a_{0,2}} + 1 = x_2$ are linearly independent

---

Algorithm 2. Construction of $2^m \times 2^m$ Involutory MDS matrix $((a_{i,j}))$ over $\mathbb{F}_{2^n}$

Input $n > 1$, the generating polynomial $\pi(x)$ of $\mathbb{F}_{2^n}$ and $m < n$.
Output Outputs a $2^m \times 2^m$ involutory MDS matrix A.

1: Set $r = 1$;
2: Select m elements labeled $a_{0,1}, a_{0,2}, a_{0,2^2} \ldots, a_{0,2^{m-1}}$ from $\mathbb{F}_{2^n}^*$ of low Hamming weight;
3: Compute m elements labeled $x_1, x_2, x_{2^2}, \ldots, x_{2^{m-1}}$, where $x_{2^i} = \frac{1}{a_{0,2^i}} + r$, $i = 0, \ldots, m - 1$;
4: Check if $x_1, x_2, x_{2^2} \ldots, x_{2^{m-1}}$ are linearly independent. If not, go to Step 2;
5: Construct G, the set of 2^m elements $x_0, x_1, x_2, x_3, \ldots, x_{2^m-1}$, where $x_i = \sum_{k=0}^{m-1} b_k x_{2^k}$, for all $0 \le i \le 2^m - 1$, $(b_{m-1}, b_{m-2}, \ldots, b_1, b_0)$ being the binary representation of i;
6: if $(r \in G)$ then go to Step 2;
7: Construct $r + G$, the set of 2^m elements $y_0, y_1, y_2, y_3, \ldots, y_{2^m-1}$, where $y_i = r + x_i$ for all $0 \le i \le 2^m - 1$;
8: Compute $c = \sum_{k=0}^{d-1} \frac{1}{y_k}$. if($c \ne 1$): go to step 2;
9: Construct the $2^m \times 2^m$ matrix $A = ((a_{i,j}))$, where $a_{i,j} = \frac{1}{x_i + y_j}$;
10: Set A as output;

---

and finally the resulting matrix is involutory MDS. If not involutory, we go for next element of higher Hamming weight for $a_{0,2}$. If no suitable candidate for $a_{0,2}$ is available, we set $a_{0,1} = 03_x = \alpha + 1$, and repeat the search of suitable candidate for $a_{0,2}$. We iterate and find the first suitable combination as $a_{0,1} = \alpha$ and $a_{0,2} = fc_x = \alpha^7 + \alpha^6 + \alpha^5 + \alpha^4 + \alpha^3 + \alpha^2$ which leads to an involutory MDS matrix. For such $a_{0,1}$, and $a_{0,2}$, we get $x_1 = \frac{1}{a_{0,1}} + 1 = \alpha^7 + \alpha^3 + \alpha^2$ and $x_2 = \frac{1}{a_{0,2}} + 1 = \alpha^7 + \alpha^6 + \alpha^3 + \alpha^2$. So we have $x_0 = 0.x_1 + 0.x_2 = 0$ and $x_3 = 1.x_1 + 1.x_2 = \alpha^6$. Thus $y_0 = 1, y_1 = \alpha^7 + \alpha^3 + \alpha^2 + 1, y_2 = \alpha^7 + \alpha^6 + \alpha^3 + \alpha^2 + 1$ and $y_3 = \alpha^6 + 1$. Finally, we get

$$A = \begin{pmatrix} 01_x & 02_x & fc_x & fe_x \\ 02_x & 01_x & fe_x & fc_x \\ fc_x & fe_x & 01_x & 02_x \\ fe_x & fc_x & 02_x & 01_x \end{pmatrix}.$$

Note that this matrix is involutory MDS. The MDS matrix A of Example 1 is more implementation friendly, but it is not involutory. Note that the matrix $\frac{1}{c} A$ of Example 1 is involutory MDS but not as efficient as the involutory MDS matrix A of Example 2.

Example 3: Here we construct $2^3 \times 2^3$ involutory MDS matrix from Algorithm 2. Let $r = 1$. Using Algorithm 2, we select $a_{0,1} = 02_x$, $a_{0,2} = 06_x$ and $a_{0,4} = 30_x$ of low Hamming weight. This generates $G = \{00_x, 8c_x, 7a_x, f6_x, 2d_x, a1_x, 57_x, db_x\}$. So we generate $r + G$ and finally the involutory MDS matrix A using Algorithm 2, first row of which is as follows: $(01_x \quad 02_x \quad 06_x \quad 8c_x \quad 30_x \quad fb_x \quad 87_x \quad c4_x)$.

Example 4: Here we construct $2^4 \times 2^4$ involutory MDS matrix from Algorithm 2. Let $r = 1$. Using Algorithm 2, we select $a_{0,1} = 03_x$, $a_{0,2} = 08_x$ and $a_{0,4} = 0d_x$ and

$a_{0,8}$ $=$ $0f_x$ of low Hamming weight. This generates $G = \{00_x, f7_x, e9_x, 1e_x, e0_x, 17_x, 09_x, fe_x, c6_x, 31_x, 2f_x, d8_x, 26_x, d1_x, cf_x, 38_x\}$. So we generate $r +$ G and finally the involutory MDS matrix A using Algorithm 2 which is as follows:

$$A = \begin{pmatrix}
01_x & 03_x & 08_x & b2_x & 0d_x & 60_x & e8_x & 1c_x & 0f_x & 2c_x & a2_x & 8b_x & c9_x & 7a_x & ac_x & 35_x \\
03_x & 01_x & b2_x & 08_x & 60_x & 0d_x & 1c_x & e8_x & 2c_x & 0f_x & 8b_x & a2_x & 7a_x & c9_x & 35_x & ac_x \\
08_x & b2_x & 01_x & 03_x & e8_x & 1c_x & 0d_x & 60_x & a2_x & 8b_x & 0f_x & 2c_x & ac_x & 35_x & c9_x & 7a_x \\
b2_x & 08_x & 03_x & 01_x & 1c_x & e8_x & 60_x & 0d_x & 8b_x & a2_x & 2c_x & 0f_x & 35_x & ac_x & 7a_x & c9_x \\
0d_x & 60_x & e8_x & 1c_x & 01_x & 03_x & 08_x & b2_x & c9_x & 7a_x & ac_x & 35_x & 0f_x & 2c_x & a2_x & 8b_x \\
60_x & 0d_x & 1c_x & e8_x & 03_x & 01_x & b2_x & 08_x & 7a_x & c9_x & 35_x & ac_x & 2c_x & 0f_x & 8b_x & a2_x \\
e8_x & 1c_x & 0d_x & 60_x & 08_x & b2_x & 01_x & 03_x & ac_x & 35_x & c9_x & 7a_x & a2_x & 8b_x & 0f_x & 2c_x \\
1c_x & e8_x & 60_x & 0d_x & b2_x & 08_x & 03_x & 01_x & 35_x & ac_x & 7a_x & c9_x & 8b_x & a2_x & 2c_x & 0f_x \\
0f_x & 2c_x & a2_x & 8b_x & c9_x & 7a_x & ac_x & 35_x & 01_x & 03_x & 08_x & b2_x & 0d_x & 60_x & e8_x & 1c_x \\
2c_x & 0f_x & 8b_x & a2_x & 7a_x & c9_x & 35_x & ac_x & 03_x & 01_x & b2_x & 08_x & 60_x & 0d_x & 1c_x & e8_x \\
a2_x & 8b_x & 0f_x & 2c_x & ac_x & 35_x & c9_x & 7a_x & 08_x & b2_x & 01_x & 03_x & e8_x & 1c_x & 0d_x & 60_x \\
8b_x & a2_x & 2c_x & 0f_x & 35_x & ac_x & 7a_x & c9_x & b2_x & 08_x & 03_x & 01_x & 1c_x & e8_x & 60_x & 0d_x \\
c9_x & 7a_x & ac_x & 35_x & 0f_x & 2c_x & a2_x & 8b_x & 0d_x & 60_x & e8_x & 1c_x & 01_x & 03_x & 08_x & b2_x \\
7a_x & c9_x & 35_x & ac_x & 2c_x & 0f_x & 8b_x & a2_x & 60_x & 0d_x & 1c_x & e8_x & 03_x & 01_x & b2_x & 08_x \\
ac_x & 35_x & c9_x & 7a_x & a2_x & 8b_x & 0f_x & 2c_x & e8_x & 1c_x & 0d_x & 60_x & 08_x & b2_x & 01_x & 03_x \\
35_x & ac_x & 7a_x & c9_x & 8b_x & a2_x & 2c_x & 0f_x & 1c_x & e8_x & 60_x & 0d_x & b2_x & 08_x & 03_x & 01_x
\end{pmatrix}.$$

Example 5: Here we construct $2^5 \times 2^5$ involutory MDS matrix from Algorithm 2. Let $r = 1$. Using Algorithm 2, we select $a_{0,1} = 02_x$, $a_{0,2} = 04_x$ and $a_{0,4} = 07_x$ and $a_{0,8} = 0b_x$ and $a_{0,16} = 0e$ of low Hamming weight. This generates $G = \{00_x, 8c_x, ca_x, 46_x, d0_x, 5c_x, 1a_x, 96_x, c1_x, 4d_x, 0b_x, 87_x, 11_x, 9d_x, db_x, 57_x, e4_x, 68_x, 2e_x, a2_x, 34_x, b8_x, fe_x, 72_x, 25_x, a9_x, ef_x, 63_x, f5_x, 79_x, 3f_x, b3_x\}$. So we generate $r + G$ and finally the involutory MDS matrix A using Algorithm 2, first row of which is as follows: $(01_x \ 02_x \ 04_x \ 69_x \ 07_x \ ec_x \ cc_x \ 72_x \ 0b_x \ 54_x \ 29_x \ be_x \ 74_x \ f9_x \ c4_x \ 87_x \ 0e_x \ 47_x \ c2_x \ c3_x \ 39_x \ 8e_x \ 1c_x \ 85_x \ 55_x \ 26_x \ 1e_x \ af_x \ 68_x \ b6_x \ 59_x \ 1f_x)$. Note that matrices in Example 2 to Example 5 are FFHadamard (see Remark 4). So $h_{i,j} = h_{0,i\oplus j}$ for all $i,j \in \{0,\ldots,31\}$.

4 FFHadamard MDS Matrices from Cauchy Based Construction and Vandermonde Based Constructions

The authors of [19] constructed FFHadamard involutory MDS matrices starting from two Special Vandermonde matrices. In this section we first show that Cauchy construction of Algorithm 1 gives FFHadamard matrices. We next show (see Theorem 5) the equivalence of Cauchy based construction and Vandermonde based construction of "FFHadamard involutory MDS matrices" of [19]. In doing so, we provide a much simpler proof (see Corollary 8) of generalization of Corollary 2 of [19]. We also prove that Cauchy based construction using Algorithm 1 is faster than the Vandermonde based construction.In the following theorem we show that the MDS matrices constructed by Algorithm 1 are FFHadamard.

Theorem 4. *Algorithm 1 generates FFHadamard Matrices.*

Proof. Let us assume that Algorithm 1 produces $2^m \times 2^m$ matrix $A = ((a_{i,j}))$. So $a_{i,j} = \frac{1}{x_i+y_j} = \frac{1}{r+x_i+x_j} = \frac{1}{r+x_{i\oplus j}}$, where x_i's and y_j's are as defined in the Algorithm 1. From Fact 9, A is FFHadamard matrix. $\square$

4.1 Equivalence of Cauchy Based Construction and Vandermonde Based Construction of Involutory MDS FFHadamard Matrices

Here we fix certain notations that will be used freely in the rest of this Section. Let $\mathbb{G} = \{\gamma_0, \gamma_1, \ldots, \gamma_{d-1}\}$ be an additive subgroup of $\mathbb{F}_{2^n}$ of order d where $\gamma_0 = 0$ and $\gamma_i + \gamma_j = \gamma_{i \oplus j}$ for $i, j \in \{0, \ldots, d-1\}$. For any two arbitrary $r_1, r_2 \in \mathbb{F}_{2^n}$, such that $r_1 + r_2 \notin \mathbb{G}$, let us define three cosets of $\mathbb{G}$ as follows: $r_1 + \mathbb{G} = \{\alpha_i : \alpha_i = r_1 + \gamma_i \text{ for } i = 0, \ldots, d-1\}$, $r_2 + \mathbb{G} = \{\beta_i : \beta_i = r_2 + \gamma_i \text{ for } i = 0, \ldots, d-1\}$ and $r_1 + r_2 + \mathbb{G} = \{\delta_i : \delta_i = r_1 + r_2 + \gamma_i \text{ for } i = 0, \ldots, d-1\}$. Let γ be the product of all nonzero elements of $\mathbb{G}$, β be the product of all elements of $r_2 + \mathbb{G}$ and δ be the product of all elements of $r_1 + r_2 + \mathbb{G}$, i.e. $\gamma = \prod_{k=1}^{d-1} \gamma_k$, $\beta = \prod_{k=0}^{d-1} \beta_k$ and $\delta = \prod_{k=0}^{d-1} \delta_k$. Also let us define two $d \times d$ Special Vandermonde matrices (SV matrices) A and B as follows: $A = van(\alpha_0, \alpha_1, \ldots, \alpha_{d-1})$ and $B = van(\beta_0, \beta_1, \ldots, \beta_{d-1})$ and let

$$B^{-1} = \begin{pmatrix} b_{0,0} & b_{0,1} & \cdots & b_{0,d-1} \\ b_{1,0} & b_{1,1} & \cdots & b_{1,d-1} \\ \vdots & \vdots & \vdots & \vdots \\ b_{d-1,0} & b_{d-1,1} & \cdots & b_{d-1,d-1} \end{pmatrix}, \text{ where } b_{i,j} \in \mathbb{F}_{2^n}.$$

We will prove in Theorem 5 the equivalence of Vandermonde based constructions (see Subsection 3.1 of [19]) and Cauchy based constructions (see Algorithm 1) of FFHadamard involutory MDS matrices. Before going into the proof, we study few properties of B and B^{-1} in Lemma 9 to Lemma 12.

Lemma 9. $det(B) = \gamma^{d/2}$.

Proof. From Fact 7, $det(B) = \prod_{k<l}(\beta_k + \beta_l) = (\prod_{k \neq l}(\beta_k + \beta_l))^{1/2} = (\prod_{k \neq l}(\gamma_k + \gamma_l))^{1/2}$. In the product $\prod_{k \neq l}(\gamma_k + \gamma_l)$, each of the terms $\gamma_1, \ldots, \gamma_{d-1}$ occurs d times. So $\prod_{k \neq l}(\gamma_k + \gamma_l) = \prod_{i=1}^{d-1} \gamma_i^d = \gamma^d$. Therefore $det(B) = \gamma^{d/2}$. $\square$

In the next lemma, we show that the elements of last row of B^{-1} i.e. $b_{d-1,j}$'s for $j = 0, \ldots, d-1$ are equal and independent of j.

Lemma 10. $b_{d-1,j} = \frac{1}{\gamma}$ for $j = 0, \ldots, d-1$.

Proof. Let $j \in \{0, 1, \ldots, d-1\}$ be arbitrary. So, $b_{d-1,j} = \frac{det(B')}{det(B)}$. Where

$$B' = \begin{pmatrix} 1 & \beta_0 & \beta_0^2 & \beta_0^3 & \cdots & \beta_0^{d-2} \\ \vdots & \vdots & \vdots & \vdots & \vdots & \vdots \\ 1 & \beta_{j-1} & \beta_{j-1}^2 & \beta_{j-1}^3 & \cdots & \beta_{j-1}^{d-2} \\ 1 & \beta_{j+1} & \beta_{j+1}^2 & \beta_{j+1}^3 & \cdots & \beta_{j+1}^{d-2} \\ \vdots & \vdots & \vdots & \vdots & \vdots & \vdots \\ 1 & \beta_{d-1} & \beta_{d-1}^2 & \beta_{d-1}^3 & \cdots & \beta_{d-1}^{d-2} \end{pmatrix}.$$

Now $(\prod_{k \neq l}(\gamma_k + \gamma_l))^{1/2} = \prod_{k<l}(\gamma_k + \gamma_l)$ and $\prod_{k \neq l, k \neq j, l \neq j}(\gamma_k + \gamma_l) = \frac{\prod_{k \neq l}(\gamma_k + \gamma_l)}{\prod_{k \neq j}(\gamma_k + \gamma_j)\prod_{l \neq j}(\gamma_j + \gamma_l)} = \frac{\prod_{k \neq l}(\gamma_k + \gamma_l)}{\prod_{k \neq 0} \gamma_k \prod_{l \neq 0} \gamma_l} = \frac{\gamma^d}{\gamma^2} = \gamma^{d-2}$. Therefore $det(B') = \prod_{k<l, k \neq j, l \neq j}(\beta_k + \beta_l) = \prod_{k<l, k \neq j, l \neq j}(\gamma_k + \gamma_l) = (\prod_{k \neq l, k \neq j, l \neq j}(\gamma_k + \gamma_l))^{1/2} = \gamma^{(d-2)/2}$.

From the relation $B^{-1} = \frac{Adj(B)^t}{det(B)}$, we get $b_{n-1,j} = \frac{det(B')}{det(B)}$. Using Lemma 9, we get $b_{n-1,j} = \frac{det(B')}{det(B)} = \frac{\gamma^{(d-2)/2}}{\gamma^{d/2}} = \frac{1}{\gamma}$. $\qquad\qquad\qquad\qquad\qquad\qquad\square$

Let us define $d-1$ degree polynomials $P_j(x) = \sum_{i=0}^{d-1} b_{i,j} x^i$ for $j = 0, \ldots, d-1$. The coefficients of $P_j(x)$ are the elements of j'th column of B^{-1}. In the next lemma we study the roots of $P_j(x)$.

Lemma 11. *The $d-1$ roots of $P_j(x)$ are $\beta_0, \ldots, \beta_{j-1}, \beta_{j+1}, \ldots, \beta_{d-1}$ for $j = 0, \ldots, d-1$.*

Proof. We know $BB^{-1} = I$, where I is the $d \times d$ identity matrix. The d elements in the j'th column of BB^{-1} are $P_j(\beta_0), P_j(\beta_1), \ldots, P_j(\beta_{d-1})$, of which only j'th element i.e. $P_j(\beta_j)$ is one and the rest $d-1$ are zero. Hence the result follows. $\square$

Corollary 6. $P_j(x) = \frac{1}{\gamma} \prod_{k \neq j} (x + \beta_k)$ *for* $j = 0, \ldots, d-1$.

Proof. From Lemma 11, roots of $P_j(x)$ are $\beta_0, \ldots, \beta_{j-1}, \beta_{j+1}, \ldots, \beta_{d-1}$. Therefore, $P_j(x) = b_{d-1,j} \prod_{k \neq j} (x - \beta_k)$. Since elements are from $\mathbb{F}_{2^n}$ which is of characteristic 2, so $P_j(x) = b_{d-1,j} \prod_{k \neq j} (x + \beta_k)$. Also from Lemma 10, $b_{d-1,j} = \frac{1}{\gamma}$ for $j = 0, \ldots, d-1$. Hence $P_j(x) = \frac{1}{\gamma} \prod_{k \neq j} (x + \beta_k)$ for $j = 0, \ldots, d-1$. $\square$

Lemma 12. $\sum_j \frac{1}{\beta_j} = \frac{\gamma}{\beta}$.

Proof. We know, $B^{-1}B = I$. So $(0,0)$'th element of $B^{-1}B$ i.e. $\sum_k b_{0,j} = 1$. Using Corollary 6, we have $P_j(0) = \frac{1}{\gamma} \prod_{k \neq j} \beta_k = \frac{\beta}{\gamma\beta_j}$. But $P_j(0) = b_{0,j}$. So $1 = \sum_j b_{0,j} = \sum_j P_j(0) = \sum_j \frac{\beta}{\gamma\beta_j} = \frac{\beta}{\gamma} \sum_j \frac{1}{\beta_j}$. Thus $\sum_j \frac{1}{\beta_j} = \frac{\gamma}{\beta}$. $\square$

Corollary 7. $\sum_j \frac{1}{\delta_j} = \frac{\gamma}{\delta}$.

Now we propose Theorem 5, which shows the equivalence between Cauchy based Construction of FFHadamard matrices (Algorithm 1) and Vandermonde based Construction of FFHadamard matrices [19]. Let $\frac{1}{c}M$ be the involutory MDS matrix produced by Algorithm 1, where $M = ((m_{i,j}))$, $m_{i,j} = \frac{1}{\gamma_i + \delta_j}$ for $i, j \in \{0, 1, \ldots, d-1\}$, $c = \sum_{k=0}^{d-1} \frac{1}{\delta_k}$. Note that in Algorithm 1, if we take G as $\mathbb{G}$, r as $r_1 + r_2$ and set $b_{Involutory} = true$, then Algorithm 1 constructs $\frac{1}{c}M$.

Theorem 5. $AB^{-1} = \frac{1}{c}M$.

Proof. Let $AB^{-1} = ((h_{i,j}))$. Now, the (i,j)'th element of AB^{-1} is $P_j(\alpha_i)$. Using Corollary 6, we have $h_{i,j} = P_j(\alpha_i) = \frac{1}{\gamma} \prod_{k \neq j} (\alpha_i + \beta_k) = \frac{1}{\gamma} \prod_{k \neq j} (r_1 + \gamma_i + r_2 + \gamma_k) = \frac{1}{\gamma} \frac{\prod_k (r_1 + r_2 + \gamma_i + \gamma_k)}{(r_1 + r_2 + \gamma_i + \gamma_j)} = \frac{1}{\gamma} \frac{\prod_k \delta_k}{(\gamma_i + \delta_j)} = \frac{\delta}{\gamma} \frac{1}{(\gamma_i + \delta_j)} = \frac{\delta}{\gamma} m_{i,j}$. Also from Corollary 7, $c = \sum_{k=0}^{d-1} \frac{1}{\delta} = \frac{\gamma}{\delta}$. Thus $h_{i,j} = \frac{1}{c} m_{i,j}$. Hence the proof. $\square$

Note that by Lemma 5 (a generalization of Corollary 2 of [19]), AB^{-1} is an FFHadamard involutory MDS matrix. The following corollary gives an alternative proof of Lemma 5.

Corollary 8. AB^{-1} *is FFHadamard involutory MDS matrix.*

Proof. Since $\frac{1}{c}M$ is FFHadamard involutory MDS (from Theorem 3 and Theorem 4), so is AB^{-1} (from Theorem 5). $\square$

4.2 Comparison of Algorithm 1 Based on Cauchy Based Construction, and Vandermonde Based Construction of [19] to Construct FFHadamard Involutory MDS Matrices

From Theorem 3, the time complexity of constructing $d \times d$ FFHadamard involutory MDS matrix $\frac{1}{c}M$ is $O(d^2)$. In the Vandermonde based Construction [19] to construct FFHadamard involutory MDS matrix AB^{-1}, it requires a multiplication of $d \times d$ matrices A and B^{-1} and the time complexity is $O(d^3)$. So, the Algorithm 1 is faster than the Vandermonde based Construction of FFHadamard involutory MDS matrix in [19].

5 The Matrix $M_{16 \times 16}$ Used in MDS-AES of [10] Is Not MDS

In [10], authors proposed 16×16 involutory MDS matrix $M_{16 \times 16}$ by Cauchy based construction with an additional restriction of allowing elements of low Hamming weights. *We checked that their method does not give MDS matrix.* It is easy to verify that the set of inverses of elements of the first row of $M_{16 \times 16}$ is not a coset of any additive subgroup of $\mathbb{F}_{2^3}$. In fact the authors of [10] did not consider the additive subgroup properly. Some authors [4, 15] recommended $M_{16 \times 16}$ to be used as a diffusion layer, but using this matrix may introduce severe weaknesses. The $M_{16 \times 16}$ matrix of [10] is given below.

$$M_{16 \times 16} = \begin{pmatrix} 01_x & 03_x & 04_x & 05_x & 06_x & 07_x & 08_x & 09_x & 0a_x & 0b_x & 0c_x & 0d_x & 0e_x & 10_x & 02_x & 1e_x \\ 03_x & 01_x & 05_x & 04_x & 07_x & 06_x & 09_x & 08_x & 0b_x & 0a_x & 0d_x & 0c_x & 10_x & 0e_x & 1e_x & 02_x \\ 04_x & 05_x & 01_x & 03_x & 08_x & 09_x & 06_x & 07_x & 0c_x & 0d_x & 0a_x & 0b_x & 02_x & 1e_x & 0e_x & 10_x \\ 05_x & 04_x & 03_x & 01_x & 09_x & 08_x & 07_x & 06_x & 0d_x & 0c_x & 0b_x & 0a_x & 1e_x & 02_x & 10_x & 0e_x \\ 06_x & 07_x & 08_x & 09_x & 01_x & 03_x & 04_x & 05_x & 0e_x & 10_x & 02_x & 1e_x & 0a_x & 0b_x & 0c_x & 0d_x \\ 07_x & 06_x & 09_x & 08_x & 03_x & 01_x & 05_x & 04_x & 10_x & 0e_x & 1e_x & 02_x & 0b_x & 0a_x & 0d_x & 0c_x \\ 08_x & 09_x & 06_x & 07_x & 04_x & 05_x & 01_x & 03_x & 02_x & 1e_x & 0e_x & 10_x & 0c_x & 0d_x & 0a_x & 0b_x \\ 09_x & 08_x & 07_x & 06_x & 05_x & 04_x & 03_x & 01_x & 1e_x & 02_x & 10_x & 0e_x & 0d_x & 0c_x & 0b_x & 0a_x \\ 0a_x & 0b_x & 0c_x & 0d_x & 0e_x & 10_x & 02_x & 1e_x & 01_x & 03_x & 04_x & 05_x & 06_x & 07_x & 08_x & 09_x \\ 0b_x & 0a_x & 0d_x & 0c_x & 10_x & 0e_x & 1e_x & 02_x & 03_x & 01_x & 05_x & 04_x & 07_x & 06_x & 09_x & 08_x \\ 0c_x & 0d_x & 0a_x & 0b_x & 02_x & 1e_x & 0e_x & 10_x & 04_x & 05_x & 01_x & 03_x & 08_x & 09_x & 06_x & 07_x \\ 0d_x & 0c_x & 0b_x & 0a_x & 1e_x & 02_x & 10_x & 0e_x & 05_x & 04_x & 03_x & 01_x & 09_x & 08_x & 07_x & 06_x \\ 0e_x & 10_x & 02_x & 1e_x & 0a_x & 0b_x & 0c_x & 0d_x & 06_x & 07_x & 08_x & 09_x & 01_x & 03_x & 04_x & 05_x \\ 10_x & 0e_x & 1e_x & 02_x & 0b_x & 0a_x & 0d_x & 0c_x & 07_x & 06_x & 09_x & 08_x & 03_x & 01_x & 05_x & 04_x \\ 02_x & 1e_x & 0e_x & 10_x & 0c_x & 0d_x & 0a_x & 0b_x & 08_x & 09_x & 06_x & 07_x & 04_x & 05_x & 01_x & 03_x \\ 1e_x & 02_x & 10_x & 0e_x & 0d_x & 0c_x & 0b_x & 0a_x & 09_x & 08_x & 07_x & 06_x & 05_x & 04_x & 03_x & 01_x \end{pmatrix}$$

The elements of $M_{16 \times 16}$ are from $\mathbb{F}_{2^8}$ and the constructing polynomial is $x^8 + x^4 + x^3 + x + 1$. Let us consider the 2×2 submatrix A of $M_{16 \times 16}$ formed by taking 0th and 2nd row and 1st and 5th column. Let α be the root of $x^8 + x^4 + x^3 + x + 1$. Then in polynomial representation,

$$A = \begin{pmatrix} 03_x & 07_x \\ 05_x & 09_x \end{pmatrix} = \begin{pmatrix} 1 + \alpha & 1 + \alpha + \alpha^2 \\ 1 + \alpha^2 & 1 + \alpha^3 \end{pmatrix}.$$

So $det(A) = (1 + \alpha)(1 + \alpha^3) + (1 + \alpha + \alpha^2)(1 + \alpha^2) = 1 + \alpha^4 + \alpha + \alpha^3 + 1 + \alpha^2 + \alpha + \alpha^3 + \alpha^2 + \alpha^4 = 0$. Thus the submatrix A is singular. So clearly from Fact 1, $M_{16 \times 16}$ is non MDS. Example 4 provides 16×16 involutory MDS matrix which can be used instead of $M_{16 \times 16}$ of [10]. Note that the matrix in Example 4 does not look as good as $M_{16 \times 16}$, in terms of Hamming weights of its elements – but $M_{16 \times 16}$ is non MDS. One can also generate different involutory MDS matrices using Algorithm 2.

6   Conclusion

In this paper, we developed techniques to construct $d \times d$ MDS matrices over $\mathbb{F}_{2^n}$. We proposed a simple algorithm (Algorithm 1) based on Lemma 6. This algorithm is a generalization of the construction proposed in [30]. We propose another algorithm (Algorithm 2) which uses Algorithm 1 iteratively to find software efficient involutory MDS matrices. We find the interesting equivalence of Cauchy based construction (Algorithm 1) and Vandermonde based construction of FFHadamard involutory MDS matrices [19]. We also prove that Cauchy based construction (Algorithm 1) is faster in terms of time complexity compared to Vandermonde based construction of FFHadamard involutory MDS matrices [19]. We have shown that the 16×16 matrix $M_{16\times16}$, used in MDS-AES of [10], is not MDS.

References

1. Barreto, P., Rijmen, V.: The Khazad Legacy-Level Block Cipher, Submission to the NESSIE Project (2000), http://cryptonessie.org
2. Barreto, P.S., Rijmen, V.: The Anubis block cipher, NESSIE Algorithm Submission (2000), http://cryptonessie.org
3. Bosma, W., Cannon, J., Playoust, C.: The Magma Algebra System I: The User Language. J. Symbolic Comput. 24(3-4), 235–265 (1997); Computational algebra and number theory (London, 1993)
4. Choy, J., Yap, H., Khoo, K., Guo, J., Peyrin, T., Poschmann, A., Tan, C.H.: SPN-Hash: Improving the Provable Resistance against Differential Collision Attacks. In: Mitrokotsa, A., Vaudenay, S. (eds.) AFRICACRYPT 2012. LNCS, vol. 7374, pp. 270–286. Springer, Heidelberg (2012)
5. Daemen, J., Knudsen, L.R., Rijmen, V.: The block cipher SQUARE. In: Biham, E. (ed.) FSE 1997. LNCS, vol. 1267, pp. 149–165. Springer, Heidelberg (1997)
6. Daemen, J., Rijmen, V.: The Design of Rijndael:AES - The Advanced Encryption Standard. Springer (2002)
7. Filho, G.D., Barreto, P., Rijmen, V.: The Maelstrom-0 Hash Function. In: Proceedings of the 6th Brazilian Symposium on Information and Computer Systems Security (2006)
8. Gauravaram, P., Knudsen, L.R., Matusiewicz, K., Mendel, F., Rechberger, C., Schlaffer, M., Thomsen, S.: Grøstl a SHA-3 Candidate. Submission to NIST (2008), http://www.groestl.info
9. Guo, J., Peyrin, T., Poschmann, A.: The PHOTON Family of Lightweight Hash Functions. In: Rogaway, P. (ed.) CRYPTO 2011. LNCS, vol. 6841, pp. 222–239. Springer, Heidelberg (2011)
10. Nakahara Jr., J., Abrahao, E.: A New Involutory MDS Matrix for the AES. International Journal of Network Security 9(2), 109–116 (2009)
11. Junod, P., Vaudenay, S.: Perfect Diffusion Primitives for Block Ciphers Building Efficient MDS Matrices. In: Handschuh, H., Hasan, M.A. (eds.) SAC 2004. LNCS, vol. 3357, pp. 84–99. Springer, Heidelberg (2004)
12. Junod, P., Vaudenay, S.: FOX: A new family of block ciphers. In: Handschuh, H., Hasan, M.A. (eds.) SAC 2004. LNCS, vol. 3357, pp. 114–129. Springer, Heidelberg (2004)

13. Junod, P., Macchetti, M.: Revisiting the IDEA philosophy. In: Dunkelman, O. (ed.) FSE 2009. LNCS, vol. 5665, pp. 277–295. Springer, Heidelberg (2009)

14. Lacan, J., Fimes, J.: Systematic MDS erasure codes based on vandermonde matrices. IEEE Trans. Commun. Lett. 8(9), 570–572 (2004)

15. Lo, J.W., Hwang, M.S., Liu, C.H.: An efficient key assignment scheme for access control in a large leaf class hierarchy. Journal of Information Sciences: An International Journal Archive 181(4), 917–925 (2011)

16. MacWilliams, F.J., Sloane, N.J.A.: The Theory of Error Correcting Codes. North Holland (1986)

17. Rao, A.R., Bhimasankaram, P.: Linear Algebra, 2nd edn. Hindustan Book Agency

18. Rijmen, V., Daemen, J., Preneel, B., Bosselaers, A., Win, E.D.: The cipher SHARK. In: Gollmann, D. (ed.) FSE 1996. LNCS, vol. 1039, pp. 99–112. Springer, Heidelberg (1996)

19. Sajadieh, M., Dakhilalian, M., Mala, H., Omoomi, B.: On construction of involutory MDS matrices from Vandermonde Matrices in $GF(2^q)$. Design, Codes Cryptography, 1–22 (2012)

20. Sajadieh, M., Dakhilalian, M., Mala, H., Sepehrdad, P.: Recursive Diffusion Layers for Block Ciphers and Hash Functions. In: Canteaut, A. (ed.) FSE 2012. LNCS, vol. 7549, pp. 385–401. Springer, Heidelberg (2012)

21. Schneier, B., Kelsey, J., Whiting, D., Wagner, D., Hall, C., Ferguson, N.: Twofish: A 128-bit block cipher. In: The first AES Candidate Conference. National Institute for Standards and Technology (1998)

22. Schneier, B., Kelsey, J., Whiting, D., Wagner, D., Hall, C., Ferguson, N.: The Twofish encryption algorithm. Wiley (1999)

23. Schnorr, C.-P., Vaudenay, S.: Black Box Cryptanalysis of Hash Networks Based on Multipermutations. In: De Santis, A. (ed.) EUROCRYPT 1994. LNCS, vol. 950, pp. 47–57. Springer, Heidelberg (1995)

24. Shannon, C.E.: Communication Theory of Secrecy Systems. Bell Syst. Technical J. 28, 656–715 (1949)

25. Sony Corporation, The 128-bit Block cipher CLEFIA Algorithm Specification (2007), http://www.sony.co.jp/Products/cryptography/clefia/download/data/clefia-spec-1.0.pdf

26. Vaudenay, S.: On the Need for Multipermutations: Cryptanalysis of MD4 and SAFER. In: Preneel, B. (ed.) FSE 1994. LNCS, vol. 1008, pp. 286–297. Springer, Heidelberg (1995)

27. Watanabe, D., Furuya, S., Yoshida, H., Takaragi, K., Preneel, B.: A new keystream generator MUGI. In: Daemen, J., Rijmen, V. (eds.) FSE 2002. LNCS, vol. 2365, pp. 179–194. Springer, Heidelberg (2002)

28. Youssef, A.M., Tavares, S.E., Heys, H.M.: A New Class of Substitution Permutation Networks. In: Workshop on Selected Areas in Cryptography, SAC 1996, Workshop Record, pp. 132–147 (1996)

29. Wu, S., Wang, M., Wu, W.: Recursive Diffusion Layers for (Lightweight) Block Ciphers and Hash Functions. In: Knudsen, L.R., Wu, H. (eds.) SAC 2012. LNCS, vol. 7707, pp. 355–371. Springer, Heidelberg (2013)

30. Youssef, A.M., Mister, S., Tavares, S.E.: On the Design of Linear Transformations for Substitution Permutation Encryption Networks. In: Workshop on Selected Areas in Cryptography, SAC 1997, pp. 40–48 (1997)

Homomorphic Encryption with Access Policies: Characterization and New Constructions

Michael Clear*, Arthur Hughes, and Hitesh Tewari

School of Computer Science and Statistics,
Trinity College Dublin

Abstract. A characterization of predicate encryption (PE) with support for homomorphic operations is presented and we describe the homomorphic properties of some existing PE constructions. Even for the special case of IBE, there are few known group-homomorphic cryptosystems. Our main construction is an XOR-homomorphic IBE scheme based on the quadratic residuosity problem (variant of the Cocks' scheme), which we show to be strongly homomorphic. We were unable to construct an anonymous variant that preserves this homomorphic property, but we achieved anonymity for a weaker notion of homomorphic encryption, which we call *non-universal*. A related security notion for this weaker primitive is formalized. Finally, some potential applications and open problems are considered.

1 Introduction

There has been much interest recently in encryption schemes with homomorphic capabilities. Traditionally, malleability was avoided to satisfy strong security definitions, but many applications have been identified for cryptosystems supporting homomorphic operations. More recently, Gentry [1] presented the first fully-homomorphic encryption (FHE) scheme, and several improvements and variants have since appeared in the literature [2–5]. There are however many applications that only require a scheme to support a single homomorphic operation. Such schemes are referred to as *partial homomorphic*. Notable examples of unbounded homomorphic cryptosystems include Goldwasser-Micali [6] (XOR), Paillier [7] and ElGamal [8].

Predicate Encryption (PE) [9] enables a sender to embed a hidden descriptor within a ciphertext that consists of attributes describing the message content. A Trusted Authority (TA) who manages the system issues secret keys to users corresponding to predicates. A user can decrypt a ciphertext containing a descriptor **a** if and only if he/she has a secret key for a predicate that evaluates to true for **a**. This construct turns out to be quite powerful, and generalizes many encryption primitives. It facilitates expressive fine-grained access control i.e. complex policies can be defined restricting the recipients who can decrypt a message. It also facilitates the evaluation of complex queries on data such as

---

* The author's work is funded by the Irish Research Council EMBARK Initiative.

A. Youssef, A. Nitaj, A.E. Hassanien (Eds.): AFRICACRYPT 2013, LNCS 7918, pp. 61–87, 2013.

range, subset and search queries. Extending the class of supported predicates for known schemes is a topic of active research at present.

PE can be viewed in two ways. It can be viewed as a means to delegate computation to a third party i.e. allow the third party to perform a precise fixed function on the encrypted data, and thus limit what the third party learns about the data. In the spirit of this viewpoint, a generalization known as Functional Encryption has been proposed [10], which allows general functions to be evaluated.

PE can also be viewed as a means to achieve more fine-grained access control. It enables a stronger separation between sender and recipient since the former must only describe the content of the message or more general conditions on its access while decryption then depends on whether a recipient's access policy matches these conditions.

Why consider homomorphic encryption in the PE setting? It is conceivable that in a multi-user environment such as a large organization, certain computations may be delegated to the cloud whose inputs depend on the work of multiple users distributed within that organization. Depending on the application, the circuit to be computed may be chosen or adapted by the cloud provider, and thus is not fixed by the delegator as in primitives such as non-interactive *verifiable computing* [11]. Furthermore, the computation may depend on data sets provided by multiple independent users. Since the data is potentially sensitive, the organization's security policy may dictate that all data must be encrypted. Accordingly, each user encrypts her data with a PE scheme using relevant attributes to describe it. She then sends the ciphertext(s) to the cloud. It is desirable that the results of the computation returned from the cloud be decryptable only by an entity whose access policy (predicate) satisfies the attributes of *all* data sets used in the computation. Of course a public-key homomorphic scheme together with a PE scheme would be sufficient if the senders were able to interact before contacting the cloud, but we would like to remove this requirement since the senders may not be aware of each other. This brings to mind the recent notion of multikey homomorphic encryption presented by López-Alt, Tromer and Vaikuntanathan [12].

Using a multikey homomorphic scheme, the senders need not interact with each other before evaluation takes place on the cloud. Instead, they must run an MPC decryption protocol to jointly decrypt the result produced by the cloud. The evaluated ciphertexts in the scheme described in [12] do not depend on the circuit size, and depend only polynomially on the security parameter and the number of parties who contribute inputs to the circuit. Therefore, the problem outlined above may be solved with a multikey fully homomorphic scheme used in conjunction with a PE scheme if we accept the evaluated ciphertext size to be polynomial in the number of parties. In this work, we are concerned with a ciphertext size that is independent of the number of parties. Naturally, this limits the composition of access policies, but if this is acceptable in an application, there may be efficiency gains over the combination of multikey FHE and PE.

In summary, homomorphic encryption in the PE setting is desirable if there is the possibility of multiple parties in a large organization (say) sending encrypted data to a semi-trusted[1] evaluator and access policies are required to appropriately limit access to the results, where the "composition" of access policies is "lossy". We assume the semi-honest model in this paper; in particular we do not consider verifiability of the computation.

The state of affairs for homomorphic encryption even for the simplest special case of PE, namely identity-based encryption (IBE), leaves open many challenges. At his talk at Crypto 2010, Naccache [13] mentioned "identity-based fully homomorphic encryption" as one of a list of theory questions. Towards this goal, it has been pointed out in [14] that some LWE-based FHE constructions can be modified to obtain a weak form of an identity-based FHE scheme using the trapdoor functions from [15]; that is, additional information is needed (beyond what can be non-interactively derived from a user's identity) in order to evaluate certain circuits and to perform bootstrapping. Therefore, the valued non-interactivity property of IBE is lost whereby no communication between encryptors and the TA is needed. To the best of our knowledge, fully-homomorphic or even "somewhat-homomorphic" IBE remains open, and a variant of the BGN-type scheme of Gentry, Halevi and Vaikuntanathan [16] is the only IBE scheme that can compactly evaluate quadratic formulae (supports 2-DNF).

As far as the authors are aware, there are no $(\mathbb{Z}_N, +)$ (like Paillier) or $(\mathbb{Z}_p^*, *)$ (like ElGamal) homomorphic IBE schemes. Many pairings-based IBE constructions admit multiplicative homomorphisms which give us a limited additive homomorphism for small ranges; that is, a discrete logarithm problem must be solved to recover the plaintext, and the complexity thereof is $O(\sqrt{M})$, where M is the size of the message space. Of a similar variety are public-key schemes such as BGN [17] and Benaloh [18]. It remains open to construct an unbounded additively homomorphic IBE scheme for a "large" range such as Paillier [7]. Possibly a fruitful step in this direction would be to look at Galbraith's variant of Paillier's cryptosystem based on elliptic curves over rings [19].

One of the contributions of this paper is to construct an additively homomorphic IBE scheme for $\mathbb{Z}_2$, which is usually referred to as XOR-homomorphic. XOR-homomorphic schemes such as Goldwasser-Micali [6] have been used in many practical applications including sealed-bid auctions, biometric authentication and as the building blocks of protocols such as private information retrieval, and it seems that an IBE XOR-homomorphic scheme may be useful in some of these scenarios.

We faced barriers however trying to make our XOR-homomorphic scheme anonymous. The main obstacle is that the homomorphism depends on the public key. We pose as an open problem the task of constructing a variant that achieves anonymity and retains the homomorphic property. Inheriting the terminology of Golle et al. [20] (who refer to re-encryption without the public key as *universal re-encryption*), we designate homomorphic evaluation in a scheme that does not require knowledge of the public key as *universal*. We introduce a weaker primitive that explicitly requires additional information to be passed to the homomorphic evaluation algorithm. Our construction can be made anonymous and retain its

---

[1] We assume all parties are semi-honest.

homomorphic property in this context; that is, if the attribute (identity in the case of IBE) is known to an evaluator. While this certainly is not ideal, it may be plausible in some scenarios that an evaluator is allowed to be privy to the attribute(s) encrypted by the ciphertexts, and it is other parties in the system to whom the attribute(s) must remain concealed. An adversary sees incoming and outgoing ciphertexts, and can potentially request evaluations on arbitrary ciphertexts. We call such a variant *non-universal*. We propose a syntax for a non-universal homomorphic primitive and formulate a security notion to capture attribute-privacy in this context.

1.1 Related Work

There have been several endeavors to characterize homomorphic encryption schemes. Gjøsteen [21] succeeded in characterizing many well-known group homomorphic cryptosystems by means of an abstract construction whose security rests on the hardness of a subgroup membership problem. More recently, Armknecht, Katzenbeisser and Peter [22] gave a more complete characterization and generalized Gjøsteen's results to the IND-CCA1 setting. However, in this work, our focus is at a higher level and not concerned with the underlying algebraic structures. In particular, we do not require the homomorphisms to be unbounded since our aim to provide a more general characterization for homomorphic encryption in the PE setting. Compactness, however, is required; that is, informally, the length of an evaluated ciphertext should be independent of the *size* of the computation.

The notion of receiver-anonymity or key-privacy was formally established by Bellare et al. [23], and the concept of universal anonymity (any user can anonymize a ciphertext) was proposed in [24]. The first universally anonymous IBE scheme appeared in [25]. Prabhakaran and Rosulek [26] consider receiver-anonymity for their definitions of homomorphic encryption.

Finally, since Cocks' IBE scheme [27] appeared, variants have been proposed ([28] and [25]) that achieve anonymity and improve space efficiency. However, the possibility of constructing a homomorphic variant has not received attention to date.

1.2 Organization

Notation and background definitions are set out in Section 2. Our characterization of homomorphic predicate encryption is specified in Section 3; the syntax, correctness conditions and security notions are established, and the properties of such schemes are analyzed. In Section 4, some instantiations are given based on inner-product PE constructions. Our main construction, XOR-homomorphic IBE, is presented in Section 5. Non-universal homomorphic encryption and the abstraction of universal anonymizers is presented in Section 6 towards realizing anonymity for our construction in a weaker setting. Conclusions and future work are presented in Section 7.

2   Preliminaries

A quantity is said to be negligible with respect to some parameter λ, written $\mathsf{negl}(\lambda)$, if it is asymptotically bounded from above by the reciprocal of all polynomials in λ.

For a probability distribution D, we denote by $x \xleftarrow{\$} D$ that x is sampled according to D. If S is a set, $y \xleftarrow{\$} S$ denotes that y is sampled from x according to the uniform distribution on S.

The support of a predicate $f : A \to \{0, 1\}$ for some domain A is denoted by $\mathsf{supp}(f)$, and is defined by the set $\{a \in A : f(a) = 1\}$.

Definition 1 (Homomorphic Encryption). *A homomorphic encryption scheme with message space M supporting a class of ℓ-input circuits $\mathbb{C} \subseteq M^\ell \to M$ is a tuple of PPT algorithms* (Gen, Enc, Dec, Eval) *satisfying the property:*

$$\forall(\mathsf{pk}, \mathsf{sk}) \leftarrow \mathsf{Gen}(1^\lambda),\ \forall C \in \mathbb{C}, \forall m_1, \ldots, m_\ell \in M$$
$$\forall c_1, \ldots, c_\ell \leftarrow \mathsf{Enc}(\mathsf{pk}, m_1), \ldots, \mathsf{Enc}(\mathsf{pk}, m_\ell)$$

$$C(m_1, \ldots, m_\ell) = \mathsf{Dec}(\mathsf{sk}, \mathsf{Eval}(\mathsf{pk}, C, c_i, \ldots, c_\ell))$$

The following definition is based on [29],

Definition 2 (Strongly Homomorphic). *Let $\mathcal{E}$ be a homomorphic encryption scheme with message space M and class of supported circuits $\mathbb{C} \subseteq \{M^\ell \to M\}$. $\mathcal{E}$ is said to be* strongly homomorphic *iff $\forall C \in \mathbb{C}$, $\forall(\mathsf{pk}, \mathsf{sk}) \leftarrow$ Gen, $\forall m_1, \ldots, m_\ell$, $\forall c_1, \ldots, c_\ell \leftarrow \mathsf{Enc}(\mathsf{pk}, m_1), \ldots, \mathsf{Enc}(\mathsf{pk}, m_\ell)$, the following distributions are statistically indistinguishable*

$$\mathsf{Enc}(\mathsf{pk}, C(m_1, \ldots, m_\ell)) \approx (\mathsf{Eval}(\mathsf{pk}, C, c_1, \ldots, c_\ell).$$

Definition 3 (Predicate Encryption (Adapted from [9] Definition 1)). *A predicate encryption (PE) scheme for the class of predicates $\mathcal{F}$ over the set of attributes A and with message space M consists of four algorithms* Setup, GenKey, Encrypt, Decrypt *such that:*

- PE.Setup *takes as input the security parameter 1^λ and outputs public parameters* PP *and master secret key* MSK.
- PE.GenKey *takes as input the master secret key* MSK *and a description of a predicate $f \in \mathcal{F}$. It outputs a key* SK_f.
- PE.Encrypt *takes as input the public parameters* PP, *a message $m \in M$ and an attribute $a \in A$. It returns a ciphertext c. We write this as $c \leftarrow$* Encrypt(PP, a, m).
- PE.Decrypt *takes as input a secret key SK_f for a predicate f and a ciphertext c. It outputs m iff $f(a) = 1$. Otherwise it outputs a distinguished symbol $\perp$ with all but negligible probability.*

Remark 1. Predicate Encryption (PE) is known by various terms in the literature. PE stems from Attribute-Based Encryption (ABE) with Key Policy, or simply KP-ABE, and differs from it in its support for attribute privacy. As a result, "ordinary" KP-ABE is sometimes known as PE with public index. Another variant of ABE is CP-ABE (ciphertext policy) where the encryptor embeds her access policy in the ciphertext and a recipient must possess sufficient attributes in order to decrypt. This is the reverse of KP-ABE. In this paper, the emphasis is placed on PE with its more standard interpretation, namely KP-ABE with attribute privacy.

3   Homomorphic Predicate Encryption

3.1   Syntax

Let M be as message space and let A be a set of attributes. Consider a set of operations $\Gamma_M \subseteq \{M^2 \to M\}$ on the message space, and a set of operations $\Gamma_A \subseteq \{A^2 \to A\}$ on the attribute space. We denote by $\gamma = \gamma_A \times \gamma_M$ for some $\gamma_A \in \Gamma_A$ and $\gamma_M \in \Gamma_M$ the operation $(A \times M)^2 \to (A \times M)$ given by $\gamma((a_1, m_1), (a_2, m_2)) = (\gamma_A(a_1, a_2), \gamma_M(m_1, m_2))$. Accordingly, we define the set of permissible "gates" $\Gamma \subseteq \{\gamma_A \times \gamma_M : \gamma_A \in \Gamma_A, \gamma_M \in \Gamma_M\} \subseteq \{(A \times M)^2 \to (A \times M)\}^2$. Thus, each operation on the plaintexts is associated with a single (potentially distinct) operation on the attributes. Finally, we can specify a class of permissible circuits $\mathbb{C}$ built from Γ.

Definition 4. *A homomorphic predicate encryption (HPE) scheme for the non-empty class of predicates $\mathcal{F}$, message space M, attribute space A, and class of ℓ-input circuits $\mathbb{C}$ consists of a tuple of five PPT algorithms* Setup, GenKey, Encrypt, Decrypt *and* Eval. *such that:*

- HPE.Setup, HPE.GenKey, HPE.Encrypt *and* HPE.Decrypt *are as specified in Definition 3.*
- HPE.Eval(PP, $C, c_1, \ldots, c_\ell$) *takes as input the public parameters* PP, *an ℓ-input circuit $C \in \mathbb{C}$, and ciphertexts* $c_1 \leftarrow$ HPE.Encrypt(PP, a_1, m_1), $\ldots, c_\ell \leftarrow$ HPE.Encrypt(PP, a_ℓ, m_ℓ).
  It outputs a ciphertext that encrypts the attribute-message pair $C((a_1, m_1), \ldots, (a_\ell, m_\ell))$.

Accordingly, the correctness criteria are defined as follows:

Correctness Conditions
For any (PP, MSK) $\leftarrow$ HPE.Setup(1^λ), $f \in \mathcal{F}$, SK$_f \leftarrow$ HPE.GenKey(PP, MSK, f), $C \in \mathbb{C}$:

1. For any $a \in A, m \in M, c \leftarrow$ HPE.Encrypt(PP, m, a):

$$\text{HPE.Decrypt}(\text{SK}_f, c) = m \iff f(a) = 1$$

---

[2] It is assumed that Γ_A and Γ_M are minimal insofar as $\forall \gamma_A \in \Gamma_A \exists \gamma_M \in \Gamma_M$ s.t. $\gamma_A \times \gamma_M \in \Gamma$ and the converse also holds. In particular, we later assume this of Γ_A.

2. $\forall m_1, \ldots, m_\ell \in M, \quad \forall a_1, \ldots, a_\ell \in A, \quad \forall c_1, \ldots, c_\ell \leftarrow$ HPE.Encrypt(PP, a_1, m_1), ..., HPE.Encrypt(PP, a_ℓ, m_ℓ) :

$\forall c' \leftarrow$ HPE.Eval(PP, $C, c_1, \ldots, c_\ell$)

(a)
$$\text{HPE.Decrypt}(\text{SK}_f, c') = m' \iff f(a') = 1$$

where $(m', a') = C((a_1, m_1), \ldots, (a_\ell, m_\ell))$

(b)
$$|c'| < L(\lambda)$$

where $L(\lambda)$ is a fixed polynomial derivable from PP.

The special case of "predicate only" encryption [9] that excludes plaintexts ("payloads") is modelled by setting $M \triangleq \{\mathbf{0}\}$ for a distinguished symbol $\mathbf{0}$, and setting $\Gamma \triangleq \{\gamma_A \times \mathbf{id_M} : \gamma_A \in \Gamma_A\}$ where $\mathbf{id_M}$ is the identity operation on M.

3.2 Security Notions

The security notions we consider carry over from the standard notions for PE. The basic requirement is IND-CPA security, which is referred to as "payload-hiding". A stronger notion is "attribute-hiding" that additionally entails indistinguishability of attributes. The definitions are game-based with non-adaptive and adaptive variants. The former prescribes that the adversary choose its target attributes at the beginning of the game before seeing the public parameters, whereas the latter allows the adversary's choice to be informed by the public parameters and secret key queries.

Definition 5. *A (H)PE scheme $\mathcal{E}$ is said to be (fully) attribute-hiding (based on Definition 2 in [9]) if an adversary $\mathcal{A}$ has negligible advantage in the following game:*

1. *In the **non-adaptive** variant, $\mathcal{A}$ outputs two attributes a_0 and a_1 at the beginning of the game.*
2. *The challenger $\mathcal{C}$ runs Setup(1^λ) and outputs (PP, MSK)*
3. ***Phase 1***
   $\mathcal{A}$ makes adaptive queries for the secret keys for predicates $f_1, \ldots, f_k \in \mathcal{F}$ subject to the constraint that $f_i(a_0) = f_i(a_1)$ for $1 \leq i \leq k$.
4. *Remark 2. In the stronger adaptive variant, $\mathcal{A}$ only chooses attributes a_0 and a_1 at this stage.*
5. *$\mathcal{A}$ outputs two messages m_0 and m_1 of equal length. It must hold that $m_0 = m_1$ if there is an i such that $f_i(a_0) = f_i(a_1) = 1$.*
6. *$\mathcal{C}$ chooses a random bit b, and outputs $c \leftarrow$ Encrypt(PP, a_b, m_b)*
7. ***Phase 2***
   A second phase is run where $\mathcal{A}$ requests secret keys for other predicates subject to the same constraint as above.
8. *Finally, $\mathcal{A}$ outputs a guess b' and is said to win if $b' = b$.*

A weaker property referred to as *weakly* attribute-hiding [9] requires that the adversary only request keys for predicates f obeying $f(a_0) = f(a_1) = 0$.

We propose another model of security for non-universal homomorphic encryption in Section 6.

3.3   Attribute Operations

We now characterize HPE schemes based on the properties of their attribute operations (elements of Γ_A).

Definition 6 (Properties of attribute operations). $\forall f \in \mathcal{F}, \quad \forall a_1, a_2 \in A, \quad \forall \gamma_A \in \Gamma_A$:

1.
$$f(\gamma_A(a_1, a_2)) \Rightarrow f(a_1) \wedge f(a_2) \tag{3.1}$$

   (Necessary condition for IND-CPA security)

2.
$$f(\gamma_A(a_1, a_1)) = f(a_1) \tag{3.2}$$

3. $\forall d \in A$:

$$f(a_1) = f(a_2) \Rightarrow f(\gamma_A(d, a_1)) = f(\gamma_A(d, a_2))$$
$$\wedge \ f(\gamma_A(a_1, d)) = f(\gamma_A(a_2, d))$$

$$\tag{3.3}$$

   (Non-monotone Indistinguishability)

4.
$$f(\gamma_A(a_1, a_2)) = f(a_1) \wedge f(a_2) \tag{3.4}$$

   (Monotone Access)

Property 3.1 is a minimal precondition for payload-hiding i.e. IND-CPA security under both adaptive and non-adaptive security definitions.

Property 3.2 preserves access under a homomorphic operation on ciphertexts with the same attribute.

Property 3.3 is a necessary condition for full attribute-hiding.

Property 3.4 enables monotone access; a user only learns a function of a plaintext if and only if that user has permission to learn the value of that plaintext. This implies that (A, γ_A) cannot be a group unless $\mathcal{F}$ is a class of constant predicates. In general, 3.4 implies that $\mathcal{F}$ is monotonic. Monotone access is equivalent to the preceding three properties collectively; that is

$$3.1 \wedge 3.2 \wedge 3.3 \iff 3.4$$

Non-Monotone Access. Non-monotone access is trickier to define and to suitably accommodate in a security definition. It can arise from policies that involve negation. As an example, suppose that it is permissible for a party to decrypt data sets designated as either "geology" or "aviation", but is not authorized to decrypt results with both designations that arise from homomorphic computations on both data sets. Of course it is then necessary to strengthen the restrictions on the adversary's choice of a_0 and a_1 in the security game. Let a_0 and a_1 be the attributes chosen by the adversary. Intuitively, the goal is to show that any sequence of transitions that leads a_0 to a an element outside the support of f, also leads a_1 to an element outside the support of f, and vice versa. Instead of explicitly imposing this non-triviality constraint on the adversary's choice of attributes, one may seek to show that there is no pair of attributes distinguishable under any γ_A and $f \in \mathcal{F}$. This is captured by the property of non-monotone indistinguishability (3.3). Trivially, the constant operations satisfy 3.3. Of more interest is an operation that limits homomorphic operations to ciphertexts with the same attribute. This captures our usual requirements for the (anonymous) IBE functionality, but it is also satisfactory for many applications of general PE where computation need only be performed on ciphertexts with matching attributes. To accomplish this, the attribute space is augmented with a (logical) absorbing element z such that $f(z) = 0 \ \forall f \in \mathcal{F}$. The attribute operation is defined as follows:

$$\delta(a_1, a_2) = \begin{cases} a_1 & \text{if } a_1 = a_2 \\ z & \text{if } a_1 \neq a_2 \end{cases} \tag{3.5}$$

δ models the inability to perform homomorphic evaluations on ciphertexts associated with unequal attributes (identities in the case of IBE). A scheme with this operation can only be fully attribute-hiding in a vacuous sense (it may be such that no restrictions are placed upon the adversary's choice of f but it is unable to find attributes a_0 and a_1 satisfying $f(a_0) = f(a_1) = 1$ for any f.) This is the case for anonymous IBE where the predicates are equality relations, and for the constant'map $(a_1, a_2) \mapsto z$ that models the absence of a homomorphic property, although this is preferably modeled by appropriately constraining the class of permissible circuits. More generally, such schemes can only be weakly attribute-hiding because their operations γ_A only satisfy a relaxation of 3.3 given as follows:

Necessary condition for weakly attribute-hiding $\forall a_1, a_2, d \in A$:

$$f(a_1) = f(a_2) = 0 \Rightarrow f(\gamma_A(d, a_1)) = f(\gamma_A(d, a_2))$$
$$\wedge \ f(\gamma_A(a_1, d)) = f(\gamma_A(a_2, d))$$

$$\tag{3.6}$$

Remark 3. In the case of general schemes not satisfying 3.3, placing constraints on the adversary's choice of attributes weakens the security definition. Furthermore, it must be possible for the challenger to efficiently check whether a pair of attributes satisfies such a condition. Given the added complications, it is tempting to move to a simulation-based definition of security. However, this is precluded by the recent impossibility results of [30] in the case of both weakly and fully attribute-hiding in the NA/AD-SIM models of security. However, for predicate encryption with public index (the attribute is not hidden), this has not been ruled out for 1-AD-SIM and many-NA-SIM where "1" and "many" refer to the number of ciphertexts seen by the adversary. See [30, 31] for more details. In the context of non-monotone access, it thus seems more reasonable to focus on predicate encryption with public index. Our main focus in this work is on schemes that facilitate attribute privacy, and therefore we restrict our attention to schemes that at least satisfy 3.6.

Delegate Predicate Encryption. A primitive presented in [32] called "Delegate Predicate Encryption" (DPE) [3] enables a user to generate an encryption key associated with a chosen attribute $a \in A$, which does not reveal anything about a. The user can distribute this to certain parties who can then encrypt messages with attribute a obliviously. The realization in [32] is similar to the widely-used technique of publishing encryptions of "zero" in a homomorphic cryptosystem, which can then be treated as a key. In fact, this technique is adopted in [33] to transform a strongly homomorphic private-key scheme into a public-key one. Generalizing from the results of [32], this corollary follows from the property of attribute-hiding

Corollary 1. *An attribute-hiding HPE scheme is a DPE as defined in [32] if there exists a $\gamma \in \Gamma$ such that $(A \times M, \gamma)$ is unital.*

4   Constructions with Attribute Aggregation

In this section, we give some meaningful examples of attribute homomorphisms (all which satisfy monotone access) for some known primitives. We begin with a special case of PE introduced by Boneh and Waters [34], which they call *Hidden Vector Encryption*. In this primitive, a ciphertext embeds a vector $\mathbf{w} \in \{0, 1\}^n$ where n is fixed in the public parameters. On the other hand, a secret key corresponds to a vector $\mathbf{v} \in V \triangleq \{*, 0, 1\}^n$ where $*$ is interpreted as a "wildcard" symbol or a "don't care" (it matches any symbol). A decryptor who has a secret key for some $\mathbf{v}$ can check whether it matches the attribute in a ciphertext.

To formulate in terms of PE, let $A = \{0, 1\}^n$ and define

$$\mathcal{F} \subseteq \{(w_1, \ldots, w_n) \mapsto \bigwedge_{i=1}^{n} (v_i = w_i \vee v_i = *) : \mathbf{v} \in V\}$$

Unfortunately, we cannot achieve a non-trivial homomorphic variant of HVE that satisfies 3.4. To see this, consider the HVE class of predicates $\mathcal{F}$ and an

---

[3] Not to be confused with the different notion of Delegatable Predicate Encryption.

operation γ_A satisfying 3.4. For any $\mathbf{x}, \mathbf{y} \in A$, let $\mathbf{z} = \gamma_A(\mathbf{x}, \mathbf{y})$. Now for 3.4 to hold, we must have that $f(\mathbf{z}) = f(\mathbf{x}) \wedge f(\mathbf{y})$ for all $f \in \mathcal{F}$. Suppose $\mathbf{x}_i \neq \mathbf{y}_i$ and $\mathbf{z}_i = \mathbf{x}_i$. Then there exists an $f \in \mathcal{F}$ with $f(\mathbf{z}) = f(\mathbf{x})$ and $f(\mathbf{z}) \neq f(\mathbf{y})$. It is necessary to restrict V. Accordingly, let $V = \{*, 1\}^n$ Setting the non-equal elements to 0 yields associativity and commutativity. Such an operation is equivalent to component-wise logical AND on the attribute vectors, and we will denote it by $\wedge^n$. $(A, \wedge^n)$ is a semilattice.

Recall that a predicate-only scheme does not incorporate a payload into ciphertexts. Even such a scheme $\mathcal{E}$ with the $\wedge^n$ attribute homomorphism might find some purpose in real-world scenarios. One particular application of $\mathcal{E}$ is secure data aggregation in Wireless Sensor Networks (WSNs), an area which has been the target of considerable research (a good survey is [35]). It is conceivable that some aggregator nodes may be authorized by the sink (base station to which packets are forwarded) to read packets matching certain criteria. An origin sensor node produces an outgoing ciphertext as follows: (1). It encrypts the attributes describing its data using $\mathcal{E}$. (2) It encrypts its sensor reading with the public key of the sink using a *separate* additively (say) homomorphic public-key cryptosystem. (3) Both ciphertexts are forwarded to the next hop.

Since an aggregator node receives packets from multiple sources, it needs to have some knowledge about how to aggregate them. To this end, the sink can authorize it to apply a particular predicate to incoming ciphertexts to check for matching candidates for aggregation. One sample policy may be ["REGION1" $\wedge$ "TEMPERATURE"']. It can then aggregate ciphertexts matching this policy. Additional aggregation can be performed by a node further along the route that has been perhaps issued a secret key for a predicate corresponding to the more permissive policy of ["TEMPERATURE"]. In the scenario above, it would be more ideal if $\mathcal{E}$ were also additively homomorphic since besides obviating the need to use another PKE cryptosystem, more control is afforded to aggregators; they receive the ability to decrypt partial sums, and therefore, to perform (more involved) statistical computations on the data.

It is possible to achieve the former case from some recent inner-product PE schemes that admit homomorphisms on both attributes and payload. We focus on two prominent constructions with different mathematical structures. Firstly, a construction is examined by Katz, Sahai and Waters (KSW) [9], which relies on non-standard assumptions on bilinear groups, assumptions that are justified by the authors in the generic group model. Secondly, we focus on a construction presented by Agrawal, Freeman and Vaikuntanathan (AFV) [36] whose security is based on the learning with errors (LWE) problem.

In both schemes, an attribute is an element of $\mathbb{Z}_m^n$ [4] and a predicate also corresponds to an element of $\mathbb{Z}_m^n$. For $\mathbf{v} \in \mathbb{Z}_m^n$, a predicate $f_\mathbf{v} : \mathbb{Z}_m^n \to \{0, 1\}$ is

---

[4] In [9], m is a product of three large primes and n is the security parameter. In [36], n is independent of the security parameter and m may be polynomial or superpolynomial in the security parameter; in the latter case m is the product of many "small" primes. We require that m be superpolynomial here.

defined by

$$f_{\mathbf{v}}(\mathbf{w}) = \begin{cases} 1 & \text{iff } \langle \mathbf{v}, \mathbf{w} \rangle \\ 0 & \text{otherwise} \end{cases}$$

Roughly speaking, in a ciphertext, all sub-attributes (in $\mathbb{Z}_m$) are blinded by the same uniformly random "blinding" element b [5]. The decryption algorithm multiplies each component by the corresponding component in the predicate vector, and the blinding element b is eliminated when the inner product evaluates to zero with all but negligible probability, which allows decryption to proceed.

Let $\mathbf{c_1}$ and $\mathbf{c_2}$ be ciphertexts that encrypt attributes a_1 and a_2 respectively. It can be easily shown that the sum $\mathbf{c}' = \mathbf{c_1} \boxplus^6 \mathbf{c_2}$ encrypts both a_1 and a_2 in a somewhat "isolated" way. The lossiness is "hidden" by the negligible probability of two non-zero inner-products summing to 0. For linear aggregation, this can be repeated a polynomial number of times (or effectively unbounded in practice) while ensuring correctness with overwhelming probability. While linear aggregation is sufficient for the WSN scenario, it is interesting to explore other circuit forms. For the KSW scheme, we observe that all circuits of polynomial depth can be evaluated with overwhelming probability. For AFV, the picture is somewhat similar to the fully homomorphic schemes based on LWE such as [4,5] but without requiring multiplicative gates.

While there are motivating scenarios for aggregation on the attributes, in many cases it is adequate or preferable to restrict evaluation to ciphertexts with matching attributes; that is, by means of the δ operation defined in Section 3.3. Among these cases is anonymous IBE. In the next section, we introduce an IBE construction that supports an unbounded XOR homomorphism, prove that it is strongly homomorphic and then investigate anonymous variants.

5   Main Construction: XOR-Homomorphic IBE

In this section, an XOR-homomorphic IBE scheme is presented whose security is based on the quadratic residuosity assumption. Therefore, it is similar in many respects to the Goldwasser-Micali (GM) cryptosystem [6], which is well-known to be XOR-homomorphic. Indeed, the GM scheme has found many practical applications due to its homomorphic property. In Section 6.3, we show how many of these applications benefit from an XOR-homomorphic scheme in the identity-based setting.

Our construction derives from the IBE scheme due to Cocks [27] which has a security reduction to the quadratic residuosity problem. To the best of our knowledge, a homomorphic variant has not been explored to date.

5.1   Background

Let m be an integer. A quadratic residue in the residue ring $\mathbb{Z}_m$ is an integer x such that $x \equiv y^2 \mod m$ for some $y \in \mathbb{Z}_m$. The set of quadratic residues in $\mathbb{Z}_m$

---

[5] a scalar in KSW and a matrix in AFV.

[6] $\boxplus$ denotes a pairwise sum of the ciphertext components in both schemes.

is denoted $\mathbb{QR}(m)$. If m is prime, it easy to determine whether any $x \in \mathbb{Z}_m$ is a quadratic residue.

Let $N = pq$ be a composite modulus where p and q are prime. Let $x \in \mathbb{Z}$. We write $\left(\dfrac{x}{N}\right)$ to denote the Jacobi symbol of $x \mod N$. The subset of integers with Jacobi symbol $+1$ (resp. -1) is denoted $\mathbb{Z}_N[+1]$ (resp. $\mathbb{Z}_N[-1]$). The quadratic residuosity problem is to determine, given input $(N, x \in \mathbb{Z}_N[+1])$, whether $x \in \mathbb{QR}(N)$, and it is believed to be intractable.

Define the encoding $\nu : \{0,1\} \rightarrow \{-1,1\}$ with $\nu(0) = 1$ and $\nu(1) = -1$. Formally, ν is a group isomorphism between $(\mathbb{Z}_2, +)$ and $(\{-1,1\}, *)$.

In this section, we build on the results of [25] and therefore attempt to maintain consistency with their notation where possible. As in [25], we let $H : \{0,1\}^* \rightarrow \mathbb{Z}_N^*[+1]$ be a full-domain hash. A message bit is mapped to an element of $\{-1,1\}$ via ν as defined earlier (0 (1 resp.) is encoded as 1 (-1 resp.)).

5.2 Original Cocks IBE Scheme

- CocksIBE.**Setup**(1^λ):
    1. Repeat: $p, q \xleftarrow{\$} \mathsf{RandPrime}(1^\lambda)$ Until: $p \equiv q \equiv 3 \pmod 4$
    2. $N \leftarrow pq$
    3. Output $(\mathsf{PP} := N, \mathsf{MSK} := (p, q))$
- CocksIBE.**KeyGen**($\mathsf{PP}, \mathsf{MSK}, \mathsf{id}$):
    1. Parse MSK as (p, q).
    2. $a \leftarrow H(\mathsf{id})$
    3. $r \leftarrow a^{\frac{N+5-p-q}{8}} \pmod N$
       ($\therefore r^2 \equiv a \pmod N$) or $r^2 \equiv -a \pmod N$))
    4. Output $\mathsf{sk_{id}} := (\mathsf{id}, r)$
- CocksIBE.**Encrypt**($\mathsf{PP}, \mathsf{id}, b$):
    1. $a \leftarrow H(\mathsf{id})$
    2. $t_1, t_2 \xleftarrow{\$} \mathbb{Z}_N^*[\nu(b)]$
    3. Output $\psi := (t_1 + at_1^{-1}, t_2 - at_2^{-1})$
- CocksIBE.**Decrypt**($\mathsf{PP}, \mathsf{sk_{id}}, \psi$):
    1. Parse ψ as (ψ_1, ψ_2)
    2. Parse $\mathsf{sk_{id}}$ as (id, r)
    3. $a \leftarrow H(\mathsf{id})$
    4. If $r^2 \equiv a \pmod N$, set $d \leftarrow \psi_1$. Else if $r^2 \equiv -a \pmod N$, set $d \leftarrow \psi_2$. Else output $\bot$ and abort.
    5. Output $\nu^{-1}\left(\left(\dfrac{d + 2r}{N}\right)\right)$

The above scheme can be shown to be adaptively secure in the random oracle model assuming the hardness of the quadratic residuosity problem.

Anonymity. Cocks' scheme is not anonymous. Boneh et al. [37] report a test due to Galbraith that enables an attacker to distinguish the identity of a ciphertext. This is achieved with overwhelming probability given multiple ciphertexts. It is shown by Ateniese and Gasti [25] that there is no "better" test for attacking anonymity. Briefly, let $a = H(\mathsf{id})$ be the public key derived from the identity $\mathsf{ID_a}$. Let c be a ciphertext in the Cocks' scheme. Galbraith's test is defined as

$$\mathsf{GT}(a, c, N) = \left(\frac{c^2 - 4a}{N}\right)$$

Now if c is a ciphertext encrypted with a, then $\mathsf{GT}(a, c, N) = +1$ with all but negligible probability. For $b \in \mathbb{Z}_N^*$ such that $b \neq a$, the value $\mathsf{GT}(b, c, N)$ is statistically close to the uniform distribution on $\{-1, 1\}$. Therefore, given multiple ciphertexts, it can be determined with overwhelming probability whether they correspond to a particular identity.

5.3   XOR-Homomorphic Construction

Recall that a ciphertext in the Cocks scheme consists of two elements in $\mathbb{Z}_N$. Thus, we have

$$(c, d) \leftarrow \mathsf{CocksIBE.Encrypt}(\mathsf{PP}, \mathsf{id}, b) \in \mathbb{Z}_N^2$$

for some identity id and bit $b \in \{0, 1\}$. Also recall that only one element is actually used for decryption depending on whether $a := H(\mathsf{id}) \in \mathbb{QR}(N)$ or $-a \in \mathbb{QR}(N)$. If the former holds, it follows that a decryptor has a secret key r satisfying $r^2 \equiv a \pmod{N}$. Otherwise, a secret key r satisfies $r^2 \equiv -a \pmod{N}$. To simplify the description of the homomorphic property, we will assume that $a \in \mathbb{QR}(N)$ and therefore omit the second "component" d from the ciphertext. In fact, the properties hold analogously for the second "component" by simply replacing a with $-a$.

In the homomorphic scheme, each "component" of the ciphertext is represented by a pair of elements in $\mathbb{Z}_N^2$ instead of a single element as in the original Cocks scheme. As mentioned, we will omit the second such pair for the moment. Consider the following encryption algorithm E_a defined by

$E_a(b : \{0, 1\}):$

    $t \xleftarrow{\$} \mathbb{Z}_N^*[\nu(b)]$

    return $(t + at^{-1}, 2) \in \mathbb{Z}_N^2$.

Furthermore, define the decryption function $D_a(\mathbf{c}) = \nu^{-1}(c_0 + rc_1)$. The homomorphic operation $\boxplus : \mathbb{Z}_N^2 \times \mathbb{Z}_N^2 \to \mathbb{Z}_N^2$ is defined as follows:

$$\mathbf{c} \boxplus \mathbf{d} = (c_0 d_0 + a c_1 d_1, c_0 d_1 + c_1 d_0) \tag{5.1}$$

It is easy to see that $D_a(\mathbf{c} \boxplus \mathbf{d}) = D_a(\mathbf{c}) \oplus D_a(\mathbf{d})$:

$$\begin{aligned}
D_a(\mathbf{c} \boxplus \mathbf{d}) &= D_a((c_0 d_0 + a c_1 d_1, c_0 d_1 + c_1 d_0)) \\
&= \nu^{-1}((c_0 d_0 + a c_1 d_1) + r(c_0 d_1 + c_1 d_0)) \\
&= \nu^{-1}(c_0 d_0 + r c_0 d_1 + r c_1 d_0 + r^2 c_1 d_1) \\
&= \nu^{-1}((c_0 + r c_1)(d_0 + r d_1)) \\
&= \nu^{-1}(c_0 + r c_1) \oplus \nu^{-1}(d_0 + r d_1) \\
&= D_a(\mathbf{c}) \oplus D_a(\mathbf{d}) \tag{5.2}
\end{aligned}$$

Let $R_a = \mathbb{Z}_N[x]/(x^2 - a)$ be a quotient of the polynomial ring $R = \mathbb{Z}_N[x]$. It is more natural and convenient to view ciphertexts as elements of R_a and the homomorphic operation as multiplication in R_a. Furthermore, decryption equates to evaluation at the point r. Thus the homomorphic evaluation of two ciphertext polynomials $c(x)$ and $d(x)$ is simply $e(x) = c(x) * d(x)$ where $*$ denotes multiplication in R_a. Decryption becomes $\nu^{-1}(e(r))$. Moreover, Galbraith's test is generalized straightforwardly to the ring R_a:

$$\mathsf{GT}(a, c(x)) = \left(\frac{c_0^2 - c_1^2 a}{N} \right).$$

We now formally describe our variant of the Cocks scheme that supports an XOR homomorphism.

Remark 4. We have presented the scheme in accordance with Definition 4 for consistency with the rest of the paper. Therefore, it uses the circuit formulation, which we would typically consider superfluous for a group homomorphic scheme.

Let $\mathbb{C} \triangleq \{\mathbf{x} \mapsto \langle \mathbf{t}, \mathbf{x} \rangle : \mathbf{t} \in \mathbb{Z}_2^\ell\} \subset \mathbb{Z}_2^\ell \to \mathbb{Z}_2$ be the class of arithmetic circuits characterized by linear functions over $\mathbb{Z}_2$ in ℓ variables. As such, we associate a representative vector $V(C) \in \mathbb{Z}_2^\ell$ to every circuit $C \in \mathbb{C}$. In order to obtain a strongly homomorphic scheme, we use the standard technique of re-randomizing the evaluated ciphertext by homomorphically adding an encryption of zero.

- xhIBE.**Encrypt**(PP, id, b):
  1. $a \leftarrow H(\text{id})$
  2. As a subroutine (used later), define
     $E(\text{PP}, a, b)$:
     (a) $t_1, t_2 \xleftarrow{\$} \mathbb{Z}_N^*[\nu(b)]$
     (b) $g_1, g_2 \xleftarrow{\$} \mathbb{Z}_N^*$
     (c) $c(x) \leftarrow (t_1 + a g_1^2 t_1^{-1}) + 2 g_1 x \in \mathbb{Z}_N[x]$
     (d) $d(x) \leftarrow (t_2 + a g_2^2 t_2^{-1}) + 2 g_2 x \in \mathbb{Z}_N[x]$
     (e) Repeat steps (a) - (d) until $\mathsf{GT}(a, c(x)) = 1$ and $\mathsf{GT}(-a, d(x)) = 1$.
     (f) Output $(c(x), d(x))$
  3. Output $\psi := (E(\text{PP}, a, b), a)$
- xhIBE.**Decrypt**(PP, sk$_{\text{id}}$, ψ):
  1. Parse ψ as $(c(x), d(x), a)$

2. Parse $\mathsf{sk_{id}}$ as (id, r)
3. If $r^2 \equiv a \pmod{N}$ and $\mathsf{GT}(a, c(x)) = 1$, set $e(x) \leftarrow c(x)$. Else if $r^2 \equiv -a$ $\pmod{N}$ and $\mathsf{GT}(-a, c(x)) = 1$, set $e(x) \leftarrow d(x)$. Else output $\perp$ and abort.
4. Output $\nu^{-1}\left(\left(\dfrac{e(r)}{N}\right)\right)$

$\mathsf{xhIBE.Eval}(\mathsf{PP}, C, \psi_1, \ldots, \psi_\ell)$:

1. Parse ψ_i as $(c_i(x), d_i(x), a_i)$ for $1 \leq i \leq \ell$
2. If $a_i \neq a_j$ for $1 \leq i, j \leq \ell$, abort with $\perp$.
3. Let $a = a_1$ and let $R_a = \mathbb{Z}_N[x]/(x^2 - a)$
4. $v \leftarrow V(C)$
5. $J \leftarrow \{1 \leq i \leq \ell : v_i = 1\}$
6. $(c'(x), d'(x)) \leftarrow (\prod_{i \in J} c_i(x) \mod (x^2 - a), \prod_{i \in I} d_i(x)) \mod (x^2 + a)$
7. $(c_z(x), d_z(x)) \leftarrow E(\mathsf{PP}, a, 0)$ (E is defined as a subroutine in the specification of $\mathsf{xhIBE.Encrypt}$)
8. Output $(c'(x) * c_z(x) \mod (x^2 - a), d'(x) * d_z(x) \mod (x^2 + a), a)$.

We now prove that our scheme is group homomorphic and strongly homomorphic. A formalization of group homomorphic public-key schemes is given in [38]. Our adapted definition for the PE setting raises some subtle points. The third requirement in [38] is more difficult to formalize for general PE; we omit it from the definition here and leave a complete formalization to the full version [7]. We remark that this property which relates to distinguishing "illegitimate ciphertexts" during decryption is not necessary to achieve IND-ID-CPA security.

Definition 7 (Adapted from Definition 1 in [38]). *Let $\mathcal{E} = (G, K, E, D)$ be a PE scheme with message space M, attribute space A, ciphertext space $\hat{\mathcal{C}}$ and class of predicates $\mathcal{F}$. The scheme $\mathcal{E}$ is group homomorphic with respect to a non-empty set of attributes $A' \subseteq A$ if for every $(\mathsf{PP}, \mathsf{MSK}) \leftarrow G(1^\lambda)$, every $f \in \mathcal{F} : A' \subseteq \mathsf{supp}(f)$, and every $\mathsf{sk}_f \leftarrow K(\mathsf{MSK}, f)$, the message space $(M, \cdot)$ is a non-trivial group, and there is a binary operation $\boxdot : \hat{\mathcal{C}}^2 \to \hat{\mathcal{C}}$ such that the following properties are satisfied for the restricted ciphertext space $\hat{\mathcal{C}}_f = \{c \in \hat{\mathcal{C}} : D_{\mathsf{sk}_f}(c) \neq \perp\}$:*

1. *The set of all encryptions $\mathcal{C} := \{c \in \hat{\mathcal{C}}_f \mid c \leftarrow E(\mathsf{PP}, a, m), a \in A', m \in M\}$ under attributes in A' is a non-trivial group under the operation $\boxdot$.*
2. *The restricted decryption $D^*_{\mathsf{sk}_f} := D_{\mathsf{sk}_f}|\mathcal{C}$ is surjective and $\forall c, c' \in \mathcal{C}$ $D_{\mathsf{sk}_f}(c \boxdot c') = D_{\mathsf{sk}_f}(c) \cdot D_{\mathsf{sk}_f}(c')$.*
3. ***IBE only (generalized in the full version)** If $\mathcal{E}$ is an IBE scheme, then $\hat{\mathcal{C}}_f$ is also required to be a group, and it is required to be computationally indistinguishable from $\mathcal{C}$; that is:*

$$\{(\mathsf{PP}, f, \mathsf{sk}_f, S, c) \mid c \xleftarrow{\$} \mathcal{C}, S \subset \{\mathsf{sk}_g \leftarrow K(g) : g \in \mathcal{F}\}\}$$

$$\underset{C}{\approx} \{(\mathsf{PP}, f, \mathsf{sk}_f, S, \hat{c}) \mid \hat{c} \xleftarrow{\$} \hat{\mathcal{C}}_f, S \subset \{\mathsf{sk}_g \leftarrow K(g) : g \in \mathcal{F}\}\}.$$

---

[7] Available at http://arxiv.org/abs/1302.1192.

Informally, the above definition is telling us that for a given subset of attributes A' satisfying a predicate f, the set of honestly generated encryptions under these attributes forms a group that is epimorphic to the plaintext group. It does not say anything about ciphertexts that are not honestly generated except in the case of IBE, where we require that all ciphertexts that do not decrypt to $\perp$ under a secret key are indistinguishable.

For the remainder of this section, we show that xhIBE fulfills the definition of a group homomorphic scheme, and that it is IND-ID-CPA secure under the quadratic residuosity assumption in the random oracle model. To simplify the presentation of the proofs, additional notation is needed. In particular, we inherit the notation from [25], and generalize it to the ring R_a.

Define the subset $G_a \subset R_a$ as follows:

$$G_a = \{c(x) \in R_a : \mathsf{GT}(a, c(x)) = 1\}$$

Define the subset $S_a \subset G_a{}^8$:

$$S_a = \{2hx + (t + ah^2t^{-1}) \in G_a \mid h, t, (t + ah^2t^{-1}) \in \mathbb{Z}_N^*\}$$

We have the following simple lemma:

Lemma 1.

1. $(G_a, *)$ is a multiplicative subgroup of R_a.
2. $(S_a, *)$ is a subgroup of G_a

Proof. We must show that G_a is closed under $*$. Let $c(x), d(x) \in G_a$, and let $e(x) = c(x) * d(x)$.

$$\mathsf{GT}(a, e(x)) = \left(\frac{e_0^2 - ae_1^2}{N}\right)$$
$$= \left(\frac{(c_0d_0 + ac_1d_1)^2 - a(c_0d_1 + c_1d_0)^2}{N}\right)$$
$$= \left(\frac{(c_0^2 - ac_1^2)(d_0^2 - ad_1^2)}{N}\right)$$
$$= \left(\frac{(c_0^2 - ac_1^2)}{N}\right)\left(\frac{(d_0^2 - ad_1^2)}{N}\right)$$
$$= \mathsf{GT}(a, c(x)) \cdot \mathsf{GT}(a, d(x))$$
$$= 1$$

Therefore, $e(x) \in G_a$.

It remains to show that every element of G_a is a unit. Let $z = c_0^2 - ac_1^2 \in \mathbb{Z}_N$. An inverse $d_1x + d_0$ of $c(x)$ can be computed by setting $d_0 = \frac{c_0}{z}$ and $d_1 = \frac{-c_1}{z}$ if it holds that z is invertible in $\mathbb{Z}_N$. Indeed such a $d_1x + d_0$ is in G_a. Now if z

---

[8] This definition is stricter than its analog in [25] in that all elements are in G_a.

is not invertible in $\mathbb{Z}_N$ then $p|z$ or $q|z$, which implies that $\left(\dfrac{z}{p}\right) = 0$ or $\left(\dfrac{z}{q}\right) = 0$. But $\mathsf{GT}(a, c(x)) = \left(\dfrac{z}{N}\right) = \left(\dfrac{z}{p}\right)\left(\dfrac{z}{q}\right) = 1$ since $c(x) \in G_a$. Therefore, z is a unit in $\mathbb{Z}_N$, and $c(x)$ is a unit in G_a.

Finally, to prove (2), note that the members of S_a are exactly the elements $c(x)$ such that $c_0^2 - c_1^2 a$ is a square, and it is easy to see that this is preserved under $*$ in R_a. □

We will also need the following corollary

Corollary 2 (Extension of Lemma 2.2 in [25]). *The distributions* $\{(N, a, t + ah^2 t^{-1}, 2h) : N \leftarrow \mathsf{Setup}(1^\lambda), a \xleftarrow{\$} \mathbb{Z}_N^*[+1], t, h \xleftarrow{\$} \mathbb{Z}_N^*)\}$ *and* $\{(N, a, z_0, z_1) : N \leftarrow \mathsf{Setup}(1^\lambda), a \xleftarrow{\$} \mathbb{Z}_N^*[+1], z_0 + z_1 x \xleftarrow{\$} G_a \backslash S_a\}$ *are indistinguishable assuming the hardness of the quadratic residuosity problem.*

Proof. The corollary follows immediately from Lemma 2.2 in [25] Let $\mathcal{A}$ be an efficient adversary that distinguishes both distributions. Lemma 2.2 in [25] shows that the distributions $d_0 := (\{(N, a, t + at^{-1}) : N \leftarrow \mathsf{Setup}(1^\lambda), a \xleftarrow{\$} \mathbb{Z}_N^*[+1], t\}$ and $d_1 := \{(N, a, z_0) : N \leftarrow \mathsf{Setup}(1^\lambda), a \xleftarrow{\$} \mathbb{Z}_N^*[+1], z_1 x + z_0 \xleftarrow{\$} G_a \backslash S_a \mid z_2 = 2\}$ are indistinguishable. Given a sample (N, a, c), the simulator generates $h \xleftarrow{\$} \mathbb{Z}_N^*$ and computes $b := h^{-2}a$. It passes the element $(N, b, c, 2h)$ to $\mathcal{A}$. The simulator aborts with the output of $\mathcal{A}$. □

Theorem 1. xhIBE *is a group homomorphic scheme with respect to the group operation of* $(\mathbb{Z}_2, +)$.

Proof. Let $a = H(\mathsf{id})$ for any valid identity string id. Assume that the secret key r satisfies $r^2 \equiv a \mod N$. The analysis holds analogously if $r^2 \equiv -a \mod N$; therefore, we omit the second component of the ciphertexts for simplicity.

By definition, $S_a = \{c(x) \in R_a \mid \psi := (c(x), d(x), a) \leftarrow$ xhIBE.Encrypt$(\mathsf{PP}, \mathsf{id}, m), m \in M\}$. By corollary 2, it holds that $S_a \approx_C G_a$ without the master secret key. The decryption algorithm only outputs $\perp$ on input $\psi := (c(x), d(x), a)$ if $c(x) \notin G_a$ or $d(x) \notin G_{-a}$. Thus, omitting the second component, we have that S_a corresponds to $\mathcal{C}$ and G_a corresponds to $\hat{\mathcal{C}}_f$ in Definition 7 (in this case f is defined as $f(\mathsf{id}') = 1$ iff $\mathsf{id}' = \mathsf{id}$). It follows that the third requirement of Definition 7 is satisfied.

By Lemma 1, G_a is a group and S_a is a non-trivial subgroup of G_a. The surjective homomorphism between $\mathcal{C} := S_a$ and $M := \mathbb{Z}_2^*$ has already been shown in the correctness derivation in equation 5.2. This completes the proof. □

Remark 5. It is straightforward to show that xhIBE also meets the criteria for a shift-type homomorphism as defined in [38].

Corollary 3. xhIBE *is strongly homomorphic.*

Proof. Any group homomorphic scheme can be turned into a strongly homomorphic scheme by rerandomizing an evaluated ciphertext. Indeed this follows from Lemma 1 in [38]. Rerandomization is achieved by multiplying the evaluated ciphertext by an encryption of the identity, as in xhIBE.Eval. Details follow for completeness.

Let id be an identity and let $a = H(\text{id})$. For any circuit $C \in \mathbb{C}$, any messages $b_1, \ldots, b_\ell$ and ciphertexts $\psi_1, \ldots, \psi_\ell \leftarrow \text{xhIBE.Encrypt}(\text{PP}, b_1, \text{id}), \ldots,$ xhIBE.Encrypt$(\text{PP}, b_\ell, \text{id})$, we have

$$(c'(x), d'(x), a) \leftarrow \text{xhIBE.\textbf{Eval}}(\text{PP}, C, \psi_1, \ldots, \psi_\ell).$$

From the last step of xhIBE.Eval, we see that $c'(x) \leftarrow c''(x) * r(x)$ where $r(x) \overset{\$}{\leftarrow} S_a^{(0)}$ and $c''(x)$ is the result of the homomorphic evaluation. Suppose that $c''(x)$ encrypts a bit b. Since S_a is a group, it follows that $c'(x)$ is uniformly distributed in the coset $S_a^{(b)}$ (of the subgroup $S_a^{(0)}$) and is thus distributed according to a "fresh" encryption of b.   □

Theorem 2. xhIBE *is IND-ID-CPA secure in the random oracle model under the quadratic residuosity assumption.*

Proof. Let $\mathcal{A}$ be an adversary that breaks the IND-ID-CPA security of xhIBE. We use $\mathcal{A}$ to construct an algorithm $\mathcal{S}$ to break the IND-ID-CPA security of the Cocks scheme with the same advantage. $\mathcal{S}$ proceeds as follows:

1. Uniformly sample an element $h \overset{\$}{\leftarrow} \mathbb{Z}_N^*$. Receive the public parameters PP from the challenger $\mathcal{C}$ and pass them to $\mathcal{A}$.
2. $\mathcal{S}$ answers a query to H for identity id with $H'(\text{id}) \cdot h^{-2}$ where H' is $\mathcal{S}$'s random oracle. The responses are uniformly distributed in $\mathbb{Z}_N[+1]$.
3. $\mathcal{S}$ answers a key generation query for id with the response $K(\text{id}) \cdot h^{-1}$ where K is its key generation oracle.
4. When $\mathcal{A}$ chooses target identity id*, $\mathcal{S}$ relays id* to $\mathcal{C}$. Assume w.l.o.g that H has been queried for id, and that $\mathcal{A}$ has not made a secret key query for id*. Further key generation requests are handled subject to the condition that id $\neq$ id* for a requested identity id.
5. Let $a = H(\text{id}^*)$. On receiving a challenge ciphertext (c, d) from $\mathcal{C}$, compute $c(x) \leftarrow 2hx + c \in R$ and $d(x) \leftarrow (2hx + d) * r(x) \in R$ where $r(x) \overset{\$}{\leftarrow} S_{-a}^{(0)}$ and $S_{-a}^{(0)}$ is the second component of the set of legal encryptions of 0. From corollary 3, $d(x)$ is uniformly distributed in $S_{-a}^{(b)}$ where the ciphertext (c, d) in the Cocks scheme encrypts the bit b. It follows that $(c(x), d(x))$ is a perfectly simulated encryption of b under identity id* in xhIBE. Give $(c(x), d(x))$ to $\mathcal{A}$.
6. Output $\mathcal{A}$'s guess b'.

Since the view of $\mathcal{A}$ in an interaction with $\mathcal{S}$ is indistinguishable from its view in the real game, we conclude that the advantage of $\mathcal{S}$ is equal to the advantage of $\mathcal{A}$.

□

In the next section, attention is drawn to obtaining an anonymous variant of our construction.

6 Anonymity

Cocks' scheme is notable as one of the few IBE schemes that do not rely on pairings. Since it appeared, there have been efforts to reduce its ciphertext size and make it anonymous. Boneh, Gentry and Hamburg [28] proposed a scheme with some elegant ideas that achieves both anonymity and a much reduced ciphertext size for multi-bit messages at the expense of performance, which is $O(n^4)$ for encryption and $O(n^3)$ for decryption (where n is the security parameter). Unfortunately the homomorphic property is lost in this construction.

As mentioned earlier (cf. Section 5.2), another approach due to Ateniese and Gasti [25] achieves anonymity and preserves performance, but its per-bit ciphertext expansion is much higher than in [28]. However, an advantage of this scheme is that it is universally anonymous (anyone can anonymize the message, not merely the encryptor [24]).

On the downside, anonymizing according to this scheme breaks the homomorphic property of our construction, which depends crucially on the public key a. More precisely, what is forfeited is the *universal homomorphic* property mentioned in the introduction (i.e. anyone can evaluate on the ciphertexts without additional information). There are applications where an evaluator is aware of the attribute(s) associated with ciphertexts, but anonymity is desirable to prevent any other parties in the system learning about such attributes. This motivates a variant of HPE, which we call non-universal HPE, denoted by $\mathsf{HPE}_{\bar{U}}$.

6.1 Non-Universal HPE

Motivation. "Non-universal" homomorphic encryption is proposed for schemes that support attribute privacy but require some information that is derivable from the public key (or attribute in the case of PE) in order to perform homomorphic evaluation. Therefore, attribute privacy must be surrendered to an evaluator. If this is acceptable for an application, while at the same time there is a requirement to hide the target recipient(s) from other entities in the system, then "non-universal" homomorphic encryption may be useful. Consider the following informal scenario. Suppose a collection of parties $P_1, \ldots, P_\ell$ outsource a computation on their encrypted data sets to an untrusted remote server S. Suppose S sends the result (encrypted) to an independent database DB from which users can retrieve the encrypted records. For privacy reasons, it may be desirable to limit the information that DB can learn about the attributes associated with the ciphertexts retrieved by certain users. Therefore, it may desirable for the encryption scheme to provide attribute privacy. However, given the asymmetric relationship between the delegators $P_1, \ldots, P_\ell$ and the target recipient(s), it might be acceptable for S to learn the target attribute(s) provided there is no collusion between S and DB. In fact, the delegators may belong to a different organization than the recipient(s).

In this paper, we introduce a syntax and security model for non-universal homomorphic IBE. The main change in syntax entails an additional input α that is supplied to the Eval algorithm. The input $\alpha \in \{0,1\}^d$ (where $d = poly(\lambda)$) models the additional information needed to compute the homomorphism(s). A description of an efficient map $Q_A : A \rightarrow \{0,1\}^d$ is included in the public parameters. We say that two attributes(i.e. identities in IBE) $a_1, a_2 \in A$ satisfying $Q_A(a_1) = Q_A(a_2)$ belong to the same attribute class.

One reason that the proposed syntax is not general enough for arbitrary PE functionalities is that it only facilitates evaluation on ciphertexts whose attributes are in the same attribute class, which suffices for (relatively) simple functionalities such as IBE.

We now formulate the security notion of attribute-hiding for non-universal homomorphic IBE. Our security model provides the adversary with an evaluation oracle whose identity-dependent input α is fixed when the challenge is produced. Accordingly, for a challenge identity id $\in A$, and binary string $\alpha = Q_A(\text{id}) \in \{0,1\}^d$, the adversary can query $\text{IBE}_{\bar{U}}.\text{Eval}(PP, \alpha, \cdot, \cdot)$ for any circuit in $\mathbb{C}$ and any ℓ-length sequence of ciphertexts.

Formally, consider the experiment

Experiment $\bar{\text{U}}\text{Priv}(\mathcal{A}_1, \mathcal{A}_2)$[9]

> $(PP, MSK) \leftarrow \text{IBE.Setup}(1^\lambda)$
> $(\text{id}_0, m_0), (\text{id}_1, m_1), \sigma \leftarrow \mathcal{A}_1^{\text{IBE}_0.\text{KeyGen}(MSK,\cdot)}(PP)$ $\triangleright \sigma$ denotes the adversary's state
> $b \xleftarrow{\$} \{0,1\}$
> $\alpha \leftarrow Q_A(\text{id}_b)$
> $c \leftarrow \text{IBE.Encrypt}(PP, \text{id}_b, m_b)$
> $b' \leftarrow \mathcal{A}_2^{\text{IBE}_0.\text{KeyGen}^*(MSK,\cdot), \text{IBE}_0.\text{Eval}(PP,\alpha,\cdot,\cdot)}(PP, c, \sigma)$
> **return** 1 iff $b' = b$ and 0 otherwise.

Define the advantage of an adversary $\mathcal{A} := (\mathcal{A}_1, \mathcal{A}_2)$ in the above experiment for a $\text{IBE}_{\bar{U}}$ scheme $\mathcal{E}$ as follows:

$$\mathbf{Adv}_\mathcal{E}^{\bar{\text{U}}\text{Priv}}(\mathcal{A}) = \Pr\left[\bar{\text{U}}\text{Priv}(\mathcal{A}) \Rightarrow 1\right] - \frac{1}{2}.$$

A $\text{IBE}_{\bar{U}}$ scheme $\mathcal{E}$ is said to be attribute-hiding if for all pairs of PPT algorithms $\mathcal{A} := (\mathcal{A}_1, \mathcal{A}_2)$, it holds that $\mathbf{Adv}_\mathcal{E}^{\bar{\text{U}}\text{Priv}}(\mathcal{A}) \leq \text{negl}(\lambda)$. Note that the above definition assumes adaptive adversaries, but can be easily modified to accommodate the non-adaptive case.

6.2 Universal Anonymizers

We now present an abstraction called a *universal anonymizer*. With its help, we can transform a universally-homomorphic, non-attribute-hiding IBE scheme

---

[9] In the random oracle model, the adversary is additionally given access to a random oracle. This is what the results in this paper will use.

$\mathcal{E}$ into a non-universally homomorphic, attribute-hiding scheme $\mathcal{E}'$. In accordance with the property of universal anonymity proposed in [24], any party can anonymize a given ciphertext.

Let $\mathcal{E} := (\mathsf{Setup}, \mathsf{KeyGen}, \mathsf{Encrypt}, \mathsf{Decrypt}, \mathsf{Eval})$ be a PE scheme parameterized with message space M, attribute space A, class of predicates $\mathcal{F}$, and class of circuits $\mathbb{C}$. Denote its ciphertext space by $\mathcal{C}$. Note that this definition of a universal anonymizer only suffices for simple functionalities such as IBE.

Definition 8. *A universal anonymizer $U_{\mathcal{E}}$ for a PE scheme $\mathcal{E}$ is a tuple $(\mathcal{G}, \mathcal{B}, \mathcal{B}^{-1}, Q_A, Q_{\mathcal{F}})$ where $\mathcal{G}$ is a deterministic algorithm, $\mathcal{B}$ and $\mathcal{B}^{-1}$ are randomized algorithms, and Q_A and $Q_{\mathcal{F}}$ are efficient maps, defined as follows:*

- *$\mathcal{G}(\mathsf{PP})$:*
  On input the public parameter of an instance of $\mathcal{E}$, output a parameters structure params. This contains a description of a modified ciphertext space $\hat{\mathcal{C}}$ as well as an integer $d = \mathsf{poly}(\lambda)$ indicating the length of binary strings representing an attribute class.
- *$\mathcal{B}(\mathsf{params}, \mathbf{c})$:*
  On input parameters params and a ciphertext $\mathbf{c} \in \mathcal{C}$, output an element of $\hat{\mathcal{C}}$.
- *$\mathcal{B}^{-1}(\mathsf{params}, \alpha, \hat{\mathbf{c}})$:*
  On input parameters params, a binary string $\alpha \in \{0,1\}^d$ and an element of $\hat{\mathcal{C}}$, output an element of $\mathcal{C}$
- *Both maps Q_A and $Q_{\mathcal{F}}$ are indexed by params: $Q_{A\,\mathsf{params}} : A \to \{0,1\}^d$ and $Q_{\mathcal{F}\,\mathsf{params}} : \mathcal{F} \to \{0,1\}^d$*

Note: params can be assumed to be an implicit input; it will not be explicitly specified to simplify notation.

The binary string α is computed by means of a map $Q_A : A \to \{0,1\}^d$. In order for a decryptor to invert $\mathcal{B}$, α must also be computable from any predicate that satisfies an attribute that maps onto α. Therefore, the map $Q_{\mathcal{F}} : \mathcal{F} \to \{0,1\}^d$ has the property that for all $a \in A$ and $f \in \mathcal{F}$:

$$f(a) = 1 \Rightarrow Q_A(a) = Q_{\mathcal{F}}(f).$$

We define an equivalence relation $\sim$ on $\mathcal{F}$ given by

$$f_1 \sim f_2 \triangleq Q_{\mathcal{F}}(f_1) = Q_{\mathcal{F}}(f_2).$$

We have that

$$f \sim g \iff \exists h_1, \dots, h_k \in \mathcal{F} \quad \mathsf{supp}(f) \cap \mathsf{supp}(h_1) \neq \emptyset \wedge \dots \wedge \mathsf{supp}(h_k) \cap \mathsf{supp}(g) \neq \emptyset.$$

It follows that each α is a representative of an equivalence class in $\mathcal{F}/\sim$. As a result, as mentioned earlier, our definition of a universal anonymizer above is only meaningful for "simple" functionalities such as IBE. For example, $|\mathcal{F}/\sim| = |\mathcal{F}|$ for an IBE scheme whose ciphertexts leak the recipient's identity.

Let c be a ciphertext associated with an attribute a. Let $\alpha = Q_A(a)$. Informally, $c' := \mathcal{B}^{-1}(\alpha, \mathcal{B}(c))$ should "behave" like c; that is, (1) it should have the same homomorphic "capacity" and (2) decryption with a secret key for any f should have the same output as that for c. A stronger requirement captured in our formal correctness criterion defined in the full version is that c and c' should be indistinguishable even when a distinguisher is given access to MSK.

A universal anonymizer is employed in the following generic transformation from a universally-homomorphic, non-attribute-hiding IBE scheme $\mathcal{E}$ to a non-universally homomorphic, attribute-hiding scheme $\mathcal{E}'$.

The transformation is achieved by setting:

- $\mathcal{E}'.\mathsf{Encrypt}(\mathsf{PP}, a, m) :=$
    $\mathcal{B}(\mathcal{E}.\mathsf{Encrypt}(\mathsf{PP}, a, m))$
- $\mathcal{E}'.\mathsf{Decrypt}(\mathsf{SK}_f, c) :=$
    $\mathcal{E}.\mathsf{Decrypt}(\mathsf{SK}_f, \mathcal{B}^{-1}(Q_{\mathcal{F}}(f), c))$
- $\mathcal{E}'.\mathsf{Eval}(\mathsf{PP}, \alpha, C, c_1, \ldots, c_\ell) :=$
    return $\mathcal{B}(\mathcal{E}.\mathsf{Eval}(\mathsf{PP}, C, \mathcal{B}^{-1}(\alpha, c_1), \ldots, \mathcal{B}^{-1}(\alpha, c_\ell)))$

Denote the above transformation by $T_{U_\mathcal{E}}(\mathcal{E})$. We leave to future work the task of establishing (generic) sufficient conditions that $\mathcal{E}$ must satisfy to ensure that $\mathcal{E}' := T_{U_\mathcal{E}}(\mathcal{E})$ is an attribute-hiding $\mathsf{HPE}_{\bar{U}}$ scheme.

An instantiation of a universal anonymizer for our XOR homomorphic scheme is given in the full version.

6.3   Applications (Brief Overview)

It turns out that XOR-homomorphic cryptosystems have been considered to play an important part in several applications. The most well-known and widely-used *unbounded* XOR-homomorphic public-key cryptosystem is Goldwasser-Micali (GM) [6], which is based on the quadratic residuosity problem. Besides being used in protocols such as private information retrieval (PIR), GM has been employed in some specific applications such as:

- Peng, Boyd and Dawson (PBD) [39] propose a sealed-bid auction system that makes extensive use of the GM cryptosystem.
- Bringer et al. [40] apply GM to biometric authentication. It is used in two primary ways: (1) to achieve PIR and (2) to assist in computing the hamming distance between a recorded biometric template and a reference one.

Perhaps in some of these applications, a group-homomorphic identity-based scheme may be of import, although the authors concede that no specific usage scenario has been identified so far.

With regard to performance, our construction requires 8 multiplications in $\mathbb{Z}_N$ for a single homomorphic operation in comparison to a single multiplication in GM. Furthermore, the construction has higher ciphertext expansion than GM by a factor of 4. Encryption involves 2 modular inverses and 6 multiplications (only 4 if the strongly homomorphic property is forfeited). In comparison, GM only requires 1.5 multiplications on average.

7   Conclusions and Future Work

We have presented a characterization of homomorphic encryption in the PE setting and classified schemes based on the properties of their attribute homomorphisms. Instantiations of certain homomorphic properties were presented for inner-product PE. However, it is clear that meaningful attribute homomorphisms are limited. We leave to future work the exploration of homomorphic encryption with access policies in a more general setting .

In this paper, we introduced a new XOR-homomorphic variant of the Cocks' IBE scheme and showed that it is strongly homomorphic. However, we failed to fully preserve the homomorphic property in anonymous variants; that is, we could not construct an anonymous universally-homomorphic variant. We leave this as an open problem. As a compromise, however, a weaker primitive (non-universal IBE) was introduced along with a related security notion. Furthermore, a transformation strategy adapted from the work of Ateniese and Gasti [25] was exploited to obtain anonymity for our XOR-homomorphic construction in this weaker primitive.

In future work, it is hoped to construct other group homomorphic IBE schemes, and possibly for more general classes of predicates than the IBE functionality.

Noteworthy problems, which we believe are still open:

1. Somewhat-homomorphic IBE scheme (even non-adaptive security in the ROM)
2. (Unbounded) Group homomorphic IBE schemes for $(\mathbb{Z}_m, +)$ where $m = O(2^\lambda)$ and $(\mathbb{Z}_p^*, *)$ for prime p. Extensions include anonymity and support for a wider class of predicates beyond the IBE functionality.

References

1. Gentry, C.: Fully homomorphic encryption using ideal lattices. In: Proceedings of the 41st Annual ACM Symposium on Symposium on Theory of Computing, STOC 2009, p. 169 (2009)
2. Smart, N.P., Vercauteren, F.: Fully Homomorphic Encryption with Relatively Small Key and Ciphertext Sizes. In: Nguyen, P.Q., Pointcheval, D. (eds.) PKC 2010. LNCS, vol. 6056, pp. 420–443. Springer, Heidelberg (2010)
3. van Dijk, M., Gentry, C., Halevi, S., Vaikuntanathan, V.: Fully Homomorphic Encryption over the Integers. In: Gilbert, H. (ed.) EUROCRYPT 2010. LNCS, vol. 6110, pp. 24–43. Springer, Heidelberg (2010)
4. Brakerski, Z., Vaikuntanathan, V.: Fully Homomorphic Encryption from Ring-LWE and Security for Key Dependent Messages. In: Rogaway, P. (ed.) CRYPTO 2011. LNCS, vol. 6841, pp. 505–524. Springer, Heidelberg (2011)

5. Brakerski, Z., Vaikuntanathan, V.: Efficient Fully Homomorphic Encryption from (Standard) LWE. Cryptology ePrint Archive, Report 2011/344 (2011), http://eprint.iacr.org/

6. Goldwasser, S., Micali, S.: Probabilistic encryption & how to play mental poker keeping secret all partial information. In: Proceedings of the Fourteenth Annual ACM Symposium on Theory of Computing, STOC 1982, pp. 365–377. ACM, New York (1982)

7. Paillier, P.: Public-key cryptosystems based on composite degree residuosity classes. In: Stern, J. (ed.) EUROCRYPT 1999. LNCS, vol. 1592, pp. 223–238. Springer, Heidelberg (1999)

8. ElGamal, T.: A public key cryptosystem and a signature scheme based on discrete logarithms 31, 469–472 (1985)

9. Katz, J., Sahai, A., Waters, B.: Predicate encryption supporting disjunctions, polynomial equations, and inner products. In: Smart, N.P. (ed.) EUROCRYPT 2008. LNCS, vol. 4965, pp. 146–162. Springer, Heidelberg (2008)

10. Boneh, D., Sahai, A., Waters, B.: Functional Encryption: Definitions and Challenges. In: Ishai, Y. (ed.) TCC 2011. LNCS, vol. 6597, pp. 253–273. Springer, Heidelberg (2011)

11. Gennaro, R., Gentry, C., Parno, B.: Non-interactive verifiable computing: Outsourcing computation to untrusted workers. In: Rabin, T. (ed.) CRYPTO 2010. LNCS, vol. 6223, pp. 465–482. Springer, Heidelberg (2010)

12. López-Alt, A., Tromer, E., Vaikuntanathan, V.: On-the-fly multiparty computation on the cloud via multikey fully homomorphic encryption. In: Proceedings of the 44th Symposium on Theory of Computing, STOC 2012, pp. 1219–1234. ACM, New York (2012)

13. Naccache, D.: Is theoretical cryptography any good in practice? Talk given at CHES 2010 and Crypto 2010 (2010)

14. Brakerski, Z., Vaikuntanathan, V.: Efficient Fully Homomorphic Encryption from (Standard) LWE. Cryptology ePrint Archive, Report 2011/344 Version: 20110627:080002 (2011), http://eprint.iacr.org/

15. Gentry, C., Peikert, C., Vaikuntanathan, V.: Trapdoors for hard lattices and new cryptographic constructions. In: Proceedings of the 40th Annual ACM Symposium on Theory of Computing, STOC 2008, pp. 197–206. ACM, New York (2008)

16. Gentry, C., Halevi, S., Vaikuntanathan, V.: A Simple BGN-Type Cryptosystem from LWE. In: Gilbert, H. (ed.) EUROCRYPT 2010. LNCS, vol. 6110, pp. 506–522. Springer, Heidelberg (2010)

17. Boneh, D., Goh, E.-J., Nissim, K.: Evaluating 2-DNF Formulas on Ciphertexts. In: Kilian, J. (ed.) TCC 2005. LNCS, vol. 3378, pp. 325–341. Springer, Heidelberg (2005)

18. Benaloh, J.: Dense probabilistic encryption. In: Proceedings of the Workshop on Selected Areas of Cryptography, pp. 120–128 (1994)

19. Galbraith, S.D.: Elliptic Curve Paillier Schemes. J. Cryptology 15, 129–138 (2002)

20. Golle, P., Jakobsson, M., Juels, A., Syverson, P.F.: Universal re-encryption for mixnets. In: Okamoto, T. (ed.) CT-RSA 2004. LNCS, vol. 2964, pp. 163–178. Springer, Heidelberg (2004)

21. Gjøsteen, K.: Homomorphic cryptosystems based on subgroup membership problems. In: Dawson, E., Vaudenay, S. (eds.) Mycrypt 2005. LNCS, vol. 3715, pp. 314–327. Springer, Heidelberg (2005)

22. Armknecht, F., Katzenbeisser, S., Peter, A.: Group homomorphic encryption: characterizations, impossibility results, and applications. Designs, Codes and Cryptography, 1–24 (2012)

23. Bellare, M., Boldyreva, A., Desai, A., Pointcheval, D.: Key-privacy in public-key encryption. In: Boyd, C. (ed.) ASIACRYPT 2001. LNCS, vol. 2248, pp. 566–582. Springer, Heidelberg (2001)

24. Hayashi, R., Tanaka, K.: Universally Anonymizable Public-Key Encryption. In: Roy, B. (ed.) ASIACRYPT 2005. LNCS, vol. 3788, pp. 293–312. Springer, Heidelberg (2005)

25. Ateniese, G., Gasti, P.: Universally anonymous IBE based on the quadratic residuosity assumption. In: Fischlin, M. (ed.) CT-RSA 2009. LNCS, vol. 5473, pp. 32–47. Springer, Heidelberg (2009)

26. Prabhakaran, M., Rosulek, M.: Homomorphic encryption with CCA security. In: Aceto, L., Damgård, I., Goldberg, L.A., Halldórsson, M.M., Ingólfsdóttir, A., Walukiewicz, I. (eds.) ICALP 2008, Part II. LNCS, vol. 5126, pp. 667–678. Springer, Heidelberg (2008)

27. Cocks, C.: An identity based encryption scheme based on quadratic residues. In: Honary, B. (ed.) Cryptography and Coding 2001. LNCS, vol. 2260, pp. 360–363. Springer, Heidelberg (2001)

28. Boneh, D., Gentry, C., Hamburg, M.: Space-efficient identity based encryption without pairings. In: FOCS, pp. 647–657. IEEE Computer Society (2007)

29. Goldwasser, S.: Lecture: Introduction to homomorophic encryption (2011), http://www.cs.bu.edu/~reyzin/teaching/s11cs937/notes-shafi-1.pdf (last Checked on March 31, 2013)

30. Agrawal, S., Gorbunov, S., Vaikuntanathan, V., Wee, H.: Functional encryption: New perspectives and lower bounds. Cryptology ePrint Archive, Report 2012/468 (2012), http://eprint.iacr.org/

31. Bellare, M., O'Neill, A.: Semantically-secure functional encryption: Possibility results, impossibility results and the quest for a general definition. Cryptology ePrint Archive, Report 2012/515 (2012), http://eprint.iacr.org/

32. Wei, R., Ye, D.: Delegate predicate encryption and its application to anonymous authentication. In: Proceedings of the 4th International Symposium on Information, Computer, and Communications Security, ASIACCS 2009, pp. 372–375. ACM, New York (2009)

33. Rothblum, R.: Homomorphic Encryption: From Private-Key to Public-Key. In: Ishai, Y. (ed.) TCC 2011. LNCS, vol. 6597, pp. 219–234. Springer, Heidelberg (2011)

34. Boneh, D., Waters, B.: Conjunctive, subset, and range queries on encrypted data. In: Vadhan, S.P. (ed.) TCC 2007. LNCS, vol. 4392, pp. 535–554. Springer, Heidelberg (2007)

35. Alzaid, H., Foo, E., Nieto, J.G.: Secure data aggregation in wireless sensor network: a survey. In: Proceedings of the Sixth Australasian Conference on Information Security, AISC 2008, vol. 81, pp. 93–105. Australian Computer Society, Inc., Darlinghurst (2008)

36. Agrawal, S., Freeman, D.M., Vaikuntanathan, V.: Functional encryption for inner product predicates from learning with errors. In: Lee, D.H., Wang, X. (eds.) ASIACRYPT 2011. LNCS, vol. 7073, pp. 21–40. Springer, Heidelberg (2011)

37. Boneh, D., Di Crescenzo, G., Ostrovsky, R., Persiano, G.: Public key encryption with keyword search. In: Cachin, C., Camenisch, J.L. (eds.) EUROCRYPT 2004. LNCS, vol. 3027, pp. 506–522. Springer, Heidelberg (2004)

38. Armknecht, F., Katzenbeisser, S., Peter, A.: Group homomorphic encryption: Characterizations, impossibility results, and applications. Cryptology ePrint Archive, Report 2010/501 (2010), http://eprint.iacr.org/
39. Peng, K., Boyd, C., Dawson, E.: A Multiplicative Homomorphic Sealed-Bid Auction Based on Goldwasser-Micali Encryption. In: Zhou, J., López, J., Deng, R.H., Bao, F. (eds.) ISC 2005. LNCS, vol. 3650, pp. 374–388. Springer, Heidelberg (2005)
40. Bringer, J., Chabanne, H., Izabachène, M., Pointcheval, D., Tang, Q., Zimmer, S.: An application of the goldwasser-micali cryptosystem to biometric authentication. In: Pieprzyk, J., Ghodosi, H., Dawson, E. (eds.) ACISP 2007. LNCS, vol. 4586, pp. 96–106. Springer, Heidelberg (2007)

Brandt's Fully Private Auction Protocol Revisited

Jannik Dreier[1], Jean-Guillaume Dumas[2], and Pascal Lafourcade[1]

[1] Université Grenoble 1, CNRS, Verimag, France
[2] Université Grenoble 1, CNRS, Laboratoire Jean Kuntzmann (LJK)
firstname.lastname@imag.fr

Abstract. Auctions have a long history, having been recorded as early as 500 B.C. [17]. Nowadays, electronic auctions have been a great success and are increasingly used. Many cryptographic protocols have been proposed to address the various security requirements of these electronic transactions, in particular to ensure privacy. Brandt [4] developed a protocol that computes the winner using homomorphic operations on a distributed ElGamal encryption of the bids. He claimed that it ensures full privacy of the bidders, i.e. no information apart from the winner and the winning price is leaked. We first show that this protocol – when using malleable interactive zero-knowledge proofs – is vulnerable to attacks by dishonest bidders. Such bidders can manipulate the publicly available data in a way that allows the seller to deduce all participants' bids. Additionally we discuss some issues with verifiability as well as attacks on non-repudiation, fairness and the privacy of individual bidders exploiting authentication problems.

1 Introduction

Auctions are a simple method to sell goods and services. Typically a *seller* offers a good or a service, and the *bidders* make offers. Depending on the type of auction, the offers might be sent using sealed envelopes which are opened simultaneously to determine the winner (the "sealed-bid" auction), or an *auctioneer* could announce prices decreasingly until one bidder is willing to pay the announced price (the "dutch auction"). Additionally there might be several rounds, or offers might be announced publicly directly (the "English" or "shout-out" auction). The winner usually is the bidder submitting the highest bid, but in some cases he might only have to pay the second highest offer as a price (the "second-price"- or "Vickrey"-Auction). In general a bidder wants to win the auction at the lowest possible price, and the seller wants to sell his good at the highest possible price. For more information on different auction methods see [17]. To address this huge variety of possible auction settings and to achieve different security and efficiency properties numerous protocols have been developed, e.g. [4,11,19,20,21,22,23] and references therein.

One of the key requirements of electronic auction (e-Auction) protocols is privacy, i.e. the bids of losing bidders remain private. Brandt proposed a first-price sealed-bid auction protocol [4,3,2] and claimed that it is fully private, i.e. it leaks no information apart from the winner, the winning bid, and what can be deduced from these two facts (*e.g.* that the other bids were lower).

A. Youssef, A. Nitaj, A.E. Hassanien (Eds.): AFRICACRYPT 2013, LNCS 7918, pp. 88–106, 2013.

Our Contributions. The protocol is based on an algorithm that computes the winner using bids encoded as bit vectors. In this paper we show that the implementation using the homomorphic property of a distributed Elgamal encryption proposed in the original paper suffers from a weakness. In fact, we prove that any two different inputs (i.e. different bids) result in different outcome values, which are only hidden using random values. We show how a dishonest participant can remove this random noise, if malleable interactive zero-knowledge proofs are used. The seller can then efficiently compute the bids of all bidders, hence completely breaking privacy. We also discuss two problems with verifiability, and how the lack of authentication enables attacks on privacy even if the above attack is prevented via non-malleable non-interactive proofs. Additionally we show attacks on non-repudiation and fairness, and propose solutions to all discovered flaws in order to recover a fully resistant protocol.

Outline. In the next section, we recall the protocol of Brandt. Then, in the following sections, we present our attacks in several steps. In Section 3, we first study the protocol using interactive zero-knowledge proofs and without noise. Then we show how a dishonest participant can remove the noise, thus mount the attack on the protocol with noise, and discuss countermeasures. Finally, in Section 4, we discuss verifiability and in Section 5 we discuss attacks on fairness, non-repudiation and privacy exploiting the lack of authentication.

2   The Protocol

The protocol of Brandt [4] was designed to ensure full privacy in a completely distributed way. It exploits the homomorphic properties of a distributed El-Gamal encryption scheme [12] for a secure multi-party computation of the winner. Then it uses zero-knowledge proofs of knowledge of discrete logarithms to ensure correctness of the bids while preserving privacy. We first give a high level description of the protocol and then present details on its main cryptographic primitives.

2.1   Informal Description

The participating n bidders and the seller communicate essentially using broadcast messages. The latter can for example be implemented using a bulletin board, i.e. an append-only memory accessible to everybody. The bids are encoded as k-bit-vectors where each entry corresponds to a price. If the bidder a wants to bid the price b_a, all entries will be 1, except the entry b_a which will be Y (a public constant). Each entry of the vector is then encrypted separately using a n-out-of-n-encryption scheme set up by all bidders. The bidders use multiplications of the encrypted values to compute values v_{aj}, exploiting the homomorphic property of the encryption scheme. Each one of this values is 1 if the bidder a wins at price j, and is a random number otherwise. The decryption of the final values takes place in a distributed way to ensure that nobody can access intermediate values.

2.2   Mathematical Description (Brandt [4])

Let $\mathbb{G}_q$ be a multiplicative subgroup of order q, prime, and g a generator of the group. We consider that $i, h \in \{1, \ldots, n\}$, $j, bid_a \in \{1, \ldots, k\}$ (where bid_a is the bid chosen by the bidder with index a), $Y \in \mathbb{G}_q \setminus \{1\}$. More precisely, the n bidders execute the following five steps of the protocol [4]:

1. **Key Generation**
   Each bidder a, whose bidding price is bid_a among $\{1, \ldots, k\}$ does the following:
   - chooses a secret $x_a \in \mathbb{Z}/q\mathbb{Z}$
   - chooses randomly m_{ij}^a and $r_{aj} \in \mathbb{Z}/q\mathbb{Z}$ for each i and j.
   - publishes $y_a = g^{x_a}$ and proves the knowledge of y_a's discrete logarithm.
   - using the published y_i then computes $y = \prod_{i=1}^n y_i$.

2. **Bid Encryption**
   Each bidder a
   - sets $b_{aj} = \begin{cases} Y & \text{if } j = bid_a \\ 1 & \text{otherwise} \end{cases}$
   - publishes $\alpha_{aj} = b_{aj} \cdot y^{r_{aj}}$ and $\beta_{aj} = g^{r_{aj}}$ for each j.
   - proves that for all j, $\log_g(\beta_{aj})$ equals $\log_y(\alpha_{aj})$ or $\log_y\left(\frac{\alpha_{aj}}{Y}\right)$, and that
   $$\log_y\left(\frac{\prod_{j=1}^k \alpha_{aj}}{Y}\right) = \log_g\left(\prod_{j=1}^k \beta_{aj}\right).$$

3. **Outcome Computation**
   - Each bidder a computes and publishes for all i and j:
   $$\gamma_{ij}^a = \left(\left(\prod_{h=1}^n \prod_{d=j+1}^k \alpha_{hd}\right) \cdot \left(\prod_{d=1}^{j-1} \alpha_{id}\right) \cdot \left(\prod_{h=1}^{i-1} \alpha_{hj}\right)\right)^{m_{ij}^a}$$
   $$\delta_{ij}^a = \left(\left(\prod_{h=1}^n \prod_{d=j+1}^k \beta_{hd}\right) \cdot \left(\prod_{d=1}^{j-1} \beta_{id}\right) \cdot \left(\prod_{h=1}^{i-1} \beta_{hj}\right)\right)^{m_{ij}^a}$$
   and proves its correctness.

4. **Outcome Decryption**
   - Each bidder a sends $\phi_{ij}^a = (\prod_{h=1}^n \delta_{ij}^h)^{x_a}$ for each i and j to the seller and proves its correctness. After having received all values, the seller publishes ϕ_{ij}^h for all i, j, and $h \neq i$.

5. **Winner determination**
   - Everybody can now compute $v_{aj} = \frac{\prod_{i=1}^n \gamma_{aj}^i}{\prod_{i=1}^n \phi_{aj}^i}$ for each j.
   - If $v_{aw} = 1$ for some w, then the bidder a wins the auction at price p_w.

2.3   Malleable Proofs of Knowledge and Discrete Logarithms

In the original paper [4] the author suggests using zero-knowledge proofs of knowledge to protect against active adversaries. The basic protocols he proposes are interactive and malleable, but can be converted into non-interactive proofs using the Fiat-Shamir heuristic [13], as advised by the author. We first recall the general idea of such proofs, then we expose the man-in-the-middle attacks on the interactive version, which we will use as part of our first attack.

Let PDL denote a *proof of knowledge of a discrete logarithm*. A first scheme for PDL was developed in 1986 by Chaum et al. [6]. In the original auction paper [4]

Brandt proposes to use a non-interactive variant of PDL as developed by Schnorr [24], which are malleable. Unfortunately, interactive malleable PDL are subject to man-in-the-middle attacks [16]. We first recall the classic Σ-protocol on a group with generator g and order q [1,5,7]. Peggy and Victor know v and g, but only Peggy knows x, so that $v = g^x$. She can prove this fact, without revealing x, by executing the following protocol:

1. Peggy chooses r at random and sends $z = g^r$ to Victor.
2. Victor chooses a challenge c at random and sends it to Peggy.
3. Peggy sends $s = (r + c \cdot x) \mod q$ to Victor.
4. Victor checks that $g^s = z \cdot v^c$.

Man-in-the-middle Attacks on Interactive PDL. Suppose Peggy possesses some secret discrete logarithm x. We present here the man-in-the-middle attack of [16], where an attacker can pretend to have knowledge of any affine combination of the secret x, even providing the associated proof of knowledge, without breaking the discrete logarithm. To prove this possession to say Victor, the attacker will start an interactive proof knowledge session with Peggy and another one with Victor. The attacker will transform Peggy's outputs and forward Victor's challenges to her. The idea is to use the proof of possession of Peggy's x, to prove possession of $1 - x$ to Victor. Indeed to prove for instance possession of just x to Victor, an attacker would only have to forward Peggy's messages to Victor and Victor's messages to Peggy. The idea of the attack is similar, except that one needs to modify the messages of Peggy. We show the example of $1 - x$ in Figure 1 since it is used in Section 3.4 to mount our attack. Upon demand by Victor to prove knowledge of $1 - x$, Mallory, the man-in-the-middle, simply starts a proof of knowledge of x with Peggy. Peggy chooses a random exponent r and sends the commitment $z = g^r$ to Mallory. Mallory simply inverts z and sends $y = z^{-1}$ to Victor. Then Victor presents a challenge c that Mallory simply forwards without modification to Peggy. Finally Peggy sends a response s that Mallory combines with c, as $u = c - s$, to provide a correct answer to Victor. This is summarized in Figure 1.

	Peggy	Mallory	Victor
Secret :	x		
Public :	$g, v = g^x$	$g, w = gv^{-1}$	g
	$z = g^r \xrightarrow{\ 1:z\ } y = z^{-1} \xrightarrow{\ 1':y\ }$		
	$\xleftarrow{\ 2:c\ } c \xleftarrow{\ 2':c\ } c$		
	$s = r + c \cdot x \xrightarrow{\ 3:s\ } u = c - s \xrightarrow{\ 3':u\ }$		
Check :		$g^s \overset{?}{==} z \cdot v^c$	$g^u \overset{?}{==} y \cdot w^c$

Fig. 1. Man-in-the-middle PDL of $1 - x$, with x an unknown discrete logarithm

Actually, the attack works in the generic settings of [5,18] or of Σ-protocols [10]. We let $f : \Gamma \to \Omega$ denote a one way homomorphic function between two commutative groups $(\Gamma, +)$ and $(\Omega, \times)$. We use this generalization to prevent possible countermeasures of our first attack in Section 3.6.

For an integral value α, $\alpha \cdot x \in \Gamma$ (resp. $y^\alpha \in \Omega$) denotes α applications of the group law $+$ (resp. $\times$). For a secret $x \in \Gamma$, and any $(h, \alpha, \beta) \in \Gamma \times \mathbb{Z}^2$, the attacker can build a proof of possession of $\alpha \cdot h + \beta \cdot x$. In the setting of the example of Figure 1, we used $f(x) = g^x$, $h = 1$, $\alpha = 1$ and $\beta = -1$.

In the general case also, upon demand of proof by Victor, Mallory starts a proof with Peggy. The secret of Peggy is x, and the associated witness v is $v = f(x)$. Then Mallory wants to prove that his witness w corresponds to any combination of x with a logarithm h that he knows. With only public knowledge and his chosen $(h, \alpha, \beta) \in \Gamma \times \mathbb{Z}^2$, Mallory is able to compute $w = f(h)^\alpha \cdot v^\beta$. For the proof of knowledge, Mallory still modifies the commitment $z = f(r)$ of Peggy to $y = z^\beta$. Mallory forwards the challenge c of Victor without modification. Finally Mallory transforms the response s of Peggy, still with only public knowledge and his chosen $(h, \alpha, \beta) \in \Gamma \times \mathbb{Z}^2$, as $u = c \cdot (\alpha \cdot h) + \beta \cdot s$. We summarize this general attack on Figure 2.

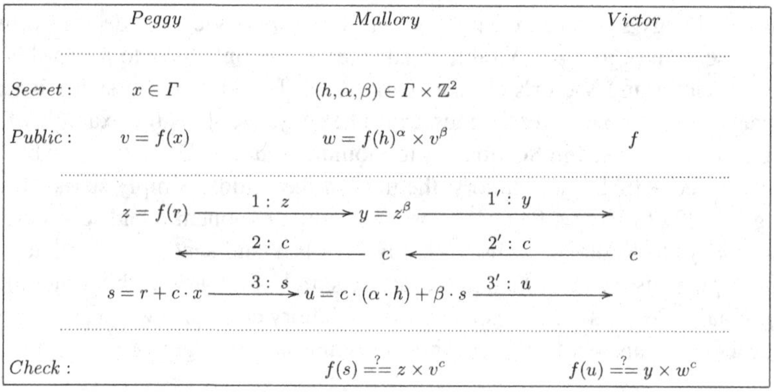

Fig. 2. Man-in-the-middle attacks proving knowledge of affine transforms of a secret discrete logarithm in the generic setting

Lemma 1. *In the man-in-the-middle attack of Figure 2 of the interactive proof of knowledge of a discrete logarithm, Victor is convinced by Mallory's proof of knowledge of* $\alpha \cdot h + \beta \cdot x$.

Proof. Indeed,

$$u = c \cdot (\alpha \cdot h) + \beta \cdot s = c \cdot (\alpha \cdot h) + \beta \cdot (r + c \cdot x) = \beta \cdot r + c \cdot (\alpha \cdot h + \beta \cdot x). \quad (1)$$

Now, since $z = f(r)$, $y = z^\beta$, $v = f(x)$ and $f(h)^\alpha \times v^\beta = w$, the latter Equation (1) proves in turn that

$$f(u) = f(r)^\beta \times f(\alpha \cdot h + \beta \cdot x)^c = z^\beta \times (f(h)^\alpha \times f(x)^\beta)^c = y \times w^c. \quad (2)$$

Now Victor has to verify the commitment-challenge-response (y, c, u) of Mallory for his witness w. Then Victor needs to checks whether $f(u)$ corresponds to $y \times w^c$, which is the case as shown by the latter Equation (2).   □

Generalizations to Equality of Discrete Logarithms. We let EQDL denote a *proof of equality of several discrete logarithms*. Any PDL can in general easily be transformed to an EQDL by applying it k times on the same witness. It is often more efficient to combine the application in one as in [8,9], or more generally as composition of Σ-protocols, here with two logarithms and two generators g_1 and g_2. Peggy wants to prove that she knows x such that $v = g_1^x$ and $w = g_2^x$:

1. Peggy chooses r at random and sends $\lambda = g_1^r$ and $\mu = g_2^r$ to Victor.
2. Victor chooses a challenge c at random and sends it to Peggy.
3. Peggy computes $s = (r + c \cdot x) \mod q$ and sends it to Victor.
4. Victor tests if $g_1^s = \lambda \cdot v^c$ and $g_2^s = \mu \cdot w^c$.

This protocol remains malleable, and the previous attacks are still valid since the response remains of the form $r + c \cdot x$.

Countermeasures. Direct countermeasures to the above attacks are to use non-interactive and/or non-malleable proofs:

- An interactive protocol can be converted into a non-interactive one using the Fiat-Shamir heuristic [13].
- Also the first PDL by [6] uses bit-flipping, and more generally non-malleable protocols like [15] could be used.

We will show in the following that if the proofs proposed in the original paper are not converted into non-interactive proofs, there is an attack on privacy. Note that even if non-interactive non-malleable zero-knowledge proofs are used, a malicious attacker in control of the network can nonetheless recover any bidder's bid as the messages are not authenticated, as we show in Section 5.

3   Attacking the Fully Private Computations

The first attack we present uses some algebraic properties of the computations performed during the protocol execution.

3.1   Analysis of the Outcome Computation

The idea is to analyze the computations done in Step 3 of the protocol. Consider the following example with three bidders and three possible prices. Then the first bidder computes

$$
\begin{aligned}
\gamma_{11}^1 &= (\ (\alpha_{12} \cdot \alpha_{13} \cdot \alpha_{22} \cdot \alpha_{23} \cdot \alpha_{32} \cdot \alpha_{33}) \cdot (1) && \cdot (1) &&)^{m_{11}^1} \\
\gamma_{12}^1 &= (\ \quad (\alpha_{13} \cdot \quad \alpha_{23} \cdot \quad \alpha_{33}) \cdot (\alpha_{11}) && \cdot (1) &&)^{m_{12}^1} \\
\gamma_{13}^1 &= (\ \quad\quad\quad (1) \cdot (\alpha_{11} \cdot \alpha_{12}) \cdot (1) &&)^{m_{13}^1} \\
\gamma_{21}^1 &= (\ (\alpha_{12} \cdot \alpha_{13} \cdot \alpha_{22} \cdot \alpha_{23} \cdot \alpha_{32} \cdot \alpha_{33}) \cdot (1) && \cdot (\alpha_{11}) &&)^{m_{21}^1} \\
\gamma_{22}^1 &= (\ \quad (\alpha_{13} \cdot \quad \alpha_{23} \cdot \quad \alpha_{33}) \cdot (\alpha_{21}) && \cdot (\alpha_{12}) &&)^{m_{22}^1} \\
\gamma_{23}^1 &= (\ \quad\quad\quad (1) \cdot (\alpha_{21} \cdot \alpha_{22}) \cdot (\alpha_{13}) &&)^{m_{23}^1} \\
\gamma_{31}^1 &= (\ (\alpha_{12} \cdot \alpha_{13} \cdot \alpha_{22} \cdot \alpha_{23} \cdot \alpha_{32} \cdot \alpha_{33}) \cdot (1) && \cdot (\alpha_{11} \cdot \alpha_{21}) &&)^{m_{31}^1} \\
\gamma_{32}^1 &= (\ \quad (\alpha_{13} \cdot \quad \alpha_{23} \cdot \quad \alpha_{33}) \cdot (\alpha_{31}) && \cdot (\alpha_{12} \cdot \alpha_{22}) &&)^{m_{32}^1} \\
\gamma_{33}^1 &= (\ \quad\quad\quad (1) \cdot (\alpha_{31} \cdot \alpha_{32}) \cdot (\alpha_{13} \cdot \alpha_{23}) &&)^{m_{33}^1}
\end{aligned}
$$

The second and third bidder do the same computations, but using different random values m_{ij}^a. Since each α_{ij} is either the encryption of 1 or Y, for example the value γ_{22}^1 will be an encryption of 1 only if

- nobody submitted a higher bid (the first block) and
- bidder 2 did not bid a lower bid (the second block) and
- no bidder with a lower index submitted the same bid (the third block).

If we ignore the exponentiation by m_{ij}^a, each γ_{ij}^a is the encryption of the product of several b_{ij}'s. Each b_{ij} can be either 1 or Y, hence $(\gamma_{ij}^a)^{-m_{ij}^a}$ will be the encryption of a value $Y^{l_{ij}}$, where $0 \le l_{ij} \le n$. The lower bound of l_{ij} is trivial, the upper bound follows from the observation that each α_{ij} will be used at most once, and that each bidder will encrypt Y at most once.

Assume for now that we know all l_{ij}. We show next that this is sufficient to obtain all bids. Consider the function f which takes as input the following vector[1]:
$b = \log_Y \left(\left(b_{11}, \ldots, b_{1k}, \quad b_{21}, \ldots, b_{2k}, \quad \ldots, \quad b_{n1}, \ldots, b_{nk} \right)^T \right)$, and returns the values l_{ij}. The input vector is thus a vector of all bid-vectors, where 1 is replaced by 0 and Y by 1. Consider our above example with three bidders and three possible prices, then we have:

$$
b = \log_Y \left(\left(b_{11}, b_{12}, b_{13}, \quad b_{21}, b_{22}, b_{23}, \quad b_{31}, b_{32}, b_{33} \right)^T \right).
$$

A particular instance where bidder 1 and 3 submit price 1, and bidder 2 submits price 2 would then look as: $b = \left(1, 0, 0, \quad 0, 1, 0, \quad 1, 0, 0 \right)^T$. Hence only the factors α_{11}, α_{22} and α_{31} are encryptions of Y, all other α's are encryptions of 1. By simply counting how often the factors α_{11}, α_{22} and α_{31} show up in each equation as described above, we can compute the following result: $f(b) = \left(1, 1, 1, \quad 2, 0, 1, \quad 2, 1, 1 \right)^T$. Note that since we chose the input of f to be a bit-vector, we have to simply count the ones (which correspond to Y's) in particular positions in b, where the positions are determined by the factors inside γ_{ij}^a. Hence we can express f as a matrix, i.e. $f(b) = M \cdot b$ for the following matrix M:

---

[1] By abuse of notation we write $\log_s (x_1, \ldots, x_n)$ for $(\log_s(x_1), \ldots, \log_s(x_n))$.

$$f(b) = M \cdot b = \begin{bmatrix} 0\ 1\ 1 & 0\ 1\ 1 & 0\ 1\ 1 \\ 1\ 0\ 1 & 0\ 0\ 1 & 0\ 0\ 1 \\ 1\ 1\ 0 & 0\ 0\ 0 & 0\ 0\ 0 \\ 1\ 1\ 1 & 0\ 1\ 1 & 0\ 1\ 1 \\ 0\ 1\ 1 & 1\ 0\ 1 & 0\ 0\ 1 \\ 0\ 0\ 1 & 1\ 1\ 0 & 0\ 0\ 0 \\ 1\ 1\ 1 & 1\ 1\ 1 & 0\ 1\ 1 \\ 0\ 1\ 1 & 0\ 1\ 1 & 1\ 0\ 1 \\ 0\ 0\ 1 & 0\ 0\ 1 & 1\ 1\ 0 \end{bmatrix} \cdot \begin{pmatrix} 1 \\ 0 \\ 0 \\ 0 \\ 1 \\ 0 \\ 1 \\ 0 \\ 0 \end{pmatrix} = \begin{pmatrix} 1 \\ 1 \\ 1 \\ 2 \\ 0 \\ 1 \\ 2 \\ 1 \\ 1 \end{pmatrix}$$

To see how the matrix M is constructed, consider for example $(\gamma_{22}^a)^{-m_{22}^a} = (\alpha_{13} \cdot \alpha_{23} \cdot \alpha_{33}) \cdot (\alpha_{21}) \cdot (\alpha_{12})$ which corresponds to the **second row** in the second vertical block:

- α_{12} and α_{13}; hence the two ones at position 2 and 3 in the first horizontal block
- α_{21} and α_{23}; hence the two ones at position 1 and 3 in the second horizontal block
- α_{33}; hence the one at position 3 in the third horizontal block

More generally, we can see that each 3×3 block consists of potentially three parts:

- An upper triangular matrix representing all bigger bids.
- On the diagonal we add a lower triangular matrix representing a lower bid by the same bidder,
- In the lower left half we add an identity matrix representing a bid at the current price by a bidder with a lower index.

This corresponds exactly to the structure of the products inside each γ_{ij}^a. It is also equivalent to formula (1) in Section 4.1.1 of the original paper [4] without the random vector R_k^*. In the following we prove that the function f is injective. We then discuss how this function can be efficiently inverted (i.e. how to compute the bids when knowing all l_{ij}'s).

3.2   Linear Algebra Toolbox

Let I_k be the $k \times k$ identity matrix; let L_k be a lower $k \times k$ triangular matrix with zeroes on the diagonal, ones in the lower part and zeroes elsewhere; and let U_k be an upper $k \times k$ triangular matrix with zeroes on the diagonal, ones in the upper part, and zeroes elsewhere:

$$I_k = \begin{bmatrix} 1 & 0 & \cdots & 0 \\ 0 & \ddots & \ddots & \vdots \\ \vdots & \ddots & \ddots & 0 \\ 0 & \cdots & 0 & 1 \end{bmatrix} \qquad L_k = \begin{bmatrix} 0 & 0 & \cdots & 0 \\ 1 & \ddots & \ddots & \vdots \\ \vdots & \ddots & \ddots & 0 \\ 1 & \cdots & 1 & 0 \end{bmatrix} \qquad U_k = \begin{bmatrix} 0 & 1 & \cdots & 1 \\ 0 & \ddots & \ddots & \vdots \\ \vdots & \ddots & \ddots & 1 \\ 0 & \cdots & 0 & 0 \end{bmatrix}$$

By abuse of notation we use I, L and U to denote respectively I_k, L_k and U_k. For a $k \times k$-matrix M_k we define $(M_k)^r = M \cdots M$ (r times) and $(M_k)^0 = I_k$. Let $(e_1, \ldots, e_k)$ be the canonical basis.

Lemma 2. *Matrices L_k and U_k have the following properties, for $0 < j \leq k$ and $r \geq 0$:* $(U_k)^r \cdot e_j = \sum_{s=1}^{j-r} e_s$ *and* $(L_k)^r \cdot e_j = \sum_{s=j+r}^{k} e_s$.

Lemma 3. *Matrices L_k and U_k are nilpotent, i.e.* $(U_k)^k = 0$ *and* $(L_k)^k = 0$.

This follows immediately from Lemma 2 by computing $(U_k)^k \cdot I_k$ and $(L_k)^k \cdot I_k$.

Lemma 4. *If $\sum_{i=1}^{k} x_i = 1$ then we have $L_k \cdot x = (1, \ldots, 1)^T - (I_k + U_k) \cdot x$.*

Proof. First note that since $\sum_{i=1}^{k} x_i = 1$,

$$
L_k \cdot x =
\begin{bmatrix}
0 & 0 & \cdots & 0 \\
1 & \ddots & \ddots & \vdots \\
\vdots & \ddots & \ddots & 0 \\
1 & \cdots & 1 & 0
\end{bmatrix}
\cdot
\begin{bmatrix}
x_1 \\
\vdots \\
x_k
\end{bmatrix}
=
\begin{bmatrix}
0 \\
x_1 \\
x_1 + x_2 \\
\vdots \\
\sum_{i=1}^{k-1} x_i
\end{bmatrix}
=
\begin{bmatrix}
1 - \sum_{i=1}^{k} x_i \\
1 - \sum_{i=2}^{k} x_i \\
\vdots \\
1 - x_k
\end{bmatrix}
$$

On the other hand, if we let $\mathbf{1} = (1, \ldots, 1)^T$, we have also:

$$
\mathbf{1} - (I_k + U_k) \cdot x = \mathbf{1} -
\begin{bmatrix}
1 & 1 & \cdots & 1 \\
0 & 1 & \ddots & \vdots \\
\vdots & \ddots & \ddots & 1 \\
0 & \cdots & 0 & 1
\end{bmatrix}
\cdot
\begin{bmatrix}
x_1 \\
\vdots \\
x_k
\end{bmatrix}
=
\begin{bmatrix}
1 - \sum_{i=1}^{k} x_i \\
1 - \sum_{i=2}^{k} x_i \\
\vdots \\
1 - x_k
\end{bmatrix}
$$

Lemma 5. $e_1^T \cdot U^{k-t-1} \cdot z = z_{k-t-1} + e_1^T \cdot U^{k-t} \cdot z$

The proof follows immediately from the fact that $e_1^T \cdot U^{k-x} = (\underbrace{0, \ldots, 0}_{k-x}, \underbrace{1, \ldots, 1}_{x})$. As

a direct consequence we obtain the following corollary.

Corollary 1. $e_1^T \cdot U^{k-t} \cdot z = z_{k-t} + e_1^T \cdot U^{k-t+1} \cdot z$

Lemma 6. *For $z = e_i - e_j$, we have that $(L_k + U_k) \cdot z = -z$.*

Proof. If $i = j$, then $z = 0$ and the results is true. Suppose w.l.o.g. that $i > j$ (otherwise we just prove the result for $-z$). Then $U_k \cdot (e_i - e_j) = \sum_{s=1}^{i-1} e_s - \sum_{s=1}^{j-1} e_s = \sum_{s=j}^{i-1} e_s$. Similarly $L_k \cdot (e_i - e_j) = \sum_{s=i+1}^{k} e_s - \sum_{s=j+1}^{k} e_s = \sum_{s=j+1}^{i} -e_s$. Therefore $(L_k + U_k) \cdot (e_i - e_j) = \sum_{s=j}^{i-1} e_s - \sum_{s=j+1}^{i} e_s = e_j - e_i = -z$.

3.3 How to Recover the Bids When Knowing the l_{ij}'s

As discussed above, we can represent the function f as a matrix multiplication. Let M be the following square matrix of size $nk \times nk$:

$$
M =
\begin{bmatrix}
(U+L) & U & \cdots & & \cdots & U \\
(U+I) & (U+L) & U & & \cdots & U \\
\vdots & \ddots & \ddots & \ddots & & \vdots \\
(U+I) & \cdots & (U+I) & (U+L) & & U \\
(U+I) & \cdots & & \cdots & (U+I) & (U+L)
\end{bmatrix}
. \text{ Then } f(b) = M \cdot b.
$$

The function takes as input a vector composed of n vectors, each of k bits. It returns the nk values l_{ij}, $1 \le i \le n$ and $1 \le j \le k$. As explained above, the structure of the matrix is defined by the formula that computes γ_{ij}^a, which consists essentially of three factors: first we multiply all α_{ij} which encode bigger bids (represented by the matrix U), then we multiply all α_{ij} which encode smaller bids by the same bidder (represented by adding the matrix L on the diagonal), and finally we multiply by all α_{ij} which encode the same bid by bidders with a smaller index (represented by adding the matrix I on the lower triangle of M). In our encoding there will be a "1" in the vector for each Y in the protocol, hence f will count how many Ys are multiplied when computing γ_{ij}^a. Using this representation we can prove the following theorem.

Theorem 1. f *is injective on valid bid vectors, i.e. for two different correct bid vectors* $u = [u_1, \ldots, u_k]^T$ *and* $v = [v_1, \ldots, v_k]^T$ *with* $u \ne v$ *we have* $M \cdot u \ne M \cdot v$.

Proof. Let u and v be two correct bid vectors such that $u \ne v$. We want to prove that $M \cdot u \ne M \cdot v$. We make a proof by contradiction, hence we assume that $M \cdot u = M \cdot v$ or that $M \cdot (u - v) = 0$. Because u and v are two correct bid vectors, each one of them is an element of the canonical basis $(e_1, \ldots, e_k)$, i.e. $u = e_i$ and $v = e_j$, as shown in Section 3.1. We denote $u - v$ by z, and consequently $z = e_i - e_j$. Knowing that $M \cdot z = 0$, we prove by induction on a that for all a the following property $P(a)$ holds:

$$P(a) : \forall l, 0 < l \le a, diag(U^{k-l}) \cdot z = 0$$

where $diag(U^{k-x})$ is a $nk \times nk$ block diagonal matrix containing only diagonal blocks of the same matrix U^{k-x}. The validity of $P(k)$ proves in particular that $diag(U^0) \cdot z_l = 0$, i.e. $z = 0$ which contradicts our hypothesis.

- Case $a = 1$: we also prove this base case by induction, i.e. for all $b \ge 1$ the property $Q(b)$ holds, where:

$$Q(b) : \forall m, 0 < m \le b, U^{k-1} \cdot z_m = 0$$

which gives us that $U^{k-1} \cdot z = 0$.
  - Base case $b = 1$: We start by looking at the multiplication of the first row of M with z. We obtain: $(L+U) \cdot z_1 + U \cdot (z_2 + \ldots + z_k) = 0$. We can multiply each side by U^{k-1}, and use Lemma 6 to obtain: $U^{k-1} \cdot [-z_1 + U^k \cdot (z_2 + \ldots + z_k)] = 0$. Since U is nilpotent, according to Lemma 3 the latter gives $-U^{k-1} \cdot z_1 = 0$. Hence we know $Q(1) : U^{k-1} \cdot z_1 = 0$, i.e. the last entry of z_1 is 0.
  - Inductive step $b + 1$: assume $Q(b)$. Consider now the multiplication of the $(b + 1)$-th row of the matrix M:
    $(U + I) \cdot z_1 + \ldots + (U + I) \cdot z_b + (L + U) \cdot z_{b+1} + U \cdot (z_{b+2} + \ldots + z_k) = 0$.
    Then by multiplying by U^{k-1} and using Lemma 6 we obtain:
    $U^{k-1} \cdot [(U + I) \cdot z_1 + \ldots + (U + I) \cdot z_b - z_{b+1} + U \cdot (z_{b+2} + \ldots + z_k)] = 0$.
    Since U is nilpotent according to Lemma 3 we have $U^{k-1} \cdot z_1 + \ldots + U^{k-1} \cdot z_b - U^{k-1} \cdot z_{b+1} = 0$. Using the fact that for all $m < b$ we have $U^{k-1} \cdot z_m = 0$, the latter gives $-U^{k-1} \cdot z_{b+1} = 0$.
- Inductive step $a + 1$: assume $P(a)$. By induction on $b \ge 1$ we will show that $Q'(b)$ holds, where

$$Q'(b) : \forall m, 0 < m \le b, U^{k-(a+1)} \cdot z_m = 0$$

which gives us that $U^{k-(a+1)} \cdot z = 0$, i.e. $P(a+1)$.

- Base case $b = 1$: Consider the multiplication of the first row with $U^{k-(a+1)}$:
  $U^{k-(a+1)} \cdot [(L+U) \cdot z_1 + U \cdot (z_2 + \ldots + z_k)] = 0$ which can be rewritten as
  $-U^{k-(a+1)} \cdot z_1 + U^{k-a} \cdot (z_2 + \ldots + z_k)] = 0$. Using $U^{k-a} \cdot z_l = 0$ for all l,
  we can conclude that $-U^{k-(a+1)} \cdot z_1 = 0$, i.e. $Q'(1)$ holds.

- Inductive step $b+1$: assume $Q'(b)$. Consider now the $(b+1)$-th row of the
  matrix M:
  $(U+I) \cdot z_1 + \ldots + (U+I) \cdot z_b + (L+U) \cdot z_{b+1} + U \cdot (z_{b+2} + \ldots + z_k) = 0$.
  Then by multiplying by $U^{k-(a+1)}$ and using Lemma 6 we obtain:
  $U^{k-(a+1)} \cdot [(U+I) \cdot z_1 + \ldots + (U+I) \cdot z_b + -z_{b+1} + U \cdot (z_{b+2} + \ldots + z_k)] = 0$. Using $U^{k-a} \cdot z_l = 0$ for all l, we can conclude that $U^{k-(a+1)} \cdot z_1 + \ldots + U^{k-(a+1)} \cdot z_b - U^{k-(a+1)} \cdot z_{b+1} = 0$. Now, for all $m < b$, we have
  $U^{k-(a+1)} \cdot z_m = 0$, so that $-U^{k-(a+1)} \cdot z_{b+1} = 0$; i.e. $Q'(b+1)$ holds.   □

This theorem shows that if there is a constellation of bids that led to certain values l_{ij}, this constellation is unique. Hence we are able to invert f on valid outputs. We will now show that this can be efficiently done.

An Efficient Algorithm. Our aim is solve the following linear system: $M \cdot x = l$. We will use the same steps we used for the proof of injectivity to solve this system efficiently. First note that

$$M \cdot x = l \Rightarrow diag(U^{k-t-1}) \cdot M \cdot x = diag(U^{k-t-1}) \cdot l.$$

Consider the r-th block of size k of the latter equality. We have $x_r = (x_{r,1}, x_{r,2}, \ldots, x_{r,k})$. When multiplying by e_1^T we obtain the first line of this block. The r-th block of $M \cdot x$ is

$$(U+I)x_1 + \ldots + (U+I)x_{r-1} + (L+U)x_r + Ux_{r+1} + \ldots + Ux_k$$
$$= U(\textstyle\sum_{i=1}^k x_i) + (\textstyle\sum_{i=1}^{r-1} x_i) + Lx_r$$

and the r-th block of l is l_r. Hence:
$e_1^T \left[U^{k-t} \left(\sum_{i=1}^k x_i \right) + U^{k-t-1} \left(\sum_{i=1}^{r-1} x_i \right) + U^{k-t-1} Lx_r \right] = e_1^T U^{k-t-1} l_r$
Using Lemma 4, we can exchange L in the latter to get:
$e_1^T \left[U^{k-t} \left(\sum_{i=1}^k x_i \right) + U^{k-t-1} \left(\sum_{i=1}^{r-1} x_i \right) + U^{k-t-1} (1 - (I_n + U_n) x_r) \right]$
$= e_1^T U^{k-t-1} l_r$. We then remark that $e_1^T U^{k-t-1} 1 = t+1$, which gives:
$e_1^T \left[U^{k-t} \left(\sum_{i=1,i\neq r}^k x_i \right) + U^{k-t-1} \left(\sum_{i=1}^{r-1} x_i \right) - U^{k-t-1} x_r \right]$
$= e_1^T U^{k-t-1} l_r - (t+1)$. Using Lemma 5, we have

$$e_1^T \left[U^{k-t} \left(\left(\sum_{i=1}^k x_i \right) - 2x_r \right) + U^{k-t-1} \left(\sum_{i=1}^{r-1} x_i \right) \right] + (t+1) - e_1^T U^{k-t-1} l_r$$
$$= x_{r,k-t-1} \quad (3)$$

Using several times Corollary 1 we have:

- $e_1^T U^{k-t} \left(\left(\sum_{i=1}^{k} x_i \right) - 2x_r \right)$
  $= e_1^T U^{k-t+1} \left(\left(\sum_{i=1}^{k} x_i \right) - 2x_r \right) + e_{k-t}^T \left(\left(\sum_{i=1}^{k} x_i \right) - 2x_r \right)$
- $e_1^T U^{k-t-1} \left(\sum_{i=1}^{r-1} x_i \right) = e_1^T U^{k-t} \left(\sum_{i=1}^{r-1} x_i \right) + e_{k-t-1}^T \left(\sum_{i=1}^{r-1} x_i \right)$
- $e_1^T U^{k-t-1} l_r = e_1^T U^{k-t} l_r + l_{r,k-t-1}$

By changing t to $t-1$ in Equation (3) we get:
$$e_1^T \left[U^{k-t+1} \left(\left(\sum_{i=1}^{k} x_i \right) - 2x_r \right) + U^{k-t} \left(\sum_{i=1}^{r-1} x_i \right) \right] + t - e_1^T U^{k-t} l_r = x_{r,k-t}.$$
Then regrouping the applications of Corollary 1 and the latter formula within Equation (3), we obtain:

$$x_{r,k-t} + e_{k-t}^T \left(\left(\sum_{i=1}^{k} x_i \right) - 2x_r \right) + e_{k-t-1} \left(\sum_{i=1}^{r-1} x_i \right) + 1 + l_{r,k-t-1} = x_{r,k-t-1} \quad (4)$$

This gives us a formula to compute the values of $x_{i,j}$, starting with the last element of the first block $x_{1,k}$. Then we can compute the last elements of all other blocks $x_{2,k}, \ldots, x_{n,k}$, and then the second to last elements $x_{1,k-1}, \ldots, x_{n,k-1}$, etc.

Complexity Analysis. To obtain all values, we have to apply the above formula for each $t \leq n$ and $r \leq k$, hence we have:

$$\sum_{t=1}^{n} \sum_{r=1}^{k} (k+r) = n \left(k^2 + \frac{k(k+1)}{2} \right) = \frac{3}{2} nk^2 + \frac{1}{2} nk \in \mathcal{O}\left(nk^2 \right)$$

This is efficient enough to be computed on a standard PC for realistic values of n (the number of bidders) and k (the number of possible bids). Those could be less than a hundred bidders with a thousand different prices, thus requiring about the order of only a hundred million arithmetic operations. It is anyway the order of magnitude of the number of operations required of each user just to compute her encrypted bids.

3.4   Attack on the Random Noise: How to Obtain the l_{ij}'s

In the previous section we showed that knowing the l_{ij}'s allows us the efficiently break the privacy of all bidders. Here is how to obtain the l_{ij}'s. The seller will learn all $v_{ij} = \left(Y^{l_{ij}} \right)^{\left(\sum_{h=1}^{n} m_{ij}^{h} \right)}$ at the end of the protocol. Since the m_{ij}^{h} are randomly chosen, this will be a random value if $l_{ij} \neq 0$. However a malicious bidder ("Mallory", of index a) can cancel out the m_{ij}^{h} as follows: in Step 3 of the protocol each bidder will compute his γ_{ij}^{a} and δ_{ij}^{a}. Mallory waits until all other bidders have published their values (the protocol does not impose any synchronization or special ordering) and then computes his values γ_{ij}^{ω} and δ_{ij}^{ω} as:

$$\gamma_{ij}^{\omega} = \left(\left(\prod_{h=1}^{n} \prod_{d=j+1}^{k} \alpha_{hd} \right) \cdot \left(\prod_{d=1}^{j-1} \alpha_{id} \right) \cdot \left(\prod_{h=1}^{i-1} \alpha_{hj} \right) \right) \cdot \left(\prod_{k \neq \omega} \gamma_{ij}^{k} \right)^{-1}$$

$$\delta_{ij}^{\omega} = \left(\left(\prod_{h=1}^{n} \prod_{d=j+1}^{k} \beta_{hd} \right) \cdot \left(\prod_{d=1}^{j-1} \beta_{id} \right) \cdot \left(\prod_{h=1}^{i-1} \beta_{hj} \right) \right) \cdot \left(\prod_{k \neq \omega} \delta_{ij}^{k} \right)^{-1}$$

The first part is a correct encryption of $Y^{l_{ij}}$, with $m_{ij}^\omega = 1$ for all i and j. The second part is the inverse of the product of all the other bidders γ_{ij}^k and δ_{ij}^k, and thus it will eliminate the random exponents. Hence after decryption the seller obtains $v_{ij} = Y^{l_{ij}}$, where $l_{ij} < n$ for a small n. He can compute l_{ij} by simply (pre-)computing all possible values Y^r and testing for equality. This allows the seller to obtain the necessary values and then to use the resolution algorithm to obtain each bidder's bid. Note that although we changed the intermediate values, the output still gives the correct result (i.e. winning bid). Therefore, the attack might even be unnoticed by the other participants. Note also that choosing a different Y_i per bidder does not prevent the attack, since all the Y_i need to be public in order to prove the correctness of the bid in Step 2 of the protocol.

However the protocol requires Mallory to prove that γ_{ij}^ω and δ_{ij}^ω have the same exponent. This is obviously the case, but Mallory does not know the exact value of this exponent. Thus it is impossible for him to execute the proposed zero-knowledge protocol directly.

In the original paper [4] the malleable interactive proof of [8], presented in Section 2.3, is used to prove the correctness of γ_{ij}^a and δ_{ij}^a in Step 3 of the protocol. If this proof is not converted into a non-interactive proof, then Mallory is able to fake it as follows.

3.5 Proof of Equality of the Presented Outcomes

Note that we can rewrite γ_{ij}^ω and δ_{ij}^ω as:

$$v = \gamma_{ij}^\omega = \underbrace{\left(\left(\prod_{h=1}^{n} \prod_{d=j+1}^{k} \alpha_{hd} \right) \cdot \left(\prod_{d=1}^{j-1} \alpha_{id} \right) \cdot \left(\prod_{h=1}^{i-1} \alpha_{hj} \right) \right)^{1-\left(\sum_{k \neq \omega} m_{ij}^k \right)}}_{g_1}$$

$$w = \delta_{ij}^\omega = \underbrace{\left(\left(\prod_{h=1}^{n} \prod_{d=j+1}^{k} \beta_{hd} \right) \cdot \left(\prod_{d=1}^{j-1} \beta_{id} \right) \cdot \left(\prod_{h=1}^{i-1} \beta_{hj} \right) \right)^{1-\left(\sum_{k \neq \omega} m_{ij}^k \right)}}_{g_2}$$

When Mallory, the bidder m, is asked by Victor for a proof of correctness of his values, he starts by asking all other bidders for proofs to initialize the man-in-the-middle attack of Figure 1. Each of them answers with values $\lambda_o = g_1^{z_o}$ and $\mu_o = g_2^{z_o}$. Mallory can then answer Victor with values $\lambda = \prod_o \lambda_o^{-1}$ and $\mu = \prod_o \mu_o^{-1}$, where $o \in ([1, n] \setminus m)$. Victor then sends a challenge c, which Mallory simply forwards to the other bidders. They answer with $r_o = z_o + c \cdot m_{ij}^o$, and Mallory sends $r = c - \sum_o r_o$ to Victor, who can check that $g_1^r = \lambda \cdot v^c$ and $g_2^r = \mu \cdot w^c$. If the other bidders did their proofs correctly, then Mallory's proof will appear valid to Victor:

$$\lambda \cdot v^c = \prod_o \lambda_o^{-1} \cdot \left(g_1^{1-\left(\sum_o m_{ij}^o \right)} \right)^c = \prod_o g_1^{-z_o} \cdot g_1^{c-c\left(\sum_o m_{ij}^o \right)} = g_1^{c-\sum_o \left(z_o + c m_{ij}^o \right)}$$

$$\mu \cdot w^c = \prod_o \mu_o^{-1} \cdot \left(g_2^{1-\left(\sum_o m_{ij}^o \right)} \right)^c = \prod_o g_2^{-z_o} \cdot g_2^{c-c\left(\sum_o m_{ij}^o \right)} = g_2^{c-\sum_o \left(z_o + c m_{ij}^o \right)}$$

Hence in the case of malleable interactive zero-knowledge proofs Mallory is able to modify the values γ_{ij}^ω and δ_{ij}^ω as necessary, and even prove the correctness using the bidders. Hence the modifications may stay undetected and the seller will be able to break privacy.

3.6 The Complete Attack and Countermeasures

Putting everything together, the attack works as follows:

1. The bidders set up the keys as described in the protocol.
2. They encrypt and publish their bids.
3. They compute γ_{ij}^h and δ_{ij}^h and publish them.
4. Mallory, who is a bidder himself, waits until all other bidders have published their values. He then computes his values as defined above, and publishes them.
5. If he is asked for a proof, he can proceed as explained above in Section 3.5.
6. The bidders (including Mallory) jointly decrypt the values.
7. The seller obtains all $Y^{l_{ij}}$'s. He can then compute the l_{ij}'s by testing at most n possibilities.
8. Once he has all values, he can invert the function f as explained above.
9. He obtains all bidders bids.

Again, note that for all honest bidders, this execution will look normal, so they might not even notice that an attack took place. To prevent this attack, one could perform the following actions:

- To counteract the removal of the noise of Section 3.4, the bidders could check whether the product of the $\gamma_{i,j}^a$ for all bidders a is equal to the product of the α_{hd} without any noise (exponent is 1). Unfortunately, the man-in-the-middle attack generalizes to any exponent as shown in Figure 2. Therefore the attacker could use a randomly chosen exponent only known to him.
- As mentioned above, another countermeasure is the use of non-interactive, non-malleable proofs of knowledge. In this case, we will show in Section 5 that it is still possible to attack a targeted bidder's privacy.

4 Attacking Verifiability

Brandt claims that the protocol is verifiable as the parties have to provide zero-knowledge proofs for their computations, however there are two problems.

4.1 Exceptional Values

First, a winning bidder cannot verify if he actually won. To achieve privacy, the protocol hides all outputs of v_{aj} except for the entry containing "1"[2]. This is done by exponentiation with random values m_{ij}^a inside all entries γ_{ij}^a and δ_{ij}^a, i.e. by computing $x_{ij}^{\sum_a m_{ij}^a}$ where x_{ij} is the product of some α_{ij} as specified in the protocol. If x_{ij} is one, x_{ij}^m

---

[2] Note that the protocol contains a mechanism to resolve ties, i.e. there should always be exactly one entry equal to 1, even in the presence of ties.

will still return one for any m, and in principle something different from one for any other value of x_{ij}. Now, the random values m_{ij}^a may add up to zero (mod q), hence the returned value will be $x_{ij}^m = x_{ij}^0 = 1$ and the bidder will conclude that he won, although he actually lost ($x_{ij} \neq 1$). Hence simply verifying the proofs is not sufficient to be convinced that the observed outcome is correct. For the same reason the seller might observe two or more "1"-values, even though all proofs are correct. In such a situation he is unable to decide which bidder actually won since he cannot determine which "1"s correspond to a real bids, and hence which bid is the highest real bid. If two "1"s correspond to real bids, he could even exploit such a situation to his advantage: he can tell both bidders that they won and take money from both, although there is only one good to sell – this is normally prohibited by the protocol's tie-breaking mechanism. If the bidders do not exchange additional data there is no way for them to discover that something went wrong, since the seller is the only party having access to all values.

A solution to this problem could work as follows: when computing the γ_{ij}^a and δ_{ij}^a, the bidders can check if the product

$$x_{ij} = \left(\prod_{h=1}^{n} \prod_{d=j+1}^{k} \alpha_{hd} \right) \cdot \left(\prod_{d=1}^{j-1} \alpha_{id} \right) \cdot \left(\prod_{h=1}^{i-1} \alpha_{hj} \right)$$

is equal to one – if yes, they restart the protocol using different keys and random values. If not, they continue, and check if $\prod_a \gamma_{ij}^a = 1$. If yes, they choose different random values m_{ij}^a and re-compute the γ_{ij}^a and δ_{ij}^a, otherwise they continue. Since the probability of the random values adding up to zero is low, this will rapidly lead to correct values.

4.2   Different Private Keys

Second, the paper does not precisely specify the proofs that have to be provided in the joint decryption phase. If the bidders only prove that they use the same private key on all decryptions *and not also that it is the one they used to generate their public key*, they may use a wrong one. This will lead to a wrong decryption where with very high probability no value is "1", as they will be random. Hence all bidders will think that they lost, thus allowing a malicious bidder to block the whole auction, as no winner is determined. Hence, if we assume that the verification test consists in verifying the proofs, a bidder trying to verify that he lost using the proofs might perform the verification successfully, although the result is incorrect and he actually won – since he would have observed a "1" if the vector had been correctly decrypted.

This problem can be addressed by requiring the bidders to also prove that they used the same private key as in the key generation phase.

5   Attacks Using the Lack of Authentication

The protocol as described in the original paper does not include any authentication of the messages. This means that an attacker in control of the network can impersonate any party, which can be exploited in many ways. However, the authors supposed in the original paper a "reliable broadcast channel, i.e. the adversary has no control of communication" [4]. Yet even under this assumption dishonest participants can impersonate other participants by submitting messages on their behalf. Additionally, this assumption is difficult to achieve in asynchronous systems [14]. In the following we consider an attacker in control of the network, however many attacks can also be executed analogously by dishonest parties (which are considered in the original paper) in the reliable broadcast setting.

5.1 Another Attack on Privacy

Our first attack on privacy only works in the case of malleable interactive proofs. If we switch to non-interactive non-malleable proofs, Mallory cannot ask the other bidders for proofs using a challenge of his choice.

However, even with non-interactive non-malleable zero-knowledge proofs, the protocol is still vulnerable to attacks on a targeted bidder's privacy if an attacker can impersonate any bidder of his choice as well as the seller, which is the case for an attacker controlling the network due to the lack of authentication. In particular, if he wants to know Alice's bid he can proceed as follows:

1. Mallory impersonates all other bidders. He starts by creating keys on their behalf and publishes the values y_i and the corresponding proofs for all of them.
2. Alice also creates her secret keyshare and publishes y_a together with a proof.
3. Alice and Mallory compute the public key y.
4. Alice encrypts her bid and publishes her α_{aj} and β_{bj} together with the proofs.
5. Mallory publishes $\alpha_{ij} = \alpha_{aj}$ and $\beta_{ij} = \beta_{aj}$ for all other bidders i and also copies Alice's proofs.
6. Alice and Mallory execute the computations described in the protocol and publish γ_{ij}^a and δ_{ij}^a.
7. They compute ϕ_{ij}^a and send it to the seller.
8. The seller publishes the ϕ_{ij}^a and computes the v_{aj}.

Since all submitted bids are equal, the seller (which might also be impersonated by Mallory) will obtain Alice's bid as the winning price, hence it is not private any more. This attack essentially simulates a whole instance of the protocol to make Alice indirectly reveal a bid that was intended for another, probably real auction. To counteract this it is not sufficient for Alice to check that the other bids are different: Mallory can produce different $\alpha_{ij} = \alpha_{aj}y^x$ together with $\beta_{ij} = \beta_{aj}g^x$ which are still correct encryptions of Alice bids.

Note that the same attack also works if dishonest bidders collude with the seller: they simply re-submit the targeted bidders bid as their own bid.

5.2 Attacking Fairness, Non-repudiation and Verifiability

The lack of authentication obviously entails that a winning bidder can claim that he did not submit his bid, hence violating non-repudiation (even in the case of reliable broadcast). Additionally, this also enables an attack on fairness: an attacker in control of the network can impersonate all bidders vis-à-vis the seller, submitting bids of his choice on their behalf and hence completely controlling the winner and winning price. This also causes another problem with verifiability: it is impossible to verify if the bids were submitted by the registered bidders or by somebody else.

5.3 Countermeasures

The solution to these problems is simple: all the messages need to be authenticated, e.g. using signatures or Message Authentication Codes (MACs) based on a trust anchor, for example a Public Key Infrastructure (PKI).

6 Conclusion

In this paper we analyze the protocol of Brandt [4] from various angles. We show that the underlying computations have a weakness which can be exploited by malicious bidders to break privacy if malleable interactive zero-knowledge proofs are used. We also identified two problems with verifiability and proposed solutions. Finally we showed how the lack of authentication can be used to mount different attacks on privacy, verifiability as well as fairness and non-repudiation. Again we suggested a solution to address the discovered flaws.

So sum up, the following countermeasures have to be implemented:

- Use of non-interactive or non-malleable zero-knowledge proofs.
- All messages have to be authenticated, e.g. using a Public-Key Infrastructure (PKI) and signatures.
- In the outcome computation step: when computing the γ_{ij}^a and δ_{ij}^a, the bidders can check if $x_{ij} = \left(\prod_{h=1}^{n} \prod_{d=j+1}^{k} \alpha_{hd} \right) \cdot \left(\prod_{d=1}^{j-1} \alpha_{id} \right) \cdot \left(\prod_{h=1}^{i-1} \alpha_{hj} \right)$ is equal to one – if yes, they restart the protocol using different keys and random values. If not, they continue, and check if $\prod_a \gamma_{ij}^a = 1$. If yes, they choose different random values m_{ij}^a and re-compute the γ_{ij}^a and δ_{ij}^a, otherwise they continue.
- In the outcome decryption step: the bidders have to prove that the value x_a they used to decrypt is the same x_a they used to generate their public key y_a in the first step.

The attacks show that properties such as authentication can be necessary to achieve other properties which might appear to be unrelated at first sight, like for instance privacy. It also points out that there is a difference between computing the winner in a fully private way, and ensuring privacy for the bidders: in the second attack we use modified inputs to break privacy even though the computations themselves are secure. Additionally our analysis highlights that the choice of interactive or non-interactive, malleable or non-malleable proofs is an important decision in any protocol design.

As for possible generalizations of our attacks, of course the linear algebra part of our first attack is specific to this protocol. Yet the man-in-the-middle attack on malleable proofs as well as the need of authentication for privacy are applicable to any protocol. Similarly, checking all exceptional cases and ensuring that the same keys are used all along the process are also valid insights for other protocols.

Acknowledgments. This work was partly supported by the ANR projects ProSe (decision ANR-2010-VERS-004-01) and HPAC (ANR-11-BS02-013).

References

1. Bangerter, E., Camenisch, J.L., Maurer, U.M.: Efficient proofs of knowledge of discrete logarithms and representations in groups with hidden order. In: Vaudenay, S. (ed.) PKC 2005. LNCS, vol. 3386, pp. 154–171. Springer, Heidelberg (2005)
2. Brandt, F.: A verifiable, bidder-resolved auction protocol. In: Falcone, R., Barber, S., Korba, L., Singh, M. (eds.) Proceedings of the 5th AAMAS Workshop on Deception, Fraud and Trust in Agent Societies, pp. 18–25 (2002)
3. Brandt, F.: Fully private auctions in a constant number of rounds. In: Wright, R.N. (ed.) FC 2003. LNCS, vol. 2742, pp. 223–238. Springer, Heidelberg (2003)
4. Brandt, F.: How to obtain full privacy in auctions. International Journal of Information Security 5, 201–216 (2006)
5. Burmester, M., Desmedt, Y.G., Piper, F., Walker, M.: A general zero-knowledge scheme. In: Quisquater, J.-J., Vandewalle, J. (eds.) EUROCRYPT 1989. LNCS, vol. 434, pp. 122–133. Springer, Heidelberg (1990)
6. Chaum, D., Evertse, J.-H., van de Graaf, J., Peralta, R.: Demonstrating possession of a discrete logarithm without revealing it. In: Odlyzko, A.M. (ed.) CRYPTO 1986. LNCS, vol. 263, pp. 200–212. Springer, Heidelberg (1987)
7. Chaum, D., Evertse, J.-H., van de Graaf, J.: An improved protocol for demonstrating possession of discrete logarithms and some generalizations. In: Price, W.L., Chaum, D. (eds.) EUROCRYPT 1987. LNCS, vol. 304, pp. 127–141. Springer, Heidelberg (1988)
8. Chaum, D., Pedersen, T.P.: Wallet databases with observers. In: Brickell, E.F. (ed.) CRYPTO 1992. LNCS, vol. 740, pp. 89–105. Springer, Heidelberg (1993)
9. Chow, S.S.M., Ma, C., Weng, J.: Zero-Knowledge Argument for Simultaneous Discrete Logarithms. In: Thai, M.T., Sahni, S. (eds.) COCOON 2010. LNCS, vol. 6196, pp. 520–529. Springer, Heidelberg (2010)
10. Cramer, R., Damgård, I.B.: Zero-Knowledge Proofs for Finite Field Arithmetic or: Can Zero-Knowledge Be for Free? In: Krawczyk, H. (ed.) CRYPTO 1998. LNCS, vol. 1462, pp. 424–441. Springer, Heidelberg (1998)
11. Curtis, B., Pieprzyk, J., Seruga, J.: An efficient eAuction protocol. In: ARES, pp. 417–421. IEEE Computer Society (2007)
12. El Gamal, T.: A public key cryptosystem and a signature scheme based on discrete logarithms. In: Blakely, G.R., Chaum, D. (eds.) CRYPTO 1984. LNCS, vol. 196, pp. 10–18. Springer, Heidelberg (1985)
13. Fiat, A., Shamir, A.: How to Prove Yourself: Practical Solutions to Identification and Signature Problems. In: Odlyzko, A.M. (ed.) CRYPTO 1986. LNCS, vol. 263, pp. 186–194. Springer, Heidelberg (1987)
14. Fischer, M.J., Lynch, N.A., Paterson, M.: Impossibility of distributed consensus with one faulty process. J. ACM 32(2), 374–382 (1985)
15. Fischlin, M., Fischlin, R.: Efficient non-malleable commitment schemes. Journal of Cryptology 22, 530–571 (2009)
16. Katz, J.: Efficient cryptographic protocols preventing "man-in-the-middle" attacks. PhD thesis, Columbia University (2002)
17. Krishna, V.: Auction Theory. Academic Press, San Diego (2002)
18. Maurer, U.: Unifying zero-knowledge proofs of knowledge. In: Preneel, B. (ed.) AFRICACRYPT 2009. LNCS, vol. 5580, pp. 272–286. Springer, Heidelberg (2009)

19. Naor, M., Pinkas, B., Sumner, R.: Privacy preserving auctions and mechanism design. In: ACM Conference on Electronic Commerce, pp. 129–139 (1999)
20. Omote, K., Miyaji, A.: A Practical English Auction with One-Time Registration. In: Varadharajan, V., Mu, Y. (eds.) ACISP 2001. LNCS, vol. 2119, pp. 221–234. Springer, Heidelberg (2001)
21. Peng, K., Boyd, C., Dawson, E., Viswanathan, K.: Robust, Privacy Protecting and Publicly Verifiable Sealed-Bid Auction. In: Deng, R.H., Qing, S., Bao, F., Zhou, J. (eds.) ICICS 2002. LNCS, vol. 2513, pp. 147–159. Springer, Heidelberg (2002)
22. Sadeghi, A.R., Schunter, M., Steinbrecher, S.: Private auctions with multiple rounds and multiple items. In: DEXA Workshops, pp. 423–427. IEEE (2002)
23. Sako, K.: An Auction Protocol Which Hides Bids of Losers. In: Imai, H., Zheng, Y. (eds.) PKC 2000. LNCS, vol. 1751, pp. 422–432. Springer, Heidelberg (2000)
24. Schnorr, C.P.: Efficient signature generation by smart cards. Journal of Cryptology 4, 161–174 (1991)

HELEN: A Public-Key Cryptosystem Based on the LPN and the Decisional Minimal Distance Problems*

Alexandre Duc** and Serge Vaudenay

Ecole Polytechnique Fédérale de Lausanne, 1015 Lausanne, Switzerland

Abstract. We propose HELEN, a code-based public-key cryptosystem whose security is based on the hardness of the Learning from Parity with Noise problem (LPN) and the decisional minimum distance problem. We show that the resulting cryptosystem achieves indistinguishability under chosen plaintext attacks (IND-CPA security). Using the Fujisaki-Okamoto generic construction, HELEN achieves IND-CCA security in the random oracle model. Our cryptosystem looks like the Alekhnovich cryptosystem. However, we carefully study its complexity and we further propose concrete optimized parameters.

Keywords: Code-based cryptosystem, learning from parity with noise problem, minimum distance problem, random linear code, public-key cryptostem.

1 Introduction

Every public-key cryptosystem relies on problems that are believed computationally hard. The two mostly used problems are the integer factorization problem [54,52] and the discrete logarithm problem [22]. However, these two problems can be solved in polynomial time on a quantum computer. It is thus important to develop new cryptosystem that are secure even on quantum computers and to correctly propose some parameters depending on the required security.

In this paper, we present HELEN, a public-key cryptosystem, the security of which relies on the hardness of the *Learning from Parity with Noise problem* (LPN) and the *minimum distance problem* which are both NP-hard.[1] The former consists in recovering an unknown vector while given access to noisy versions of its scalar product with random vectors. There is also no known polynomial-time algorithm on quantum computers. In short, the keys in HELEN consists in a low-weight parity check equation h (the private key) which is hidden in a random matrix G (the public key) such that it is indistinguishable from a totally random matrix. The matrix G spans a linear code. Our cryptosystem looks like the Alekhnovich cryptosystem [1]. However, we carefully study its

---

* This paper is an extended version of [19].
** Supported by a grant of the Swiss National Science Foundation, 200021_143899/1.
[1] HELEN stands for Hidden Equation for Linear Encryption with Noise.

A. Youssef, A. Nitaj, A.E. Hassanien (Eds.): AFRICACRYPT 2013, LNCS 7918, pp. 107–126, 2013.

complexity, we further propose concrete and optimized parameters, and we make incorrectness small.

We encrypt a duplicated bit by hiding it using a random linear codeword as well as a random biased noise vector. For decryption, the random linear codeword is removed by multiplying the ciphertext with h. The noise is removed by majority logic decoding. With a proper parameter choice, the probability of decrypting erroneously the message is small. We show in a further section how to reduce this probability of error as well as how to encrypt multiple bits at the same time using HELEN.

Related Work. The LPN problem is well studied in the cryptographic community. There is an authentication protocol based on the LPN problem named HB by Hopper and Blum [34]. This protocol was later improved into the HB$^+$ protocol by Juels and Weis [36]. However, HB$^+$ was shown vulnerable to man-in-the-middle attacks [28]. Several variants were proposed [12,21,47] but all of them suffer from the same vulnerability [29]. A new variant HB$^\#$ was proposed by Gilbert, Robshaw and Seurin [30] to improve the transmission cost of the protocol and its securtiy against man-in-the-middle attacks but an attack was also found in this variant [49]. Two more recent versions were introduced based on the hardness of some variant of the LPN problem, namely Ring-LPN [32] and subspace LPN [38].

Among other work based on the LPN problem, a PRNG is presented by Blum et al. in [10] along with a one-way function and a private-key encryption scheme based on some hard learning problems. A private-key encryption scheme named LPN-C was proposed by Gilbert, Robshaw and Seurin [31]. LPN-C was shown IND-CPA secure.

The construction of HELEN [19] presents some similarities with the trapdoor cipher TCHo [20,3,24] by Aumasson et al. which similarly encrypts a message by adding some random biased noise and some contribution from a linear code. In TCHo, this noise is introduced using an LFSR whose feedback polynomial has a multiple of low weight.

A class of lattice-based cryptosystems introduced by Regev is based on the worst-case complexity of the *learning with errors* (LWE) problem [53,50,43,57], which is a generalisation of the LPN problem on fields $\mathbb{F}_q$ with $q > 2$. The last two introduce the *ring-LWE* problem, an algebraic variant of the LWE problem. According to the authors, it is the first truly practical lattice-based cryptosystem based on the LWE problem.

Other well-known post-quantum cryptosystems include the McEliece cryptosystem [46] and its dual the Niederreiter cryptosystem [48], which are code-based making use of Goppa codes. In lattice-based cryptosystem, one has to mention NTRU [33] based on the hardness of the shortest vector problem in a particular class of lattices. We refer the reader to [7] for a more exhaustive survey on post-quantum cryptosystems.

More closely related cryptosystems were proposed. Gentry et al. proposed an LWE-based cryptosystem [27] in which users share a common random matrix and whose private key (resp. public key) consists in a random error vector (resp. its syndrome). Extensions to $p = 2$ have been open so far. Our procedure is different from theirs in the sense that we hide a low-parity check equation in a matrix so that this matrix looks random, whereas they pick a totally random matrix. Similarly, Alekhnovich proposed a scheme based on problem to distinguish $(A, Ax + e)$ with x following uniform distribution and e either in $\binom{n}{n^\delta}$ or $\binom{n}{n^\delta + 1}$ with $\delta < 1/2$ which he conjectures to be hard [1]. Our scheme differs with the scheme proposed in [1] in the following ways. First, we encode the bit so that decryption is correct with constant probability ϕ and which is independent from the encrypted bit b (in [1], this probability is just known to be close to one for $b = 0$ and $1/2$ for $b = 1$). Finally, we propose concrete parameters and asymptotic parameters for our scheme. Applebaum et al. proposed a scheme, which is very similar to ours but which uses sparse matrices instead of random ones. Thus, the security reduces to the less-studied 3LIN problem instead of LPN. This problem is similar to the LPN problem except that queries are done with vectors of weight 3 instead of random vectors. Also, the authors do not provide any concrete parameters [2]. n Asiacrypt 2012, Döttling et al. presented an IND-CCA secure cryptosystem based on Alekhnovich's scheme, but again, no concrete parameters are given [18]. IND-CCA security is obtained using a technique by Dolev et al. [17] based on one-time signatures and a tool by Rosen and Segev [55]. So, to the best of our knowledge, we propose for the first time a *concrete PKC* whose security is based on LPN.

2 Preliminaries

We denote by log the logarithm in base two. The concatenation of two bitstrings x and y is written $x \| y$. We consider vectors as row vectors. The transpose of a vector v is denoted by v^t. We denote the Hamming weight of a bitstring x by $\mathsf{wt}(x)$. We write $x \xleftarrow{U} \mathcal{D}$ if an element x is drawn uniformly at random in a domain $\mathcal{D}$. A function $f(\lambda)$ is *negligible* if for all $d \in \mathbb{R}$ we have $f(\lambda) = O(\lambda^{-d})$. We denote the Bernoulli distribution with parameter p by $\mathrm{Ber}(p)$, i.e., if $x \leftarrow \mathrm{Ber}(p)$, we have $\Pr[x = 1] = p$ and $\Pr[x = 0] = 1 - p$. We write S_p^n to denote the sequence of n independent Bernoulli trials with parameter p. We write $\mathsf{S}_p^n(r)$ when we need to specify the seed r used to generate this sequence. Given a permutation σ in $\mathfrak{S}_n$, the group of all permutations over n elements, and given $h \in \{0, 1\}^n$, we write $\sigma \star h$ when we apply σ on the bits of h. That is, $(\sigma \star h)_i = h_{\sigma^{-1}(i)}$. Given a $k \times n$ matrix G, we write $\sigma \star G$ when we apply σ on the columns of G, i.e., $(\sigma \star G)_{i,j} = G_{i, \sigma^{-1}(j)}$.

Notation. Given some initial parameters Π and a predicate P, we write

$$\Pr \left[P(v_1, \ldots, v_m; r_p) : \begin{array}{l} v_1 \leftarrow f_1(\Pi; r_1) \\ \vdots \\ v_m \leftarrow f_m(\Pi, v_1, \ldots, v_{m-1}; r_m) \end{array} \right]$$

to denote the probability (over the randomnesses $r_1, \ldots, r_m, r_p$) that there exist $v_1 \leftarrow f_1(\Pi; r_1), \ldots, v_m \leftarrow f_m(\Pi, v_1, \ldots, v_m; r_m)$ such that $P(v_1, \ldots, v_m; r_p)$.

2.1   Security Notions

Definition 1 (Public-key Encryption Scheme). *Given a function $\varphi(\lambda)$, a $\varphi(\lambda)$-cryptosystem over a given message space $\mathcal{M}$ and random coin space $\mathcal{R}$ consists of three polynomial-time algorithms:*

- *a probabilistic key-generation algorithm $\mathsf{Gen}(1^\lambda; \rho_g)$ taking as input some security parameter 1^λ in unary representation and some random coins ρ_g, and producing a secret key K_s and a public key K_p;*
- *a probabilistic encryption algorithm $\mathsf{Enc}(K_p, m; r)$ taking as input a public key K_p and a message $m \in \mathcal{M}$ with some random coins $r \in \mathcal{R}$, and producing a ciphertext y in the ciphertext space $\mathcal{C}$;*
- *a deterministic decryption algorithm $\mathsf{Dec}(K_s, c)$ taking as input a secret key K_s and a ciphertext $c \in \mathcal{C}$, and producing a message or an error.*

The cryptosystem must satisfy the following correctness property:

$$\max_{m \in \mathcal{M}} \Pr\left[\mathsf{Dec}(K_s, \mathsf{Enc}(K_p, m; \rho)) \neq m : \quad (K_s, K_p) \leftarrow \mathsf{Gen}(1^\lambda; \rho_g)\right] \leq \varphi(\lambda) .$$

We will also use the following security notions and acronyms. Adaptive Chosen Ciphertext Attack is denoted CCA, Chosen Plaintext Attack CPA, Indistinguishability IND and one-wayness OW.

Definition 2 (IND-CPA-security). *A cryptosystem is said (t, ε)-IND-CPA-secure or (t, ε)-semantically secure against chosen plaintext attacks if no adversary $\mathcal{A} = (\mathcal{A}_1, \mathcal{A}_2)$ with running time bounded by t can distinguish the encryption of two different plaintexts m_0 and m_1 with a probability higher than ε.[2] More formally, for all $\mathcal{A}$ bounded by t,*

$$\Pr\left[\mathcal{A}_2(K_p, c; \rho) = b : \begin{array}{c} (K_s, K_p) \leftarrow \mathsf{Gen}(1^\lambda; \rho_g) \\ m_0, m_1 \leftarrow \mathcal{A}_1(K_p; \rho) \quad (\star) \\ r \xleftarrow{U} \mathcal{R}; \ b \xleftarrow{U} \{0, 1\} \\ c \leftarrow \mathsf{Enc}(K_p, m_b; r) \end{array}\right] \leq \frac{1}{2} + \varepsilon . \tag{1}$$

Asymptotically, a cryptosystem is IND-CPA-secure if for any polynomial $t(\lambda)$ there exists a negligible function $\varepsilon(\lambda)$ such that it is $(t(\lambda), \varepsilon(\lambda))$-IND-CPA-secure.

IND-CPA-security can also be represented in the simple real-or-random game model [6,5].[3]

Definition 3 (Simple real-or-random IND-CPA game security). *A cryptosystem is (t, ε)-real-or-random-IND-CPA-secure if in Definition 2, line $(\star)$ in (1) is replaced by $m_0 \leftarrow \mathcal{A}_1(K_p; \rho); \ m_1 \xleftarrow{U} \mathcal{M}$*

---

[2] We include in the running time the size of the code of $\mathcal{A}$ in a fixed RAM model of computation to avoid trivial adversaries.

[3] In our definition of real-or-random game model, we consider only *simple* adversaries, i.e., adversaries who can query the oracle once. This definition is enough to prove the IND-CPA-security of our scheme.

A (t, ε)-real-or-random-IND-CPA-secure system is $(t, 2\varepsilon)$-IND-CPA-secure [5]. Conversely, a (t, ε)-IND-CPA-secure system is (t, ε)-real-or-random-IND-CPA-secure. Asymptotically, both models are equivalent.

Definition 4 (IND-CCA-security). *A cryptosystem is said (t, ε)-IND-CCA-secure or (t, ε)-secure against adaptive chosen ciphertext attacks if no adversary $\mathcal{A} = (\mathcal{A}_1, \mathcal{A}_2)$, with access to a decryption oracle $\mathcal{O}_{K_s}$ and with running time bounded by t can distinguish the encryption of two different plaintexts m_0 and m_1 with a probability higher than ε. More formally, for all $\mathcal{A}$ bounded by t,*

$$\Pr\left[\mathcal{A}_2^{\mathcal{O}_{K_s}}(K_p, c; \rho) = b : \begin{array}{c} (K_s, K_p) \leftarrow \mathsf{Gen}(1^\lambda; \rho_g) \\ m_0, m_1 \leftarrow \mathcal{A}_1^{\mathcal{O}_{K_s}}(K_p; \rho) \\ r \xleftarrow{U} \mathcal{R};\ b \xleftarrow{U} \{0, 1\} \\ c \leftarrow \mathsf{Enc}(K_p, m_b; r) \end{array}\right] \leq \frac{1}{2} + \varepsilon,$$

where $\mathcal{O}_{K_s,c}(y) = \mathsf{Dec}(K_s, y)$ for $y \neq c$ and $\mathcal{O}_{K_s,c}(c) = \bot$. Asymptotically, a cryptosystem is IND-CCA-secure if for any polynomial $t(\lambda)$ there exists a negligible function $\varepsilon(\lambda)$ such that it is $(t(\lambda), \varepsilon(\lambda))$-IND-CCA-secure.

Definition 5 (Statistical distance). *Given two discrete distributions $\mathcal{D}_0$ and $\mathcal{D}_1$ over a set $\mathcal{Z}$, we define the* statistical distance *between $\mathcal{D}_0$ and $\mathcal{D}_1$ by*

$$d(\mathcal{D}_0, \mathcal{D}_1) := \frac{1}{2} \sum_{z \in \mathcal{Z}} |\mathcal{D}_1(z) - \mathcal{D}_0(z)| .$$

Definition 6. *Given two distributions $\mathcal{D}_0$ and $\mathcal{D}_1$, a* distinguisher *between them is an algorithm $\mathcal{A}$ that takes as input one sample x from either $\mathcal{D}_0$ or $\mathcal{D}_1$ and has to decide which distribution was used. Its* advantage *is*

$$\mathrm{Adv}_\mathcal{A}(\mathcal{D}_0, \mathcal{D}_1) = \Pr\left[\mathcal{A}(x) = 1 : x \leftarrow \mathcal{D}_1\right] - \Pr\left[\mathcal{A}(x) = 1 : x \leftarrow \mathcal{D}_0\right] .$$

We know that for all $\mathcal{A}$, $\mathrm{Adv}_\mathcal{A}(\mathcal{D}_0, \mathcal{D}_1) \leq d(\mathcal{D}_0, \mathcal{D}_1)$. Equality is reached for $\mathcal{A}$ defined by $\mathcal{A}(x) = 1$ iff $\mathcal{D}_1(x) \geq \mathcal{D}_0(x)$.

We say that $\mathcal{D}_0$ and $\mathcal{D}_1$ are ϵ-statistically indistinguishable if $d(\mathcal{D}_0, \mathcal{D}_1) \leq \epsilon$.

We say that the two distributions are (t, ε)-computationally indistinguishable if for any distinguisher $\mathcal{A}$ with running time bounded by t,

$$|\mathrm{Adv}_\mathcal{A}(\mathcal{D}_0, \mathcal{D}_1)| \leq \varepsilon .$$

Asymptotically, two distributions depending on a parameter λ are computationally indistinguishable if for any polynomial $t(\lambda)$ there exists a negligible function $\varepsilon(\lambda)$ such that, they are $(t(\lambda), \varepsilon(\lambda))$-computationally indistinguishable.

2.2 The Learning from Parity with Noise Problem

The *Learning from Parity with Noise* (LPN) problem has been well studied both in learning theory and in cryptography. The goal of this problem is to find out an unknown vector $\boldsymbol{u}$, given some noisy versions of its scalar product with some known random vector. More formally

Definition 7 (LPN Oracle). *An LPN oracle $\Pi_{u,p}$ for a hidden vector $u \in \{0,1\}^k$ and $0 < p < \frac{1}{2}$ is an oracle returning vectors of the form*

$$\langle a \xleftarrow{U} \{0,1\}^k , a \cdot u \oplus \nu \rangle ,$$

where, $\nu \leftarrow \text{Ber}(p)$. Note that the output is a $k+1$-bit vector.

Problem 8 (Learning from Parity with Noise Problem). The (k,p)-Learning from Parity with Noise Problem $((k,p)$-LPN) consists, given an LPN Oracle $\Pi_{u,p}$, to recover the hidden vector u.

We say that an algorithm $\mathcal{A}$ (t,n,δ)-solves the (k,p)-LPN problem if $\mathcal{A}$ runs in time at most t, makes at most n oracle queries and

$$\Pr\left[u \xleftarrow{U} \{0,1\}^k : \mathcal{A}^{\Pi_{u,p}}(1^k) = u \right] \geq \delta .$$

The Decisional LPN Problem. The LPN problem has also a decisional form. The problem is the following: let U_{k+1} be an oracle returning random $k+1$-bit vectors. Then, an algorithm $\mathcal{A}$ (t,n,δ)-solves the (k,p)-*decisional LPN problem* (D-LPN) if $\mathcal{A}$ runs in time at most t, makes at most n oracle queries and

$$\left| \Pr\left[u \xleftarrow{U} \{0,1\}^k : \mathcal{A}^{\Pi_{u,p}}(1^k) = 1 \right] - \Pr\left[\mathcal{A}^{U_{k+1}}(1^k) = 1 \right] \right| \geq \delta .$$

It is shown [37,53] that if there exists an algorithm $\mathcal{A}$ that (t,n,δ)-solves the (k,p)-D-LPN problem, then there is an algorithm $\mathcal{A}'$ that $(t',n',\delta/4)$-solves the (k,p)-LPN problem, with $t' := O\left(t \cdot k\delta^{-2} \log k\right)$ and $n' := O\left(n \cdot \delta^{-2} \log k\right)$. Thus, the hardness of the LPN problem implies that the output of the LPN vector oracle is indistinguishable from a random source.

We say that the (k,p)-D-LPN problem is (t,ϵ)-hard, if there is no algorithm solving it with running time bounded by t and advantage higher than ϵ.

Algorithms that Solve the LPN Problem. The first subexponential algorithm to solve the LPN problem was given by Blum, Kalai, and Wasserman in [11] and they estimated its complexity to $2^{O(k/\log k)}$. We denote this algorithm by BKW algorithm.

The idea of the BKW algorithm is to first query the LPN oracle to obtain a large amount of LPN vectors. It searches then for basis vectors e_j by finding a low amount of vectors that xor to e_j. If the number of vectors that xor to e_j is small, the noise for this vector will be small as well. Using different independent instances that xor to the same e_j, one can recover the jth bit of u with good probability. All this procedure can be done using a large amount of queries.

The BKW algorithm was analyzed in details and improved in [40,25]. We give here the complexity of the improvement given in [40] that we will use as a security bound in our cryptosystem.

Theorem 9 ([40], Theorem 2). *For $b \geq 1$, let $a := k/b$ and $q := (8b+200) \times (1-2p)^{-2^a} + (a-1) \times 2^b$. There exists an algorithm that $(kaq, q, \frac{1}{2})$-solves the (k,p)-LPN problem.*

Some parameters along with their security are given in [40, Section 5.2]. This algorithm requires a subexponential (in k) number of samples. When the number of samples is polynomial (as it is in our case), Lyubashevsky showed that one can scramble randomly the samples to get more of them with a higher noise level [42]. Then, the problem is solvable in $2^{O(k/\log\log k)}$. More precisely, one can transform the (k,p)-LPN problem with $k^{1+\epsilon}$ samples in the (k,p')-LPN problem with enough samples to use the BKW algorithm and with

$$p' = \frac{1}{2} - \frac{1}{2}\left(\frac{1}{4} - \frac{p}{2}\right)^{\frac{2k}{\epsilon\log k}}. \tag{2}$$

Combining this idea with Theorem 9, we get the following time complexity ($\mathsf{T_{LPN}}$) for solving LPN and we will use it as a security bound.

Theorem 10 (LPN with limited number of queries). *For $b \geq 1$, let $q := k^{1+\epsilon}$, and let*

$$\mathsf{T_{LPN}} := \min_{0<a\leq k}\left(k \times a \times \left(\left(\frac{8k}{a} + 200\right) \times (1 - 2p')^{-2^a} + (a-1) \times 2^{\frac{k}{a}}\right)\right), \tag{3}$$

where p' is given in Equation (2). There exists an algorithm that $(\mathsf{T_{LPN}}, q, \frac{1}{2})$-solves the (k,p)-LPN problem.

2.3   Finding a Low-weight Codeword in a Random Linear Code

In our security proof, we will also need to bound the complexity of finding a low-weight parity-check equation in a random linear code which is the same as finding a low-weight codeword in the dual code. This problem of finding a low-weight codeword is also called the minimum distance problem.

Problem 11 (Minimum Distance Problem (MDP)). The (n, k, w)-*decisional minimum distance problem* is the following. Given an $(n - k) \times n$ matrix H drawn uniformly and given $w \in \mathbb{N}, w \geq 0$, is there a *non-zero* $\boldsymbol{x} \in \mathbb{F}_2^n$ with $\mathsf{wt}(\boldsymbol{x}) \leq w$ such that $\boldsymbol{x}H^t = \boldsymbol{0}$?

The computation counterpart of this problem consists in finding such an x.

Its hardness remained open for a long time. It was even set the "open problem of the month" in [35]. It was finally shown to be NP-hard by Vardy [59] using a reduction from the decisional syndrome decoding problem. Many algorithms solving this problem were developed (e.g. [39,58,13,14,15,23].)

Finally, a general lower-bound on the complexity of the information set decoding algorithm was derived by Finiasz and Sendrier [23] using idealized algorithms. However, it was shown in [9,45] and very recently in [4] that it is possible to do better than this bound.

A new lower-bound for information set decoding is proposed in [9]. This bound is much simpler and we give it in Assumption 12.

Assumption 12 ([9]). Let $r := n - k$. Given an $[n, k]$-code and given a weight w, if $\binom{n}{w} \leq 2^r$, the cost of finding a parity-check equation of weight w is lower-bounded by

$$\mathsf{T_{MDP}}(w, n, k) := \min_i \frac{\binom{n}{w}}{2\binom{k}{w-i}\sqrt{\binom{r}{i}}}, \tag{4}$$

bit operations, with $r = n - k$.

We will assume this lower-bound for our cryptosystem. Note that a similar analysis for linear codes over a general field $\mathbb{F}_q$ is presented in [51].

3   The Cryptosystem

We will first consider how to encrypt one single bit b. Hence, our message space is $\mathcal{M} = \{0, 1\}$. We denote the cryptosystem by HELEN. We generalize the encryption to multiple bits in Section 6.

HELEN uses the following parameters which are described below: n, k, p, w, c, and $\mathcal{H}$. We encode first our message bit b with a binary $[n, 1]$-error-correcting code C_1, for $n \in \mathbb{N}$. The goal of this code is to be able to recover b when errors occur. Let $c \in \{0, 1\}^n$ be the generating matrix of this code (in fact, it is a vector). We encode b as $b \cdot c$. This message is hidden by a random codeword from a random binary linear $[n, k]$-code C_2 which has a low-weight parity-check equation $h \in \{0, 1\}^n$ and a generator matrix $G \in \{0, 1\}^{k \times n}$. The parameter $k \in \mathbb{N}$ determines the dimension of the codeword space in C_2 and needs to be tuned so that the system has the required security. The parity-check equation h will be the *private key* of our system while G will be the *public key*. Since h is a parity check equation of the code generated by G, we have $h \cdot G^t = 0$. We denote the weight of h by w and the set of all possible h by $\mathcal{H}$. We require $\mathcal{H}$ to verify the following property: there should exist a subgroup P of $\mathfrak{S}_n$ such that for any $\sigma \in P$ and any $h \in \mathcal{H}$, $\sigma \star h \in \mathcal{H}$. The group P defines a *group action* on the set $\mathcal{H}$. We require P to be a *transitive* group action, i.e, for any two $h, h' \in \mathcal{H}$, there exists a $\sigma \in P$ such that $\sigma \star h = h'$. In the following, $\mathcal{H}$ will be the set of all vectors of weight w and dimension n but we keep this more general $\mathcal{H}$ for further improvements. We also hide then the message further by adding some low weight random noise vector $\nu \in \{0, 1\}^n$ produced by a source S_p.

For correct decryption, we require also that $h \cdot c^t = 1$ for all $h \in \mathcal{H}$. When $\mathcal{H}$ contains all the vectors of weight w, this condition implies $c = (1, \ldots, 1)$ (see (5) below).

In the following, we describe more precisely the cryptosystem. All algorithms are summarized in Algorithm 1.

3.1   Encryption

A bit $b \in \mathcal{M}$ is encrypted as $\mathsf{BEnc}(G, b; r_1 \| r_2) = b \cdot c \oplus r_1 G \oplus \nu$, where c is the generator vector for C_1, G is the generator matrix for C_2, $r_1 \in \{0, 1\}^k$ is random and $\nu := S_p^n(r_2)$, i.e., it is the n first bits generated by the source S_p with random seed r_2. The ciphertext space is, thus, $\mathcal{C} = \{0,1\}^n$. The complexity of encryption is $O(kn)$.

3.2 Decryption

We define $b' := \mathsf{BDec}(h, y) = h \cdot y^t$. Given a ciphertext $y \in \{0,1\}^n$, we recover the original message by first removing the noise due to C_2. This is done by applying h on y since $h \cdot G^t = 0$. Hence, we get $b' := \mathsf{BDec}(h, y) = h \cdot y^t = (h \cdot c^t \cdot b^t) \oplus \nu'$, for $\nu' := h \cdot \nu$ a noise with

$$\Pr[\nu' = 1] = \frac{1 - (1 - 2p)^w}{2}$$

by Lemma 14. Note that it is necessary that

$$h \cdot c^t = 1 \tag{5}$$

for all vector $h \in \mathcal{H}$ if one wants to be able to recover b. When $\mathcal{H}$ includes all vectors of weight w, this condition is equivalent to setting c to the all-one vector and w to an odd number. The resulting bit b' is then different from b with probability φ, which is given in the following theorem.

Theorem 13. *HELEN is a φ-cryptosystem, where $\varphi := (1 - (1 - 2p)^w)/2$.*

Note that the complexity of decryption is $O(n)$.

Lemma 14. *Let X be a random variable defined as the sum modulo 2 of w iid Bernoulli random variables equals to 1 with probability p and to 0 else. Then*

$$\Pr[X = 1] = \frac{1 - (1 - 2p)^w}{2} .$$

Proof. We have $1 - 2\Pr[X = 1] = \mathbb{E}\left[(-1)^X\right] = (1 - 2p)^w$. □

3.3 Key Generation

We need now to generate a code that is indistinguishable from a random code but that contains a known secret parity-check equation h of low weight. Let w be the required weight of h and let $\mathcal{H}$ be the set of all possible private keys. The key generation algorithm is given in Algorithm 1.

The resulting public key size is $k \times n$ bits, since we have to store the matrix G. The private key is $w \log n$ bits long. The key generation complexity is $O(k \times n)$. Note that we have $hG^t = 0$.

4 Security Analysis

We will reduce the security of our scheme to the LPN problem presented in Section 2.2. To do this, we will proceed in two steps. First, we show that the code we construct for C_2 is computationally indistinguishable from a random matrix.

Algorithm 1. Algorithm to generate keys, to encrypt, and to decrypt.

Key Generation:
Input: Lengths k, n and a set $\mathcal{H}$.
Output: A private key h and a public key G.
 1: Draw a random vector h of length n in the set $\mathcal{H}$.
 2: Let $0 < u \leq n$ be any index of h such that $h_i = 1$, e.g., $\max\{i : h_i = 1\}$.
 3: Let $g_{ij} \leftarrow \mathrm{Ber}(\frac{1}{2})$, for $1 \leq i \leq k$ and $1 \leq j \leq n$, $j \neq u$.
 4: Let

$$g_{iu} = \sum_{\substack{1 \leq j \leq n \\ j \neq u}} g_{ij} h_j$$

for $1 \leq i \leq k$, where the sum is taken over $\mathbb{F}_2$.
 5: **return** the matrix $G := [g_{ij}]_{\substack{1 \leq i \leq k \\ 1 \leq j \leq n}}$ and the vector h.

Encryption:
Input: A bit b to encrypt, a public key G, two random seeds r_1 and r_2, a length n, an n-bit vector c, and a noise parameter p.
Output: A ciphertext y encrypted under the public key G.
 1: Let $\nu := S_p^n(r_2)$.
 2: **return** $y \leftarrow b \cdot c \oplus r_1 G \oplus \nu$.

Decryption:
Input: A ciphertext y and a private key h.
Output: The original plaintext b with probability φ defined in Theorem 13.
 1: **return** $b' \leftarrow h \cdot y^t$.

4.1   Link to Random Codes

We will compare the distributions of the output of different generators and show that their statistical distance is negligible using various lemmas. We conclude in Theorem 15. The first generator is our key generation algorithm.

Generator A: Run the key generation algorithm to obtain G and h and return $A := G$.

Generator G_1: Run generator A until the resulting matrix G has only one parity check equation in $\mathcal{H}$ and return $G_1 := G$.

Generator G_2: Draw a random $k \times n$ matrix G_2 until it has *a single* parity check equation in $\mathcal{H}$ and return G_2.

Generator G_3: Draw a random $k \times n$ matrix G_3 until it has *at least one* parity check equation in $\mathcal{H}$ and return G_3.

Generator B: Return a random $k \times n$ matrix B.

In the following, we show that the statistical distance between A and G_3 is negligible for suitable parameters.

Theorem 15. *Assume that there exists a subgroup P of $\mathfrak{S}_n$ that acts transitively on $\mathcal{H}$. Then,*

$$d(A, G_3) \leq \frac{(\#\mathcal{H} - 1)(\#\mathcal{H} + 2)}{2^{k+1}} =: \mathsf{D}_{\mathsf{A},\mathsf{G}_3} . \qquad (6)$$

Proof. We do the proof in three steps.

– We have $d(G_1, G_2) = 0$.

Recall that there exists a subgroup P of $\mathfrak{S}_n$ that acts transitively on $\mathcal{H}$. Clearly, G_2 generates a uniform distribution among all G's which have a unique parity check equation in $\mathcal{H}$. So, we just have to prove that G_1 has the same distribution. Clearly, $hG^t = 0$ if and only if $(\sigma \star h) \times (\sigma \star G)^t = 0$. Also, A generates uniformly a pair (h, G) with $h \in \mathcal{H}$ and G such that $hG^t = 0$. Let $\mathcal{G}_h$ be the set of all G's for which h is the only element of $\mathcal{H}$ satisfying $hG^t = 0$. For any $h \in \mathcal{H}$ and any $G \in \mathcal{G}_h$, we have

$$\Pr[G_1 \to G] = \frac{1}{\#\mathcal{H} \times \#\mathcal{G}_h}$$

Due to the above property on the action $\star$, any σ induces a permutation from $\mathcal{G}_h$ to $\mathcal{G}_{\sigma \star h}$. Since the action is further transitive, all $\mathcal{G}_h$'s have same cardinality. Hence, G_1 generates a uniform distribution among all the G's which have a unique parity check equation in $\mathcal{H}$.

– We have $d(G_2, G_3) \leq \frac{(\#\mathcal{H}-1)\#\mathcal{H}}{2^{k+1}}$.

Let $p_1(G_3)$ denote the probability that generator G_3 has exactly one parity-check equation in $\mathcal{H}$. The best distinguisher between G_2 and G_3 outputs 1 if and only if the generated matrix has two or more parity-check equations in $\mathcal{H}$. So, $d(G_2, G_3) = 1 - p_1(G_3)$.

Let a (resp. b) be the probability that a random matrix verifies at least one (resp. two) parity-check equations in $\mathcal{H}$. Then $a \geq 2^{-k}$, since any parity-check equation is verified with probability exactly 2^{-k}. Similarly,

$$b \leq \frac{(\#\mathcal{H})(\#\mathcal{H} - 1)}{2} \times 2^{-2k}$$

Thus,

$$d(G_2, G_3) = 1 - p_1(G_3) = \frac{b}{a} \leq \frac{(\#\mathcal{H} - 1)\#\mathcal{H}}{2} \times 2^{-k} .$$

– We have $d(A, G_1) \leq \frac{\#\mathcal{H}-1}{2^k}$.

Let $p_1(A)$ denote the probability that the output of generator A has exactly one parity-check equation in $\mathcal{H}$. The best distinguisher between A and G_1 checks if the generated matrix has only one parity-check equation. So, $d(A, G_1) = 1 - p_1(A) \leq \frac{\#\mathcal{H}-1}{2^k}$ since we are looking for a second parity-check equation in a random matrix which has already one of them.

Using triangular inequality, we get the wanted result. □

We want now to link this distribution with the distribution of an uniformly distributed $k \times n$ matrix, i.e., a matrix produced by generator B. We will need suitable parameters such that G_3 is *computationally indistinguishable* from B.

The best distinguisher between G_3 and B consists in deciding whether the output of the unknown generator has a parity-check equation in $\mathcal{H}$ or not. As discussed, the decisional problem is believed as hard as the computational problem. Hence, we extend Assumption 12 to the following one.

Assumption 16. For any distinguisher between G_3 and B, the complexity over advantage ratio is lower bounded by $\mathsf{T}_{\mathsf{MDP}}(w, n, k)$, which is defined in (4).

So, by selecting parameters such that the right-hand side of (6) is negligible and such that $\mathsf{T}_{\mathsf{MDP}}(w, n, k) \geq 2^\lambda$, for a security parameter λ, any game involving our cryptosystem produces a computationally indistinguishable outcome when the key generator is replaced by B.

4.2 Semantic Security

Now that we have B computationally indistinguishable from A, we can link our cryptosystem with the LPN problem.

Theorem 17. *Let $\varepsilon_0 := d(A, G_3)$ as defined in Theorem 15. If the (n, k, w)-decisional minimum distance problem is (t_1, ε_1)-computationally unsolvable, and if the (k, p)-decisional LPN problem is (t_2, ε_2)-hard, then there exists a constant τ such that our cryptosystem is*

$$(\min\{t_1, t_2 - \tau k n\}, 2(\varepsilon_0 + \varepsilon_1 + \varepsilon_2)) \text{-IND-CPA-secure}.$$

Proof. We introduce the following three games Γ_0, Γ_1 and Γ_2. Γ_0 is the IND-CPA game for our cryptosystem in the simple real-or-random model. Γ_1 is the IND-CPA game in the same model but using generator B instead of A. Γ_2 is the (k, p)-D-LPN game.

By the assumptions, we know that the best advantage between Γ_0 and Γ_1 is $\varepsilon_1 + \varepsilon_2$.

For the best advantage between Γ_1 and Γ_2, we do the following. Recall that in the simple real-and-random game this model, the adversary submits first a chosen plaintext b using an algorithm $\mathcal{A}_1^{\mathsf{ror}}(G)$. Then, given a n-bit word u, has to decide using an algorithm $\mathcal{A}_2^{\mathsf{ror}}(G, u)$, whether u is the encryption of b or is a random bitstring. Let $(\mathcal{A}_1^{\mathsf{ror}}(G), \mathcal{A}_2^{\mathsf{ror}}(G, u))$ be an IND-CPA adversary for our cryptosystem when G is generated using generator B.

We show that using this adversary, we can solve the D-LPN problem. We query first the unknown oracle of the D-LPN problem n times to obtain n-vectors $\alpha_1, \ldots, \alpha_n$. Note that each of these α_i has exactly $k + 1$ bits. We create now the $k \times n$ matrix $\tilde{G}$ using the first k bits of α_i as column i, for $1 \leq i \leq n$. Using $\mathcal{A}_1^{\mathsf{ror}}(\tilde{G})$, we recover a plaintext b. Let $z := b \cdot c \oplus (\alpha_{1|k+1} \| \ldots \| \alpha_{n|k+1})$, where $\alpha_{i|k+1}$ denotes the $k+1$-th bit of α_i. If the unknown oracle returns random bitstrings, then z will be random as well. However, if it is an LPN oracle, then z is a valid ciphertext of b using the public key $\tilde{G}$. Note also that the matrix $\tilde{G}$ follows the same distribution as the output of generator B.

Hence, using $\mathcal{A}_2^{ror}(\tilde{G}, z)$, we can decide whether z is a ciphertext corresponding to b or not. The complexity of this simulation is τkn for a constant $\tau > 0$ large enough. Thus, the advantage between game Γ_1 and Γ_2 is zero.

Since the D-LPN problem is supposed (t_2, ε_2)-hard, we get that our cryptosystem when we use generator B is $(t_2 - \tau kn, \varepsilon_2)$-IND-CPA-secure in the simple real-or-random model. Similarly, we get that the original cryptosystem is $(\min\{t_1, t_2 - \tau kn\}, \varepsilon_0 + \varepsilon_1 + \varepsilon_2)$-IND-CPA-secure in the simple real-or-random model. Thus, our cryptosystem is $(\min\{t_1, t_2 - \tau kn\}, 2(\varepsilon_0 + \varepsilon_1 + \varepsilon_2))$-IND-CPA-secure in the standard model [6]. □

Hence, we reduced the semantic security of our cryptosystem to the hardness of the decisional LPN problem with n queries and noise parameter p.

Note that since we encrypt one single bit, an IND-CPA adversary has to distinguish $\mathsf{BEnc}(G, 0)$ from $\mathsf{BEnc}(G, 1)$ which is equivalent to OW-CPA security.

5   Selection of Parameters

To summarize, we need to tune the following security parameters for HELEN:

- The dimension k of the code C_2 generated by G,
- The ciphertext length n (also the length of the codewords in C_2),
- The weight w of the secret key, and
- The noise probability p.

For our cryptosystem to be semantically secure, we need the parameters to verify Theorem 17. In particular, this implies that the D-LPN problem should be hard, that finding a low-weight parity-check equation in the code is hard as well, i.e., that $\mathsf{T}_{\mathsf{MDP}}(w, n, k) \geq 2^\lambda$ and that the statistical distance $\mathsf{D}_{\mathsf{A}, \mathsf{G}_3}$ defined in Theorem 15 is lower than $2^{-\lambda}$. We need also w to be odd. For the LPN problem, we want $\mathsf{T}_{\mathsf{LPN}} \geq 2^\lambda$, where $\mathsf{T}_{\mathsf{LPN}}$ is given in Equation (3).

Recall that the probability of decrypting incorrectly a bit is

$$P_{\mathsf{error}} := \frac{1 - (1 - 2p)^w}{2}. \tag{7}$$

Hence, to compare different parameters, we will normalize them with the capacity of a binary symmetric channel (BSC) with parameter P_{error}. Recall that the capacity of the BSC is $C := 1 - H_2(P_{\mathsf{error}})$ with $H_2(p) := -p \log(p) - (1 - p) \log(1 - p)$. We normalize by this factor, since we know that such a rate is achievable by the channel coding theorem. This gives us a good way of comparing the parameters.

We propose two sets of parameters. Some (I) which minimizes the n/C ratio to minimize the number of transmitted bits and some (II) with a smaller kn/C ratio to minimize the encryption/decryption complexity. We give in Table 1 concrete parameters for different security parameters λ.

Table 1. Parameters for our cryptosystem

	λ	k	n	w	p	kn	n/C	kn/C	$\mathsf{T}_{\mathsf{MDP}}$	$\mathsf{D}_{\mathsf{A},\mathsf{G}_3}$	$\mathsf{T}_{\mathsf{LPN}}$	C
I	64	4 500	18 000	33	0.01	$2^{26.3}$	$2^{16.4}$	$2^{28.6}$	$2^{65.3}$	2^{-3813}	$\geq 2^k$	0.20
II	64	2 200	16 000	23	0.02	$2^{25.0}$	$2^{17.1}$	$2^{28.2}$	$2^{64.7}$	2^{-1707}	$\geq 2^k$	0.11
I	80	5 600	28 000	35	0.01	$2^{27.2}$	$2^{17.2}$	$2^{29.7}$	$2^{80.5}$	2^{-4832}	$\geq 2^k$	0.18
II	80	2 800	27 000	25	0.02	$2^{26.2}$	$2^{18.1}$	$2^{29.6}$	$2^{80.4}$	2^{-2232}	$\geq 2^k$	0.10

In Table 2, we compare for concrete parameters HELEN with the code-based McEliece cryptosystem [46] and with an LWE-based cryptosystem [41]. Note that for encryption and decryption time, we neglect the cost of encoding and decoding.

We propose the following asymptotic parameters for our system:

$$k = \Theta\left(\lambda^2\right) \qquad n = \Theta\left(\lambda^2\right) \qquad w = \Theta\left(\lambda\right) \qquad p = \Theta\left(1/\lambda\right).$$

Indeed, we obtain $\mathsf{T}_{\mathsf{MDP}}$ and $\mathsf{T}_{\mathsf{LPN}} \geq 2^\lambda$, $\mathsf{D}_{\mathsf{A},\mathsf{G}_3} \leq 2^{-\lambda}$, $P_{\text{error}} = \frac{1}{2} - \frac{1}{e^{O(1)}}$, and $C > 0$. In Table 3, we compare the asymptotic parameters.

Table 2. Comparison with other cryptosystems

Name	λ	Message expansion	Pub key size	Encryption time	Decryption time
HELEN I	80	$2^{17.2}$	$2^{27.2}$	$O\left(2^{29.7}\right)$	$O\left(2^{17.2}\right)$
McEliece [8]	80	1.29	$2^{18.8}$	$O\left(2^{21.0}\right)$	$O\left(2^{21.3}\right)$
LWE [41]	128	22	$2^{17.5}$	$O\left(2^{24}\right)$	$O\left(2^{18.5}\right)$
Ring-LWE [41]	128	22	$\approx 2^{10}$	$O\left(2^{24}\right)$	$O\left(2^{18.5}\right)$

Table 3. Asymptotic comparison with other cryptosystems. The $\Theta\left(.\right)$'s have been omitted.

Name	Message expansion	Public key size	Private key size	Key generation	Encryption	Decryption
HELEN	λ^2	λ^4	$\lambda \log \lambda$	λ^4	λ^4	λ^2
TCHo	λ^2	λ^2	$\lambda \log \lambda$	$\lambda^6 \log \lambda \log \log \lambda$	λ^5	λ^4
McEliece	1	λ^2	λ^2	λ^3	λ^2	$\lambda^2 \log \lambda$
RSA	1	λ^3	λ^3	λ^{12}	λ^6	λ^9
NTRU	1	λ	λ	λ^3	λ^2	λ^2

6   Encrypting More than One Bit

In this section, we show how to encrypt more than one bit using HELEN. Taking advantage of an efficient coding scheme, we can also improve the probability of decrypting correctly the message. In addition to the previous parameters n, k, p, w and $\mathcal{H}$ we add a $[\mu, \kappa]$-error-correcting code. Let Encode be this $[\mu, \kappa]$-error-correcting code. Let also Decode be an efficient decoding algorithm corresponding to this code.

Encryption: We encrypt a plaintext $m \in \{0, 1\}^{\kappa}$ in two steps. First we compute $b_1 \| \ldots \| b_\mu := \mathsf{Encode}(m)$. The ciphertext c is then $\mathsf{BEnc}(G, b_1) \| \ldots \| \mathsf{BEnc}(G, b_\mu)$. The complexity of encryption is $O\left(\mu k n + T_{\mathsf{Encode}}\right)$, where T_{Encode} is the complexity of the encoding algorithm.

Decryption: To decrypt, we first decrypt each block of n bits using BDec to recover $b_1' \| \ldots \| b_\mu'$, where each $b_i' \neq b_i$ with probability $(1 - (1/2p)^w)/2 =: P_{\mathsf{error}}$. The complexity of decryption is $O\left(\mu n + T_{\mathsf{Decode}}\right)$, where T_{Decode} is the complexity of the decoding algorithm. Let ρ be the maximum number of errors the error-correcting code can correct. Then, using a Chernoff bound, the probability of decrypting incorrectly the message is

$$\sum_{i=\rho+1}^{\mu} \binom{\mu}{i} (P_{\mathsf{error}})^i (1 - P_{\mathsf{error}})^{\mu-i} \leq \exp\left[-2\mu\left(\frac{\rho}{\mu} - P_{\mathsf{error}}\right)^2\right] =: \phi . \quad (8)$$

Theorem 18. *HELEN with parameter μ, κ is a ϕ-cryptosystem, where ϕ is given in (8).*

Theorem 19. *Let ε_b be the* IND-CPA *advantage for the elementary cryptosystem HELEN with $\mu = \kappa = 1$. Then, the advantage of an* IND-CPA *adversary against the full cryptosystem HELEN with parameter μ and κ is smaller than $\mu\varepsilon_b$.*

Proof. Let $\mathcal{A} := (\mathcal{A}_1, \mathcal{A}_2)$ be an IND-CPA adversary HELEN with parameter μ, κ. Given $i \in \{1, \ldots, \mu\}$, we define $\mathcal{B}_i := (\mathcal{B}_{i,1}(G), \mathcal{B}_{i,2}(G, c))$ as follows.

$\mathcal{B}_{i,1}(G)$:

1. Let $m_0, m_1 \leftarrow \mathcal{A}_1(G)$
2. Let $b_1^0 \| \ldots \| b_\mu^0 \leftarrow \mathsf{Encode}(m_0)$, the encoding of m_0
3. Let $b_1^1 \| \ldots \| b_\mu^1 \leftarrow \mathsf{Encode}(m_1)$, the encoding of m_1
4. Return b_i^0, b_i^1.

$\mathcal{B}_{i,2}(G, c)$:

1. Compute $c_1 \leftarrow \mathsf{BEnc}(G, b_1^1), \ldots,$ $c_{i-1} \leftarrow \mathsf{BEnc}(G, b_{i-1}^1)$.
2. Let $c_i = c$
3. Compute $c_{i+1} \leftarrow \mathsf{BEnc}(G, b_{i+1}^0), \ldots,$ $c_\mu \leftarrow \mathsf{BEnc}(G, b_\mu^0)$.
4. Set $y := c_1 \| \ldots \| c_\mu$
5. return $\mathcal{A}_2(G, y)$

We know that $\text{Adv}\,\mathcal{B}_i \leq \varepsilon_b$. We have

$$\Pr[\mathcal{A} \to 0 \mid m_0 \text{ encrypted}] = \Pr[\mathcal{B}_1 \to 0 \mid b_1^0 \text{ encrypted}]$$

and

$$\Pr[\mathcal{A} \to 0 \mid m_1 \text{ encrypted}] = \Pr[\mathcal{B}_\mu \to 0 \mid b_\mu^1 \text{ encrypted}]\ .$$

Also,

$$\Pr[\mathcal{B}_i \to 0 \mid b_i^1 \text{ encrypted}] = \Pr[\mathcal{B}_{i+1} \to 0 \mid b_{i+1}^0 \text{ encrypted}]\ .$$

Hence,

$$\text{Adv}\,\mathcal{A} = (\Pr[\mathcal{A} \to 0 \mid m_0 \text{ encrypted}] - \Pr[\mathcal{A} \to 0 \mid m_1 \text{ encrypted}])$$
$$= \sum_{i=1}^{\mu} (\Pr[\mathcal{B} \to 0 \mid b_i^0 \text{ encrypted}] - \Pr[\mathcal{B} \to 0 \mid b_i^1 \text{ encrypted}]) \leq \mu\varepsilon_b\ .$$

$$\square$$

Obviously HELEN is not IND-CCA-secure, since it is clearly malleable. It suffices to change one single bit of the ciphertext and to submit it to the decryption oracle to decrypt the plaintext with good probability. To achieve IND-CCA security, one can use well-known construction like the Fujisaki-Okamoto hybrid construction [26]. This construction uses two random oracles H_1 and H_2 as well as a symmetric encryption scheme. However, such a construction work only if the cryptosystem is Γ-uniform.

Definition 20 (Γ-uniformity). *Let* Enc *be an asymmetric encryption scheme, with key generation algorithm* $\text{Gen}(1^\lambda)$ *and encryption algorithm* $\text{Enc}(K_p, m; r)$ *over the message space* $\mathcal{M}$ *and the random coins space* $\mathcal{R}$. Enc *is* Γ-uniform *if for any plaintext* $m \in \mathcal{M}$, *for any keys drawn by* Gen *and for any* $y \in \{0,1\}^*$, *we have*

$$\Pr\left[h \xleftarrow{U} \mathcal{R} : y = \text{Enc}(K_p, m; h)\right] \leq \Gamma\ ,$$

i.e., the probability that a plaintext and a ciphertext match is bounded.

Lemma 21. *HELEN is* $(1-p)^n$*-uniform.*

Proof. Recall that the HELEN encryption of b is $y = b \cdot c \oplus r_1 G \oplus S_p^n(r_2)$, for random coins r_1 and r_2. We need to bound the probability (taken over r_1 and r_2) that a given plaintext x and ciphertext y match. Since in HELEN we consider only $p < \frac{1}{2}$, the most probable ciphertext corresponds to $y = b \cdot c \oplus r_1 G$, i.e., when S_p^n is the zero bitstring. This happens with probability $(1-p)^n$. When we take the average over the possible r_1, this probability can only decrease. Hence, HELEN is $(1-p)^n$-uniform. $\square$

Theorem 22. *Let* q_1*(resp.* q_2*) be the number of queries an adversary makes to* H_1 *(resp.* H_2*). Let* q_d *be the number of queries performed to the decryption oracle. Then, if HELEN is* (t, ϵ)*-IND-CPA-secure, the Fujisaki-Okamoto hybrid* *construction using a one-time pad for symmetric encryption with key length ℓ is (t_1, ϵ_1)-IND-CCA-secure in the random oracle model, where*

$$t_1 := t - O\left((q_1 + q_2) \times (k + \ell)\right)$$
$$\epsilon_1 := (2(q_1 + q_2)\epsilon + 1)(1 - (1 - p)^n - 2^{-\ell})^{-q_d} - 1 .$$

Proof. Since HELEN is OW-CPA secure and $(1 - p)^n$-uniform, the result follows from [26, Theorem 14]. □

7 Conclusion

Further Work. HELEN can be extended in multiple ways. A first idea is to use different $\mathcal{H}$ to reduce the probability of error and, hence, to reduce the transmission overhead. This implies also to verify that Assumption 16 holds for this new $\mathcal{H}$. Another idea would be to encrypt a message in $\mathbb{F}_q$ for $q > 2$. The codes C_1 and C_2 described in Section 3 need then to be modified accordingly as well as the noise we add. This new extension could then be linked to the learning with error (LWE) problem [53], a generalization of the LPN problem over a finite field $\mathbb{F}_q$. Finally, the LPN problem deserves some more analysis in particular when p is not fixed.

In conclusion, HELEN is a code-based public-key cryptosystem based on the hardness of some well-known problems. Since its margin of progression is still large, HELEN can become a competitive cryptosystem with truly practical parameters.

References

1. Alekhnovich, M.: More on Average Case vs Approximation Complexity. In: FOCS, pp. 298–307. IEEE Computer Society (2003)
2. Applebaum, B., Barak, B., Wigderson, A.: Public-key cryptography from different assumptions. In: Schulman (ed.) [56], pp. 171–180
3. Aumasson, J.-P., Finiasz, M., Meier, W., Vaudenay, S.: TCHo: A Hardware-Oriented Trapdoor Cipher. In: Pieprzyk, J., Ghodosi, H., Dawson, E. (eds.) ACISP 2007. LNCS, vol. 4586, pp. 184–199. Springer, Heidelberg (2007)
4. Becker, A., Joux, A., May, A., Meurer, A.: Decoding Random Binary Linear Codes in $2^{n/20}$: How $1 + 1 = 0$ Improves Information Set Decoding. In: Pointcheval, D., Johansson, T. (eds.) EUROCRYPT 2012. LNCS, vol. 7237, pp. 520–536. Springer, Heidelberg (2012)
5. Bellare, M., Desai, A., Jokipii, E., Rogaway, P.: A Concrete Security Treatment of Symmetric Encryption: Analysis of the DES Modes of Operation (Full Version) (1997), http://cseweb.ucsd.edu/users/mihir
6. Bellare, M., Desai, A., Jokipii, E., Rogaway, P.: A Concrete Security Treatment of Symmetric Encryption (Extended Abstract). In: FOCS, pp. 394–403 (1997)

7. Bernstein, D.J.: Introduction to post-quantum cryptography. In: Bernstein, D.J., Buchmann, J., Dahmen, E. (eds.) Post-Quantum Cryptography, pp. 1–14. Springer (2009)

8. Bernstein, D.J., Lange, T., Peters, C.: Attacking and Defending the McEliece Cryptosystem. In: Buchmann, J., Ding, J. (eds.) PQCrypto 2008. LNCS, vol. 5299, pp. 31–46. Springer, Heidelberg (2008)

9. Bernstein, D.J., Lange, T., Peters, C.: Smaller Decoding Exponents: Ball-Collision Decoding. In: Rogaway, P. (ed.) CRYPTO 2011. LNCS, vol. 6841, pp. 743–760. Springer, Heidelberg (2011)

10. Blum, A., Furst, M.L., Kearns, M., Lipton, R.J.: Cryptographic Primitives Based on Hard Learning Problems. In: Stinson, D.R. (ed.) CRYPTO 1993. LNCS, vol. 773, pp. 278–291. Springer, Heidelberg (1994)

11. Blum, A., Kalai, A., Wasserman, H.: Noise-Tolerant Learning, the Parity Problem, and the Statistical Query Model. J. ACM 50(4), 506–519 (2003)

12. Bringer, J., Chabanne, H., Dottax, E.: HB^{++}: a Lightweight Authentication Protocol Secure against Some Attacks. In: SecPerU, pp. 28–33. IEEE Computer Society (2006)

13. Canteaut, A., Chabanne, H.: A Further Improvement of the Work Factor in an Attempt at Breaking McEliece's Cryptosystem. In: Charpin, P. (ed.) EUROCODE (1994)

14. Canteaut, A., Chabaud, F.: A New Algorithm for Finding Minimum-Weight Words in a Linear Code: Application to McEliece's Cryptosystem and to Narrow-Sense BCH Codes of Length 511. IEEE Transactions on Information Theory 44(1), 367–378 (1998)

15. Canteaut, A., Sendrier, N.: Cryptanalysis of the Original McEliece Cryptosystem. In: Ohta, K., Pei, D. (eds.) ASIACRYPT 1998. LNCS, vol. 1514, pp. 187–199. Springer, Heidelberg (1998)

16. Chekuri, C., Jansen, K., Rolim, J.D.P., Trevisan, L. (eds.): APPROX/RANDOM 2005. LNCS, vol. 3624. Springer, Heidelberg (2005)

17. Dolev, D., Dwork, C., Naor, M.: Non-Malleable Cryptography (Extended Abstract). In: Koutsougeras, C., Vitter, J.S. (eds.) STOC, pp. 542–552. ACM (1991)

18. Döttling, N., Müller-Quade, J., Nascimento, A.C.A.: IND-CCA Secure Cryptography Based on a Variant of the LPN Problem. In: Wang, X., Sako, K. (eds.) ASIACRYPT 2012. LNCS, vol. 7658, pp. 485–503. Springer, Heidelberg (2012)

19. Duc, A., Vaudenay, S.: HELEN: a Public-key Cryptosystem Based on the LPN and the Decisional Minimal Distance Problems (Extended Abstract). In: Yet Another Conference on Cryptography (2012)

20. Duc, A., Vaudenay, S.: TCHo: A Code-Based Cryptosystem. In: Kranakis, E. (ed.) Advances in Network Analysis and its Applications, Mathematics in Industry, vol. 18, pp. 149–179. Springer, Heidelberg (2013)

21. Duc, D.N., Kim, K.: Securing HB$^+$ against GRS man-in-the-middle attack. In: Institute of Electronics, Information and Communication Engineers, Symposium on Cryptography and Information Security (2007)

22. El Gamal, T.: A Public Key Cryptosystem and a Signature Scheme Based on Discrete Logarithms. In: Blakely, G.R., Chaum, D. (eds.) CRYPTO 1984. LNCS, vol. 196, pp. 10–18. Springer, Heidelberg (1985)

23. Finiasz, M., Sendrier, N.: Security Bounds for the Design of Code-Based Cryptosystems. In: Matsui (ed.) [44], pp. 88–105

24. Finiasz, M., Vaudenay, S.: When Stream Cipher Analysis Meets Public-Key Cryptography. In: Biham, E., Youssef, A.M. (eds.) SAC 2006. LNCS, vol. 4356, pp. 266–284. Springer, Heidelberg (2007)

25. Fossorier, M.P.C., Mihaljević, M.J., Imai, H., Cui, Y., Matsuura, K.: An Algorithm for Solving the LPN Problem and Its Application to Security Evaluation of the HB Protocols for RFID Authentication. In: Barua, R., Lange, T. (eds.) INDOCRYPT 2006. LNCS, vol. 4329, pp. 48–62. Springer, Heidelberg (2006)

26. Fujisaki, E., Okamoto, T.: Secure Integration of Asymmetric and Symmetric Encryption Schemes. In: Wiener, M. (ed.) CRYPTO 1999. LNCS, vol. 1666, pp. 537–554. Springer, Heidelberg (1999)

27. Gentry, C., Peikert, C., Vaikuntanathan, V.: Trapdoors for hard lattices and new cryptographic constructions. In: Dwork, C. (ed.) STOC, pp. 197–206. ACM (2008)

28. Gilbert, H., Robshaw, M., Sibert, H.: Active attack against HB$^+$: a provably secure lightweight authentication protocol. Electronics Letters 41(21), 1169–1170 (2005)

29. Gilbert, H., Robshaw, M., Seurin, Y.: Good Variants of HB$^+$ Are Hard to Find. In: Tsudik, G. (ed.) FC 2008. LNCS, vol. 5143, pp. 156–170. Springer, Heidelberg (2008)

30. Gilbert, H., Robshaw, M., Seurin, Y.: HB$^\#$ Increasing the Security and Efficiency of HB$^+$. In: Smart, N.P. (ed.) EUROCRYPT 2008. LNCS, vol. 4965, pp. 361–378. Springer, Heidelberg (2008)

31. Gilbert, H., Robshaw, M., Seurin, Y.: How to Encrypt with the LPN Problem. In: Aceto, L., Damgård, I., Goldberg, L.A., Halldórsson, M.M., Ingólfsdóttir, A., Walukiewicz, I. (eds.) ICALP 2008, Part II. LNCS, vol. 5126, pp. 679–690. Springer, Heidelberg (2008)

32. Heyse, S., Kiltz, E., Lyubashesvky, V., Paar, C., Pietrzak, K.: An Efficient Authentication Protocol Based on Ring-LPN. In: ECRYPT Workshop on Lightweight Cryptography 2007 (2011), http://www.uclouvain.be/crypto/ecrypt_lc11/static/pre_proceedings_2.pdf

33. Hoffstein, J., Pipher, J., Silverman, J.H.: NTRU: A Ring-Based Public Key Cryptosystem. In: Buhler, J.P. (ed.) ANTS 1998. LNCS, vol. 1423, pp. 267–288. Springer, Heidelberg (1998)

34. Hopper, N.J., Blum, M.: Secure Human Identification Protocols. In: Boyd, C. (ed.) ASIACRYPT 2001. LNCS, vol. 2248, pp. 52–66. Springer, Heidelberg (2001)

35. Johnson, D.S.: The NP-Completeness Column: An Ongoing Guide. J. Algorithms 3(2), 182–195 (1982)

36. Juels, A., Weis, S.A.: Authenticating Pervasive Devices with Human Protocols. In: Shoup, V. (ed.) CRYPTO 2005. LNCS, vol. 3621, pp. 293–308. Springer, Heidelberg (2005)

37. Katz, J., Shin, J.S.: Parallel and Concurrent Security of the HB and HB$^+$ Protocols. In: Vaudenay, S. (ed.) EUROCRYPT 2006. LNCS, vol. 4004, pp. 73–87. Springer, Heidelberg (2006)

38. Kiltz, E., Pietrzak, K., Cash, D., Jain, A., Venturi, D.: Efficient Authentication from Hard Learning Problems. In: Paterson, K.G. (ed.) EUROCRYPT 2011. LNCS, vol. 6632, pp. 7–26. Springer, Heidelberg (2011)

39. Lee, P.J., Brickell, E.F.: An Observation on the Security of McEliece's Public-Key Cryptosystem. In: Günther, C.G. (ed.) EUROCRYPT 1988. LNCS, vol. 330, pp. 275–280. Springer, Heidelberg (1988)

40. Levieil, É., Fouque, P.-A.: An Improved LPN Algorithm. In: De Prisco, R., Yung, M. (eds.) SCN 2006. LNCS, vol. 4116, pp. 348–359. Springer, Heidelberg (2006)

41. Lindner, R., Peikert, C.: Better Key Sizes (and Attacks) for LWE-Based Encryption. In: Kiayias, A. (ed.) CT-RSA 2011. LNCS, vol. 6558, pp. 319–339. Springer, Heidelberg (2011)

42. Lyubashevsky, V.: The Parity Problem in the Presence of Noise, Decoding Random Linear Codes, and the Subset Sum Problem. In: Chekuri, et al. (eds.) [16], pp. 378–389

43. Lyubashevsky, V., Peikert, C., Regev, O.: On Ideal Lattices and Learning with Errors over Rings. In: Gilbert, H. (ed.) EUROCRYPT 2010. LNCS, vol. 6110, pp. 1–23. Springer, Heidelberg (2010)

44. Matsui, M. (ed.): ASIACRYPT 2009. LNCS, vol. 5912. Springer, Heidelberg (2009)

45. May, A., Meurer, A., Thomae, E.: Decoding Random Linear Codes in $\tilde{O}(2^{0.054n})$. In: Lee, D.H., Wang, X. (eds.) ASIACRYPT 2011. LNCS, vol. 7073, pp. 107–124. Springer, Heidelberg (2011)

46. McEliece, R.J.: A public-key cryptosystem based on algebraic coding theory. DSN Progress Report 42(44), 114–116 (1978)

47. Munilla, J., Peinado, A.: HB-MP: A further step in the HB-family of lightweight authentication protocols. Computer Networks 51(9), 2262–2267 (2007)

48. Niederreiter, H.: Knapsack-type cryptosystems and algebraic coding theory. Problems of Control and Information Theory 15(2), 159–166 (1986)

49. Ouafi, K., Overbeck, R., Vaudenay, S.: On the Security of $HB^{\#}$ against a Man-in-the-Middle Attack. In: Pieprzyk, J. (ed.) ASIACRYPT 2008. LNCS, vol. 5350, pp. 108–124. Springer, Heidelberg (2008)

50. Peikert, C.: Public-key cryptosystems from the worst-case shortest vector problem: extended abstract. In: Mitzenmacher, M. (ed.) STOC, pp. 333–342. ACM (2009)

51. Peters, C.: Information-Set Decoding for Linear Codes over $\mathbf{F}_q$. In: Sendrier, N. (ed.) PQCrypto 2010. LNCS, vol. 6061, pp. 81–94. Springer, Heidelberg (2010)

52. Rabin, M.: Digitalized signatures and public-key functions as intractable as factorization (1979)

53. Regev, O.: On lattices, learning with errors, random linear codes, and cryptography. In: Gabow, H.N., Fagin, R. (eds.) STOC, pp. 84–93. ACM (2005)

54. Rivest, R.L., Shamir, A., Adleman, L.: A method for obtaining digital signatures and public-key cryptosystems. Communications of the ACM 21(2), 120–126 (1978)

55. Rosen, A., Segev, G.: Chosen-Ciphertext Security via Correlated Products. In: Reingold, O. (ed.) TCC 2009. LNCS, vol. 5444, pp. 419–436. Springer, Heidelberg (2009)

56. Schulman, L.J. (ed.): Proceedings of the 42nd ACM Symposium on Theory of Computing, STOC 2010, Cambridge, Massachusetts, USA, June 5-8. ACM (2010)

57. Stehlé, D., Steinfeld, R., Tanaka, K., Xagawa, K.: Efficient Public Key Encryption Based on Ideal Lattices. In: Matsui (ed.) [44], pp. 617–635

58. Stern, J.: A method for finding codewords of small weight. In: Wolfmann, J., Cohen, G. (eds.) Coding Theory 1988. LNCS, vol. 388, pp. 106–113. Springer, Heidelberg (1989)

59. Vardy, A.: The Intractability of Computing the Minimum Distance of a Code. IEEE Transactions on Information Theory 43(6), 1757–1766 (1997)

Attacking AES Using Bernstein's Attack on Modern Processors

Hassan Aly* and Mohammed ElGayyar

Department of Mathematics, Faculty of Science,
Cairo University, Giza 12613, Egypt
{hassan.aly,melgayar}@sci.cu.edu.eg

Abstract. The Advanced Encryption Standard (AES) was selected by NIST due to its heavy resistance against classical cryptanalysis like differential and linear cryptanalysis. Even after the appearance of the modern side-channel attacks like timing and power consumption side-channel attacks, NIST claimed that AES is not vulnerable to timing attacks. In 2005, Bernstein [6] has successfully attacked the OpenSSL AES implementation on a Pentium III processor and completely retrieved the full AES key using his cache timing side-channel attack. This paper reproduces Bernstein's attack on Pentium Dual-Core and Core 2 Duo processors. We have successfully attacked the AES implemented in the latest OpenSSL release 1.0.1c using the most recent GCC compiler 4.7.0 running on both Windows and Linux in some seconds by sending 2^{22} plaintexts at most. We improved Bernstein's first round attack by using 2 way measurements. Instead of using only the above average timing information, we added the above minimum timing information which significantly improved the results.

Keywords: AES, timing attack, Bernstein's attack, cache memory attack, side-channel attack, cryptanalysis.

1 Introduction

For a long time, attacking cryptographic systems was relying only on its mathematical basis like the case in differential and linear cryptanalysis. To conduct such attacks you have to know either a number of ciphertexts or pairs of ciphertexts and plaintexts. Nowadays, several attacks are based on the information revealed from the encryption devices. Since this information is not the ciphertext or the plaintext, so it is often called side-channel information [5]. This information may be revealed by measuring the power consumption, heat consumption, cache access or time elapsed during processing.

The timing attacks can be considered as the most popular attacks that have greatly developed during the last ten years. Measuring the time taken to access cache memory helps identifying cache hits and misses which in turn is considered

---

* Current address: Department of Computer Science and Information, College of Science, Majmaah University, AzZulfi 11932, P.O. Box 1712, Kingdom of Saudi Arabia.

A. Youssef, A. Nitaj, A.E. Hassanien (Eds.): AFRICACRYPT 2013, LNCS 7918, pp. 127–139, 2013.

a dangerous information to be revealed. NIST has stated in [20] (section 3.6.2) that "Table lookup: not vulnerable to timing attacks; relatively easy to effect a defense against power attacks by software balancing of the lookup address.". However, the most powerful timing attacks on AES depend on measuring table lookup access times, which reveals most of the AES key bits.

2 AES Implementation Background

AES is the most widely used secret key block cipher. It is a substitution-permutation network (SPN), it has an iterative structure that works in rounds. That is, both encryption and decryption consists of number of iterations, called rounds, each round consists of a fixed set of operations, namely (SubBytes, ShiftRows, MixColumns, and AddRoundKey). The operations are applied to the input plaintext block, called state. After iterating a predefined number of rounds, (10, 12, and 14 rounds for 128-bit, 192-bit, and 256-bit keys, respectively), the final state is the ciphertext block.

Although the optimized ANSI C code submitted with Rijndael proposal uses 5 T-tables [31], high performance implementations like OpenSSL [25], uses only four 256-entry 32-bit T-tables T_0, T_1, T_2 and T_3 during encryption. This implementation precomputes original S-boxes and stores every lookup vector in a different T-table each of 1024-byte size. Each round state word $(x_{4i}^{(r)}, x_{4i+1}^{(r)}, x_{4i+2}^{(r)}, x_{4i+3}^{(r)})$, $i = 0, 1, 2, 3$, is generated as:

$$(x_0^{(r)}, x_1^{(r)}, x_2^{(r)}, x_3^{(r)}) = T_0[x_0^{(r-1)}] \oplus T_1[x_5^{(r-1)}] \oplus T_2[x_{10}^{(r-1)}] \oplus T_3[x_{15}^{(r-1)}] \oplus (k_0^{(r-1)}, k_1^{(r-1)}, k_2^{(r-1)}, k_3^{(r-1)})$$

$$(x_4^{(r)}, x_5^{(r)}, x_6^{(r)}, x_7^{(r)}) = T_0[x_4^{(r-1)}] \oplus T_1[x_9^{(r-1)}] \oplus T_2[x_{14}^{(r-1)}] \oplus T_3[x_3^{(r-1)}] \oplus (k_4^{(r-1)}, k_5^{(r-1)}, k_6^{(r-1)}, k_7^{(r-1)})$$

$$(x_8^{(r)}, x_9^{(r)}, x_{10}^{(r)}, x_{11}^{(r)}) = T_0[x_8^{(r-1)}] \oplus T_1[x_{13}^{(r-1)}] \oplus T_2[x_2^{(r-1)}] \oplus T_3[x_7^{(r-1)}] \oplus (k_8^{(r-1)}, k_9^{(r-1)}, k_{10}^{(r-1)}, k_{11}^{(r-1)})$$

$$(x_{12}^{(r)}, x_{13}^{(r)}, x_{14}^{(r)}, x_{15}^{(r)}) = T_0[x_{12}^{(r-1)}] \oplus T_1[x_1^{(r-1)}] \oplus T_2[x_6^{(r-1)}] \oplus T_3[x_{11}^{(r-1)}] \oplus (k_{12}^{(r-1)}, k_{13}^{(r-1)}, k_{14}^{(r-1)}, k_{15}^{(r-1)})$$

Here, T_0, T_1, T_2, T_3 are four lookup tables with 1 byte input and 4 bytes output and $(k_{4i}^{(r)}, k_{4i+1}^{(r)}, k_{4i+2}^{(r)}, k_{4i+3}^{(r)})$, $i = 0, 1, 2, 3$ is the i-th word of the r-th round key.

3 Cache Based Side Channel Attacks Background

Cache based side channel attacks, cache attacks, can be classified into three major types of attacks: time-driven, access-driven, and trace-driven.

In a trace-driven attack, the attacker should be able to monitor and collect the cache activity including every memory access during an encryption. This data collection process is meant to create a profile (trace) of cache hits and misses for a single encryption. The quality of the attack depends on how many traces are needed to begin the analysis phase, a better attack requires less traces.

In a time-driven attack, the attacker do not need to have the ability of collecting data of every memory access. Instead, the attack relies on a value that can be used to describe or approximate the total number of cache hits and misses. In most of time-driven attacks, the attacker collects only the total execution time of an encryption as many times as needed, then use mathematical tools to analyze the collected data. The quality of the attack depends on how many encryptions are needed to begin the analysis phase, a better attack requires less encryptions.

In an access-driven attack, the attacker can detect which cache sets were modified by the encryption process, which leads the attacker to know which lookup table entries are accessed during the encryption. Then, uses elimination and non-elimination techniques to detect the right key candidates. For more information about microarchitectural attacks, consult [2].

4   Related Work

In this section, we introduce the history of cache based side channel attacks and timing attacks related to Bernstein's attack.

In 1996, Kocher [16] first introduced timing attacks on implementations of Diffie-Hellman, RSA, DSS, and other systems. In 1998, Kelsey et al. [15] mentioned the "attacks based on cache hit ratio in large S-box ciphers" prospect. In 2002, Page [27] expanded the idea proposed by Kelsey et al., of cache memory being used as a side-channel which leaks information during the run of DES. As well as describing and simulating the theoretical attack. He discussed how hardware and algorithmic alterations can be used to defend against such techniques in [28] in 2003. Later on, Tsunoo et al. [34,35] implemented the first practical cache timing cryptanalysis of DES, 3DES, MISTY1, Camellia and AES.

In 2003, Brumley and Boneh [10] devised and implemented a remote timing attack against unprotected OpenSSL implementation of RSA over a local area network.

In 2005, Acıiçmez et al. [4] improved the efficiency of Brumley and Boneh timing attack on unprotected SSL implementations of RSA-CRT by a factor of more than 10. Earlier in the same year, Bertoni et al. [7] introduced the first trace-driven attack on AES based on induced cache misses. Also they proposed a simple countermeasure against the attack.

Bernstein [6] derived an attack on AES which depends only on calculating the encryption time information caused by cache memory hits and misses then comparing timing data using statistical methods. Bernstein's attack is a time-driven cache attack. It is performed as a template attack where at first a profile under a known key is generated with the same platform as the later attacked one. The real attack is performed in a second phase where a profile of an unknown key is generated. Those two profiles are then correlated and the key space for the unknown key is reduced. In a last phase for full key recovery, a brute-force of the remaining key space is performed. This attack is not affected by cache architecture or active manipulation, it only depends on the similarity between reference and target machines. Later, Percival [29] was the first to use access-driven attack against RSA, and demonstrated that the shared access to cache memory provides an easily used covert channel between threads, allowing in many cases for theft of cryptographic keys.

Osvik et al. [26,33] led the work on attacking AES using access-driven cache memory attacks, and described several software side-channel attacks based on inter-process leakage through the state of the CPU's cache memory. The authors discussed an attack called *synchronous attack*, which requires knowledge of either the plaintext or the ciphertext. The synchronous attacker can operate synchronously with the encryption on the same processor. Moreover, they demonstrated an extremely strong type of attack called *asynchronous attack*, which does not require any knowledge of plaintexts or ciphertexts. The asynchronous attacker will execute his own program on the same processor as the encryption program without any explicit interaction, depending on the knowledge of the non-uniform distribution of the plaintexts or ciphertexts. They also experimentally demonstrated their applicability to real systems, such as OpenSSL and presented a variety of countermeasures which can be used to mitigate such attacks. However, they did not give a description of how to perform a full asynchronous attack. Lauradoux [17] proposed some countermeasures against these attacks, and Canteaut et al. [11] followed him in 2006.

In 2006 also, Neve et al. [23,22] presented a thorough analysis of Bernstein's attack, reproducing the attack and demonstrating results of important experiments practically. They answered a lot of open questions about the attack like, what if there is no learning phase? Can this attack be a real remote threat or not? and more. Then they extended the attack with a second round attack to reveal other key bits that could not be revealed by Bernstein's first round attack.

In the same year, Neve et al. [21] introduced an access-driven attack on AES, by demonstrating how a spy process running on the same single threaded CPU can measure the number of accessed cache lines by another process running on the same CPU.

Another cache memory attack was introduced in 2006 by Bonneau et al. [9], in this cache collision attack, they aimed to predict cache collisions timing variation using a simplified cache model. Their most powerful attack recovered a full 128-bit AES key with an improvement of almost four orders of magnitude over Bernstein's attack.

A cache based remote timing attack followed by Acıiçmez et al. [3], they described an expanded second round attack that can be used to obtain secret keys of remote cryptosystems. Their attack requires hyper threading enabled system with a large enough workload. In 2006, Acıiçmez et al. [1] presented a trace-driven attack on AES. They described a first two rounds attack and a last round attack as well. At the end of their work, they show the trade-off between the online and offline cost of the attack in details. Later in the same year, Bonneau [8] described a final round trace-driven attack on AES, building off of previous work by Acıiçmez and Koç [1]. Bonneau introduced an algorithm that reduces the problem of attacking AES given a small set of cache traces, to a simple constraint satisfaction problem.

In 2007, Tiri et al. [32] have proposed an analytical model for time-driven cache attacks. They presented a tool to help us evaluate the security of symmetric key ciphers against against such attacks.

In 2008, Zhao et al. [37] introduced a first two rounds access-driven attack on AES. Introducing the elimination technique in guessing the key bytes.

They succeeded in recovering the full 128-bit AES key through the first round attack using about 350 samples, and two rounds attack using about 80 samples in a few seconds.

In 2010, Rebeiro et al. [30] justified that cache timing attacks on AES are unable to force hits in the third round and concluded that a similar third round cache timing attack does not work. Hence, protecting only the first two AES rounds prohibits cache based timing attacks. Zhao and Wang [36] presented an improved trace-driven attack on AES and CLEFIA by considering S-box mis-alignment, and due to this feature, about 200 samples are enough to obtain full 128-bit AES key within seconds. Bogdanov et al. introduced a novel differential collision attack based on the MDS properties of AES on embedded CPUs. Their experiments show that efficient attacks on embedded systems implementing AES are not theoretical any more.

In 2011, Gallais et al. [12] introduced an improved adaptive plaintext, and presented a new known plaintext trace-driven cache-collision attacks against embedded AES implementations. Their experiments show that with approximately 30 known plaintexts, the key space of AES 128-bit is reduced to 2^{30}. Gullasch et al. [13] improved over prior work [26,21,33] by providing a first practical access-driven cache attack on AES in the asynchronous model. They introduced a novel approach by using neural networks to handle noise surrounding key candidates. Their experiments shows that performing only 100 encryptions is enough to find the key in average of 3 minutes including key search phase. They mentioned a way to transfer the offline phase to another machine by downloading 62.5 KB only per attack.

In 2012, Mowery et al. [19] proved that any cache timing attack against x86 processors that does not somehow subvert the prefetcher, physical indexing, and massive memory requirements of modern programs is doomed to fail.

5   Bernstein's Attack on AES

Bernstein's attack is a first round cache timing attack, it consists of two online phases, namely profiling and attacking phases, and two offline phases, namely correlation and key search phase. During profiling and attacking phases, it measures and collects total execution time of a single encryption, thousands or millions of times. Then a mathematical correlation phase matches between results of the earlier online stages and generates a list of possible key candidates. The last phase is the key search phase, it is a brute force attack, searching for the unknown key in the reduced key space generated by the correlation phase.

The biggest advantage of this attack, it is a generic attack and it can be performed mostly without any knowledge about many processor details.

Bernstein succeeded to attack an OpenSSL implementation of AES, which makes use of four T-tables only, and utilizes a total of four kilobytes (4096 bytes) of memory. The idea is that for the first round, the table lookup indices $x_i^{(0)}$ are each related to only one key byte $k_i^{(0)}$ and one plaintext byte p_i:

$$x_i^{(0)} = p_i \oplus k_i^{(0)}, \, i = 0, 1, 2, ..., 15.$$

So, at the profiling phase we know the i-th key byte $k_i^{(0)}$ and the i-th plaintext byte p_i, which leads directly to the table lookup index $x_i^{(0)}$. On the other hand, at the attacking phase we know both p_i and $x_i^{(0)}$, which reveals the unknown i-th key byte $\hat{k}_i^{(0)}$:

$$\hat{k}_i^{(0)} = p_i \oplus x_i^{(0)}, i = 0, 1, 2, ..., 15.$$

Osvik et al. [26,33] mentioned a lot of important shortcomings about Bernstein's attack:

- it requires reference measurements of encryption under known key in an identical configuration, and these are often not readily available (e.g., a user may be able to write data to an encrypted file system, but creating a reference file system with a known key is a privileged operation).
- it relies on timing the encryption and thus, it seems impractical on many real systems due to excessively low signal-to-noise ratio.
- even when this attack works, it requires high number of analyzed encryptions.

To work around these shortcomings, Neve et al. [23,22] suggested that instead of using another reference machine, attacking two different keys on the same machine might recover some bits by comparing similar byte signatures. Also, we tried to compare between two different attacking stages for the same key, but we failed to get any good results.

However, in some cases like a shared computer with different users accounts, the attacker has access to his own account on the machine in which he can collect required information about his own known key.

6   Our Work

Bernstein in [6] presented his attack against the OpenSSL 0.9.7a of AES implementation on an 850MHz Pentium III desktop computer running FreeBSD 4.8. O'Hanlon and Tonge [24] failed to collect any useful data about the key by attacking a Pentium IV running GCC 4.0.0 and OpenSSL 0.9.7f. They succeeded with a Pentium III running GCC 2.95.3 against the MIRACL [18] implementation of AES. Followed by Canteaut et al., they attacked a Pentium IV processor [11] trying to modify the cache state before the attack by removing system calls. Also recently, Jayasinghe et al. [14] succeeded to implement Bernstein's attack on the original configuration used by Bernstein at 2005. Table 1 outlines all these implementation of the attack.

While through our experiments we tested over 100 random keys, we decided to attack a fixed key $k = \{2b, a8, 62, a3, 4d, 42, e2, 44, 27, 89, a4, 4a, c6, 7e, cd, eb\}$, through the rest of this paper.

After testing the attack on Ubuntu 9.10, OpenSSL version 0.9.8g and GCC 4.4.1 on a Pentium Dual-Core processor as shown in Table (2), we noticed that smaller packet sizes are giving better results. Our best results were achieved by sending 2^{27} plaintexts of size 100 bytes.

---

[1] This attack was performed on a 32-bit Windows7 Ultimate sp1.

[2] This attack was performed on a 64-bit Windows7 Home Premium sp1.

Table 1. This table shows attackers hardware and software configurations

Attacker	CPU model	GCC	AES software	#Packets
Bernstein[6]	Pentium III	2.95.4	OpenSSL 0.9.7a	2^{27}
O'Hanlon et al.[24]	Pentium III	2.95.3	MIRACL	$2^{30}/3$
Canteaut et al.[11]	Pentium III	3.2.2	Original [31]	2^{30}
	Pentium IV	3.2.2	Original [31]	2^{26}
Jayasinghe et al.[14]	Pentium III	2.95.4	OpenSSL 0.9.7a	2^{27}
Our attack	Pentium Dual-Core	4.4.1	OpenSSL 0.9.8g	2^{26}
	Pentium Dual-Core[1]	4.7.0	OpenSSL 1.0.1c	2^{20}
	Pentium Core 2 Duo[2]	4.7.0	OpenSSL 1.0.1c	2^{20}

Table 2. This table shows the attacked processor model and cache size

CPU model	Level	Cache line size	Cache sets	Associativity	Total size
Pentium Dual-Core T2060	L1	64 B	2×64	8	2×32 KB
	L2	64 B	4096	4	1 MB
Pentium Core 2 Duo P7550	L1	64 B	2×64	8	2×32 KB
	L2	64 B	4096	12	3 MB

After testing the attack on a similar configuration to Bernstein's, we started porting the attack to test on Windows7 32-bit and 64-bit operating systems. We chose to port the attack to MinGW, not Cygwin, so the attack is not restricted to machines running Cygwin only. Porting the attack to MinGW was a hard job, due to the differences between sockets implemented in Linux and Windows.

Performing the first attack on Windows 32-bit, OpenSSL version 0.9.8g and GCC 4.7.0 on a Pentium Dual-Core processor as shown in Table (2) was successful, and some key candidates appeared after sending 2^{26} samples of size 100. After this we followed Neve et al. [22] by removing the network delay from the attack, since the execution time is measured on the server. We merged the original study and server programs to create the new ServerNoNetwork program which runs locally. Also, we fixed the samples length to 16 as Neve et al. advised, since only the first 16 bytes are encrypted even if the packet size is 1000.

Performing the attack again with the new attack program on our testing platforms lead to better results in less time. Our best results began to appear at 2^{25} samples. This attack took less than 30 minutes and the results included 4 accurate peaks for k_1, k_5, k_9, k_{13}. By repeating the attack several times with the same configuration, the same 4 peaks kept appearing and no other candidates were observed.

Since the main concept used by Bernstein depends on touching the cache memory before and after AES encryption, we restored back the ability to send different sizes of samples again. At the first glance, we tried to speedup the attack by removing unused code like calling rand() function, which fills the whole packet array while we need only the first 16 entries to be randomized, also we added a command line argument for the limit to stop whenever it reaches that limit.

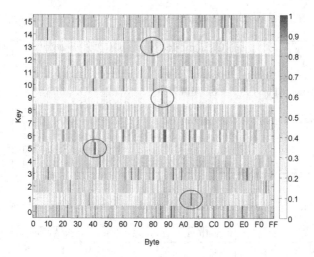

Fig. 1. Single peaks for group $g_1 = \{a8, 42, 89, 7e\}$ are marked with red. x-axis represents byte values in hexadecimal and y-axis represents the 16 key bytes

By repeating the attack ten times, we discovered that results are mostly appearing as single peaks on this configuration see Fig. 1. Another interesting behavior is the relation between results of the same attack, it appears that results are grouped in 4 groups, g_0, g_1, g_2, and g_3, each group contains 4 key candidates: $g_i = \{k_i, k_{i+4}, k_{i+8}, k_{i+12}\}, i = 0, 1, 2, 3..$ The first group appeared while sending samples of size 16 was g_1, other groups appeared later with larger samples.

After improving and optimizing the speed of the merged program, a successful attack on Windows7, GCC 4.71, and OpenSSL 1.0.1c, required sending only 1M of samples with different sizes, and we recovered the full key without a brute force attack in less than 20 seconds. We didn't need to use the brute force attack, since all key candidates were found as single peaks.

The reason behind this grouping is how OpenSSL implemented AES and how the MinGW GCC compiler assembles the implementation on Windows. Every group corresponds to a set of certain table lookup indices, i.e g_0 corresponds to T_0 lookup indices and so on, recalling that:

$$\left(x_{4i}^{(r)}, x_{4i+1}^{(r)}, x_{4i+2}^{(r)}, x_{4i+3}^{(r)}\right) = T_0\left[x_{4i}^{(r-1)}\right] \oplus T_1\left[x_{4i+5}^{(r-1)}\right] \oplus T_2\left[x_{4i+10}^{(r-1)}\right] \oplus T_3\left[x_{4i+15}^{(r-1)}\right] \oplus \left(k_{4i}^{(r-1)}, k_{4i+1}^{(r-1)}, k_{4i+2}^{(r-1)}, k_{4i+3}^{(r-1)}\right)$$

where $i = 0, 1, 2, 3$ and all sum operations are done modulo 16.

While analyzing the results in Table 3, we found that peaks are changing approximately every extra 300-350 bytes to the sample. It appears that the attacked operating system and/or MinGW GCC compiler are partitioning the T-tables into 256 byte chunks (4 cache lines per table), and due to misalignment of the tables in cache, an extra cache line is used which means $256 + 64 = 320$ bytes in cache.

Following this theory, we found that 320, 640, 960, and 1280 bytes are very special sample sizes that reveals more than 8 key candidates at once in some

Table 3. This table shows the relation between revealed groups and sample size by sending only 2^{20} samples

Sample Size in Bytes	Recovered Key Indices	Group	Time in Seconds	Improved Time
16	$1, 5, 9, 13$	g_1	1.6	0.9
100-300	$5, 9, 13$	g_1	2-10	1.1-1.5
350	$0, 4, 8, 12$	g_0	12	1.6
400-600	$0, 4, 8$	g_0	13-19	1.7-2.0
650	$3, 7, 11, 15$	g_3	21	2.1
700-950	$7, 11, 15$	g_3	22.5-30	2.2-2.7
1000	$2, 6, 10, 14$	g_2	32.5	2.8

cases. Figure 2 shows the evolution of the correct key candidate against the number of samples, the figure shows the results of measuring 1M samples of size 320 Byte, as you can notice, the correct key candidates were very clear at the level of 256K and in some cases at 64K samples only.

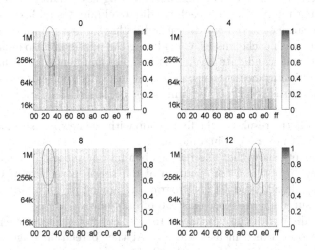

Fig. 2. Single peaks for group $g_0 = \{2b, 4d, 27, c6\}$ are marked with red. In this grouped correlation graph, the x-axis represents the byte value and the y-axis represents number of samples.

The last part of our work is the most exciting part. We tried to modify the core of the attack by replacing the main measurement criteria by another one. The core measurement criteria in the attack is $u[j][b] - taverage$ where

$$taverage := \frac{\sum_{i=0}^{total\#packets} timing_i}{total\#packets}$$

is the total overall average time and

$$u[j][b] := \frac{\sum\limits_{i=0}^{tnum[j][b]} timing[j][b]_i}{tnum[j][b]},$$

$tnum[j][b]$ is the total number of samples for the j-th key candidate and the plaintext byte b. We aimed to replace this complicated formula with a simpler one to help gain more speed and accuracy in our attack program, our formula is $umin[j][b] - tmin$ where

$$tmin := min_i(timing_i)$$

is the overall minimum timing and

$$umin[j][b] := min_i(timing[j][b]_i)$$

is the local minimum for the j-th key candidate and the plaintext byte b.

We were not surprised when this new measure succeeded to recover the same groups as the original measure with the same configurations, since the minimum is used in access-driven attacks to eliminate wrong key candidates. Calculating minimum timing information for each candidate eliminates all low value candidates easily because candidates with cache hits have a lower timing information than candidates with cache misses. Calculating the minimum also eliminates the noise; if a particular candidate is affected by noise and hence had a high timing value, using minimum means that at the first instant of noise absence, the real timing information is recorded and will never be raised again even if the noise is back.

Figure 3 shows the results of the first group with the new measurement, which looks more clear than the original measure.

Our experiments show that the reasons behind our success in recovering the full 128-bit key so fast is a combination of the simple structure of GCC 4.7.0 for MinGW, optimizing the attack program, removing all redundant code, and using two measurement criterions instead of only one.

After succeeding with the new measure, we kept both measurements, so the attacker can choose which measure to use. At the correlation stage, both measurements data are kept together in the file for double checking.

At this point, using the merged program ServerNoNetwork, with the original and new measurements, we succeeded to recover the full AES key, without a brute force search in less than 20 seconds, using less than 1M random plaintexts.

7   Future Work

After attacking OpenSSL successfully we plan to attack other cryptographic libraries. Our first trial was to attack MIRACL version 5.5.4 [18] which implements AES in a small slow implementation and another high performance one, using 5 lookup T-tables: 4 for all rounds and one for the final round. Attacking this implementation failed to extract more than 2 or 3 key bytes for the small implementation, while we succeeded to get a better chance with the fast one.

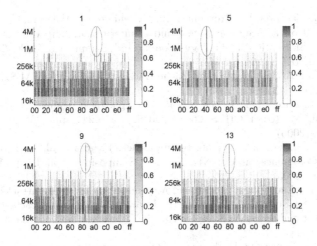

Fig. 3. Single peaks for group $g_1 = \{a8, 42, 89, 7e\}$ are marked with red. In this grouped correlation graph, the x-axis represents the byte value and the y-axis represents number of samples.

8 Conclusion

We succeeded to attack the latest OpenSSL implementation of AES using Bernstein's cache timing attack on a different testing environment from those used earlier. We replaced the original "above average" measure with a simple "above minimum" one. Our experiments shows that GCC 4.7.0 for MinGW might be the reason behind recovering the 128-bit key in seconds using either the original or new measurements.

References

1. Acıiçmez, O., Koç, Ç.: Trace-driven cache attacks on AES (short paper). Information and Communications Security, 112–121 (2006)
2. Acıiçmez, O., Koç, K.: Microarchitectural attacks and countermeasures. Cryptographic Engineering, 475–504 (2009)
3. Acıiçmez, O., Schindler, W., Koç, Ç.K.: Cache based remote timing attack on the AES. In: Abe, M. (ed.) CT-RSA 2007. LNCS, vol. 4377, pp. 271–286. Springer, Heidelberg (2006)
4. Acıiçmez, O., Schindler, W., Koç, Ç.: Improving Brumley and Boneh timing attack on unprotected SSL implementations. In: Proceedings of the 12th ACM Conference on Computer and Communications Security, pp. 139–146. ACM (2005)
5. Bar-El, H.: Introduction to side channel attacks, vol. 43. Discretix Technologies Ltd. (2003)
6. Bernstein, D.: Cache-timing attacks on AES (2005),
   http://cr.yp.to/antiforgery/cachetiming-20050414.pdf

7. Bertoni, G., Zaccaria, V., Breveglieri, L., Monchiero, M., Palermo, G.: AES power attack based on induced cache miss and countermeasure. In: International Conference on Information Technology: Coding and Computing, ITCC 2005, vol. 1, pp. 586–591. IEEE (2005)
8. Bonneau, J.: Robust final-round cache-trace attacks against AES. Tech. rep., Citeseer (2006)
9. Bonneau, J., Mironov, I.: Cache-collision timing attacks against AES. In: Goubin, L., Matsui, M. (eds.) CHES 2006. LNCS, vol. 4249, pp. 201–215. Springer, Heidelberg (2006)
10. Brumley, D., Boneh, D.: Remote timing attacks are practical. In: Proceedings of the 12th Conference on USENIX Security Symposium, vol. 12, p. 1. USENIX Association (2003)
11. Canteaut, A., Lauradoux, C., Seznec, A.: Understanding cache attacks (2006)
12. Gallais, J., Kizhvatov, I., Tunstall, M.: Improved trace-driven cache-collision attacks against embedded AES implementations. Information Security Applications, 243–257 (2011)
13. Gullasch, D., Bangerter, E., Krenn, S.: Cache games–bringing access-based cache attacks on AES to practice. In: 2011 IEEE Symposium on Security and Privacy (SP), pp. 490–505. IEEE (2011)
14. Jayasinghe, D., Fernando, J., Herath, R., Ragel, R.: Remote cache timing attack on Advanced Encryption Standard and countermeasures. In: 2010 5th International Conference on Information and Automation for Sustainability (ICIAFs), pp. 177–182. IEEE (2010)
15. Kelsey, J., Schneier, B., Wagner, D., Hall, C.: Side channel cryptanalysis of product ciphers. In: Quisquater, J.-J., Deswarte, Y., Meadows, C., Gollmann, D. (eds.) ESORICS 1998. LNCS, vol. 1485, pp. 97–110. Springer, Heidelberg (1998)
16. Kocher, P.C.: Timing attacks on implementations of diffie-hellman, RSA, DSS, and other systems. In: Koblitz, N. (ed.) CRYPTO 1996. LNCS, vol. 1109, pp. 104–113. Springer, Heidelberg (1996)
17. Lauradoux, C.: Collision attacks on processors with cache and countermeasures. In: Western European Workshop on Research in Cryptology WEWoRC, vol. 5, pp. 76–85 (2005)
18. MIRACL: Multiprecision Integer and Rational Arithmetic C/C++ Library. Shamus Software Ltd., Dublin, http://www.shamus.ie
19. Mowery, K., Keelveedhi, S., Shacham, H.: Are AES x86 cache timing attacks still feasible? In: Proceedings of the 2012 ACM Workshop on Cloud Computing Security Workshop, pp. 19–24. ACM (2012)
20. Nechvatal, J., Barker, E., Bassham, L., Burr, W., Dworkin, M., Foti, J., Roback, E.: Report on the development of the Advanced Encryption Standard (AES). Journal of Research of the National Institute of Standards and Technology 106(3) (2001), http://archive.org/details/jresv106n3p511
21. Neve, M., Seifert, J.-P.: Advances on access-driven cache attacks on AES. In: Biham, E., Youssef, A.M. (eds.) SAC 2006. LNCS, vol. 4356, pp. 147–162. Springer, Heidelberg (2007)
22. Neve, M., Seifert, J., Wang, Z.: Cache time-behavior analysis on AES. In: Selected Area of Cryptology (2006)
23. Neve, M., Seifert, J., Wang, Z.: A refined look at Bernstein's AES side-channel analysis. In: Proceedings of the 2006 ACM Symposium on Information, Computer and Communications security. pp. 369–369. ACM (2006)
24. O'Hanlon, M., Tonge, A.: Investigation of cache timing attacks on AES. School of Computing, Dublin City University (2005)

25. OpenSSL: The open source toolkit for SSL/TLS, http://www.openssl.org
26. Osvik, D.A., Shamir, A., Tromer, E.: Cache attacks and countermeasures: The case of AES. In: Pointcheval, D. (ed.) CT-RSA 2006. LNCS, vol. 3860, pp. 1–20. Springer, Heidelberg (2006)
27. Page, D.: Theoretical use of cache memory as a cryptanalytic side-channel. Tech. rep., Citeseer (2002)
28. Page, D.: Defending against cache-based side-channel attacks. Information Security Technical Report 8(1), 30–44 (2003)
29. Percival, C.: Cache missing for fun and profit. In: BSDCan 2005 (2005)
30. Rebeiro, C., Mondal, M., Mukhopadhyay, D.: Pinpointing cache timing attacks on AES. In: 23rd International Conference on VLSI Design, VLSID 2010, pp. 306–311. IEEE (2010)
31. Rijmen, V., Bosselaers, A., Barreto, P.: Optimised ANSI C code for the Rijndael cipher (now AES). Public domain software (2000), http://fastcrypto.org/front/misc/rijndael-alg-fst.c
32. Tiri, K., Acıiçmez, O., Neve, M., Andersen, F.: An analytical model for time-driven cache attacks. In: Biryukov, A. (ed.) FSE 2007. LNCS, vol. 4593, pp. 399–413. Springer, Heidelberg (2007)
33. Tromer, E., Osvik, D., Shamir, A.: Efficient cache attacks on AES, and countermeasures. Journal of Cryptology 23(1), 37–71 (2009)
34. Tsunoo, Y.: Cryptanalysis of block ciphers implemented on computers with cache. In: Preproceedings of ISITA 2002 (2002)
35. Tsunoo, Y., Saito, T., Suzaki, T., Shigeri, M., Miyauchi, H.: Cryptanalysis of DES implemented on computers with cache. In: Walter, C.D., Koç, Ç.K., Paar, C. (eds.) CHES 2003. LNCS, vol. 2779, pp. 62–76. Springer, Heidelberg (2003)
36. Zhao, X., Wang, T.: Improved cache trace attack on AES and CLEFIA by considering cache miss and S-box misalignment. Tech. rep., Cryptology ePrint Archive, Report 2010/056 (2010)
37. Zhao, X., Wang, T., Dong, M., Yuanyuan, Z., Zhaoyang, L.: Robust first two rounds access driven cache timing attack on AES. In: 2008 International Conference on Computer Science and Software Engineering, vol. 3, pp. 785–788. IEEE (2008)

Optimal Public Key Traitor Tracing Scheme in Non-Black Box Model

Philippe Guillot[1], Abdelkrim Nimour[2], Duong Hieu Phan[1], and Viet Cuong Trinh[1]

[1] Université Paris 8, LAGA, CNRS, (UMR 7539), Université Paris 13, Sorbonne Paris Cité 2 rue de la liberté, 93526 Saint-Denis, Cedex
[2] Canal+, Paris, France

Abstract. In the context of secure content distribution, the content is encrypted and then broadcasted in a public channel, each legitimate user is provided a decoder and a secret key for decrypting the received signals. One of the main threat for such a system is that the decoder can be cloned and then sold out with the pirate secret keys. Traitor tracing allows the authority to identify the malicious users (are then called traitors) who successfully collude to build pirate decoders and pirate secret keys. This primitive is introduced by Chor, Fiat and Naor in '94 and a breakthrough in construction is given by Boneh and Franklin at Crypto '99 in which they consider three models of traitor tracing: *non-black-box tracing* model, *single-key black box tracing* model, and *general black box tracing* model.

Beside the most important open problem of obtimizing the black-box tracing, Boneh-Franklin also left an open problem concerning non-black-box tracing, by mentioning: "it seems reasonable to believe that there exists an efficient public key traitor tracing scheme that is completely collusion resistant. In such a scheme, any number of private keys cannot be combined to form a new key. Similarly, the complexity of encryption and decryption is independent of the size of the coalition under the pirate's control. An efficient construction for such a scheme will provide a useful solution to the public key traitor tracing problem".

As far as we know, this problem is still open. In this paper, we resolve this question in the affirmative way, by constructing a very efficient scheme with all parameters are of constant size and in which the full collusion of traitors cannot produce a new key. Our proposed scheme is moreover dynamic.

Keywords: traitor tracing, non-black-box tracing, full collusion, pairings.

1 Introduction

Traitor tracing, introduced in [12], is an important cryptographic primitive in the context of secure content distribution. Traitor tracing is a main ingredient in many practical applications of global networking such as pay-per-view television,

A. Youssef, A. Nitaj, A.E. Hassanien (Eds.): AFRICACRYPT 2013, LNCS 7918, pp. 140–155, 2013.

satellite transmission. In secure content distribution, the content is encrypted and broadcasted in a public channel, each legitimate user is provided a decoder and a secret key for decrypting the received signals. The main threat in this context is that the decoder can be cloned or be produced and then sold out with the pirate secret keys. Traitor tracing allows the authority to identify the malicious users (are then called traitors) who successfully collude to build pirate decoders and pirate secret keys.

A breakthrough was proposed by Boneh-Franklin in [6] in which an efficient public key traitor tracing scheme was introduced. They considered three following tracing models:

1. *Non-black-box tracing* model considers the situation where the collusion of t traitors can derive a new valid secret key. The tracing algorithm takes as inputs this new valid secret key and outputs at least a traitor in the collusion.
2. *Single-key black box tracing* model extends a bit the non-black-box tracing model. It always considers the scenario that the collusion of t traitors can derive a new valid secret key and then this new valid secret key is embedded in a pirate decoder. The tracing algorithm takes as inputs the pirate decoder and should be able to output the identity of one of the traitors.
3. *General black box tracing* model is the strongest model of tracing in which the tracer cannot open the pirate decoder and only interact with it in a black box manner by sending the ciphertext and observing the output of the pirate decoder. It is required that whenever the pirate can decrypt the ciphertext, the tracer should be able to trace back one of the traitors.

1.1   Non-Black-Box Tracing vs. General Black Box Tracing

The *general black box tracing* is evidently the most desired model as it covers all the possible strategies of the pirate. However, all the schemes in this model are still quite impractical. The most efficient black box traitor tracing are code based schemes [15,2,8,17]. However, the main weakness of code based schemes is that the user's secret key is long (at least $O(t^2 \log N)$ where t, N are the number of traitors and of users in the system) and thus it cannot be highly protected as one cannot put a long key in a tamper-resistant memory in a smart-card. Moreover, the leakage of some small part of the key can be efficiently used in the attack as shown in Pirates 2.0 [3]. Therefore, these schemes are still far to be applicable in practice. Algebraic schemes achieve the *general black box tracing* [6,16,9,10] in inefficient ways: either the tracing algorithm is of exponential time complexity [6,16], or the ciphertext size is still large (*i.e.*, $O(\sqrt{N})$) and the constructions make use of bilinear maps in groups of composite order [9,10]. These two last schemes are very interesting in the sense that they can deal with full collusion.

While it seems a very difficult and challenging problem to achieve a practical *general black box tracing*, it's of practical interest in considering the weaker models of the non-black-box tracing and the single-key black box tracing. Moreover, these models are also very practical, there are many scenarios that these models are suitable, as also discussed in [19,14].

Let us explain some details in the context of pay-TV. In the majority of the existing systems, each user has been provided a Set-Top box (STB) and a smartcard. The secret key of the user is stored in the smartcard which has the role of decrypting the session key for every crypto-period (between 2 and 10 seconds), this session key is then transmitted to the STB for decrypting the content. The pirate always wants to minimize the cost of distribution of his solution and in practice, he really wants to try to produce new pirate smartcard to be used in already deployed STBs. It is thus necessary that these pirate smartcard are compatible with the STBs in the fields (including the legitimate STBs). As a consequence, the smartcard should preserve the functionality of the legitimate smartcard and it has to embed a pirate but valid key in the memory. It is often in reality that the authority can reverse this key in the memory of the pirate smartcard and the scenario exactly falls in the non-black-box tracing model. Even if the tracer cannot reserve the memory of the pirate smartcard, we argue that the single-key pirate tracing model is suitable. Indeed, in the modern CAS (Conditional Access Systems), the session key is delivered at the last moment so that there is only a small delay between the time the smartcard decrypts the session key and the time the decoder receive the encrypted content. Therefore, if the pirate card (which is evidently cannot more performant than a legitimate smartcard) always try to decrypt the session key with different, say two, keys, it will fails to decrypt the content in time and will give the STB the session key after the encrypted content arrive for that crypto-period. One could wonder what happens if the pirate decoder only try to detect the presence of tracing algorithm from time to time. Fortunately, the single-key black box tracing algorithms, as in Boneh-Franklin schemes and in our scheme, only need to ask just one query and the decoder is resettable in practice, this strategy of pirate does not work. All in all, we would like to argue that the non-black-box tracing model and the single-key black box tracing model, though much weaker than the general black box tracing model and cannot thus cover all the strategies of the pirate, are still very practical. In fact, there are quite a lot of interesting works that only concentrate on these models, namely [19,14,1].

In a theoretical point of view, it's also a very interesting problem to consider non-black-box tracing because there is still no optimal solution, far from that, in spite of many efforts. Indeed, the Boneh-Fraklin is efficient with respect to the non-black-box tracing and single-key black box tracing but its ciphertext size is still linear in the number of traitors. The Tonien-Safavi scheme [19] and the Junod-Karlov-Lenstra sheme [14] managed to improve the tracing algorithm but the ciphertext size is always linear in the number of traitors. A side effect of this high ciphertext size in the number of traitors is that these schemes cannot be used with full collusion because in this later case, these schemes are worse than the trivial scheme of assigning each user an independent key. Agrawal *et. al.* [1] go one step further by achieving an intermediate level between bounded tracing (when one assumes a maxmixum t number of traitors) and full collusion: they allow the pirate to collect up to t keys and get some bounded partial information about the others keys. We notice that the authors in [1] only considers the non-black-box tracing model and therefore a full collusion resistant scheme in the non-black-box tracing model satisfies immediately their security notion proposed. All in all, there is still an important gap between the efficiency of all these schemes and an optimal solution: the ciphertext size depends on the number of traitors and none of them can deal with full collusion. Our objective is to close this gap.

1.2 Our Contributions

We consider the *non-black-box tracing* and the *single-key black box tracing* models for which we propose an optimal scheme in the sense that all the parameters including private key size, public key size, ciphertext size, encryption and decryption time complexity are constant. In addition, our scheme also achieves two interesting properties of a public key traitor tracing scheme: it is fully collusion resistant and dynamic where there is no need to update any parameter when a user joins the system. We also highly improve the time complexity in tracing algorithms, in particular we achieve $O(1)-$time non-black-box tracing. Regarding the single-key black box tracing, we consider both the full access model (where the decoder pirate has to return the correct message for any valid ciphertext) and the minimal access model (where the pirate decoder only needs to return a single bit signifying whether the ciphertext is valid or not). We then design a $O(\log N)-$time full access single-key black box tracing and a $O(N)-$time minimal access single-key black box tracing.

The detailed comparison between our scheme and other schemes is given in the full version of this paper [18]. We notice that our scheme is the only scheme that allows minimal access single-key black box tracing.

The main weakness in our scheme is that the security for the tracing problem is based on a type of q-assumption. However, we notice that these types of assumptions have been widely used in security proofs, for example in [4,5], [7,11]. We also prove that the proposed assumptions hold in the generic group.

2 Preliminaries

2.1 Traitor Tracing Scheme

We refine the definition of a non-black-box public key traitor tracing scheme from [6]. Formally, a non-black-box public key traitor tracing encryption scheme is made up of the following algorithms:

Setup(λ): Takes as input the security parameter λ, it returns a master key msk and a public key mpk.

Joint(i, msk): Takes as inputs a user's index i, together with the master key, and outputs a user's secret key sk_i.

Encrypt(M, mpk): Takes as inputs a message M, together with the public key, and outputs a ciphertext C.

Decrypt(sk_i, mpk, C): Takes as inputs a secret key sk_i, public key, and a ciphertext C, outputs the corresponding message M.

Trace$(\mathcal{D}, sk^*, \text{tracing} - \text{key}) \rightarrow i$: Takes as input the public key mpk, the tracing $-$ key, a pirate decoder $\mathcal{D}$ and some valid secret key sk^* embedded in $\mathcal{D}$ and outputs an index i corresponding to an accused traitor.

When the knowledge of the tracer about the pirate decoder is more restricted, one can get the stronger following notions, which were discussed in [6]:

- in the single-key black box tracing model, the tracing algorithm only takes as inputs the public key mpk, the tracing $-$ key, and interact with a pirate decoder $\mathcal{D}$ with the assumption that the pirate decoder only embed a single valid key sk^*.
- in the general black box tracing model, there is no any assumption on the pirate decoder and the tracer can only interact with it. It is however required that $\mathcal{D}$ can decrypt the well-form ciphertexts with a non-negligible probability because otherwise the pirate decoder is useless.

For correctness, we require that for all $i \in \mathbb{N}$, if $(\text{msk}, \text{mpk}) \leftarrow \textbf{Setup}(\lambda)$, $sk_i \leftarrow \textbf{Joint}(i, \text{msk})$ and $C \leftarrow \textbf{Encrypt}(M, \text{mpk})$ then one should get $M = \textbf{Decrypt}(sk_i, C)$.

The security of the scheme is defined in terms of two properties: semantic security and tracing security.

Semantic security. The standard notion of semantic security requires that, for any PPT $\mathcal{A}$, we have $|\Pr[\mathcal{A}\ wins] - 1/2|$ is negligible in the following game:

- In the setup phase, the challenger runs $\textbf{Setup}(\lambda)$ algorithm to get a master key msk and a public key mpk. It then gives mpk to $\mathcal{A}$.
- In the challenge phase, $\mathcal{A}$ outputs two messages M_0, M_1. The challenger then chooses a bit $b \in \{0, 1\}$ at random, sets $C \leftarrow \textbf{Encrypt}(M_b, \text{mpk})$, and gives C to $\mathcal{A}$.
- In the guess phase, the attacker $\mathcal{A}$ outputs a bit b'. We say $\mathcal{A}$ wins if $b' = b$.

non-black-box tracing security. We say that a secret key sk is a valid secret key iff there exists some message M in message-domain such that if $C = \textbf{Encrypt}(M, \text{mpk})$ then one should get $M = \textbf{Decrypt}(sk, C)$ with probability at least $\frac{1}{2}$.

We say that non-black-box tracing security holds if, for any PPT $\mathcal{A}$, we have $|\Pr[challenger\ wins] - 1/2|$ is considerable in the following game:

- In the setup phase, the challenger runs $\textbf{Setup}(\lambda)$ algorithm to get a master key msk and a public key mpk. It then gives mpk to $\mathcal{A}$.
- In the query phase, $\mathcal{A}$ may adaptively ask corrupt query for user index i and gets sk_i.
- At some point $\mathcal{A}$ outputs some sk^* and a pirate decoder $\mathcal{D}$ in which sk^* is embedded in. The challenger then runs $\textbf{Trace}(\mathcal{D}, sk^*, \text{tracing} - \text{key}) \rightarrow i$. We say that the challenger wins if the secret key sk^* is a valid secret key and the traced index i is in the set of corrupted indexes.

In the single-key black box tracing security, $\mathcal{A}$ only outputs a decoder $\mathcal{D}$ in which only sk^*, mpk are embedded in it. In the general black box tracing security, $\mathcal{A}$ only outputs a decoder $\mathcal{D}$ with a requirement that $\mathcal{D}$ can decrypt the well-form ciphertexts with a non-negligible probability because otherwise the pirate decoder is useless.

Full access black box tracing vs Minimal access black box tracing. These two types of models are discussed in [6].

1. In the full access black box tracing model, the tracer can query the pirate decoder on a ciphertext C, if C is a well-form ciphertext, he will always receive the corresponding plantext M. Otherwise, the pirate decoder can return an arbitrary output (it can return a signal indicating that the ciphertext C is invalid or can maliciously choose a random message M' and return M').
2. In the minimal access black box tracing model, the tracer queries the pirate decoder on a pair (C, M) and only receives a signal: *valid* if the ciphertex C is a valid encryption of M, *invalid* if not.

Dynamic public key traitor tracing scheme. We adapt the definition of a dynamic broadcast encryption in [13] for a public key traitor tracing scheme, note that our definition is in the *strongest* sense because it requires no any update in the parameters of the systems. Indeed:

1. the system setup as well as the ciphertext size are fully independent from the number of users in the system. The number of users in the system is flexible,
2. a new user can join the system at anytime without implying a modification of preexisting user decryption keys and of the encryption key.

2.2   Bilinear Maps

Our scheme employs bilinear maps and related assumptions, which we now recall.

Let $\mathbb{G}$ and $\mathbb{G}_T$ denote two finite multiplicative abelian groups of large prime order $p > 2^\lambda$ where λ is the security parameter. Let g be a generator of $\mathbb{G}$. We assume that there exists an admissible bilinear map $e : \mathbb{G} \times \mathbb{G} \to \mathbb{G}_T$, meaning that for all $a, b \in \mathbb{Z}_p$

(1) $e(g^a, g^b) = e(g, g)^{ab}$,
(2) $e(g^a, g^b) = 1$ iff $a = 0$ or $b = 0$,
(3) $e(g^a, g^b)$ is efficiently computable.

$(p, \mathbb{G}, \mathbb{G}_T, e(\cdot, \cdot))$ is then called a bilinear map group system. We now recall the generalization of the Diffie-Hellman exponent assumption in [5] on bilinear map group system.

Let $(p, \mathbb{G}, \mathbb{G}_T, e(\cdot, \cdot))$ a bilinear map group system and $g \in \mathbb{G}$ be a generator of $\mathbb{G}$, and set $g_T = e(g, g) \in \mathbb{G}_T$. Let s, n be positive integers and $P, Q \in \mathbb{F}_p[X_1, \ldots, X_n]^s$ be two s-tuples of n-variate polynomials over $\mathbb{F}_p$. Thus, P and

Q are just two lists containing s multivariate polynomials each. We write $P = (p_1, p_2, \ldots, p_s)$ and $Q = (q_1, q_2, \ldots, q_s)$ and impose that $p_1 = q_1 = 1$. For any function $h : \mathbb{F}_p \to \Omega$ and vector $(x_1, \ldots, x_n) \in \mathbb{F}_p^n$, $h(P(x_1, \ldots, x_n))$ stands for $(h(p_1(x_1, \ldots, x_n)), \ldots, h(p_s(x_1, \ldots, x_n))) \in \Omega^s$. We use a similar notation for the s-tuple Q. Let $f \in \mathbb{F}_p[X_1, \ldots, X_n]$. It is said that f depends on (P, Q), which denotes $f \in \langle P, Q \rangle$, when there exists a linear decomposition

$$f = \sum_{1 \leq i,j \leq s} a_{i,j} \cdot p_i \cdot p_j + \sum_{1 \leq i \leq s} b_i \cdot q_i, \qquad a_{i,j}, b_i \in \mathbb{Z}_p$$

Let P, Q be as above and $f \in \mathbb{F}_p[X_1, \ldots, X_n]$. The $(P, Q, f) - \text{GDDHE}$ problem is defined as follows.

Definition 1. $((P, Q, f) - \text{GDDHE})$ [5].
Given $H(x_1, \ldots, x_n) \in \mathbb{G}^s \times \mathbb{G}_T^s$ as above and $T \in \mathbb{G}_T$ decide whether $T = g_T^{f(x_1, \ldots, x_n)}$.

The $(P, Q, f) - \text{GDDHE}$ assumption says that it is hard to solve the $(P, Q, f) - \text{GDDHE}$ problem if f is independent of (P, Q). In this paper, we will prove our scheme is semantically secure under this assumption.

3   Construction

Let $(p, \mathbb{G}, \mathbb{G}_T, e(\cdot, \cdot))$ a bilinear map group system and $g \in \mathbb{G}$ be a generator of $\mathbb{G}$, our scheme is constructed as follows:

Setup(λ). The algorithm chooses $e_1, e_2, v \xleftarrow{\$} \mathbb{Z}_p$ then sets $d_1 = e_1^{-1}, d_2 = e_2^{-1}$
The master key msk is (e_1, e_2, v). The system public keys mpk is:

$$(g^{d_1}, e(g, g)^{d_2}, g^{d_1 \cdot d_2}, e(g, g), e(g, g)^v, e(g, g)^{d_2 \cdot v})$$

Joint(i, msk). For each user i chooses $a_i \xleftarrow{\$} \mathbb{Z}_p$ such that $a_i \neq -1, -v, d_2 - 1$. The secret key for user i is set as: $A_i = g^{e_1(a_i + v)}, B_i = \frac{1}{(a_i + 1)} - e_2$. We call the secret keys in the case $a_i = -v$ or $a_i = d_2 - 1$ are special keys. The users in the system can be assigned to all secret keys in the secret key space except these special keys. Note that the special key, in the case $a_i = -v$, is not useful for decryption.

Encrypt(M, mpk). Encryptor picks a random k in $\mathbb{Z}_p$, then computes:

$$C_1 = g^{d_1 \cdot k}, C_2 = e(g, g)^{d_2 \cdot k}, C_3 = g^{d_1 \cdot d_2 \cdot k},$$

$$C_4 = e(g, g)^{k \cdot v}, C_5 = e(g, g)^{d_2 \cdot k \cdot v}, C_6 = e(g, g)^{-k} \cdot M$$

Finally, outputs $C = (C_1, C_2, C_3, C_4, C_5, C_6)$.

Decrypt(A_i, B_i, C). User $i'th$ first computes:

$$\frac{e(A_i, C_1)}{C_4} \cdot C_2^{B_i-1} \cdot \left(\frac{e(A_i, C_3)}{C_5}\right)^{B_i} = \frac{e(g^{e_1(a_i+v)}, g^{d_1 \cdot k})}{e(g,g)^{k \cdot v}} \cdot e(g,g)^{d_2 \cdot k \cdot (\frac{1}{(a_i+1)} - e_2 - 1)}.$$

$$\cdot \left(\frac{e(g^{e_1(a_i+v)}, g^{d_1 \cdot d_2 \cdot k})}{e(g,g)^{d_2 \cdot k \cdot v}}\right)^{\frac{1}{(a_i+1)} - e_2} =$$

$$= e(g,g)^{k \cdot a_i} \cdot e(g,g)^{\frac{d_2 \cdot k}{(a_i+1)}} \cdot e(g,g)^{-k} \cdot e(g,g)^{-d_2 \cdot k} \cdot e(g,g)^{k \cdot a_i \cdot (\frac{d_2}{(a_i+1)} - 1)} = e(g,g)^{-k}.$$

then outputs $M = C_6/e(g,g)^{-k}$.

Intuition about our construction. In the decryption, we emphasize that the crucial element is $\left(\frac{e(A_i, C_3)}{C_5}\right)^{B_i}$. We remark that, though a pirate can perform a linear combination on the elements A_i in his collected keys, there is no way for the pirate to exploit the combination of his keys to do a linear combination for the elements $\left(\frac{e(A_i, C_3)}{C_5}\right)$ because C_5 is changed for each encryption. Therefore the well-known pirate's strategy of making a linear combination on the collected keys do not work for our scheme. The next section is devoted for formal analysis of security.

4   Security

Definition 2 (GDDHE$_1$ Assumption). *The (t, ε) – GDDHE$_1$ assumption says that for any t-time adversary $\mathcal{A}$ that is given* input $= (g, g^x, g^y, g^{xy}, g^{kx}, g^{ky}, g^{kxy})$ *cannot distinguish between a value $e(g,g)^k \in \mathbb{G}_T$ or a random value $T \in \mathbb{G}_T$, where $x, y, k \in \mathbb{Z}_p, g \in \mathbb{G}$, with advantage greater than ε:*

$$\mathbf{Adv}^{\mathsf{GDDHE_1}}(\mathcal{A}) = \left| \begin{array}{l} \Pr[\mathcal{A}(\mathsf{input}, e(g,g)^k) = 1] \\ - \Pr[\mathcal{A}(\mathsf{input}, T) = 1] \end{array} \right| \leq \epsilon.$$

It is not hard to see that GDDHE$_1$ assumption is a special case of (P, Q, f) – GDDHE assumption. Indeed, we set $P = (p_1 = 1, p_2 = X, p_3 = Y, p_4 = XY, p_5 = KX, p_6 = KY, p_7 = KXY), Q = (q_1 = 1), f = K$. Suppose that f is not independent to $\langle P, Q \rangle$, i.e., one can find $a_8 \neq 0$ such that the following equation holds for all $X, Y, K \in \mathbb{Z}_p$

$$a_8 f = \sum_{1 \leq i,j \leq 7} a_{i,j} \cdot p_i \cdot p_j + b_1 \cdot q_1$$

$$\iff a_8 K = (KX + KY + KXY)(a_1 + a_2 X + a_3 Y + a_4 XY + a_5 KX + a_6 KY + a_7 KXY)$$

$$\iff a_8 = (X + Y + XY)(a_1 + a_2 X + a_3 Y + a_4 XY + a_5 KX + a_6 KY + a_7 KXY)$$

$$\iff (X + Y + XY)(a_1 + a_2 X + a_3 Y + a_4 XY + a_5 KX + a_6 KY + a_7 KXY) - a_8 = 0$$

This implies that the constant term $a_8 = 0$ which is a contradiction with the requirement that $a_8 \neq 0$. Therefore, f is independent to $\langle P, Q \rangle$.

Theorem 1. *Under the GDDHE$_1$ assumption, our scheme is semantically secure.*

Proof. Assume that there exists an adversary $\mathcal{B}$ who is successful in breaking the semantic security of our scheme, we prove that there also exists an adversary $\mathcal{A}$ which attacks the GDDHE$_1$ assumption with the same advantage.

We show that $\mathcal{A}$ can simulate the interaction with $\mathcal{B}$ and then use the output of $\mathcal{B}$ to break the GDDHE$_1$ assumption as follow:

In the setup, $\mathcal{A}$ receives the inputs from his challenger:

$$(g, g^x, g^y, g^{xy}, g^{kx}, g^{ky}, g^{kxy}, T)$$

and needs to distinguish T is either $e(g,g)^k$ or a random value in $\mathbb{G}_T$.

In the next step, $\mathcal{A}$ provides the inputs for $\mathcal{B}$ as follow:

He chooses randomly $z \in \mathbb{Z}_p$, implicitly sets $d_1 = zy, d_2 = x, v = y$, then computes the public key:

$$g^{d_1} = (g^y)^z, e(g,g)^{d_2} = e(g, g^x), g^{d_1 \cdot d_2} = (g^{xy})^z, e(g,g), e(g,g)^v = e(g, g^y),$$

$$e(g,g)^{d_2 \cdot v} = e(g, g^{xy})$$

In the challenge phase, $\mathcal{B}$ outputs two messages M_0 and M_1. $\mathcal{A}$ chooses randomly a bit $b \in \{0,1\}$ then computes the challenge ciphertext as follow:

$$C_1 = (g^{ky})^z = g^{d_1 \cdot k}, C_2 = e(g, g^{kx}) = e(g,g)^{d_2 \cdot k}, C_3 = (g^{kxy})^z = g^{d_1 \cdot d_2 \cdot k},$$
$$C_4 = e(g, g^{ky}) = e(g,g)^{k \cdot v}, C_5 = e(g, g^{kxy}) = e(g,g)^{d_2 \cdot k \cdot v}, C_6 = \frac{1}{T} \cdot M_b$$

then gives it to $\mathcal{B}$.

$\mathcal{B}$ outputs its guess b' for b. If $b' = b$ the algorithm $\mathcal{A}$ outputs 0 (indicating that $T = e(g,g)^k$). Otherwise, it outputs 1 (indicating that T is random in $\mathbb{G}_T$).

As the simulation of $\mathcal{A}$ is perfect, $\mathcal{A}$ can thus break GDDHE$_1$ assumption with the same advantage that $\mathcal{B}$ can break the semantic security.

5 Traitor Tracing

5.1 Non-Black-Box Tracing

Definition 3 (GDDHE$_2$ Assumption). *The* $(t, \varepsilon) - $ GDDHE$_2$ *assumption says that for any t-time adversary $\mathcal{A}$ that is given $(b_1, \ldots, b_l, \mathsf{input})$ in which $b_1, \ldots, b_l$ are random in $\mathbb{Z}_p$ and $\neq 0$,*

$$\mathsf{input} = \left(g^{d_1}, g^{d_1 d_2}, g^{\frac{1}{d_1}}, g^{\frac{d_2 - b_1 d_2 - 1}{d_1(b_1 d_2 + 1)}}, \ldots, g^{\frac{d_2 - b_l d_2 - 1}{d_1(b_l d_2 + 1)}} \right)$$

its probability to output a value $g^{\frac{d_2}{d_1}} \in \mathbb{G}$, where $d_1, d_2 \in \mathbb{Z}_p, g \in \mathbb{G}$, is bounded by ε:

$$\mathbf{Succ}^{\mathsf{GDDHE_2}}(\mathcal{A}) = \Pr[\mathcal{A}(b_1, \ldots, b_l, \mathsf{input}) = g^{\frac{d_2}{d_1}}] \leq \varepsilon.$$

We show that this assumption holds in the generic group, the details can be found in the full version of this paper [18]. Next, we recall the definition of Modified$-l-$SDH assumption from [11].

Definition 4 (Modified−l−SDH Assumption).
Given $g, g^\alpha \in \mathbb{G}$ and $l-1$ pairs $\langle w_j, g^{1/(\alpha+w_j)} \rangle \in \mathbb{Z}_p \times \mathbb{G}$ for a fixed parameter $l \in \mathbb{N}$.
Output another pair $\langle w, g^{1/(\alpha+w)} \rangle \in \mathbb{Z}_p \times \mathbb{G}$.

Theorem 2. *Under the GDDHE$_2$ assumption and Modified−l−SDH assumption, our scheme is secure in the non-black-box tracing model.*

Proof. It is sufficient for us to show that the collusion of any number of traitors cannot derive a new valid secret key. Then, the proof is automatically followed since at least a traitor's key must be embedded in the pirate decoder and when the tracer reverse this key, the identity of the corresponding traitor is revealed.

To prove that the collusion of any number of traitors cannot derive a new valid secret key, we first prove that they cannot derive a special key A, B in which $a = d_2 - 1$, we then prove that they also cannot derive any new valid secret key that differs from this special key.

Lemma 1. *Under the GDDHE$_2$ assumption, the collusion of any number of traitors cannot derive a special key A, B in which $a = d_2 - 1$.*

Proof. Assume that there is an adversary $\mathcal{B}$ which takes as inputs l traitors' keys, for any number l, the system public key, and successfully derive a special key A, B in which $a = d_2 - 1$. We construct an algorithm $\mathcal{A}$ which can simulate the interaction with $\mathcal{B}$ and then use the output of $\mathcal{B}$ to break the GDDHE$_2$ assumption as follow:

In the setup, $\mathcal{A}$ receives the inputs from his challenger:

$$b_1, \ldots, b_l, g^{d_1}, g^{d_1 d_2}, g^{\frac{1}{d_1}}, g^{\frac{d_2 - b_1 d_2 - 1}{d_1(b_1 d_2 + 1)}}, \ldots, g^{\frac{d_2 - b_l d_2 - 1}{d_1(b_l d_2 + 1)}}$$

And needs to output the value $g^{\frac{d_2}{d_1}}$.

In the next step, $\mathcal{A}$ first chooses randomly $v \in \mathbb{Z}_p$, then provides the inputs for $\mathcal{B}$ as follow:

- $\mathcal{A}$ provides a secret key $A_i, B_i, i = 1, \ldots, l$ for $\mathcal{B}$ by setting $B_i = b_i = \frac{1}{a_i + 1} - e_2$, therefore implicitly $a_i = \frac{d_2 - b_i d_2 - 1}{(b_i d_2 + 1)}$, then computes

$$A_i = g^{\frac{d_2 - b_i d_2 - 1}{d_1(b_i d_2 + 1)}} \cdot g^{\frac{v}{d_1}} = g^{\frac{a_i}{d_1}} \cdot g^{\frac{v}{d_1}} = g^{e_1(a_i + v)}$$

  where $e_1 = d_1^{-1}, e_2 = d_2^{-1}$. Note that because b_i, d_1, d_2, v are randomly chosen in $\mathbb{Z}_p$, the resulted secret key is also chosen in the same distribution as in the joint algorithm.
- For the public key, $\mathcal{A}$ computes:

$$g^{d_1}, e(g, g)^{d_2} = e(g^{d_1 d_2}, g^{\frac{1}{d_1}}), g^{d_1 d_2}, e(g, g) = e(g^{d_1}, g^{\frac{1}{d_1}}), e(g, g)^v,$$

$$e(g, g)^{v \cdot d_2} = e(g^{d_1 d_2}, g^{\frac{v}{d_1}})$$

When $\mathcal{B}$ outputs the special secret key A, B in which $a = d_2 - 1$

$$A = g^{e_1(d_2+v-1)}, B = 0$$

then $\mathcal{A}$ outputs

$$\frac{A \cdot g^{\frac{1}{d_1}}}{g^{\frac{v}{d_1}}} = g^{\frac{d_2}{d_1}}$$

As a result, the probability that the collusion of any number of traitors can derive a special key A, B in which $a = d_2 - 1$ is the same as the probability that a t-time adversary $\mathcal{A}$ who breaks the security of the GDDHE$_2$ assumption.

Lemma 2. *Under the* Modified$-l-$SDH *assumption, the collusion of any number of traitors cannot derive any new valid secret key that differs from the special key above.*

Proof. Assume that there is an adversary $\mathcal{B}$ which takes as inputs $l - 2$ traitors' keys, for any number l, the system public key, and successfully derive a new valid secret key which is different from these $l - 2$ traitors' keys and the special key above. We construct an algorithm $\mathcal{A}$ which can simulate the interaction with $\mathcal{B}$ and then use the output of $\mathcal{B}$ to break the Modified$-l-$SDH assumption as follow:

In the setup, $\mathcal{A}$ receives the inputs from his challenger:

$$(w_1, \ldots, w_{l-1}, g, g^\alpha, g^{\frac{1}{\alpha+w_1}}, \ldots, g^{\frac{1}{\alpha+w_{l-1}}})$$

In the next step, $\mathcal{A}$ provides the inputs for $\mathcal{B}$ as follow:

He first chooses randomly $e_1, v \in \mathbb{Z}_p$, then implicitly sets $e_2 = \alpha + w_1$ thus $g^{\frac{1}{\alpha+w_1}} = g^{\frac{1}{e_2}} = g^{d_2}$. $\mathcal{A}$ can easily compute the system public keys and gives them to $\mathcal{B}$.

To compute $A_i, B_i, i = 2, \ldots, l - 1$, $\mathcal{A}$ sets $B_i = \frac{1}{a_i+1} - e_2 = w_i - w_1$ thus $a_i = \frac{1}{e_2+w_i-w_1} - 1$ and

$$A_i = (g^{\frac{1}{\alpha+w_i}})^{e_1} \cdot g^{e_1(v-1)} = g^{e_1(\frac{1}{e_2+w_i-w_1}+v-1)} = g^{e_1(a_i+v)}$$

Note that $\alpha = e_2 - w_1$.

When $\mathcal{B}$ outputs a new secret key

$$A = g^{e_1(a+v)}, B = \frac{1}{(a+1)} - e_2$$

where $a \neq -1, d_2 - 1, a_2, \ldots, a_{l-1}$, then $\mathcal{A}$ outputs $w = B + w_1 = \frac{1}{(a+1)} - e_2 + w_1$ thus $a = \frac{1}{e_2+w-w_1} - 1$, and

$$g^{\frac{1}{\alpha+w}} = (\frac{A}{g^{e_1(v-1)}})^{\frac{1}{e_1}} = \frac{g^{a+v}}{g^{v-1}} = g^{a+1} = g^{\frac{1}{e_2+w-w_1}-1+1} = g^{\frac{1}{e_2+w-w_1}} = g^{\frac{1}{\alpha+w}}$$

Note that $a \neq -1, d_2 - 1, a_2, \ldots, a_{l-1}$ thus $w \neq w_1, \ldots, w_{l-1}$.

As the simulation of $\mathcal{A}$ is perfect, $\mathcal{A}$ can thus break Modified$-l-$SDH assumption with the same advantage that $\mathcal{B}$ can successfully derive a new valid secret key.

5.2 Single-Key Black Box Tracing

Definition 5 (GDDHE$_3$ Assumption). *The $(t, \varepsilon) -$ GDDHE$_3$ assumption says that for any t-time adversary $\mathcal{A}$ that is given a pair (b, input) in which $b \neq 0$ is random in $\mathbb{Z}_p$ and*

$$\mathsf{input} = \left(g, g^{d_1}, g^{d_1 d_2}, g^{\frac{v}{d_1}}, g^{\frac{d_2 - bd_2 - 1}{d_1(bd_2+1)}}, g^{kd_1}, g^{kd_1 d_2}, e(g,g)^{d_2}, e(g,g)^k \right)$$

cannot distinguish between a value $e(g,g)^{kd_2} \in \mathbb{G}_T$ and a random value $T \in \mathbb{G}_T$, where $d_1, d_2, v, k \in \mathbb{Z}_p, g \in \mathbb{G}$, with an advantage greater than ε:

$$\mathbf{Adv}^{\mathsf{GDDHE}_3}(\mathcal{A}) = \left| \begin{matrix} \Pr[\mathcal{A}(b, \mathsf{input}, e(g,g)^{kd_2}) = 1] \\ - \Pr[\mathcal{A}(b, \mathsf{input}, T) = 1] \end{matrix} \right| \leq \epsilon$$

We notice that, unlike the Modified$-l-$SDH assumption, this is a static assumption. We show that this assumption holds in the generic group, the details can be found in the full version of this paper [18].

Theorem 3. *Under the* GDDHE$_3$ *assumption, our scheme is secure in the single-key black box tracing model.*

Proof. We note that in the single-key black box tracing model, there are two separate functions which are called the *key-builder* and the *box-builder*. In the first one, the traitors will collude to derive a new valid secret key. In the second one, one receives this new secret key and build a pirate decoder based on it.

In our proof we first prove that the pirate decoder takes as inputs a secret key and the public key, cannot distinguish a probe ciphertext and a well-form ciphertext, therefore it will run the decryption algorithm normally. Finally, we present a tracing algorithm in which the tracer creates a probe ciphertext and then queries the pirate decoder on this probe ciphertext. After the pirate decoder outputs the answer, the tracer can identify the secret key that pirate decoder is using to decrypt.

Assume that there is a pirates decoder $\mathcal{B}$, on inputs a secret key and the public key, can successfully distinguish a probe ciphertext and a well-form ciphertext. We show that $\mathcal{A}$ can simulate the interaction with $\mathcal{B}$ and then use the output of $\mathcal{B}$ to break the GDDHE$_3$ assumption:

In the setup, $\mathcal{A}$ receives the inputs from his challenger:

$$b, g, g^{d_1}, g^{d_1 d_2}, g^{\frac{v}{d_1}}, g^{\frac{d_2 - bd_2 - 1}{d_1(bd_2+1)}}, g^{kd_1}, g^{kd_1 d_2}, e(g,g)^{d_2}, e(g,g)^k, T$$

with b, d_1, d_2, v, k are randomly chosen in $\mathbb{Z}_p$, and needs to distinguish T is $e(g,g)^{kd_2}$ or not.

In the next step, $\mathcal{A}$ provides the inputs for $\mathcal{B}$ as follow:

- $\mathcal{A}$ provides a secret key for $\mathcal{B}$ by setting $B = b = \frac{1}{a+1} - e_2$, therefore implicitly $a = \frac{d_2 - bd_2 - 1}{(bd_2+1)}$, then computes

$$A = g^{\frac{d_2 - bd_2 - 1}{d_1(bd_2+1)}} \cdot g^{\frac{v}{d_1}} = g^{\frac{a}{d_1}} \cdot g^{\frac{v}{d_1}} = g^{e_1(a+v)}$$

where $e_1 = d_1^{-1}, e_2 = d_2^{-1}$. Note that because b, d_1, d_2, v are randomly chosen in $\mathbb{Z}_p$, the resulted secret key is also chosen in the same distribution as in the joint algorithm.
- For the public key, $\mathcal{A}$ computes:

$$g^{d_1}, e(g,g)^{d_2}, g^{d_1 \cdot d_2}, e(g,g), e(g,g)^v = e(g^{d_1}, g^{\frac{v}{d_1}}), e(g,g)^{v \cdot d_2} = e(g^{d_1 d_2}, g^{\frac{v}{d_1}})$$

$\mathcal{A}$ next chooses a random message M and uses T to compute the challenge ciphertext and passes it to $\mathcal{B}$:

$$g^{kd_1}, T, g^{kd_1 d_2}, e(g^{\frac{v}{d_1}}, g^{kd_1}), e(g^{\frac{v}{d_1}}, g^{kd_1 d_2}), e(g,g)^{-k} \cdot M$$

In the guess phase, if $\mathcal{B}$ outputs 0 (indicating that this is well-form ciphertext) then $\mathcal{A}$ outputs 0 (indicating that T is $e(g,g)^{kd_2}$), and otherwise if $\mathcal{B}$ outputs 1 (indicating that this is probe ciphertext) then $\mathcal{A}$ also outputs 1 (indicating that T is a random element).

We also note that $\mathcal{B}$ can maliciously output a random message M' in the case he knows the challenge ciphertext is a probe ciphertext, however $\mathcal{A}$ still knows the right answer of $\mathcal{B}$ because he knows the real message M.

As the simulation of $\mathcal{A}$ is perfect, $\mathcal{A}$ can thus break GDDHE_3 assumption with the same advantage that $\mathcal{B}$ can successfully distinguish a probe ciphertext and a well-form ciphertext. We can thus construct a single-key black box tracing algorithm as follow:

Full Access Single-Key Black Box Tracing Algorithm: When a user j joins the system, the tracer computes and stores the pair $(j, e(g,g)^{B_j})$ in a sorted table Tab. The tracing algorithm then works as follow:
1. The tracer picks random $k, r \in \mathbb{Z}_p$ then creates a probe ciphertext:

$$C_1 = g^{kd_1}, C_2 = e(g,g)^{kd_2+r}, C_3 = g^{kd_1 d_2}, C_4 = e(g,g)^{kv},$$

$$C_5 = e(g,g)^{kvd_2}, C_6 = M'$$

2. Assume the decryption key A_i, B_i is embedded in the pirate decoder. Then the tracer queries the pirate decoder on this probe ciphertext. The pirate decoder will compute:

$$K = \frac{e(A_i, C_1)}{C_4} \cdot C_2^{B_i - 1} \cdot \left(\frac{e(A_i, C_3)}{C_5}\right)^{B_i} = e(g,g)^{\frac{kd_2 a_i}{a_i + 1}} \cdot e(g,g)^{(kd_2+r)(\frac{1}{a_i+1} - \frac{1}{d_2} - 1)}$$

$$= e(g,g)^{-k} \cdot e(g,g)^{r(B_i - 1)}$$

Then outputs:

$$C_6/K$$

3. The tracer first recovers K then computes $e(g,g)^{B_i}$ since it knows k, r. Then the tracer simply verifies if the element $e(g,g)^{B_i}$ is in the table Tab and eventually outputs the traitor. It is easy to see that our tracing algorithm never accuses any innocent user and the time complexity of our tracing security is $O(\log N)$. We also notice that, in our system, N is the effective number of the actual users in the system.

Minimal Access Single-key Black Box Tracing Algorithm: In the setup phase, the tracer picks random $k, r \in \mathbb{Z}_p$ and a message M, then creates:

$$C_1 = g^{kd_1}, C_2 = e(g,g)^{kd_2+r}, C_3 = g^{kd_1d_2}, C_4 = e(g,g)^{kv}, C_5 = e(g,g)^{kvd_2}$$

and store these values in a table Tab.

When a user j joins the system, the tracer computes

$$C_{6,j} = e(g,g)^{-k} \cdot e(g,g)^{r(B_j-1)} \cdot M$$

and stores the pair $(j, C_{6,j})$ in the table Tab.

The tracing algorithm then works as follow:

1. For each user's indices j, the tracer queries the pirate decoder on a pair

$$(C = (C_1, C_2, C_3, C_4, C_5, C_{6,j}), M)$$

2. Assume the decryption key A_i, B_i is embedded in the pirate decoder. The pirate decoder will compute:

$$K = \frac{e(A_i, C_1)}{C_4} \cdot C_2^{B_i-1} \cdot \left(\frac{e(A_i, C_3)}{C_5}\right)^{B_i} = e(g,g)^{\frac{kd_2a_i}{a_i+1}} \cdot e(g,g)^{(kd_2+r)(\frac{1}{a_i+1}-\frac{1}{d_2}-1)}$$

$$= e(g,g)^{-k} \cdot e(g,g)^{r(B_i-1)}$$

Then computes:

$$M' = C_{6,j}/K$$

3. At user's indices j, if the tracer receives a signal *valid* which indicates that C is a valid encryption of M, then the tracer outputs user's indices j is a traitor. It is easy to see that our tracing algorithm never accuses any innocent user and the time complexity of our tracing security is $O(N)$. We also notice that, in our system, N is the effective number of the actual users in the system.

6   Conclusion

In this paper, we restrict ourselves to the non-black-box tracing and the single-key black box tracing models and proposed an optimal and practical scheme in these models. As far as we know, this is the first practical fully collusion resistant traitor tracing scheme. However the most important open problem in traitor tracing remains the construction of a practical fully collusion resistant traitor tracing scheme in the general black box tracing model. The schemes in [2,8] has constant ciphertext size but when considering the full collusion, the secret key size of user is $O(N^2)$ which is impractical. The most relevant schemes in [9] and in [10] still have large ciphertext size of $O(\sqrt{N})$ and require the use of bilinear maps in groups of composite order. We also recall that, non-black-box tracing and the single- key black box tracing models deal with pirates who are required to implement a key that has the form of the keys distributed to the users (this consideration is justified and discussed in the introduction) and do not consider pirates who can produce new form of key that can help to decrypt ciphertexts. One of the promising direction is to consider a model between the single-key black box tracing and the general black box tracing model in which one can still achieve a practical scheme.

Acknowledgments. This work was partially supported by the French ANR-09-VERSO-016 BEST Project and partially conducted within the context of the International Associated Laboratory Formath Vietnam (LIAFV).

References

1. Agrawal, S., Dodis, Y., Vaikuntanathan, V., Wichs, D.: On continual leakage of discrete log representations. Cryptology ePrint Archive, Report 2012/367 (2012), http://eprint.iacr.org/2012/367
2. Billet, O., Phan, D.H.: Efficient Traitor Tracing from Collusion Secure Codes. In: Safavi-Naini, R. (ed.) ICITS 2008. LNCS, vol. 5155, pp. 171–182. Springer, Heidelberg (2008)
3. Billet, O., Phan, D.H.: Traitors collaborating in public: Pirates 2.0. In: Joux, A. (ed.) EUROCRYPT 2009. LNCS, vol. 5479, pp. 189–205. Springer, Heidelberg (2009)
4. Boneh, D., Boyen, X.: Short signatures without random oracles. In: Cachin, C., Camenisch, J.L. (eds.) EUROCRYPT 2004. LNCS, vol. 3027, pp. 56–73. Springer, Heidelberg (2004)
5. Boneh, D., Boyen, X., Goh, E.-J.: Hierarchical identity based encryption with constant size ciphertext. In: Cramer, R. (ed.) EUROCRYPT 2005. LNCS, vol. 3494, pp. 440–456. Springer, Heidelberg (2005)
6. Boneh, D., Franklin, M.K.: An efficient public key traitor scheme (Extended abstract). In: Wiener, M. (ed.) CRYPTO 1999. LNCS, vol. 1666, pp. 338–353. Springer, Heidelberg (1999)
7. Boneh, D., Gentry, C., Waters, B.: Collusion resistant broadcast encryption with short ciphertexts and private keys. In: Shoup, V. (ed.) CRYPTO 2005. LNCS, vol. 3621, pp. 258–275. Springer, Heidelberg (2005)
8. Boneh, D., Naor, M.: Traitor tracing with constant size ciphertext. In: Ning, P., Syverson, P.F., Jha, S. (eds.) ACM CCS 2008, pp. 501–510. ACM Press (October 2008)
9. Boneh, D., Sahai, A., Waters, B.: Fully collusion resistant traitor tracing with short ciphertexts and private keys. In: Vaudenay, S. (ed.) EUROCRYPT 2006. LNCS, vol. 4004, pp. 573–592. Springer, Heidelberg (2006)
10. Boneh, D., Waters, B.: A fully collusion resistant broadcast, trace, and revoke system. In: Juels, A., Wright, R.N., Vimercati, S. (eds.) ACM CCS 2006, pp. 211–220. ACM Press (October/ November 2006)
11. Boyen, X.: Mesh signatures. In: Naor, M. (ed.) EUROCRYPT 2007. LNCS, vol. 4515, pp. 210–227. Springer, Heidelberg (2007)
12. Chor, B., Fiat, A., Naor, M.: Tracing traitors. In: Desmedt, Y.G. (ed.) CRYPTO 1994. LNCS, vol. 839, pp. 257–270. Springer, Heidelberg (1994)
13. Delerablée, C., Paillier, P., Pointcheval, D.: Fully collusion secure dynamic broadcast encryption with constant-size ciphertexts or decryption keys. In: Takagi, T., Okamoto, T., Okamoto, E., Okamoto, T. (eds.) Pairing 2007. LNCS, vol. 4575, pp. 39–59. Springer, Heidelberg (2007)
14. Junod, P., Karlov, A., Lenstra, A.K.: Improving the boneh-franklin traitor tracing scheme. In: Jarecki, S., Tsudik, G. (eds.) PKC 2009. LNCS, vol. 5443, pp. 88–104. Springer, Heidelberg (2009)
15. Kiayias, A., Yung, M.: Traitor tracing with constant transmission rate. In: Knudsen, L.R. (ed.) EUROCRYPT 2002. LNCS, vol. 2332, pp. 450–465. Springer, Heidelberg (2002)

16. Naor, M., Pinkas, B.: Efficient trace and revoke schemes. In: Frankel, Y. (ed.) FC 2000. LNCS, vol. 1962, pp. 1–20. Springer, Heidelberg (2001)
17. Nuida, K.: A general conversion method of fingerprint codes to (More) robust fingerprint codes against bit erasure. In: Kurosawa, K. (ed.) ICITS 2009. LNCS, vol. 5973, pp. 194–212. Springer, Heidelberg (2010)
18. Guillot, P., Nimour, A., Phan, D.H., Trinh, V.C.: Optimal Public Key Traitor Tracing Scheme in Non-Black Box Model. Full version available at, http://www.di.ens.fr/users/phan/2013-africa-a.pdf
19. Tonien, D., Safavi-Naini, R.: An efficient single-key pirates tracing scheme using cover-free families. In: Zhou, J., Yung, M., Bao, F. (eds.) ACNS 2006. LNCS, vol. 3989, pp. 82–97. Springer, Heidelberg (2006)

NaCl on 8-Bit AVR Microcontrollers*

Michael Hutter[1] and Peter Schwabe[2]

[1] Graz University of Technology
Institute for Applied Information Processing and Communications (IAIK)
Inffeldgasse 16a, 8010, Graz, Austria
michael.hutter@iaik.tugraz.at
[2] Radboud University Nijmegen
Digital Security Group
P.O. Box 9010, 6500GL Nijmegen, The Netherlands
peter@cryptojedi.org

Abstract. This paper presents first results of the Networking and Cryptography library (NaCl) on the 8-bit AVR family of microcontrollers. We show that NaCl, which has so far been optimized mainly for different desktop and server platforms, is feasible on resource-constrained devices while being very fast and memory efficient. Our implementation shows that encryption using Salsa20 requires 268 cycles/byte, authentication using Poly1305 needs 195 cycles/byte, a Curve25519 scalar multiplication needs 22 791 579 cycles, signing of data using Ed25519 needs 23 216 241 cycles, and verification can be done within 32 634 713 cycles. All implemented primitives provide at least 128-bit security, run in constant time, do not use secret-data-dependent branch conditions, and are open to the public domain (no usage restrictions).

Keywords: Elliptic-curve cryptography, Edwards curves, Curve25519, Ed25519, Salsa20, Poly1305, AVR, ATmega.

1   Introduction

This paper describes implementations of the Networking and Cryptography library (NaCl) [4] on 8-bit AVR microcontrollers. More specifically, we describe two different approaches, one aiming at higher speed, one aiming at smaller memory requirements, of porting NaCl to the AVR ATmega family of microcontrollers. The aim of the high-speed implementation is not to achieve the highest possible speed at all (memory-)costs for all primitives. Similarly, the aim of the low-memory implementation is not to obtain the smallest possible footprint without any performance considerations. The two implementations are rather

---

* This work was supported by the Austrian Science Found (FWF) under the grant number TRP251-N23. Part of this work was done while Peter Schwabe was employed by the Research Center for Information Technology Innovation, Academia Sinica, Taiwan. Permanent ID of this document: cd4aad485407c33ece17e509622eb554. Date: April 1, 2013

A. Youssef, A. Nitaj, A.E. Hassanien (Eds.): AFRICACRYPT 2013, LNCS 7918, pp. 156–172, 2013.

two example tradeoffs between speed and memory footprint that we consider reasonable and useful for various applications and different microcontrollers in the ATmega family.

Previous NaCl optimization focused on large general-purpose server and desktop CPUs; the "smallest" architecture targeted by previous NaCl optimization is ARMv7 CPUs with the NEON vector-instruction set [10]. Despite this focus on large processors, the NaCl designers claim in [4, Section 4] that

> "all of the cryptographic primitives in NaCl can fit onto much smaller CPUs: there are no requirements for large tables or complicated code"

This paper shows that this claim is actually correct.

The cryptographic primitives used by default in NaCl to provide public-key authenticated encryption are the Curve25519 elliptic-curve Diffie-Hellman key-exchange protocol [2], the Poly1305 authenticator [5], and the Salsa20 stream cipher [3]. The designers of NaCl announced, that the next release of NaCl will use the Ed25519 elliptic-curve signature scheme [7,8] to provide cryptographic signatures. This signature scheme—as described in the original paper and as implemented in this paper—uses the SHA-512 hash function [28].

We will put all software described in this paper into the public domain to maximize reusability of our results[1]. We will furthermore discuss possibilities for public benchmarking with the editors of eBACS [9] and XBX [35]. Currently eBACS does not support benchmarking on AVR microcontrollers; XBX only supports benchmarking of hash functions.

Main contribution. There exists an extensive literature describing implementations of cryptographic *primitives* on AVR microcontrollers and other embedded processors. Some of them have been integrated into libraries that offer a set of cryptographic functionalities, e.g., AVR-Crypto-Lib [15], TinyECC [24], NanoECC [32], or the AVR Cryptolibrary from Efton s.r.o. [13]. These libraries are specifically tailored to match the specific restricted environment of the AVR.

This paper is the first to describe implementations of the entire NaCl library on AVR microcontrollers. These include the cryptographic primitives Salsa20 [3], Poly1305 [5], Curve25519 [2], and Ed25519 [8]. All primitives are based—in contrast to existing AVR libraries—on at least 128-bit security and provide new speed records for that level of security. In addition, all functions run in constant time and do not contain secret-data-dependent branch conditions. This is important to provide a certain level of security against basic implementation attacks [22,25]. In particular the implementation is protected against *remote* side-channel attacks. Other cryptographic libraries for AVR do not address this issue. Moreover, the entire library is very small in size and requires only 17366 bytes of code, no static RAM, and less than 1350 bytes of stack memory; it therefore fits into very resource-constrained devices such as the very small ATmega family of microcontrollers, e.g., the ATmega32, ATmega328, and ATmega324A. Last but not least, we present new speed records for Salsa20 on AVRs and give first

---

[1] The software is available online at http://cryptojedi.org/crypto/#avrnacl

results of scalar multiplication for Curve25519 and signing and verifying using Ed25519 on AVR.

Roadmap. The paper is organized as follows. In Section 2, we briefly describe the AVR family of microcontrollers. Section 3 describes the NaCl library and the general approach to porting it to AVR. In Section 4, we describe the implementation of Salsa20. In Section 5, we describe the implementation of Poly1305. Section 6 presents the implementations of Curve25519 and Ed25519 (including SHA-512). Results, a comparison with previous work, and a discussion are given in Section 7.

2   The 8-Bit Family of AVR Microcontrollers

Atmel offers a wide range of 8-bit microcontrollers that can be mainly separated into three groups. High-end devices with high performance (ATxmega), mid-range devices featuring most functionality needed for the majority of applications (ATmega), and low-end devices with limited memory and processing power (ATtiny). Typical use cases of those devices are embedded systems such as motor control, sensor nodes, smart cards, networking, metering, medical applications, etc.

All those devices process data on 8-bit words. There are 32 general-purpose registers available, R0-R31, which can be freely used by implementations. Some of them have special features like R26-R31, which are register pairs used to address 16-bit addresses in SRAM, i.e., X (R27:R26), Y (R29:R28), and Z (R31:R30). Some of those registers (R0-R15) can also only be accessed by a limited set of instructions (in fact only those that do not have an immediate value as one operand).

The instruction set offers up to 90 instructions which are equal for all AVR devices. For devices with more memory or enhanced cores, it is extended by more than 30 additional instructions. The most important instruction for (public-key) cryptography is multiplication. It is not available for minimal cores such as the ATtiny or AT90Sxxxx family. But for enhanced cores like most of the ATmega and also all ATxmega cores, it allows (signed or unsigned) multiplication of two 8-bit words within two clock cycles. The 16-bit result of the multiplication is always stored in the register pair R1:R0. The software described in this paper makes use of these multipliers and does therefore not support the low-end ATtiny and AT90Sxxxx devices.

ATmega example configurations. We perform all benchmarks on an ATmega2560 which has a maximal clock frequency of 16 MHz, a flash storage of 256 KB and 8 KB of RAM. Other typical configurations of ATmega microcontrollers are, for example, the ATmega128 with a maximal clock frequency of 16 MHz, 128 KB of flash storage and 4 KB of RAM and the ATmega328 with a maximal clock frequency of 20 MHz, 32 KB of flash storage and 2 KB of RAM.

Radix-2^8 representation. The typical representation of integers of size larger than 8 bits on an 8-bit architecture is to split integers into byte arrays using radix 2^8. In other words, an m-bit integer x is represented as $n = \lceil m/8 \rceil$ bytes $(x_0, x_1, \ldots, x_{n-1})$ such that $x = \sum_{i=0}^{n} x_i 2^{8*i}$. We use this representation for all integers and elements of finite fields.

3   The NaCl Library

The Networking and Cryptography library (short: NaCl; pronounced: "salt") is a cryptographic library for securing Internet communication [4]. It was developed as one deliverable of the project CACE (Computer Aided Cryptography Engineering) funded by the European Commission. After CACE ended in December 2010, development of NaCl continued within the VAMPIRE virtual lab [23] of the European Network of Excellence in Cryptology, ECRYPT II [12]. The main features of the library are the following:

Easy usability. The library provides a high-level API for public-key authenticated encryption through one function call to `crypto_box`. The receiver of the message verifies the authentication and recovers the message through one function call to `crypto_box_open`. A pair of a public and a private key is generated through `cryto_box_keypair`. A similarly easy-to-use API is offered for cryptographic signatures: A function call to `crypto_sign` signs a message, `crypto_sign_open` verifies the signature and recovers the message, `crypto_sign_keypair` generates a keypair for use with this signature scheme. Implementors of information-security systems obtain high-security cryptographic protection without having to bother with the details of the underlying primitives and parameters. Those are chosen by the NaCl designers.

High security. The key sizes are chosen such that the security level of the primitives is at least 128 bits. Furthermore, NaCl is the only cryptographic library that systematically protects against timing attacks by avoiding loads from addresses that depend on secret data and avoiding branch conditions that depend on secret data. For further security features of NaCl see the extensive discussion in [4, Section 3].

High speed. The cryptographic primitives chosen for NaCl allow very fast implementations on a large variety of architectures.

No usage restrictions. The library is free of copyright restrictions. It is in the public domain. Furthermore the library avoids all patents that the authors are aware of. NaCl is free for download at `http://nacl.cr.yp.to/`.

3.1   Porting NaCl to AVRs

Reusing code. Porting a whole cryptographic library to a memory-restricted and storage-restricted environment such as AVR microcontrollers is different from porting each primitive in the library separately. To minimize code size we can use some functionalities (such as big-integer arithmetic) in multiple primitives. Sometimes this requires optimizing algorithm choices *across primitives.*

For example, the Poly1305 authenticator described in Section 5 needs multiplication of 130-bit numbers; the Curve25519 key-exchange and Ed25519 signatures described in Section 6 need fast multiplication of 256-bit (or at least 255-bit) numbers. With the Karatsuba technique [20] we decompose the 256-bit (32 × 32-byte) multiplication into two 16 × 16-byte multiplications and one 17 × 17-byte multiplication. The latter one can directly be used for the Poly1305 authenticator.

Secret load addresses. On all architectures targeted in previous NaCl optimization, loading data from an address that depends on secret data causes timing variation that can be used by an attacker to mount a timing attack. The reason is that memory access on all these architectures uses a hierarchy of transparent caches; the time required for a load operation depends on whether the requested data is in cache (*cache hit*) or not (*cache miss*). Memory access on the AVR microcontroller is not cached, it takes a constant amount of time. Loading data from a secret position on an AVR will not leak timing information. Avoiding loads from secret positions incurs performance penalties, we therefore decided to *not* avoid loads from secret addresses on the AVR.

Secret branch conditions. Conditional branches are an even more obvious source for timing variation than data loads. Even if both possible branches take the same amount of time to execute, branch conditions that depend on secret data will leak timing information on most architectures. The reason is that most processors use branch-prediction techniques to avoid pipeline stalls. If a branch is predicted correctly, the branch will incur only a small or no penalty; a mispredicted branch typically takes much more time.

AVR microcontrollers do not use any branch-prediction techniques so in principle one can write software that *does* use secret branch conditions and still runs in constant time. However, it is very tedious to review such code for constant-time behavior and the performance benefits are relatively small. We therefore follow the strategy of all other NaCl optimizations and avoid all data flow from secret data to branch conditions.

Randomness generation. NaCl uses the operating-system's random-number generator and reads random bytes from /dev/urandom (see [4, Section 3, "Centralizing randomness"]). This is not possible on the AVR microcontroller. Our implementation of NaCl does not contain any cryptographically secure randomness generator. To test the key-generation functions that require randomness we used the deterministic randombytes function from the try-anything program of the SUPERCOP benchmarking suite. There are two different ways to address randomness generation on the AVR: One can use NaCl in a way that does not require randomness by computing key pairs on an external device and transferring them to the AVR. In NaCl, all operations except key-generation are deterministic. See [4, Section 3, "Avoiding unnecessary randomness"].

If one needs to generate keys on an AVR microcontroller it is necessary to include cryptographically secure randomness generation. One possible source of randomness is, for example, the jitter of the RC oscillator as described in [18].

Message lengths. In the C interface of NaCl, message lengths are passed as 64-bit unsigned integers (datatype `unsigned long long`). Addresses on the AVR ATmega microcontrollers have only 16 bits; we therefore omit expensive arithmetic on 64-bit integers to support messages of a length that would anyway not fit into the addressable memory.

Benchmarking. The cycle-count numbers of the various primitives presented in this paper have been obtained as follows. The numbers given in the following sections are the results of cycle-accurate simulations for an ATmega2560 microcontroller. The results given in the Section 7 (the results given in Table 1 in particular), are obtained through actual measurements on the same targeted microcontroller. For this purpose we re-implemented the 64-bit resolution `cpucycles` cycle counter included in NaCl and the eBACS benchmarking suite SUPERCOP [9] for AVR. We combine the 8-bit and the 16-bit cycle counters into one 24-bit cycle counter and increase the overall count by 2^{24} for an overflow interrupt of the higher counter. The cycle counts include an 247-cycle overhead (284-cycle overhead for the low-area variant) for function call and reading the 64-bit cycle count; this is reported as "empty" benchmark in Table 1. We measured this overhead by subsequently calling an empty function and reading the cycle counter many times and computing the differences of the measurements. We also measured the overhead for reading the cycle counter without the overhead of function calls by computing differences of subsequent readings of the cycle counter. This overhead is 230 cycles (274 cycles for the low-area variant); it is reported as "nothing" benchmark in Table 1.

4   Implementation of Salsa20

Salsa20 is a stream cipher which has been proposed in 2005 [3]. It has been included in the final portfolio of the eSTREAM project initiated in 2004 by the European Network of Excellence for Cryptology (ECRYPT). The cipher consists of 20 rounds[2] where an internal state is modified by various (logical and arithmetic) transformations. To encrypt a message, a 32-byte key is used.

4.1   High-Speed Implementation

The Salsa20 stream cipher is implemented in the library functions `crypto_stream` and `crypto_stream_xor`. The function `crypto_stream` only generates a pseudorandom bitstream, the function `crypto_stream_xor` generates this stream and xors it to a message to produce a ciphertext. The pseudorandom stream is generated in blocks of 64 bytes, each block is generated by the function `crypto_core`. This function first initializes a 32-byte state and starts the round calculation afterwards. We implemented both the initialization and `crypto_core` in assembly to improve the performance in Salsa20. The functions `crypto_stream` and

---

[2] Note that there also exist round-reduced versions of Salsa20, e.g., Salsa20/12 applying 12 rounds instead of 20.

crypto_stream_xor are written in C; to save code size we implemented crypto_stream as a call to crypto_stream_xor with an all zero-message.

Initialization of the state. The function init_core mainly consists of 7 loop iterations where the state x (and a copy of the state j which is later added to the cipher output) gets initialized with the 32-byte key, the 64-byte input, and a 16-byte nonce. The initialization takes 642 clock cycles in total.

Round calculation. The round-calculation function provides the most promising potential to increase the speed of Salsa20. It treats 64-byte blocks as 4×4 matrix of 32-bit words and transforms this state matrix through ten loop iterations consisting of 8 quarterround function calls each (thus 80 function calls in total). Within one quarterround function, three different 32-bit operations (addition, bitwise addition, and rotations) are performed on either the rows or the columns of the state x.

We implemented the following optimizations. First, we used all 32 available registers of the AVR to avoid unnecessary storing and loading from the stack which is costly in terms of memory and speed. For this, we passed the addresses of the current row or column of the state in the registers R18-R25. The values of the state are then loaded into the registers R0-R15. The register pair R17:R16 is reserved to store the 16-bit base address. It will not be modified within the quarterround function. The remaining address registers R26-R31 are used for fast addressing during the round transformations. They allow to implicitly decrement the addresses before or after a ST (store) or LD (load) instruction. Second, the state variables are modified in-place. This means that the state is directly modified without needing extra variables and copy instructions. Third, we implemented shifts by 7 and 9 as cheap logical shift (LSR and LSL) and rotate-through-carry instructions (ROR and ROL). Shifts by 13 and 18 are performed as multiplications (MUL instruction) with the constants $2^5 = 32$ and $2^2 = 4$.

One quarterround function call requires 176 clock cycles in total. The entire round calculation needs 15 763 clock cycles. The entire crypto_stream_xor function needs 17 787 clock cycles to encrypt a 64-byte message. The code size of Salsa20 is 1 556 bytes, including crypto_stream and crypto_stream_xor.

4.2   Low-Area Implementation

For the low-area version, we looped the final addition of j at the end of the quarterround function. The remaining assembly parts are already optimized in terms of low area. We also used the -Os compiler flag to optimize for small code size. With these modifications, the performance is slightly reduced by 159 clock cycles, resulting in 17 893 clock cycles for crypto_stream_xor; the code size is reduced by 426 bytes to only 1 130 bytes, i.e., by 27.38%.

5   Implementation of Poly1305

Poly1305 is a message authentication code (MAC) proposed in 2005 [5]. The name is related to the underlying finite field $2^{130} - 5$. A message m with variable size n is authenticated using a (random) 32-byte one-time secret key s (and a 16-byte nonce). The secret key s consists of two parts, each 16-bytes in length, i.e., $s = (k, r)$. First, the message m is split into 16-byte blocks where each block is padded with a 1. The resulting 17-byte chunks c_i, where $i \in [1, q]$ and $q = \lceil n/16 \rceil$, are then represented as unsigned little-endian integers. After that, one addition and one modular multiplication is performed for each chunk c resulting in the 16-byte authenticator h, i.e.,

$$h = (((c_1 \cdot r^q + c_2 \cdot r^{q-1} + ... + c_q \cdot r^1) \bmod 2^{130} - 5) + s) \bmod 2^{128}.$$

5.1 High-Speed Implementation

The most time-consuming operation in Poly1305 is modular multiplication in the field $2^{130} - 5$. In order to obtain high speeds, we implemented both multiplication and reduction in assembly. To save code size, we implemented a 2^{136}-bit multiplier that is also (re)used by the Karatsuba-multiplier implementation for Curve25519 and Ed25519 as described in Section 6.

17 × 17-byte multiplication. There exist various ways to implement large-integer multiplication, for example, the widely used schoolbook or Comba multiplication. On AVRs, it has been shown by various papers that a combination of both techniques significantly helps in speeding up the computation. See, for example, [16,24,32,34].

We followed a similar approach by breaking the 136-bit multiplication into 8×8-byte, 9×9-byte, and 9×8-byte multiplications and combine the partial results within each block in a conventional schoolbook approach. The 17×17-byte multiplication takes 1 882 cycles (excluding function call overhead). The code size of the fully unrolled implementation is 2 944 bytes.

Reduction mod $2^{130} - 5$ on AVR. We implemented modular reduction as follows. Since the prime $p = 2^{130} - 5$ is a Mersenne-like prime, we can apply fast reduction by using simple shifts and additions only which are relatively cheap on AVRs. Consider the integer $X \in [0, p^2)$ and let $X = X_1 \cdot 2^{130} + X_0$ be the result of the multiplication. Then, we can exploit the congruence $2^{130} \equiv 5$ and we can add $x_1 \cdot 5$ to the lower part x_0, i.e., $X \equiv x_1 \cdot 5 + x_0 = x_1 + (x_1 \ll 2) + x_0 \pmod{p}$. Note that one of the two input operands to the multiplication in Poly1305 has only 124 bits. Even if we assume that the other argument has full 17 bytes, i.e., 136 bits, the result of the multiplication has at most 260 bits. After adding $x_1 \cdot 5$ to x_0 we obtain a number of at most 133 bits; addition with a 128-bit number during processing of the next block yields at most 134 bits which fits into 17 bytes and is thus safe to use as input for the following multiplication. We therefore do not have to reduce further after adding $x_1 \cdot 5$ to x_0.

We optimize the reduction by exploiting the gap between 2^{128} and $2^{130} - 5$ on the AVR. Since we operate on radix-2^8, the integer X is represented as $X = X_1' \cdot 2^{128} + X_0'$ where X_0' is a 128-bit integer represented as 16-byte array and X_1' is an integer represented as a 17-byte array. Let $X_1'' = 4 \cdot \lfloor X_1/4 \rfloor$, i.e., X_1' with the two lowest bits set to zero. Note that $4X_1 = X_1''$. We compute the reduction as $X_0 + X_1'' + X_1''/4 = X_0 + X_1'' + (X_1'' \gg 2)$. Shifting X_1'' right by one bit can done in two clock cycles per byte through a logical-shift-right (LSR) instruction (which shifts the LSB to the carry register) and a rotate-right-through-carry (ROR) instruction which rotates a byte by shifting the carry into the MSB. Shifting by two bits means performing this shift twice.

5.2 Low-Area Implementation

For the low-area version of Poly1305, we implemented three operations in a loop, i.e., two initializations of intermediate variables and the addition operation. For the latter operation we simply re-used the function bigint_add, which is also used for scalar arithmetic in Ed25519. These modifications have only a slight impact in performance (13 270 clock cycles are needed for a 64-byte message instead of 12 525) but the code size is reduced from 1 153 bytes to only 729, i.e., by 36.77 %.

6   Curve25519 and Ed25519

In 2006, Bernstein introduced the Curve25519 elliptic-curve Diffie-Hellman key-exchange primitive and the corresponding high-speed software for various x86 CPUs [2]. Curve25519 uses the elliptic curve defined by the equation $E : y^2 = x^3 + 486662x^2 + x$ over the field $\mathbb{F}_{2^{255}-19}$. The scalar multiplication performed in Curve25519 uses the x-coordinate-based differential addition introduced by Montgomery in [27, Section 10]. The main computational effort for the scalar multiplication are 255 so called ladder steps, 255 conditional swaps, each based on one bit of the scalar, and one inversion in $2^{2^{255}-19}$. Each of the laddersteps consists of 5 multiplications, 4 squarings, 1 multiplication with the constant 121666, 4 additions, and 4 subtractions in $\mathbb{F}_{2^{255}-19}$.

In 2011, Bernstein, Lange, Duif, Schwabe, and Yang introduced the Ed25519 elliptic-curve digital-signature scheme and presented corresponding high-speed software for Intel Nehalem/Westmere processors [7,8]. The signatures are based on arithmetic on the twisted Edwards curve [6] defined by the equation $E : x^2 + y^2 = 1 - \frac{121665}{121666}x^2y^2$ over $\mathbb{F}_{2^{255}-19}$. This curve is birationally equivalent to the Montgomery curve used in the Curve25519 key-exchange software. The main computational effort for Ed25519 key-pair generation and signing is one fixed-base-point scalar multiplication with a secret scalar. The main computational effort for signature verification is one point decompression (Ed25519 stores only the y coordinate and one bit of the x coordinate of public keys) and one double-point scalar multiplication with public scalars. One of the two points involved in this double-point scalar multiplication is the fixed-base-point also used in key-pair generation and signing.

6.1 High-Speed Implementation

Arithmetic in $\mathbb{F}_{2^{255}-19}$. The computations of both Curve25519 key exchange and Ed25519 signatures break down to operations in the field $\mathbb{F}_{2^{255}-19}$. The most speed-critical operations are multiplications and squarings. We decided to not specialize squarings to save code size.

Multiplication is implemented as one level of Karatsuba multiplication, that breaks the 32×32-byte multiplication into two 16×16-byte multiplications and one 17×17-byte multiplication. Note that the latter multiplication is also used for the Poly1305 authenticator described in Section 5. On top these multiplications, we need two 16-byte additions, two 33-byte additions, and two 33-byte subtractions to accumulate the intermediate results. The entire 32×32-byte Karatsuba multiplication takes 6 868 cycles; this is slightly slower than the current state of the art presented at CHES 2011 [19] (6 208 cycles); but we save in code size, especially for the low-area variant as described later. For the completely unrolled high-speed version of the 32×32-byte multiplication, 7 184 bytes of code are required.

Throughout the whole computation we do not reduce modulo $2^{255} - 19$, but instead only modulo $2^{256} - 38$. Only at the very end we "freeze" the values modulo $2^{255} - 19$. To perform modular reduction after a multiplication or squaring we multiply the upper 32 bytes of the 64-byte result by 38 and then add those to the lower 32 bytes. This will leave us with a 33-bit value. We multiply the highest byte again by 38 and add the 2-byte result to the lowest two bytes and ripple the carry through all 32 bytes. This may again produce a carry which we multiply by 38, add to the lowest byte and carry to the second byte. Note that this final addition of the carry bit can not produce a carry. After an addition or subtraction we simply multiply the final carry bit by 38 and add to (or subtract from) the lowest byte; then ripple through the carry and again multiply the carry by 38 and add to the lowest byte. These reductions after multiplication and addition use fully unrolled loops. We use a separate function call to the modular reduction after multiplication and squaring. This way we are able to reuse the 32×32-byte multiplication for arithmetic on scalars in Ed25519 signature verification. Addition and subtraction in $\mathbb{F}_{2^{255}-19}$ do not use separate function calls to reduction. They have been also fully unrolled.

Curve25519. Our Curve25519 software uses the same sequence of 255 Montgomery ladder steps and 255 conditional swaps as previous optimized implementations of Curve25519 [10,2]. The conditional swaps neither use lookups from secret addresses nor (as previously explained) secret branch conditions; a conditional swap between two values a and b depending on one secret bit s is computed as two conditional moves; each conditional move is computed by first expanding the secret bit s to an all-one or all-zero mask $\bar{s}$ and then computing $a \leftarrow a$ XOR ($\bar{s}$ AND (a XOR b)).

The final inversion in $\mathbb{F}_{2^{255}-19}$ is computed as exponentiation with $2^{255} - 21$ using the same sequence of 254 squarings and 11 multiplications as [2]. We implemented this sequence of function calls in C and used the compiler flags `-mmcu=atmega2560 -Os -mcall-prologues` to translate it.

Ed25519 key-pair generation and signing. The fixed-base-point scalar multiplication in key-pair generation and signing is implemented through a signed-fixed-window scalar multiplication with window size 4. The elliptic-curve arithmetic uses the extended coordinates introduced in [17]. In total the fixed-base-point scalar multiplication requires 64 table lookups, 63 additions of a precomputed multiple of the basepoint to a point in extended coordinates, and 252 doublings in extended coordinates. At the end of this computation we need one inversion and two multiplications in $\mathbb{F}_{2^{255}-19}$ to convert to affine coordinates. The precomputed multiples of the base point are in an array marked as PROGMEM. This way they do not occupy space in the data segment in RAM but only in the (much larger) flash memory. Before performing the fixed-base-point scalar multiplication we copy this table of precomputed points into a space on the stack to avoid (secretly indexed) lookups from flash memory.

Ed25519 verification. We perform point decompression of the public key in the same way as explained in [8, Section 5]. We implement the required exponentiation by $2^{252} - 3$ the same way as the inversion: A sequence of function calls to multiplications and squarings implemented in C and compiled with the flags -mmcu=atmega2560 -Os -mcall-prologues.

For double-scalar multiplication we apply Straus' algorithm [31] with window-size 1, a special case that is sometimes referred to as "Shamir's trick". For the multiplication of 256-bit scalars modulo the group order we use the 32×32-byte multiplication and subsequent Barrett reduction [1].

SHA-512. Ed25519 signatures need a 512-bit-output hash function; the original paper [8] uses SHA-512 but the authors comment that they "will not hesitate to recommend Ed25519-SHA-3 after SHA-3 is standardized". In order to provide a compatible implementation to the Ed25519 implementations currently included in SUPERCOP [9] we also use Ed25519-SHA-512. We implemented all speed-critical low-level functions, in particular arithmetic on 64-bit integers, in assembly. This assembly implementation unrolls all length-8 loops. Calls to the low-level assembly functionalities are implemented in C. Compiling this SHA-512 C code with the -O3 flag, which we use for most files in the high-speed version, results in unacceptably large code; for SHA-512 we therefore use compiler flags -mmcu=atmega2560 -Os -mcall-prologues.

6.2   Low-Area Implementation

Arithmetic in $\mathbb{F}_{2^{255}-19}$. The main difference in the implementation of finite-field arithmetic for the low-area implementation is that we get rid of the 16×16-byte multiplication. Instead we copy the arguments to 17-byte arrays with leading zero byte and use the 17×17-byte multiplication. The resulting assembly implementation of 32×32-byte multiplication that performs 3 calls to 17×17 byte multiplication and all necessary additions and copies for the Karatsuba multiplication has a size of 3 358 bytes (53.25 % less code size compared to the high-speed version). The runtime is increased to 8 322 clock cycles.

Aside from that change we do not unroll the loops in the modular reduction after multiplication, addition, and subtraction to further reduce code size.

Curve25519. The high-level implementation of Curve25519 is the same for the small-area implementation as for the high-speed implementation.

Ed25519 key-pair generation and signing. For the fixed-base-point scalar multiplication we also use a signed-fixed-window scalar-multiplication algorithm. Instead of window size 4 (as in the high-speed implementation) we use a window size of only 2 to save space in flash and RAM.

Ed25519 verification. The high-level implementation of verification is the same for the small-area implementation as for the high-speed implementation.

SHA-512. SHA-512 uses almost the same code as same in the high-speed implementation. The only difference is that we do not unroll the 3 length-8 loops in the σ-transformation of SHA-256. This change slightly shrinks the code size without significantly hurting performance.

7   Results

In this section we report benchmarks of our software and give a comparison with previous results. As described in Subsection 3.1, the benchmarks are not obtained in a simulator but by measuring cycles on an actual ATmega2560 microcontroller clocked at 16 MHz (on the Arduino Mega 2560 development board). Measuring cycles incurs a certain overhead; we give this overhead as a "nothing" benchmark, i.e., simply differences of subsequent readings to the cycle counter. The reported numbers are the median of the cycle counts of 20 runs of the respective primitive.

We compiled all C software with avr-gcc version 4.7.2. For the high-speed implementation we used compiler flags -mmcu=atmega2560 -O3 where not otherwise reported; for the low-area implementation we used the compiler flags -mmcu=atmega2560 -Os -mcall-prologues. Our implementation does not use any space in the data segment and no dynamic memory allocation; so RAM is only used by the stack[3]. We measured stack space by writing a canary value to the whole stack before running the actual function; then reading later how many of the canary bytes have been overwritten. Reporting code sizes for individual primitives does not make much sense because of large portions of code that is shared between the primitives (for example Curve25519 and Ed25519 share the code for field arithmetic in $\mathbb{F}_{2^{255}-19}$). Instead, we report the code size (i.e. required space in the flash memory) for both implementations of the whole library. These sizes were obtained with avr-size from GNU binutils version 2.20.1.20100303. Our results are summarized in Table 1.

Comparison with related work. To the authors' knowledge, there exist three resources that present results of Salsa20 on AVR microcontrollers. Meiser et al. [26] and Eisenbarth et al. [14] reported results of Salsa20 implemented in C and assembly. Their fastest design needs 17 812 clock cycles for one 64-byte message block needing 2 984 bytes of code. Their low-area variant needs 18 400 clock cycles and 1 452 bytes of code. Both implementations need 280 bytes of RAM.

---

[3] We observed that earlier versions of avr-gcc, for example, avr-gcc 4.5, place some constants in the data segment; gcc-4.7 stores those constants in program memory.

Table 1. Benchmark results of NaCl on the AVR ATmega2560 microcontroller

Primitive		Message bytes	Cycles	Stack bytes
nothing	high-speed		230	
	low-area		274	
empty	high-speed		247	
	low-area		284	
Salsa20	high-speed	8	17 076	268
		64	17 787	
		576	155 195	
		1024	275 427	
		2048	550 243	
	low-area	8	17 202	273
		64	17 893	
		576	155 981	
		1024	276 808	
		2048	552 984	
Poly1305	high-speed	8	4 411	148
		64	12 525	
		576	98 477	
		1024	173 685	
		2048	345 588	
	low-area	8	4 773	148
		64	13 270	
		576	103 286	
		1024	182 050	
		2048	362 081	
SHA-512	high-speed	8	536 133	689
		64	535 945	
		576	2 656 525	
		1024	4 777 297	
		2048	9 018 552	
	low-area	8	607 082	669
		64	606 916	
		576	3 012 120	
		1024	5 417 516	
		2048	10 228 019	

Primitive		Operation	Cycles	Stack bytes
Curve25519	high-speed	crypto_scalarmult_base	22 791 580	677
		crypto_scalarmult	22 791 579	677
	low-area	crypto_scalarmult_base	27 926 288	917
		crypto_scalarmult	27 926 278	920
Ed25519	high-speed	crypto_sign_keypair	21 928 751	1 566
		crypto_sign	23 216 241	1 642
		crypto_sign_open	32 634 713	1 315
	low-area	crypto_sign_keypair	32 870 759	1 282
		crypto_sign	34 303 972	1 289
		crypto_sign_open	40 083 281	1 346

NaCl implementation	Code size (in bytes)
high-speed	27 962
low-area	17 366

There is also a C implementation of Salsa20 in the AVR-Crypto-Lib [15] written by Daniel Otte. His implementation requires 723 clock cycles for initializing the state and 94 476 clock cycles for encryption.

In view of elliptic-curve implementations on AVR, there exist many results presented for example in [16,21,33,34]. Most of these results are hard to compare since the implementations differ in various ways such as in the size of the underlying finite field, the used ECC group formulas, the multiplication technique (both in terms of group and field arithmetic), and additionally implemented higher-level protocols (e.g., hash functions, signing and verifying of messages, random number generation, ...). For example, one of the first who reported the performance of ECC on an ATmega128 are Gura et al. [16] who presented their results at CHES 2004. They implemented ECC using the NIST standardized curves over the prime fields $\mathbb{F}_{p160}$, $\mathbb{F}_{p192}$, and $\mathbb{F}_{p224}$. Their implementation needs 17.52 million clock cycles for a single scalar multiplication on the curve over $\mathbb{F}_{p224}$. Uhsadel et al. [33] reported around 10 million cycles for a 160-bit scalar multiplication.

One of the few AVR libraries that support also higher-level protocols are TinyECC, NanoECC, or CRS-AVR010X-ECC. TinyECC has been presented by Liu et al. [24] in 2008. The library implements ECDSA, ECDH, and ECIES on the SECG curves over $\mathbb{F}_{p128}$, $\mathbb{F}_{p160}$[4], and $\mathbb{F}_{p192}$. Signing using ECDSA-SECP160r1 needs 16 million clock cycles and 27 million cycles in addition to precompute the base-point multiples of the implemented sliding window scalar-multiplication method. The entire library needs between 15 492 and 19 308 bytes of code (depending on the used multiplication method) and around 1 500 bytes of RAM. The low-area variant needs 10 180 bytes of code and 152 bytes of RAM. NanoECC has been proposed by Szczechowiak et al. [32]. The library implements the NIST-K163 Koblitz curve over $\mathbb{F}_{p160}$. They reported 9.37 million clock cycles for one scalar multiplication and the code size of the library is 46 100 bytes[5] and the RAM usage is 1 800 bytes. There exist also another library called CRS-AVR010X-ECC [29] that implements ECDSA and ECDH on SECG curves over $\mathbb{F}_{p160}$, $\mathbb{F}_{p192}$, $\mathbb{F}_{p224}$, and $\mathbb{F}_{p256}$. The implementation on the curve over $\mathbb{F}_{p256}$ needs 5 to 8 kB of code and 750 to 900 bytes of RAM. Signing using ECDSA requires 76.8 million cycles. Their high-speed implementation requires only 27.2 million cycles with an additional memory of 16 384 bytes.

Recently, Chu et al. [11] set new speed records for a single scalar multiplication on Twisted Edwards curves on AVRs. Their implementation needs only 5.9 million clock cycles for a 160-bit curve on an ATmega128. However, the authors aimed for high-speed without considering implementation attacks, e.g., they implemented the conventional double-and-add method and used data-dependent branch conditions which can be exploited in implementation attacks [22,25].

---

[4] Curve *secp160r1* has been used in [24] for evaluating the performance of TinyECC.
[5] NanoECC is based on the MIRACLE (Multi-precision Integer and Rational Arithmetic C/C++ Library) [30], which provides many functions and tools to implement higher-level protocols.

Discussion. As explained in the introduction, our implementation of NaCl does not aim at highest speed at all costs. Instead we aimed at good speeds with a moderate RAM and ROM usage. With this paper we are hoping for feedback from potential users of AVR NaCl telling us what the specific requirements of their application are. For applications that require higher speeds for a specific primitive there are various possibilities for speedups, in particular in Curve25519 and Ed25519:

- Arithmetic in $\mathbb{F}_{2^{255}-19}$ does not use special code for squarings but instead uses calls to the multiplication. A specialized squaring implementation would speed up both Curve25519 and Ed25519.
- The Karatsuba multiplier used for multiplication in $\mathbb{F}_{2^{255}-19}$ is only slightly slower than the operand-caching multiplication presented in [19]; however, switching to operand-caching multiplication would offer further speedups for Curve25519 and Ed25519.
- The multiplication with the small constant 121 666 in Curve25519 is not specialized; again we are using a call to the full multiplication. A specialized function for multiplication with this constant would speed up Curve25519.
- Ed25519 signature verification uses Straus' algorithm with window size 1 instead of, for example, a sliding-window algorithm that would require significantly more RAM. If RAM usage is not a critical limitation we could thus easily speed up signature verification.
- We do not expect users of AVR NaCl to have any use for the fast batch verification of signatures; processing *many* signatures in short time is not exactly the typical domain for embedded microcontrollers. If applications benefit from fast batch verification and are willing to spend some space in RAM, we could also include the fast batch verification based on the Bos-Coster multi-scalar-multiplication algorithm described in [8, Section 5].

References

1. Barrett, P.: Implementing the Rivest Shamir and Adleman Public Key Encryption Algorithm on a Standard Digital Signal Processor. In: Odlyzko, A.M. (ed.) CRYPTO 1986. LNCS, vol. 263, pp. 311–323. Springer, Heidelberg (1987)
2. Bernstein, D.J.: Curve25519: New Diffie-Hellman Speed Records. In: Yung, M., Dodis, Y., Kiayias, A., Malkin, T. (eds.) PKC 2006. LNCS, vol. 3958, pp. 207–228. Springer, Heidelberg (2006), http://cr.yp.to/papers.html#curve25519
3. Bernstein, D.J.: The Salsa20 family of stream ciphers. In: Robshaw, M., Billet, O. (eds.) New Stream Cipher Designs. LNCS, vol. 4986, pp. 84–97. Springer, Heidelberg (2008), http://cr.yp.to/papers.html#salsafamily
4. Bernstein, D.J., Lange, T., Schwabe, P.: The Security Impact of a New Cryptographic Library. In: Hevia, A., Neven, G. (eds.) LATINCRYPT 2012. LNCS, vol. 7533, pp. 159–176. Springer, Heidelberg (2012), http://cryptojedi.org/papers/#coolnacl
5. Bernstein, D.J.: The Poly1305-AES Message-Authentication Code. In: Gilbert, H., Handschuh, H. (eds.) FSE 2005. LNCS, vol. 3557, pp. 32–49. Springer, Heidelberg (2005), http://cr.yp.to/papers.html#poly1305

6. Bernstein, D.J., Birkner, P., Joye, M., Lange, T., Peters, C.: Twisted Edwards Curves. In: Vaudenay, S. (ed.) AFRICACRYPT 2008. LNCS, vol. 5023, pp. 389–405. Springer, Heidelberg (2008), http://cr.yp.to/papers.html#twisted

7. Bernstein, D.J., Duif, N., Lange, T., Schwabe, P., Yang, B.-Y.: High-speed high-security signatures. In: Preneel, B., Takagi, T. (eds.) CHES 2011. LNCS, vol. 6917, pp. 124–142. Springer, Heidelberg (2011); see also full version [8]

8. Bernstein, D.J., Duif, N., Lange, T., Schwabe, P., Yang, B.-Y.: High-speed high-security signatures. Journal of Cryptographic Engineering 2(2), 77–89 (2012) see also short version [7], http://cryptojedi.org/papers/#ed25519

9. Bernstein, D.J., Lange, T.: eBACS: ECRYPT benchmarking of cryptographic systems, http://bench.cr.yp.to (accessed January 31, 2013)

10. Bernstein, D.J., Schwabe, P.: NEON crypto. In: Prouff, E., Schaumont, P. (eds.) CHES 2012. LNCS, vol. 7428, pp. 320–339. Springer, Heidelberg (2012), http://cryptojedi.org/papers/#neoncrypto

11. Chu, D., Großschädl, J., Liu, Z.: Twisted Edwards-Form Elliptic Curve Cryptography for 8-bit AVR-based Sensor Nodes. Cryptology ePrint Archive: Report 2012/730 (2012)

12. European Network of Excellence in Cryptology II, http://www.ecrypt.eu.org/index.html (accessed January 18, 2013)

13. Efton. 8051 and AVR Cryptolibrary, http://www.efton.sk/crypt/index.htm

14. Eisenbarth, T., Kumar, S., Paar, C., Poschmann, A., Uhsadel, L.: A Survey of Lightweight-Cryptography Implementations. IEEE Design & Test of Computers - Design and Test of ICs for Secure Embedded Computing 24(6), 522–533 (2007) ISSN 0740-7475

15. Das Labor e.V. AVR-Crypto-Lib, http://www.das-labor.org/wiki/AVR-Crypto-Lib/en

16. Gura, N., Patel, A., Wander, A., Eberle, H., Shantz, S.C.: Comparing Elliptic Curve Cryptography and RSA on 8-bit CPUs. In: Joye, M., Quisquater, J.-J. (eds.) CHES 2004. LNCS, vol. 3156, pp. 119–132. Springer, Heidelberg (2004)

17. Hisil, H., Wong, K.K.-H., Carter, G., Dawson, E.: Twisted edwards curves revisited. In: Pieprzyk, J. (ed.) ASIACRYPT 2008. LNCS, vol. 5350, pp. 326–343. Springer, Heidelberg (2008), http://eprint.iacr.org/2008/522/

18. Hlaváč, J., Lórencz, R., Hadáček, M.: True random number generation on an Atmel AVR microcontroller. In: 2010 2nd International Conference on Computer Engineering and Technology (ICCET), vol. 2, pp. 493–495. IEEE (2010)

19. Hutter, M., Wenger, E.: Fast multi-precision multiplication for public-key cryptography on embedded microprocessors. In: Preneel, B., Takagi, T. (eds.) CHES 2011. LNCS, vol. 6917, pp. 459–474. Springer, Heidelberg (2011)

20. Karatsuba, A., Ofman, Y.: Multiplication of Multidigit Numbers on Automata. Soviet Physics-Doklady 7, 595–596 (1963); Translated from Doklady Akademii Nauk SSSR 145(2), 293–294 (July 1962)

21. Kargl, A., Pyka, S., Seuschek, H.: Fast Arithmetic on ATmega128 for Elliptic Curve Cryptography. IACR Cryptology ePrint Archive, report 2008/442 (October 2008), http://eprint.iacr.org/2008/442

22. Kocher, P.C.: Timing Attacks on Implementations of Diffie-Hellman, RSA, DSS, and Other Systems. In: Koblitz, N. (ed.) CRYPTO 1996. LNCS, vol. 1109, pp. 104–113. Springer, Heidelberg (1996)

23. Lange, T.: Vampire – virtual applications and implementations research lab (2007), http://hyperelliptic.org/ECRYPTII/vampire/ (accessed January 28, 2013)

24. Liu, A., Ning, P.: TinyECC: A Configurable Library for Elliptic Curve Cryptography in Wireless Sensor Networks. In: Proceedings of International Conference on Information Processing in Sensor Networks, IPSN 2008, St. Louis, Missouri, USA, MO, April 22–24, pp. 245–256 (April 2008)

25. Mangard, S., Oswald, E., Popp, T.: Power Analysis Attacks – Revealing the Secrets of Smart Cards. Springer (2007) ISBN 978-0-387-30857-9

26. Meiser, G., Eisenbarth, T., Lemke-Rust, K., Paar, C.: Efficient Implementation of eSTREAM Ciphers on 8-bit AVR Microcontrollers. In: International Symposium on Industrial Embedded Systems, SIES 2008, pp. 58–66 (June 2008)

27. Montgomery, P.L.: Speeding the Pollard and elliptic curve methods of factorization. Mathematics of Computation 48(177), 243–264 (1987),
http://www.ams.org/journals/mcom/1987-48-177/S0025-5718-1987-0866113-7/S0025-5718-1987-0866113-7.pdf

28. National Institute of Standards and Technology (NIST). FIPS-180-3: Secure Hash Standard (October 2008), http://www.itl.nist.gov/fipspubs/

29. Center of Mathematical Modeling Sigma. CRS-AVR010X-ECC,
http://www.cmmsigma.eu/products/crypto/crs_avr010x.en.html

30. Scott, M.: MIRACLE – A Multiprecision Integer and Rational Arithmetic C/C++ Library (2003), http://www.shamus.ie

31. Straus, E.G.: Addition chains of vectors (problem 5125). American Mathematical Monthly 70, 806–808 (1964), http://cr.yp.to/bib/1964/straus.html

32. Szczechowiak, P., Oliveira, L.B., Scott, M., Collier, M., Dahab, R.: NanoECC: Testing the Limits of Elliptic Curve Cryptography in Sensor Networks. In: Verdone, R. (ed.) EWSN 2008. LNCS, vol. 4913, pp. 305–320. Springer, Heidelberg (2008)

33. Uhsadel, L., Poschmann, A., Paar, C.: Enabling Full-Size Public-Key Algorithms on 8-Bit Sensor Nodes. In: Stajano, F., Meadows, C., Capkun, S., Moore, T. (eds.) ESAS 2007. LNCS, vol. 4572, pp. 73–86. Springer, Heidelberg (2007)

34. Wang, H., Li, Q.: Efficient Implementation of Public Key Cryptosystems on Mote Sensors (Short Paper). In: Ning, P., Qing, S., Li, N. (eds.) ICICS 2006. LNCS, vol. 4307, pp. 519–528. Springer, Heidelberg (2006)

35. Wenzel-Benner, C., Gräf, J.: XBX: eXternal Benchmarking eXtension,
http://xbx.das-labor.org/trac/wiki/WikiStart (accessed January 31, 2013)

W-OTS+– Shorter Signatures
for Hash-Based Signature Schemes

Andreas Hülsing*

Cryptography and Computeralgebra
Department of Computer Science
TU Darmstadt
huelsing@cdc.informatik.tu-darmstadt.de

Abstract. We present W-OTS+, a Winternitz type one-time signature scheme (W-OTS). We prove that W-OTS+ is strongly unforgeable under chosen message attacks in the standard model. Our proof is exact and tight. The first property allows us to compute the security of the scheme for given parameters. The second property allows for shorter signatures than previous proposals without lowering the security. This improvement in signature size directly carries over to all recent hash-based signature schemes. I.e. we can reduce the signature size by more than 50% for XMSS+ at a security level of 80 bits. As the main drawback of hash-based signature schemes is assumed to be the signature size, this is a further step in making hash-based signatures practical.

Keywords: digital signatures, one-time signature schemes, hash-based signatures, provable security, hash functions.

1 Introduction

Digital signatures are among the most important cryptographic primitives in practice. They have many applications, including the use in SSL/TLS and securing software updates. Hash-based or Merkle signature schemes (MSS) are an interesting alternative to the signature schemes used today, not only because they are assumed to resist quantum computer aided attacks, but also because of their fast signature generation and verification times as well as their strong security guarantees. Most MSS come with a standard model security proof and outperform RSA in many settings regarding runtimes. The main drawback of MSS is the signature size which to a large extent depends on the used one-time signature scheme (OTS). Recent MSS proposals [BDH11, HBB13] use a variant of the Winternitz OTS (W-OTS) introduced in [BDE+11]. The main reason for this choice is the reduced signature size. Using W-OTS, a MSS signature does not have to contain the OTS public key as it can be computed given the W-OTS signature. Moreover, W-OTS type signature schemes allow for a trade-off between signature size and runtime.

---

* Supported by grant no. BU 630/19-1 of the German Research Foundation (www.dfg.de).

A. Youssef, A. Nitaj, A.E. Hassanien (Eds.): AFRICACRYPT 2013, LNCS 7918, pp. 173–188, 2013.

In this work we introduce W-OTS$^+$, a W-OTS type OTS that allows to reduce the signature size more than previous W-OTS variants and reaches a higher level of security. We prove that W-OTS$^+$ is strongly unforgeable under adaptive chosen message attacks (SU-CMA) in the standard model, if the used hash function is second-preimage resistant, undetectable and one-way (Indeed, we only present the proof for EU-CMA security in this extended abstract). Previous proposals require non-standard assumptions to achieve SU-CMA security (i.e. „key-collision resistance" in case of [BDE$^+$11]). Besides the SU-CMA secure variants there exist W-OTS that achieve EU-CMA security, either using a collision resistant, undetectable hash function [HM02, DSS05] or a pseudorandom function family [BDE$^+$11]. The first security requirement is strictly stronger than that of W-OTS$^+$. While the second is comparable, the corresponding proof is less tight. However, both cases result in larger signatures.

Besides provable security we are also concerned with the practical performance of the scheme. We show how to use the exact security proof to compute the security level of W-OTS$^+$ for a given set of parameters. Moreover we discuss how to instantiate W-OTS$^+$ in practice and present parameter sizes for recent MSS (XMSS [BDH11], XMSS$^+$ [HBB13]) when instantiated with W-OTS$^+$.

Organization. We start by introducing W-OTS$^+$ in Section 2. Afterwards we state our main result about the security of W-OTS$^+$ and prove it in Section 3. In Section 4 we discuss possible instantiations and compare W-OTS$^+$ with previous proposals. Finally, we conclude in Section 5.

2   The Winternitz One-Time Signature Scheme

In this section we describe W-OTS$^+$. The core idea of all W-OTS is to use a certain number of function chains starting from random inputs. These random inputs are the secret key. The public key consists of the final outputs of the chains, i.e. the end of each chain. A signature is computed by mapping the message to one intermediate value of each function chain. All previous variants of W-OTS constructed the function chains as plain iteration of the used function (or function family in case of [BDE$^+$11]). In contrast, for W-OTS$^+$ we use a special mode of iteration which enables the tight security proof without requiring the used hash function family to be collision resistant. We start with some preliminaries. Afterwards we present W-OTS$^+$.

2.1   Signature Schemes

We now fix some notation and define digital signature schemes and existential unforgeability under adaptive chosen message attacks (EU-CMA). Through out the paper we write $x \xleftarrow{\$} \mathcal{X}$ if x is randomly chosen from the set $\mathcal{X}$ using the uniform distribution. We further write log for $\log_2$.

Digital Signature Schemes. Let $\mathcal{M}$ be the message space. A digital signature scheme Dss $=$ (Kg, Sign, Vf) is a triple of probabilistic polynomial time algorithms:

- Kg(1^n) on input of a security parameter 1^n outputs a private signing key sk and a public verification key pk;
- Sign(sk, M) outputs a signature σ under sk for message M, if $M \in \mathcal{M}$;
- Vf(pk, σ, M) outputs 1 iff σ is a valid signature on M under pk;

such that $\forall$(pk, sk) $\longleftarrow$ Kg(1^n), $\forall (M \in \mathcal{M})$: Vf(pk, Sign(sk, M), M) = 1.

EU-CMA Security. The standard security notion for digital signature schemes is existential unforgeability under adaptive chosen message attacks (EU-CMA), which is defined using the following experiment. By Dss(1^n) we denote a signature scheme with security parameter n.

Experiment $\mathsf{Exp}^{\text{EU-CMA}}_{\text{Dss}(1^n)}(\mathcal{A})$
   (sk, pk) $\longleftarrow$ Kg(1^n)
   $(M^\star, \sigma^\star) \longleftarrow \mathcal{A}^{\mathsf{Sign(sk,\cdot)}}$(pk)
   Let $\{(M_i, \sigma_i)\}_1^q$ be the query-answer pairs of Sign(sk, $\cdot$).
   Return 1 iff Vf(pk, $M^\star, \sigma^\star$) = 1 and $M^\star \notin \{M_i\}_1^q$.

For the success probability of an adversary $\mathcal{A}$ in the above experiment we write

$$\mathsf{Succ}^{\text{EU-CMA}}_{\text{Dss}(1^n)}(\mathcal{A}) = \Pr\left[\mathsf{Exp}^{\text{EU-CMA}}_{\text{Dss}(1^n)}(\mathcal{A}) = 1\right].$$

Using this, we define EU-CMA the following way.

Definition 1 (EU-CMA). *Let* $n, t, q \in \mathbb{N}$, $t, q = \text{poly}(n)$, Dss *a digital signature scheme. We call* Dss *EU-CMA-secure, if the maximum success probability* InSec$^{\text{EU-CMA}}$ (Dss(1^n); t, q) *of all possibly probabilistic adversaries* $\mathcal{A}$, *running in time* $\leq t$, *making at most q queries to* Sign *in the above experiment, is negligible in n:*

$$\text{InSec}^{\text{EU-CMA}}\left(\text{Dss}(1^n); t, q\right) \stackrel{def}{=} \max_{\mathcal{A}}\{\mathsf{Succ}^{\text{EU-CMA}}_{\text{Dss}(1^n)}(\mathcal{A})\} = negl(n).$$

An EU-CMA secure one-time signature scheme (OTS) is a Dss that is EU-CMA secure as long as the number of oracle queries of the adversary is limited to one, i.e. $q = 1$.

2.2 W-OTS$^+$

Now we present W-OTS$^+$. Like all previous variants of W-OTS, W-OTS$^+$ is parameterized by security parameter $n \in \mathbb{N}$, the message length m and the Winternitz parameter $w \in \mathbb{N}, w > 1$, which determines the time-memory trade-off. The last two parameters are used to compute

$$\ell_1 = \left\lceil \frac{m}{\log(w)} \right\rceil, \quad \ell_2 = \left\lfloor \frac{\log(\ell_1(w-1))}{\log(w)} \right\rfloor + 1, \quad \ell = \ell_1 + \ell_2.$$

Furthermore, W-OTS$^+$ uses a family of functions $\mathcal{F}_n : \{f_k : \{0,1\}^n \to \{0,1\}^n |$ $k \in \mathcal{K}_n\}$ with key space $\mathcal{K}_n$. The reader might think of it as a cryptographic hash function family that is non-compressing. Using $\mathcal{F}_n$ we define the following chaining function.

$c_k^i(x, \mathbf{r})$: On input of value $x \in \{0,1\}^n$, iteration counter $i \in \mathbb{N}$, key $k \in \mathcal{K}$ and randomization elements $\mathbf{r} = (r_1, \ldots, r_j) \in \{0,1\}^{n \times j}$ with $j \geq i$, the chaining function works the following way. In case $i = 0$, c returns x ($c_k^0(x, \mathbf{r}) = x$). For $i > 0$ we define c recursively as

$$c_k^i(x, \mathbf{r}) = f_k(c_k^{i-1}(x, \mathbf{r}) \oplus r_i),$$

i.e. in every round, the function first takes the bitwise xor of the intermediate value and bitmask r and evaluates f_k on the result afterwards. We write $\mathbf{r}_{a,b}$ for the subset $r_a, \ldots, r_b$ of $\mathbf{r}$. In case $b < a$ we define $\mathbf{r}_{a,b}$ to be the empty string. We assume that the parameters m, w and the function family $\mathcal{F}_n$ are publicly known. Now we describe the three algorithms of W-OTS$^+$:

Key Generation Algorithm ($\mathsf{Kg}(1^n)$): On input of security parameter n in unary the key generation algorithm choses $\ell + w - 1$ n-bit strings uniformly at random. The secret key $\mathsf{sk} = (\mathsf{sk}_1, \ldots, \mathsf{sk}_\ell)$ consists of the first ℓ random bit strings. The remaining $w - 1$ bit strings are used as the randomization elements $\mathbf{r} = (r_1, \ldots, r_{w-1})$ for c. Next, Kg chooses a function key $k \xleftarrow{\$} \mathcal{K}$ uniformly at random. The public verification key pk is computed as

$$\mathsf{pk} = (\mathsf{pk}_0, \mathsf{pk}_1, \ldots, \mathsf{pk}_\ell) = ((\mathbf{r}, k), c_k^{w-1}(\mathsf{sk}_1, \mathbf{r}), \ldots, c_k^{w-1}(\mathsf{sk}_\ell, \mathbf{r})).$$

Signature Algorithm ($\mathsf{Sign}(M, \mathsf{sk}, \mathbf{r})$): On input of a m bit message M, secret signing key sk and the randomization elements $\mathbf{r}$, the signature algorithm first computes a base w representation of M: $M = (M_1 \ldots M_{\ell_1})$, $M_i \in \{0, \ldots, w-1\}$. Therefor, M is treated as the binary representation of a natural number x and then the w-ary representation of x is computed. Next it computes the checksum

$$C = \sum_{i=1}^{\ell_1}(w - 1 - M_i)$$

and its base w representation $C = (C_1, \ldots, C_{\ell_2})$. The length of the base w representation of C is at most ℓ_2 since $C \leq \ell_1(w - 1)$. We set $B = (b_1, \ldots, b_\ell) = M \parallel C$, the concatenation of the base w representations of M and C. The signature is computed as

$$\sigma = (\sigma_1, \ldots, \sigma_\ell) = (c_k^{b_1}(\mathsf{sk}_1, \mathbf{r}), \ldots, c_k^{b_\ell}(\mathsf{sk}_\ell, \mathbf{r})).$$

Please note that the checksum guarantees that given the b_i, $0 < i \leq \ell$ corresponding to one message, the b_i' corresponding to any other message include at least one $b_i' < b_i$.

Verification Algorithm ($\mathsf{Vf}(1^n, M, \sigma, \mathsf{pk})$)*:* On input of message M of binary length m, a signature σ and a public verification key pk, the verification algorithm first computes the b_i, $1 \le i \le \ell$ as described above. Then it does the following comparison:

$$\mathsf{pk} = (\mathsf{pk}_0, \mathsf{pk}_1, \ldots, \mathsf{pk}_\ell)$$

$$\overset{?}{=} ((\mathbf{r}, k), c_k^{w-1-b_1}(\sigma_1, \mathbf{r}_{b_1+1, w-1}), \ldots, c_k^{w-1-b_\ell}(\sigma_\ell, \mathbf{r}_{b_\ell+1, w-1}))$$

If the comparison holds, it returns **true** and **false** otherwise.

The runtime of all three algorithms is bounded by ℓw evaluations of f_k. The size of a signature and the secret key is $|\sigma| = |\mathsf{sk}| = \ell n$ bits. The public key size is $(\ell + w - 1)n + |k|$ bits, where $|k|$ denotes the number of bits required to represent any element of $\mathcal{K}$.

3   Security of W-OTS$^+$

In this section we analyze the security of W-OTS$^+$. We prove W-OTS$^+$ is existentially unforgeable under chosen message attacks, if the used function family is a second-preimage resistant family of undetectable one-way functions. More precisely, we prove the following theorem:

Theorem 1. *Let* $n, w, m \in \mathbb{N}$, $w, m = poly(n)$, $\mathcal{F}_n : \{f_k : \{0,1\}^n \to \{0,1\}^n | k \in \mathcal{K}_n\}$ *a second preimage resistant, undetectable one-way function family. Then,* $\mathrm{InSec}^{\mathrm{EU\text{-}CMA}}\left(W\text{-}OTS^+(1^n, w, m); t, 1\right)$, *the insecurity of* $W\text{-}OTS^+$ *against an EU-CMA attack is bounded by*

$$\mathrm{InSec}^{\mathrm{EU\text{-}CMA}}\left(W\text{-}OTS^+(1^n, w, m); t, 1\right)$$
$$\le w \cdot \mathrm{InSec}^{\mathrm{UD}}\left(\mathcal{F}_n; t^\star\right) + w\ell \cdot \max\left\{\mathrm{InSec}^{\mathrm{OW}}\left(\mathcal{F}_n; t'\right), w \cdot \mathrm{InSec}^{\mathrm{SPR}}\left(\mathcal{F}_n; t'\right)\right\}$$

with $t' = t + 3\ell w$ *and* $t^\star = t + 3\ell w + w - 1$.

It seems natural to assume that the existence of a function that combines these properties is equivalent to the existence of a one-way function. As the function has to be one-way itself, the one direction is trivial. On the other hand, we know that second-preimage resistant functions exist if a one-way function exists [Rom90] and we know the same for undetectable functions, i.e. pseudorandom generators [HILL99]. We leave the question if this also implies the existence of a function family that combines all three properties for future work. If this was the case, it would mean that W-OTS$^+$ has minimal security requirements. The practical implications of the proof are discussed in the next section.

In this extended abstract we only prove that W-OTS$^+$ is EU-CMA secure. In fact it also fulfills the stronger notion of SU-CMA, where the adversary is also allowed to return a new signature on the message send to the signature oracle. The claimed bound in Theorem 1 holds for the SU-CMA case, too. We present the EU-CMA proof, because it contains all important ideas but has less different cases to handle. Before we present the proof we give some preliminaries. At the end of this sections we show how to compute the security level of W-OTS$^+$.

3.1   Preliminaries

In this subsection we provide some more notation and formal definitions. We denote the uniform distribution over bit strings of length n by $\mathcal{U}_n$. In our proofs, we measure all runtimes counting the evaluations of elements from $\mathcal{F}_n$. In some proofs and definitions we use the (distinguishing) advantage of an adversary which we now define.

Definition 2 (Advantage). *Given two distributions $\mathcal{X}$ and $\mathcal{Y}$, we define the advantage $\mathrm{Adv}_{\mathcal{X},\mathcal{Y}}(\mathcal{A})$ of an adversary $\mathcal{A}$ in distinguishing between these two distributions as*

$$\mathrm{Adv}_{\mathcal{X},\mathcal{Y}}(\mathcal{A}) = |\Pr\left[1 \longleftarrow \mathcal{A}(\mathcal{X})\right] - \Pr\left[1 \longleftarrow \mathcal{A}(\mathcal{Y})\right]|.$$

Functions. We now define three properties for families of functions that we use. In what follows, we only consider families $\mathcal{F}_n$ as defined in the last section. We require that it is possible given $n \in \mathbb{N}$ to sample a key k from key space $\mathcal{K}_n$ using the uniform distribution in polynomial time. Furthermore we require that all functions from $\mathcal{F}_n$ can be evaluated in polynomial time. We first recall the definitions of one-wayness (OW) and second preimage resistance (SPR).
   The success probability of an adversary against the one-wayness of $\mathcal{F}_n$ is:

$$\mathrm{Succ}_{\mathcal{F}_n}^{\mathrm{OW}}(\mathcal{A}) = \Pr[\, k \xleftarrow{\$} \mathcal{K}_n; x \xleftarrow{\$} \{0,1\}^n, y \longleftarrow f_k(x),$$
$$x' \xleftarrow{\$} \mathcal{A}(k,y) : y = f_k(x')] \tag{1}$$

The success probability of an adversary against the second preimage resistance of $\mathcal{F}_n$ is:

$$\mathrm{Succ}_{\mathcal{F}_n}^{\mathrm{SPR}}(\mathcal{A}) = \Pr[\, k \xleftarrow{\$} \mathcal{K}_n; x \xleftarrow{\$} \{0,1\}^n, x' \longleftarrow \mathcal{A}(k,x) :$$
$$(x \neq x') \wedge (f_k(x) = f_k(x'))] \tag{2}$$

We call a function family $\mathcal{F}_n$ one-way (second preimage resistant, resp.) if the respective success probability given above of any PPT adversary is negligible in n.
   Besides SPR and OW, we require $\mathcal{F}_n$ to provide another property called undetectability to proof W-OTS$^+$ secure. Intuitively, a function family is undetectable if its outputs can not be distinguished from uniformly random values. This is what we require from a pseudorandom generator, which in contrast to $\mathcal{F}_n$ has to be length expanding.
   To define undetectability, assume the following two distributions over $\{0,1\}^n \times \mathcal{K}$. A sample (u, k) from the first distribution $\mathcal{D}_{\mathrm{UD},\mathcal{U}}$ is obtained by sampling $u \longleftarrow \mathcal{U}_n$ and $k \xleftarrow{\$} \mathcal{K}$ uniformly at random from the respective domain. A sample (u, k) from the second distribution $\mathcal{D}_{\mathrm{UD},\mathcal{F}}$ is obtained by sampling $k \xleftarrow{\$} \mathcal{K}$ and then evaluating f_k on a uniformly random bit string, i.e. $u \longleftarrow f_k(\mathcal{U}_n)$. The advantage of an adversary $\mathcal{A}$ against the undetectability of $\mathcal{F}_n$ is simply the distinguishing advantage for these two distributions:

$$\mathrm{Adv}_{\mathcal{F}_n}^{\mathrm{UD}}(\mathcal{A}) = \mathrm{Adv}_{\mathcal{D}_{\mathrm{UD},\mathcal{U}},\mathcal{D}_{\mathrm{UD},\mathcal{F}}}(\mathcal{A})$$

Using this we define undetectability as:

Definition 3 (Undetectability (UD)). *Let* $n \in \mathbb{N}$, $\mathcal{F}_n$ *a family of functions as described above. We call* $\mathcal{F}_n$ *undetectable, if* $\mathrm{InSec}^{\mathrm{UD}}(\mathcal{F}_n; t)$ *the advantage of any adversary* $\mathcal{A}$ *against the undetectability of* $\mathcal{F}_n$ *running in time less or equal* t *is negligible:*

$$\mathrm{InSec}^{\mathrm{UD}}(\mathcal{F}_n; t) \stackrel{def}{=} \max_{\mathcal{A}} \{\mathrm{Adv}_{\mathcal{F}_n}^{\mathrm{UD}}(\mathcal{A})\} = negl(n).$$

Undetectability was already used by Dods et al. [DSS05] to prove a former version of W-OTS secure.

3.2 Security Proof

We now present the proof of Theorem 1. The general idea is, that because of the checksum, a successful forgery must contain at least one intermediate value x for one chain α, that is closer to the start value of chain α than the value σ_α contained in the answer to the signature query. We try to guess the position of σ_α and place our preimage challenge y_c there. So we can answer the signature query and hopefully extract a preimage given x. We also include a second preimage challenge in the same chain α, manipulating the randomization elements. This is necessary, as x must lead to the same public key value pk_α than y_c but the chain continued from x does not need to contain y_c as an intermediate value. But in this case it contains a second preimage which we try to extract.

Manipulating the public key to place our challenges, we slightly change the distribution of the key. In the second part of the proof we show that this does not significantly change the success probability of the adversary using the undetectability of $\mathcal{F}_n$.

Proof (of Theorem 1). For the sake of contradiction assume there exists an adversary $\mathcal{A}$ that can produce existential forgeries for W-OTS$^+$($1^n, w, m$) running an adaptive chosen message attack in time $\leq t$ and with success probability $\epsilon_{\mathcal{A}} = \mathrm{Succ}_{\mathrm{W\text{-}OTS}(1^n, w, m)}^{\mathrm{EU\text{-}CMA}}(\mathcal{A})$ greater than the claimed bound $\mathrm{InSec}^{\mathrm{EU\text{-}CMA}}(\mathrm{W\text{-}OTS}^+$ ($1^n, w, m$); $t, 1$). We first show how to construct an oracle machine $\mathcal{M}^{\mathcal{A}}$ that either breaks the second preimage resistance or one-wayness of $\mathcal{F}_n$ using $\mathcal{A}$ with a possibly different input distribution. A pseudo-code description of $\mathcal{M}^{\mathcal{A}}$ is given as Algorithm 1.

The oracle machine $\mathcal{M}^{\mathcal{A}}$ first runs the W-OTS$^+$ key generation to obtain a key pair (sk, pk). Then, $\mathcal{M}^{\mathcal{A}}$ selects the positions to place its challenges in the public key. Therefor it selects a random function chain choosing the index α. Second it chooses an index β to select a random intermediate value of this chain. $\mathcal{M}^{\mathcal{A}}$ places the preimage challenge at this position. This is done, setting y_c as the βth intermediate value of the chain. If $\beta < w - 1$, i.e. $\mathcal{M}^{\mathcal{A}}$ did not sample the last position in the chain, another intermediate value between β and the end of the chain is selected, sampling γ. $\mathcal{M}^{\mathcal{A}}$ places the second preimage challenge at the input of the γth evaluation of the chain continued from y_c, replacing

Algorithm 1. $\mathcal{M}^{\mathcal{A}}$

Input: Security parameter n, function key k, one-way challenge y_c and second preimage resistance challenge x_c.

Output: A value x that is either a preimage of y_c or a second preimage for x_c under f_k or fail.

1. Run $\mathsf{Kg}(1^n)$ to generate W-OTS$^+$ key pair $(\mathsf{sk}, \mathsf{pk})$
2. Choose indices $\alpha \xleftarrow{\$} \{1, ..., \ell\}, \beta \xleftarrow{\$} \{1, ..., w-1\}$ uniformly at random
3. **If** $\beta = w - 1$ **then** set $\mathbf{r}' = \mathbf{r}$
4. **Else**
    (a) Choose index $\gamma \xleftarrow{\$} \{\beta+1, ..., w-1\}$ uniformly at random
    (b) Obtain $\mathbf{r}'$ from $\mathbf{r}$, replacing r_γ by $c_k^{\gamma-\beta-1}(y_c, \mathbf{r}_{\beta+1,l}) \oplus x_c$.
5. Obtain pk' by setting $\mathsf{pk}'_i = c_k^{w-1}(\mathsf{sk}_i, \mathbf{r}'), 0 < i \le \ell, i \ne \alpha$,
    $\mathsf{pk}'_\alpha = c_k^{w-1-\beta}(y_c, \mathbf{r}'_{\beta+1,w-1})$ and $\mathsf{pk}_0 = (\mathbf{r}', k)$
6. Run $\mathcal{A}^{\mathsf{Sign}(\mathsf{sk}, \cdot)}(\mathsf{pk}')$
7. **If** $\mathcal{A}^{\mathsf{Sign}(\mathsf{sk}, \cdot)}(\mathsf{pk}')$ queries Sign with message M **then**
    (a) compute $B = (b_1, ..., b_\ell)$
    (b) **If** $b_\alpha < \beta$ **then return** fail
    (c) Generate signature σ of M:
        i. Run $\sigma = (\sigma_1, ..., \sigma_\ell) \longleftarrow \mathsf{Sign}(M, \mathsf{sk}, \mathbf{r}')$
        ii. Set $\sigma_\alpha = c_k^{b_\alpha - \beta}(y_c, \mathbf{r}'_{\beta+1,w-1})$
    (d) Reply to query using σ
8. **If** $\mathcal{A}^{\mathsf{Sign}(\mathsf{sk}, \cdot)}(\mathsf{pk})$ returns valid (σ', M') **then**
    (a) Compute $B' = (b'_1, ..., b'_\ell)$
    (b) **If** $b'_\alpha \ge \beta$ **return** fail
    (c) **If** $\beta = w - 1$
        i. **Return** preimage $c_k^{w-1-b'_\alpha-1}(\sigma'_\alpha, \mathbf{r}'_{b'_\alpha+1,w-1}) \oplus r_{w-1}$
    (d) **Else**
        i. **If** $c_k^{\beta-b'_\alpha}(\sigma'_\alpha, \mathbf{r}'_{b'_\alpha+1,w-1}) = y_c$ **then**
            return preimage $c_k^{\beta-b'_\alpha-1}(\sigma'_\alpha, \mathbf{r}'_{b'_\alpha+1,w-1}) \oplus r_\beta$
        ii. **Else if** $x' = c_k^{\gamma-b'_\alpha-1}(\sigma'_\alpha, \mathbf{r}'_{b'_\alpha+1,w-1}) \oplus r_\gamma \ne x_c$ **and** $c_k^{\gamma-b'_\alpha}(\sigma'_\alpha, \mathbf{r}'_{b'_\alpha+1,w-1}) = c_k^{\gamma-\beta}(y_c, \mathbf{r}_{\beta+1,w-1})$ **return** second preimage x'
9. In any other case **return** fail

---

the randomization element r_γ (Line 4b). A manipulated public key pk' is computed using the new set of randomization elements. The αth value of pk' is computed continuing the chain from y_c at position β (Line 5). Then $\mathcal{M}^{\mathcal{A}}$ runs $\mathcal{A}$ on input pk'.

W.l.o.g. we assume that $\mathcal{A}$ asks for the signature on one message M (Line 7). So $\mathcal{M}^{\mathcal{A}}$ computes the b_i as described in the signature algorithm. $\mathcal{M}^{\mathcal{A}}$ knows the secret key value sk_i for all chains with exception of chain α. For chain α $\mathcal{M}^{\mathcal{A}}$ only knows the βth intermediate value. Hence, $\mathcal{M}^{\mathcal{A}}$ can answer the query if $b_\alpha \ge \beta$ as all intermediate values $\ge \beta$ of the αth chain can be computed using y_c. If this is not the case, $\mathcal{M}^{\mathcal{A}}$ aborts.

If $\mathcal{A}$ returns an existential forgery (σ', M'), $\mathcal{M}^{\mathcal{A}}$ computes the b_i'. The forgery is only useful if $b_\alpha' < \beta$. If this is not the case, $\mathcal{M}^{\mathcal{A}}$ returns fail. Now, there are two mutually exclusive cases. If $\beta = w-1$, i.e. we selected the end of chain α, the forgery contains a preimage of y_c. This is the case because σ'_α is an intermediate value of chain alpha that ends in y_c. So, $\mathcal{M}^{\mathcal{A}}$ extracts the preimage and returns it (Line 8(c)i). Otherwise, there are again two mutually exclusive cases. The chain continued from σ'_α either has y_c as the βth intermediate value or it has not. In the first case, again a preimage can be extracted (Line 8(d)i). In the second case, the chains continued from y_c and σ'_α must collide at some position between $\beta + 1$ and $w - 1$ according to the pigeonhole principle. If they collide at position γ for the first time, a second preimage for x_c can be extracted (Line 8(d)ii). Otherwise $\mathcal{M}^{\mathcal{A}}$ aborts.

Now we compute the success probability of $\mathcal{M}^{\mathcal{A}}$. To make it easier, we only compute the probability for a certain success case. We assume that the b_α obtained from $\mathcal{A}$'s query equals β. This happens with probability w^{-1} as β was chosen uniformly at random. As our modifications might have changed the input distribution of $\mathcal{A}$, it does not necessarily succeed with probability $\epsilon_{\mathcal{A}}$. For the moment we only denote the probability that $\mathcal{A}$ returns a valid forgery when run by $\mathcal{M}^{\mathcal{A}}$ as $\epsilon'_{\mathcal{A}}$. Because of the construction of the check sum, M' leads to at least one $b_i' < b_i, 0 < i \le \ell$. With probability ℓ^{-1} this happens for $i = \alpha$ and the condition in line 8b is fulfilled. At this point there are two mutually exclusive cases, so one of them occurs with probability p and the other one with probability $(1 - p)$.

Case 1: Either $\beta = w - 1$ or the chain continued from σ'_α has y_c as the βth intermediate value. In this case, $\mathcal{M}^{\mathcal{A}}$ returns a preimage for y_c with probability 1.

Case 2: $\beta < w - 1$ and the chain continued from σ'_α does not have y_c as the βth intermediate value. In this case, $\mathcal{M}^{\mathcal{A}}$ returns a second preimage for x_c if the two chains collide for the first time at position γ. This happens with probability greater w^{-1} as gamma was chosen uniformly at random from within the interval $[\beta + 1, w - 1]$.

Using the assumptions about the one-wayness and second preimage resistance of $\mathcal{F}_n$ we can bound the success probability of $\mathcal{A}$ if called by $\mathcal{M}^{\mathcal{A}}$:

$$\epsilon'_{\mathcal{A}} \le w\ell \cdot \max\left\{\text{InSec}^{\text{OW}}\left(\mathcal{F}_n; t'\right), w \cdot \text{InSec}^{\text{SPR}}\left(\mathcal{F}_n; t'\right)\right\} \tag{3}$$

where the time $t' = t + 3\ell w$ is an upper bound for the runtime of $\mathcal{A}$ plus the time needed to run each algorithm of W-OTS$^+$ once.

As the second step, we bound the difference between the success probability $\epsilon'_{\mathcal{A}}$ of $\mathcal{A}$ when called by $\mathcal{M}^{\mathcal{A}}$ and its success probability $\epsilon_{\mathcal{A}}$ in the original experiment. If the first is greater than the latter we already have a contradiction. Hence we assume $\epsilon_{\mathcal{A}} \ge \epsilon'_{\mathcal{A}}$ in what follows. Please note, that among the elements of pk$'$ only the distribution of pk$'_\alpha$ might differ from the distribution of a public key generated by Kg. r_γ is uniformly distributed in $\{0,1\}^n$, because x_c is uniformly distributed in $\{0,1\}^n$. We define two distributions $\mathcal{D}_{\mathcal{M}}$ and $\mathcal{D}_{\text{Kg}}$ over $\{0, \ldots, w-1\} \times \{0,1\}^n \times \{0,1\}^{(n \times w-1)} \times \mathcal{K}$. A sample $(\beta, u, \mathbf{r}, k)$ follows $\mathcal{D}_{\mathcal{M}}$ if the entries

$\beta \xleftarrow{\$} \{0, \ldots, w-1\}$, $u \xleftarrow{\$} \{0,1\}^n$, $\mathbf{r} \xleftarrow{\$} \{0,1\}^{n \times w-1}$ and $k \xleftarrow{\$} \mathcal{K}$ are chosen uniformly at random. A sample $(\beta, u, \mathbf{r}_{1,i}, k)$ follows $\mathcal{D}_{\mathsf{Kg}}$ if $\beta \xleftarrow{\$} \{0, \ldots, w-1\}$, $\mathbf{r} \xleftarrow{\$} \{0,1\}^{n \times w-1}$ and $k \xleftarrow{\$} \mathcal{K}$ are chosen uniformly at random and $u = c_k^\beta(\mathcal{U}_n, \mathbf{r})$. So the two distributions only differ in the way u is chosen. We now construct an oracle machine $\mathcal{M}'^{\mathcal{A}}$ that uses the possibly different behavior of $\mathcal{A}$ when given differently distributed inputs, to distinguish between $\mathcal{D}_{\mathsf{Kg}}$ and $\mathcal{D}_{\mathcal{M}}$. Using $\mathcal{M}'^{\mathcal{A}}$ we can then upper bound $\epsilon_{\mathcal{A}}$ by a function of the distinguishing advantage of $\mathcal{M}'^{\mathcal{A}}$ and $\epsilon'_{\mathcal{A}}$. Afterwards we use a hybrid argument to bound the distinguishing advantage of $\mathcal{M}'^{\mathcal{A}}$ using the undetectability of $\mathcal{F}_n$.

The oracle machine $\mathcal{M}'^{\mathcal{A}}$ works the following way. On input of a sample $(\beta, u, \mathbf{r}, k)$ that is either chosen from $\mathcal{D}_{\mathcal{M}}$ or from $\mathcal{D}_{\mathsf{Kg}}$, $\mathcal{M}'^{\mathcal{A}}$ generates a W-OTS$^+$ key pair. Instead of using Kg, $\mathcal{M}'^{\mathcal{A}}$ samples a secret key $\mathsf{sk} \xleftarrow{\$} \{0,1\}^{n \times \ell}$ and an index $\alpha \xleftarrow{\$} \{1, \ldots, \ell\}$ uniformly at random. It computes the public key pk as $\mathsf{pk}_0 = (\mathbf{r}, k)$ and

$$\mathsf{pk}_i = \begin{cases} c_k^{w-1}(\mathsf{sk}_i, \mathbf{r}) & \text{, if } 1 \leq i \leq \ell \text{ and } i \neq \alpha \\ c_k^{w-1-\beta}(u, \mathbf{r}_{\beta+1,w-1}) & \text{, if } i = \alpha. \end{cases}$$

Then $\mathcal{M}'^{\mathcal{A}}$ runs $\mathcal{A}$ on input pk. If $\mathcal{A}$ queries $\mathcal{M}'^{\mathcal{A}}$ for the signature on a message M, $\mathcal{M}'^{\mathcal{A}}$ behaves the same way as $\mathcal{M}^{\mathcal{A}}$. If $b_\alpha \geq \beta$, $\mathcal{M}'^{\mathcal{A}}$ uses sk and u to compute the signature, otherwise it aborts. If $\mathcal{A}$ returns a valid forgery, $\mathcal{M}'^{\mathcal{A}}$ returns 1 and otherwise 0. The runtime of $\mathcal{M}'^{\mathcal{A}}$ is bounded by the runtime of $\mathcal{A}$ plus one evaluation of each algorithm of W-OTS$^+$. So we get $t'' = t + 3\ell w$ as an upper bound.

Now, we compute the distinguishing advantage $\mathrm{Adv}_{\mathcal{D}_{\mathcal{M}}, \mathcal{D}_{\mathsf{Kg}}}(\mathcal{M}'^{\mathcal{A}})$ of $\mathcal{M}'^{\mathcal{A}}$. If the sample is taken from $\mathcal{D}_{\mathcal{M}}$, the distribution of the public keys pk generated by $\mathcal{M}'^{\mathcal{A}}$ is the same as the distribution of the public keys pk' generated by $\mathcal{M}^{\mathcal{A}}$. Hence $\mathcal{M}'^{\mathcal{A}}$ outputs 1 with probability

$$\Pr\left[(\beta, u, \mathbf{r}, k) \longleftarrow \mathcal{D}_{\mathcal{M}} : 1 \longleftarrow \mathcal{M}'^{\mathcal{A}}(\beta, u, \mathbf{r}, k)\right] = \epsilon'_{\mathcal{A}}.$$

If the sample was taken from $\mathcal{D}_{\mathsf{Kg}}$, the public keys pk generated by $\mathcal{M}'^{\mathcal{A}}$ follow the same distribution than those generated by Kg and so $\mathcal{M}'$ outputs 1 with probability

$$\Pr\left[(\beta, u, \mathbf{r}, k) \longleftarrow \mathcal{D}_{\mathsf{Kg}} : 1 \longleftarrow \mathcal{M}'^{\mathcal{A}}(\beta, u, \mathbf{r}, k)\right] = \epsilon_{\mathcal{A}}.$$

So the distinguishing advantage of $\mathcal{M}'^{\mathcal{A}}$ is

$$\mathrm{Adv}_{\mathcal{D}_{\mathsf{Kg}}, \mathcal{D}_{\mathcal{M}}}(\mathcal{M}'^{\mathcal{A}}) = |\epsilon_{\mathcal{A}} - \epsilon'_{\mathcal{A}}|.$$

As mentioned above, we only have to consider the case $\epsilon_{\mathcal{A}} \geq \epsilon'_{\mathcal{A}}$. So we obtain the following bound on $\epsilon_{\mathcal{A}}$:

$$\epsilon_{\mathcal{A}} = \mathrm{Adv}_{\mathcal{D}_{\mathsf{Kg}}, \mathcal{D}_{\mathcal{M}}}(\mathcal{M}'^{\mathcal{A}}) + \epsilon'_{\mathcal{A}} \tag{4}$$

We now limit the distinguishing advantage of $\mathcal{M}'^{\mathcal{A}}$ in our last step. We use a hybrid argument to show that this advantage is bound by the undetectability of $\mathcal{F}_n$.

For a given $\beta \in \{0, \ldots, w-1\}$, we define the hybrids $H_j = (\beta, c_k^{\beta-j}(\mathcal{U}_n, \mathbf{r}_{j+1,w-1}),$ $\mathbf{r}, k)$ with $\mathbf{r} \xleftarrow{\$} \{0,1\}^{n \times w-1}, k \xleftarrow{\$} \mathcal{K}$ for $0 \leq j \leq \beta$. Given an adversary $\mathcal{B}$ that can distinguish between H_0 and H_β with advantage $\epsilon_\mathcal{B}$, a hybrid argument leads that there must exist two consecutive hybrids that $\mathcal{B}$ distinguishes with advantage $\geq \epsilon_\mathcal{B}/\beta$. Assume these two hybrids are H_α and $H_{\alpha+1}$. Then we can construct an oracle machine $\mathcal{M}''^\mathcal{B}$ that uses $\mathcal{B}$ to distinguish between $\mathcal{D}_{\mathrm{UD},\mathcal{U}}$ and $\mathcal{D}_{\mathrm{UD},\mathcal{F}}$ as defined in the preliminaries and thereby attacking the undetectability of $\mathcal{F}_n$. Given a distinguishing challenge (u, k), $\mathcal{M}''^\mathcal{B}$ selects $\mathbf{r} \xleftarrow{} \mathcal{U}_n^{w-1}$, computes $x = c^{\beta-(\alpha+1)}(u, \mathbf{r}_{\alpha+2,w-1})$, runs $b \xleftarrow{} \mathcal{B}(\beta, x, \mathbf{r}, k)$ and outputs b.

Let's analyze the advantage $\mathrm{Adv}_{\mathcal{F}_n}^{\mathrm{UD}}(\mathcal{M}''^\mathcal{B})$ of $\mathcal{M}''^\mathcal{B}$. If the sample is taken from $\mathcal{D}_{\mathrm{UD},\mathcal{U}}$, u is uniformly random and $x = c^{\beta-(\alpha+1)}(u, \mathbf{r}_{\alpha+2,w-1})$ is distributed exactly like the second element of $H_{\alpha+1}$. Otherwise, if the sample is taken from $\mathcal{D}_{\mathrm{UD},\mathcal{F}}$, then $u \xleftarrow{} f_k(\mathcal{U}_n)$ is an output of f_k and we get

$$x = c^{\beta-(\alpha+1)}(f_k(\mathcal{U}_n), \mathbf{r}_{\alpha+2,w-1}) = c^{\beta-(\alpha+1)+1}(\mathcal{U}_n \oplus r_{\alpha+1}, \mathbf{r}_{\alpha+1,w-1})$$

$$= c^{\beta-\alpha}(\mathcal{U}_n, \mathbf{r}_{\alpha+1,w-1}) = H_{\alpha(2)}$$

where $H_{\alpha(2)}$ denotes the second element of H_α. Here we used the fact, that the xor of a uniformly distributed variable and a fixed value leads again to a uniformly distributed variable. Summing up, the input of $\mathcal{B}$, produced by $\mathcal{M}''^\mathcal{B}$ is either distributed like H_α or like $H_{\alpha+1}$, depending on $\mathcal{M}''^\mathcal{B}$s distinguishing challenge. Hence, the advantage of $\mathcal{M}''^\mathcal{B}$ is exactly that of $\mathcal{B}$ distinguishing between these two hybrids. So we get

$$\mathrm{Adv}_{\mathcal{F}_n}^{\mathrm{UD}}(\mathcal{M}''^\mathcal{B}) \geq \epsilon_D/\beta.$$

As the advantage of $\mathcal{M}''^\mathcal{B}$ is bounded by the undetectability of $\mathcal{F}_n$ per assumption, $\mathcal{M}'^\mathcal{A}$ does exactly what we assume $\mathcal{B}$ to do and the runtime of $\mathcal{M}''^\mathcal{B}$ is that of $\mathcal{B}$ plus at most $w-1$ evaluations of elements from $\mathcal{F}_n$, we get

$$\mathrm{InSec}^{\mathrm{UD}}(\mathcal{F}_n; t^\star) \geq \mathrm{Adv}_{\mathcal{F}_n}^{\mathrm{UD}}(\mathcal{M}''^\mathcal{B}) \geq \frac{\epsilon_\mathcal{B}}{i} = \frac{\mathrm{Adv}_{\mathcal{D}_{\mathrm{Kg}},\mathcal{D}_\mathcal{M}}(\mathcal{M}'^\mathcal{A})}{\beta}$$

where $t^\star = t'' + w - 1 = t + 3\ell w + w - 1$ is the runtime of $\mathcal{M}''^\mathcal{B}$. As $\beta \in \{0, \ldots, w-1\}$, we obtain the following bound on the advantage of $\mathcal{M}'^\mathcal{A}$:

$$\mathrm{Adv}_{\mathcal{D}_{\mathrm{Kg}},\mathcal{D}_\mathcal{M}}(\mathcal{M}'^\mathcal{A}) \leq w \cdot \mathrm{InSec}^{\mathrm{UD}}(\mathcal{F}_n; t^\star). \tag{5}$$

Putting equations (3), (4) and (5) together we obtain a final bound on $\epsilon_\mathcal{A}$ which leads the required contradiction:

$$\epsilon_\mathcal{A} \leq w \cdot \mathrm{InSec}^{\mathrm{UD}}(\mathcal{F}_n; t^\star) + w\ell \cdot \max\{\mathrm{InSec}^{\mathrm{OW}}(\mathcal{F}_n; t'), w \cdot \mathrm{InSec}^{\mathrm{SPR}}(\mathcal{F}_n; t')\}$$

with $t' = t + 3\ell w$ and $t^\star = t + 3\ell w + w - 1$. $\qquad\square$

3.3   Security Level of W-OTS$^+$

Given Theorem 1, we can compute the security level in the sense of [Len04]. This allows a comparison of the security of W-OTS$^+$ with the security of a symmetric primitive like a block cipher for given security parameters. Following [Len04], we say that W-OTS$^+$ has security level b if a successful attack on the scheme can be expected to require 2^{b-1} evaluations of functions from $\mathcal{F}_n$ on average. We can compute the security level, finding a lower bound for t s.th. $1/2 \leq \text{InSec}^{\text{EU-CMA}} (\text{W-OTS}(1^n, w, m); t, 1)$. According to the proof of Theorem 1, W-OTS$^+$ can only be attacked by either attacking the second preimage resistance, one-wayness or undetectability of $\mathcal{F}_n$. Following the reasoning in [Len04], we only take into account generic attacks on $\mathcal{F}_n$.

Regarding the insecurity of $\mathcal{F}(n)$ under generic attacks we assume $\text{InSec}^{\text{SPR}} (\mathcal{F}(n); t) = \text{InSec}^{\text{OW}} (\mathcal{F}(n); t) = \frac{t}{2^n}$ which corresponds to a brute force search for (second-)preimages. For the insecurity regarding undetectability we assume $\text{InSec}^{\text{UD}} (\mathcal{F}(n); t) = \frac{t}{2^n}$ following [DSS05]. In the following we assume that the small additive increase of the attack runtime coming from the reduction is negligible, compared to the value of t for any practical attack. So we assume $t = t' = t^\star$. We compute the lower bound on t.

$$\frac{1}{2} \leq \text{InSec}^{\text{EU-CMA}} (\text{W-OTS}(1^n, w, m); t, 1)$$

$$\leq w\frac{t}{2^n} + w\ell \cdot \max\left\{\frac{t}{2^n}, w \cdot \frac{t}{2^n}\right\} = \frac{tw}{2^n} + \frac{tw^2\ell}{2^n} = \frac{t(w^2\ell + w)}{2^n}$$

Solving this for t gives us

$$t \geq \frac{1}{2} \cdot \frac{2^n}{w^2\ell + w} = 2^{n-1-\log(w^2\ell+w)}.$$

So, for the security level b we obtain $b \geq n - \log(w^2\ell + w)$.

4   W-OTS$^+$ in Practice

In this section we discuss the practical implications of our result. We first present practical instantiations of W-OTS$^+$. Then we discuss the implications of the new security proof, comparing W-OTS$^+$ to other W-OTS type OTS and present results for XMSS and XMSS$^+$ when instantiated using W-OTS$^+$.

4.1   Instantiations

To use W-OTS$^+$ in practice $\mathcal{F}_n$ has to be instantiated. We propose two different instantiations. The first and most obvious way to instantiate $\mathcal{F}_n$ is to simply use a cryptographic hash function like SHA2 or SHA3. These functions are assumed to fulfill all the properties we require $\mathcal{F}_n$ to provide. In case the input length of the function is bigger then the output length, we pad the inputs using the required number of zeros. As we do not allow arbitrary length messages, we do not need a more involved padding.

Another way is to use a block cipher. It is well known that a cryptographic hash function can be constructed using a block cipher. This is very useful, as many smart cards and CPUs provide hardware acceleration for AES. To construct $\mathcal{F}_n$ using a block cipher, we apply the Matyas-Meyer-Oseas (MMO) construction [MMO85] in a manner similar to [BDH11]. The MMO construction was shown to be secure by Black et al. [BRS02]. Assume we have a block cipher $E_n : \{0,1\}^n \times \{0,1\}^n \to \{0,1\}^n$ with block and key size n. Then we construct $\mathcal{F}_n$ with key space $\mathcal{K} = \{0,1\}^n$ defining the elements of $\mathcal{F}_n$ as $f_k(x) = E_k(x) \oplus x$ where $E_K(M)$ denotes an evaluation of E using key K and message M. So, one evaluation of f_k takes either one evaluation of the used hash function or one evaluation of the underlying block cipher.

4.2   Performance Comparison

We now compare the performance of W-OTS+ with that of the schemes from [DSS05] and [BDE+11] which we call W-OTS^{CR} and W-OTS^{PRF}, respectively. Comparing W-OTS+ with W-OTS^{CR}, the most important point is, that W-OTS^{CR} requires an undetectable collision resistant hash function. While this is a strictly stronger security requirement, it also has practical implications. Namely, collision resistance is threatened by birthday attacks. Hence, to achieve a security level of b bits, a hash function with $n = 2b$ bits output size is required. This leads to larger signatures and slows down the scheme, as in general hash functions get slower with increased output size. It is possible to reach the same signature size as for W-OTS+ using a greater w, but this further slows down the scheme. On the other hand, the W-OTS+ public key is bigger than that of W-OTS^{CR} which has only ℓn bits. This is because of the randomization elements. But as we will show later, this is of no relevance in many practical scenarios as we can reuse randomness.

Comparing W-OTS+ with W-OTS^{PRF}, the differences are more subtle. First, looking at the instantiations, when using a hash function H to instantiate W-OTS^{PRF}, two evaluations of H are needed per evaluation of $\mathcal{F}_n$ (see [BDH11]) in contrast to one for W-OTS+. So the runtimes are doubled in this case. For a block cipher based instantiation the runtimes are the same. Second, at a first glance the sizes of both schemes are the same, only the W-OTS+ public key contains the additional randomization elements. But the bit security of W-OTS^{PRF} is $n - w - 1 - 2\log(\ell w)$, i.e. it contains w as a negative linear term while the bit security of W-OTS+ only looses a term logarithmic in w. In practice, the consequence of this difference is that the possible choices for w are limited if we target a certain bit security. This is best illustrated in the following example. Table 1 shows sizes and runtimes for a signature size below 1kB at a security level of 100 bit or more. Using W-OTS^{PRF} it is simply impossible to achieve a signature size below 1kB at 100 bit security. For W-OTS^{CR} it is theoretically possible, but one needs more than 10 times the number of evaluations of $\mathcal{F}_n$ which are also slower because of the bigger n.

Table 1. Parameters for signatures below 1kB for message length $m = 256$ and security level $b \geq 100$. For W-OTS^{PRF} this is impossible so we give the best possible signature size for $b \geq 100$. Runtime is given in number of evaluations of $\mathcal{F}_n$. As key generation, signature and verification times are the same, we only included the signature time t_{Sign}.

	n	w	$|\sigma|$	t_{Sign}	b
W-OTS$^+$	128	21	992	1,302	113
W-OTS^{CR}	256	455	992	14,105	128
W-OTS^{PRF}	128	8	1,440	720	100

4.3   Impact on XMSS and XMSS$^+$

OTS have numerous applications. The application that motivated this work is usage in hash-based signature schemes. Current hash-based signature schemes like XMSS [BDH11] and XMSS$^+$ [HBB13] are based on W-OTS^{PRF} which turned out to be the best choice for an OTS so far. In the following we will shortly discuss what happens if we replace W-OTS^{PRF} by W-OTS$^+$. We do not describe XMSS and XMSS$^+$ in detail due to the constrained space and refer the reader to the original papers. Table 2 shows a table from [HBB13] where we recomputed the results for the case that W-OTS$^+$ is used. Where the values changed, we included the old values for W-OTS^{PRF} in brackets. The table shows, that in most cases the public key of the overall scheme does not change. The reason is that XMSS and XMSS$^+$ public keys already contain public randomization elements that can be reused. There is only one case where randomization elements have to be added. We assume that the runtimes do not change. The W-OTS^{PRF}

Table 2. Results for XMSS and XMSS$^+$ using W-OTS$^+$ for message length $m = 256$ on an Infineon SLE78. We use the same k and w for both trees. b denotes the security level in bits. The signature times are worst case times. Numbers in brackets are the values when using W-OTS^{PRF}.

Scheme	h	k	w	Timings (ms)			Sizes (byte)			b
				KeyGen	Sign	Verify	Secret key	Public key	Signature	
XMSS$^+$	16	2	4	5,600	106	25	3,760	544	3,476	96 (85)
XMSS$^+$	16	2	8	5,800	105	21	3,376	512	2,436	95 (81)
XMSS$^+$	16	2	16	6,700	118	22	3,200	512	1,892	93 (71)
XMSS$^+$	16	2	32	10,500	173	28	3,056	544 (480)	1,588	92 (54)
XMSS$^+$	20	4	4	22,200	106	25	4,303	608	3,540	92 (81)
XMSS$^+$	20	4	8	22,800	105	21	3,920	576	2,500	91 (77)
XMSS$^+$	20	4	16	28,300	124	22	3,744	576	1,956	89 (67)
XMSS$^+$	20	4	32	41,500	176	28	3,600	544	1,652	88 (50)
XMSS	10	4	4	14,600	86	22	1,680	608	2,292	103 (92)
XMSS	10	4	16	18,800	100	17	1,648	576	1,236	100 (78)
XMSS	16	4	4	925,400	134	23	2,448	800	2,388	97 (86)
XMSS	16	4	16	1,199,100	159	18	2,416	768	1,332	94 (72)

function chains were implemented using one AES encryption per iteration. As shown above the same can be done for W-OTS$^+$, requiring one additional xor operation per AES evaluation. This should not lead any recognizable overhead. Moreover, the table shows that certain parameter sets — those with small signatures — have a very low level of security when using W-OTS^{PRF}. In practice a scheme has to provide at least a security level of 80 bits. Hence, these parameter sets could not be used before. Using W-OTS$^+$, the same parameter sets now lead to a level of security above 80 bits. Hence, they can now be used in practice.

5 Conclusion

In this work we introduced W-OTS$^+$. We proved its security, showed how to compute the security level of a given parameter set and discussed possible practical instantiations. As shown in the last section, W-OTS$^+$ can be used to decrease the signature size of hash-based signature schemes significantly without lowering the security of the scheme. I.e. we can decrease the signature size by 50% for XMSS$^+$ at a security level of 80 bits. Hopefully this leads to a broader acceptance of hash-based signature schemes, as the signature size was so far assumed to be the main drawback of these schemes. The only drawback of W-OTS$^+$ compared to previous W-OTS variants is the increased public key size. As for the case of hash-based signature schemes, it might be possible to reuse public randomness in other scenarios to mitigate this, too. An interesting question we left open is whether the existance of a one-way function implies the existence of a second-preimage resistant family of undetectable one-way functions.

References

[BDE⁺11] Buchmann, J., Dahmen, E., Ereth, S., Hülsing, A., Rückert, M.: On the Security of the Winternitz One-Time Signature Scheme. In: Nitaj, A., Pointcheval, D. (eds.) AFRICACRYPT 2011. LNCS, vol. 6737, pp. 363–378. Springer, Heidelberg (2011)

[BDH11] Buchmann, J., Dahmen, E., Hülsing, A.: XMSS - A Practical Forward Secure Signature Scheme Based on Minimal Security Assumptions. In: Yang, B.-Y. (ed.) PQCrypto 2011. LNCS, vol. 7071, pp. 117–129. Springer, Heidelberg (2011)

[BRS02] Black, J.A., Rogaway, P., Shrimpton, T.: Black-box analysis of the block-cipher-based hash-function constructions from PGV. In: Yung, M. (ed.) CRYPTO 2002. LNCS, vol. 2442, pp. 320–335. Springer, Heidelberg (2002)

[DSS05] Dods, C., Smart, N.P., Stam, M.: Hash Based Digital Signature Schemes. In: Smart, N.P. (ed.) Cryptography and Coding 2005. LNCS, vol. 3796, pp. 96–115. Springer, Heidelberg (2005)

[HBB13] Hülsing, A., Busold, C., Buchmann, J.: Forward Secure Signatures on Smart Cards. In: Knudsen, L.R., Wu, H. (eds.) SAC 2012. LNCS, vol. 7707, pp. 66–80. Springer, Heidelberg (2013)

[HILL99] Håstad, J., Impagliazzo, R., Levin, L.A., Luby, M.: A pseudorandom generator from any one-way function. SIAM J. Comput. 28, 1364–1396 (1999)

[HM02]   Hevia, A., Micciancio, D.: The provable security of graph-based one-time signatures and extensions to algebraic signature schemes. In: Zheng, Y. (ed.) ASIACRYPT 2002. LNCS, vol. 2501, pp. 379–396. Springer, Heidelberg (2002)

[Len04]   Lenstra, A.K.: Key lengths. In: Contribution to The Handbook of Information Security (2004)

[MMO85]   Matyas, S., Meyer, C., Oseas, J.: Generating strong one-way functions with cryptographic algorithms. IBM Technical Disclosure Bulletin 27, 5658–5659 (1985)

[Rom90]   Rompel, J.: One-way functions are necessary and sufficient for secure signatures. In: STOC 1990: Proceedings of the Twenty-Second Annual ACM Symposium on Theory of Computing, pp. 387–394. ACM Press, New York (1990)

New Speed Records for Salsa20 Stream Cipher Using an Autotuning Framework on GPUs[*]

Ayesha Khalid[1], Goutam Paul[2], and Anupam Chattopadhyay[1]

[1] Institute for Communication Technologies and Embedded Systems,
RWTH Aachen University, Aachen 52074, Germany
{ayesha.khalid,anupam.chattopadhyay}@ice.rwth-aachen.de
[2] Department of Computer Science and Engineering,
Jadavpur University, Kolkata 700 032, India
goutam.paul@ieee.org

Abstract. Since the introduction of the CUDA programming model, GPUs are considered a viable platform for accelerating non-graphical applications. Many cryptographic algorithms have been reported to achieve remarkable performance speedups, especially block ciphers. For stream ciphers, however, the lack of reported GPU acceleration endeavors is due to their inherent iterative structures that prohibit parallelization. In this paper, we propose an efficient implementation methodology for data-parallel cryptographic functions in a batch processing fashion on modern GPUs in general and optimizations for Salsa20 in particular. We present an autotuning framework to reach the most optimized set of device and application parameters for Salsa20 kernel variants with throughput maximization as a figure of merit. The peak performance achieved by our implementation for Salsa20/12 is 2.7 GBps and 43.44 GBps with and without memory transfers respectively on NVIDIA GeForce GTX 590. These figures beat the fastest reported GPU implementation of any stream cipher in the eSTREAM portfolio including Salsa20/12, as well as the block cipher AES optimized by hand-tuning, and thus, to the best of our knowledge set a new speed record.

Keywords: CUDA, eSTREAM, GPU, Salsa20, Salsa20/r, stream cipher.

1 Introduction and Motivation

Performance enhancement on a GPU is a function of the extent of parallelism within the application. In case of symmetric block ciphers, for the encryption of long messages, the plaintext is first partitioned into chunks of the cipher's *blocksize* and then encrypted. For avoiding the weakness of generating identical ciphertexts for identical plaintext blocks, chaining dependencies between adjacent plaintext blocks are added, defined by *modes of operations*. The ciphertext C_i for the ith plaintext block P_i under different modes of operations is

---

[*] This work was done in part while the second author was visiting RWTH Aachen, Germany as an Alexander von Humboldt Fellow.

A. Youssef, A. Nitaj, A.E. Hassanien (Eds.): AFRICACRYPT 2013, LNCS 7918, pp. 189–207, 2013.
© Springer-Verlag Berlin Heidelberg 2013

given below. Here IV stands for the Initialization Vector and E_k stands for the encryption function parametrized by the secret key k.

Operation Mode	C_i
Electronic codebook (ECB)	$E_k(P_i)$
Counter (CTR)	$P_i \oplus E_k(nonce, counter)$
Cipher block chaining (CBC)	$E_k(P_i \oplus C_{i-1}), C_0 = IV$
Propagating CBC (PCBC)	$E_k(P_i \oplus P_{i-1} \oplus C_{i-1}), P_0 \oplus C_0 = IV$
Cipher feedback (CFB)	$E_k(C_{i-1}) \oplus P_i, C_0 = IV$

Observe that all the modes of operation of block ciphers are not parallelizable. The ECB and CTR modes of operation pose encryption as a massively parallel problem with all the plaintext blocks being available for simultaneous encryption without any dependency. However, in CBC, PCBC and CFB modes, due to the dependency or a "carry over" from the previous block, the encryption must progress sequentially on a block by block basis. Consequently, almost all the results of GPU based acceleration undertake block cipher encryption or decryption in Electronic Codebook (ECB) or Counter (CTR) mode for which the inter-dependency between data blocks does not exist and a parallel encryption of blocks of plaintext data is possible.

The use of stream ciphers is best suited for applications requiring high throughput and where the amount of data is either unknown, or continuous. A block cipher is generic in nature and can be used as a stream cipher in CTR mode. In comparison to block ciphers, stream ciphers are simpler and faster and typically execute at a higher speed than block ciphers [3]. Using a block cipher as a stream ciphers is therefore an *overkill* and consequently, results in a lower throughput of encryption in comparison. For example, on a Core 2 Intel processor, 20 rounds of Salsa20 stream cipher run at 3.93 cycles/byte, while 10 rounds of AES block cipher are reported to run more than twice as slow at 9.2 cycles/byte for long data streams (bitsliced AES-CTR) [2]. In this paper, we focus on the GPU implementation of Salsa20 series [4] of stream ciphers.

1.1   Why Salsa20?

The eSTREAM [1] competition was created to attract stream ciphers in two separate profiles, namely for software and hardware platforms. Out of 34 initial submissions, four software stream ciphers, namely, HC-128, Rabbit, Salsa20/12, SOSEMANUK and three hardware stream ciphers, namely, Grain v1, MICKEY 2.0 and Trivium made into the final portfolio [1]. Unlike the parallelizable modes of operations defined for block ciphers, most stream ciphers do not have the liberty of employing the "divide and rule" policy on chunks of plaintext and exhort parallelism on GPUs. Their highly iterative structures have inter-dependencies on subsequent keystream values generated.

For example, in case of HC-128 [14], the limitation on parallelization of the keystream generation routine from two S-boxes P and Q is severe. This is because there are inter-S-box as well as intra-S-box dependencies. The update of the values in S-boxes is a function of previous index values in the array (update of $Q[j]$ requires $Q[j \boxminus 3]$, $Q[j \boxminus 10]$ and $Q[j \boxminus 511]$, where $\boxminus$ is subtraction modulo 512). This limitation renders no more than 3 parallel threads deployment to ensure correctness of results [8]. The other five eSTREAM finalists are also no different. The update of the internal states for generation of next block of keystream is dependent on its previous values.

Salsa20 [4] has an edge over the rest of the stream ciphers for mapping on GPUs, since it has no chaining or dependence between blocks of data during encryption / decryption. Hence a large number of parallel homogeneous threads can be subjected to plaintext data chunks enabling instruction execution in a Single Instruction Multiple Thread (SIMT) fashion exploiting well the parallelism offered by many-core architecture of GPUs. Each block takes a nonce, a secret key, constants and a counter and combines them to generate a block of keystream. For additive stream ciphers the keystream generation is independent of the plaintext. For generating ciphertext, keystream is simply XOR-ed with the plaintext. This property motivated us to take up and report an efficient implementation of Salsa20 stream ciphers on recent graphics hardware.

1.2   Why Autotuning?

Another motivation of the work was the development of an autotuning framework for cryptographic kernels with optimization of throughput performance in mind. The recommended autotuning framework can tune optimally to other and newer devices of NVIDIA GPUs and can be extended to other cryptographic algorithms. *The need of such autotuners is emphasized by the fact that considering device occupancy as a figure of merit is not guaranteed to achieve maximized throughput.* Extensive experimentation is recommended with variation of factors like register usage, thread-block sizes, loop unroll factor etc. The implementation results after autotuning stand out in performance compared to hand-tuned codes for Salsa20.

Autotuning methodologies for multi-core devices are gaining popularity since hand-tuning a large number of parameters optimally for an algorithm on a machine is hard. Most of these autotuning efforts are limited to either a class of similar algorithms, a family of similar devices or an optimization strategy of one parameter for performance enhancement. Murthy *et al.* studied the effect of loop unrolling on various GPGPU programs and claimed 70 percent better throughput by optimally unrolling iterations [11]. A class of algorithms extensively undertaken for autotuning is General Matrix Multiply (GEMM), a part of Basic Linear Algebra Subprograms (BLAS) for matrix multiplication [18]. Kurzak *et al.* presented the optimized choice of tiling and thread arrangement for various versions of GEMM mapped on Fermi family of NVIDIA GPUs [9]. Our autotuning framework is similar in spirit to their work, however specialized for symmetric cryptographic schemes, for which no autotuning endeavors have been reported so far.

1.3   Our Contributions

The major contributions of our work are summarized as follows.

(1) We introduce a batch processing framework for processing any parallelizable cryptographic task in a hybrid CPU-GPU environment.
(2) We recommend a better memory hierarchy utilization for Salsa20 kernels, i.e., use of constant memory instead of shared memory for keeping the initial state vector for Salsa20 (boosting throughput considerably).
(3) We propose an autotuning framework for Salsa20 kernels with two goals in mind: fast device portability and selection of application-specific, device-specific and compiler-specific optimization parameters for throughput maximization. Performance tuning by various parameter search space generation and pruning is generic enough to be extended to other cryptographic schemes, for which no autotuning framework has been reported.
(4) Throughput curves for very long message streams encrypted by Salsa20/r, with and without memory transfers are presented. We hereby report so far the fastest implementation results for Salsa20 variants mapped on any GPU.

2   Parallelism Opportunities of Salsa20 in GPUs

We begin with a functional description of the Salsa20 stream cipher, followed by an overview of the CUDA programming model for NVIDIA GPUs. Then we connect these two by critically analyzing the parallelization opportunities of Salsa20 in GPU.

2.1   Description of Salsa20

Salsa20 accepts four types of inputs, each consisting of 32-bit words: an input key of either 256-bit $(k_0, k_1, ..., k_7)$ or 128-bit $(k_4 = k_0, ..., k_7 = k_3)$ size, a 64-bit nonce (n_0, n_1), a 64-bit counter (t_0, t_1) and four words of pre-defined constants ϕ_i, whose values are dependent upon the key size.

Initialization. These inputs are arranged in a predefined order into a 4x4 state vector X, as follows.

$$X = \begin{pmatrix} x_0 & x_1 & x_2 & x_3 \\ x_4 & x_5 & x_6 & x_7 \\ x_8 & x_9 & x_{10} & x_{11} \\ x_{12} & x_{13} & x_{14} & x_{15} \end{pmatrix} = \begin{pmatrix} \phi_0 & k_0 & k_1 & k_2 \\ k_3 & \phi_1 & n_0 & n_1 \\ t_0 & t_1 & \phi_2 & k_4 \\ k_5 & k_6 & k_7 & \phi_3 \end{pmatrix}.$$

Keystream Generation. The state vector is subjected to a series of rounds composed of additions, cyclic rotations and XORs, to achieve a random permutation. Originally, the number of rounds was set to 20 (Salsa20/r, r=20); however, the version included in the eSTREAM portfolio [1], it was reduced to 12 rounds, for performance reasons.

Then Salsa20/r function for keystream generation can be represented mathematically as:

$Salsa20_k(X) = DoubleRound^{r/2}(X) + X$, with $DoubleRound(X) = RowRound(ColumnRound(X))$.

Each double round comprises of four *QuarterRounds* (in short, *QR*) performed first on the columns of the state vector X and then on the rows of the output.

$Y = (y_0, y_1, ..., y_{15}) = ColumnRound(X)$, and $Z = (z_0, z_1, ..., z_{15}) = RowRound(Y)$, where
$$(y_0, y_4, y_8, y_{12}) = QR(x_0, x_4, x_8, x_{12}),\quad (y_5, y_9, y_{13}, y_1) = QR(x_5, x_9, x_{13}, x_1),$$
$$(y_{10}, y_{14}, y_2, y_6) = QR(x_{10}, x_{14}, x_2, x_6),\quad (y_{15}, y_3, y_7, y_{11}) = QR(x_{15}, x_3, x_7, x_{11}),$$
$$(z_0, z_1, z_2, z_3) = QR(y_0, y_1, y_2, y_3),\quad (z_5, z_6, z_7, z_4) = QR(y_5, y_6, y_7, y_4),$$
$$(z_{10}, z_{11}, z_8, z_9) = QR(y_{10}, y_{11}, y_8, y_9),\ (z_{15}, z_{12}, z_{13}, z_{14}) = QR(y_{15}, y_{12}, y_{13}, y_{14}).$$

Each *QuarterRound*(a, b, c, d) consists of four *ARXrounds*, comprising of additions (A), cyclic rotations (R) and XOR (X) operations only, as below:

$$b = b \oplus ((a + d) \lll 7),\ c = c \oplus ((b + a) \lll 9),\ d = d \oplus ((c + b) \lll 13),\ a = a \oplus ((d + c) \lll 18).$$

Encryption and Decryption. A 16-word ciphertext block C is calculated simply by bitwise XOR-ing a 16-word plaintext block P with the 16-word keystream block S. On the receiver side, the same keystream, when bitwise XOR-ed with the ciphertext C, reproduces the plaintext P.

2.2   CUDA Programming Model Overview

CUDA defines a convenient programming model for heterogeneous computing environment for a CPU *host* and GPU *device*. This section briefly presents NVIDIA GPU architecture and its programming environment. The reader is kindly referred to CUDA C programming guide [16] and Fermi Architecture manual [15] for more information.

Execution Model. CUDA device execution model is depicted in Fig. 1. Parallel portions of an application, executed on the device are called *kernels*. A Kernel call launches a number of *threads*, each executing the same code but having a unique *threadID*. Threads are forwarded to the CUDA device in groups called *warps* for execution. A *threadblock* is a batch of threads that may or may not cooperate with each other by sharing data or by synchronizing their execution. Threads from different threadblocks cannot cooperate.

Kernels are launched in *grids* for execution, comprising of one or more threadblocks. The grid dimensions are specified by *blocksPerGrid* and *threadsPerBlock*. A CUDA device consists of several *Streaming Multiprocessors* (*SM*), each responsible for handling one or more blocks in a grid. Threads in a block are not divided across multiple SMs.

Memory Model. CUDA provides explicit methods to organize memory architecture. Local variables within a kernel reside in registers (*regs*) or in the off-chip local memory (*lmem*). Shared memory (*shmem*) is shared by each threadblock. Global memory (*gmem*) is accessible by all threads as well as host. The lifetime of global memory is from allocation to de-allocation by the host. However, for the other memories mentioned, the lifetime is only during the kernel execution. Other than these memories, each thread within a grid can access *read-only*, *constant* and *texture* memories. These memories can be modified from the host only,

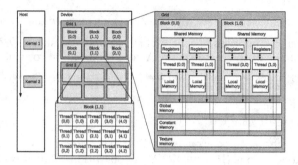

Fig. 1. CUDA GPU execution model [16]

and are useful for storing immutable data structures like lookup tables. The performance of any algorithmic implementation on the GPUs depends heavily on the proper utilization of this memory hierarchy.

2.3   Analyzing Parallelism Opportunities of Salsa20

The immense parallelism offered by the GPUs for acceleration can be better harnessed by a careful study of the parallelism opportunities offered by the application intended for mapping on it. The degree of parallelism also effects the potential throughput performance achievable after mapping. For Salsa20, we observe two categories of parallelism.

Functional Parallelism. As evident from Section 2.1, each block of 64 bytes of Salsa20 keystream can be independently generated and mixed with data to get the ciphertext. Salsa20/r has $r/2$ *DoubleRounds*, each comprising of a *ColumnRound* and a *RowRound*. These *Column* or *Row-Rounds* undergo 4 *ARXrounds* for each row/column. Hence a total of $16 \times r$ invertible *ARXrounds* complete the keystream generation for one block of Salsa20/r. A CUDA compatible device, capable of launching t parallel threads, each undertaking one data block of plaintext, will give a throughput of $(t \times 64)/(16 \times r \times \alpha)$ Bytes/sec if α is the time taken for one *ARXround* as depicted in Fig. 2. We ignore the final addition of *DoubleRound* output with the state vector for keystream generation since its overhead is negligible in comparison.

Data Parallelism. In Salsa20, each *QuarterRound* operates on either a row or a column of the 4x4 array. Each of the four *ARXrounds* constitutes of a *QuarterRound*, modifying exactly one value of that row or column. Hence 4 parallel *QuarterRounds* can be executed due to absence of inter-column/row dependence. Consequently, $16 \times r$ transformations of one Salsa20 block can be broken down as $4 \times r$ transformations mapped on 4 parallel threads giving a throughput of $(t \times 64)/(4 \times r \times \alpha)$ Bytes/sec, or 4 times higher than a single thread per block mapping. Exploiting further parallelism within one *ARXround* is not possible due to dependence of XOR (X) operation on the output of rotation (R) and addition (A) as shown in Fig. 2.

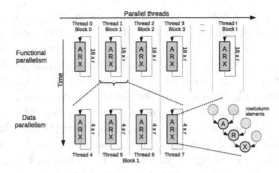

Fig. 2. Parallelism in Salsa20/r

For coding Salsa20 kernels employing functional parallelism (one thread per data block), internal registers and shared memory were used for storing results of *DoubleRounds* and X respectively. For manipulating data parallelism (four threads per data block) inter-thread communication is required within the threads of a threadblock. Therefore the results of *DoubleRounds* are also held in shared memory. For this implementation the need of thread-synchronization makes it lag behind in performance compared to single thread per block implementation. Experimentation of mapping AES on GPUs with different granularities also conform to our findings, as the best throughput performance is achieved when no synchronization is required between different threads [10,7]. Another reason for avoiding intra-block synchronization is that for most GPU devices is as follows. *The limited number of shared memory limits instruction-level parallelism by restricting the number of threads launched, lowering occupancy. Hence, for the rest of the discussion, we consider only the single thread per data block implementation due to its superior throughput performance.*

3   Batch Processing Framework

Salsa20 algorithm is a classic case of a parallelizable application, for which performance is dependent on the amount of parallel work received. For all such applications, a batch processing framework of operation is recommended. It is termed *batch* processing, since a batch of threads work simultaneously to encrypt one data block each and iterating in a loop for encryption of more plaintext. The batch of active threads die when all the data to be encrypted is exhausted.

3.1   CPU-GPU Interaction

Algorithm 1 explains the batch processing framework for encryption or decryption in a hybrid environment (CPU-GPU). We consider plaintext (P), given as 1-D data to be the input to the application. Inputs to the framework for Salsa20 encryption have already been explained in the functional description of the algorithm in Section 2.1. The byte-length of a data block for encryption or decryption is called the *blocksize*. The initial state vector (X) is set up at the host machine using algorithm specific *Initialization* routine and transferred to the global memory (*gmem*). Assuming encryption of P having size larger than global memory (*gmem*) size, P is divided into chunks equal to size of *gmem*, termed as P_k. Every k^{th} iteration encrypts a portion of plaintext P_k into an equal sized ciphertext C_k (line no. 3). For simplicity, we assume the total size of plaintext to be a multiple of the size of *gmem*, in case of non-conformity, the number of data blocks forwarded to kernel for encryption is changed to the residue after division with $size(gmem)$ in the last iteration. For Salsa20, X is a 16-word array and its subscript represents its existence location, i.e., h, g, s, r representing host, global memory, shared memory and registers respectively. After the transfer of P_k to device's *gmem*, launch of kernel is kick-started in an iterative fashion. One batch of threads or *threadsPerGrid*, executed in parallel on device, is *blocksPerGrid* $\times$ *threadsPerBlock*. In every iteration, when the kernel call is terminated, *gmem* contains the cipher text, that must be read out by the host (line no. 7) before writing the next plaintext chunk into the device memory (line no. 4).

Input: $key(k)$, $nonce(n)$, $counter(t)$, $constants(\phi_i)$, $rounds$, $blocksize$, $plaintext(P)$
Output: *ciphertext* (C)
1 $X_h = Initialization(key, constants, counter, nonce)$;
2 $X_h : host \Rightarrow gmem$;
3 **for** $k=1$ **to** $\lceil \frac{size(P)}{size(gmem)} \rceil$ **step** 1 **do**
4 | $P_k : host \Rightarrow gmem$;
5 | $Salsa20\_kernel \lll blocksPerGrid, threadsPerBlock \ggg$
   | $(rounds, size(gmem)/blocksize)$;
6 | $X_g : gmem \Rightarrow host$;
7 | $C_k : gmem \Rightarrow host$;
   end

Algorithm 1. Batch processing for a cryptographic kernel

3.2   The CUDA Kernel

Algorithm 2 is the CUDA kernel call and is executed on the GPU device. Although CUDA kernel functions do not have any output, the algorithm represents a pseudo-code and the output specified is not the output of the kernel function. Two local variables called *counter* and *batch* are declared and initialized, containing the unique threadID and the total number of threads in a batch respectively. Variable *counter* is used to update the counter in the state vector of Salsa20, incremented by the variable *batch* after every iteration. When a thread finishes encrypting a block, it encrypts again the block corresponding to that thread index plus the total number of active threads running (*batch*), which is constant and device dependent.

The state vector, residing in global memory, is first copied to faster *shmem*. As the size of global memory of newer NVIDIA GPUs is in GBs, a single batch of parallel threads each encrypting one data block, will not finish up the P_k,

```
    Input: rounds, dataBlocks, plaintext(P_k)
    Output: ciphertext (C_k)
 1  counter = blockDim.x * blockIdx.x + threadIdx.x;
 2  batch = gridDim.x × blockDim.x;
 3  X_g : gmem ⇒ shmem;
 4  for i=1 to dataBlocks step (dataBlocks/batch) do
 5  │   X_s = X_s + counter;
 6  │   X_s : shmem ⇒ regs;
 7  │   for j=1 to rounds step 2 do
 8  │   │   state_r = DoubleRound(state_r);
    │   end
 9  │   S_i = X_s ⊕ state_r;
10  │   P_ki : gmem ⇒ regs;
11  │   C_ki = P_ki ⊕ S_i;
12  │   C_ki : regs ⇒ gmem;
13  │   counter+ = batch
    end
14  X_s : shmem ⇒ gmem;
```

Algorithm 2. Salsa20 kernel

requiring iterations over variable i, as given in line no. 4. Here too, for the sake of simplicity, we consider the number of *dataBlocks* forwarded to the kernel for encryption or decryption to be a multiple of *batch* of threads. In case of nonconformity, the pseudo code can be modified to launch lesser number of threads in the batch in the last iteration. The state vector is updated with the counter value as given in line no. 5. Since threadID is different for each thread in a batch, all threads get a different state vector. The variable $state_r$ refers to the register copy of the state vector (it is copied from *shmem* to *regs* in line no. 6).

The value of *rounds* is either 8, 12 or 20 for various flavors of Salsa20/r. A copy of $state_r$ in thread-local registers apply *DoubleRound* transformations for $\frac{rounds}{2}$ times. One *block* of keystream, generated by XOR-ing the state vector with its transformed copy in local registers (line no. 9), is held in S_i.

The last step is the encryption of the plaintext with the generated keystream. Plaintext is read from *gmem*, one *block* at a time (P_{ki}), XOR-ed with the generated keystream to produce a block of ciphertext (C_{ki}) and then is written back to *gmem*. Saving of state vector into *gmem* is required before exiting the kernel, since its lifetime in shared memory only lasts as long as threadblock's lifetime.

3.3   Programming Recommendations

CUDA programmers are recommended to follow the guides [16,17] to achieve the best performance. We summarize some more relevant recommendations for good throughput performance when Algorithm 1 and 2 are mapped onto a GPU.

Avoiding threadBlock Switching Overhead. Each kernel launch on the device bears overheads of a kernel call, memory allocation and argument copying into the device. If the amount of work per kernel is small in comparison to the total workload, the run-time of the application is dominated by these overheads instead of the actual computation time. In order to decrease these overheads, the amount of work per kernel call should be increased. Hence we resort to iterations computed inside a kernel call (loop indexed with i in Algorithm 2) to continue as long as the entire workload is finished instead of launching a new batch of threads. This strategy amortizes the overhead of multiple kernel calls across more computation and boosts throughput.

Reuse Memory. For cryptographic applications, the plaintext P_k once handed over to the device is not needed back by the host device. A prudent decision is to *overwrite* the plaintext with ciphertext in the *gmem*. It saves the iterations of loop indexed by k by half in Algorithm 1.

Data Coalescing: Global memory accesses incur a 100x access penalty compared to kernel local registers [16]. If these accesses are close to each other and dispatched in a group, they are *coalesced* as a single access. The device can read 4-byte, 8-byte, or 16-byte words from global memory into registers in a single instruction. Mixing of plaintext for generating ciphertext requires reading, XORing with keystream and writing back into the global memory as given by line no.s 10, 11 and 12 respectively (in Algorithm 2). Maximum memory coalescing that the device supports gives good saving in access time.

Autotune. The choice of grid dimensions, *blocksPerGrid* and *threadsPerBlock* is critical since it affects the throughput. It is discussed in detail in Section 4.

3.4   Optimization for Salsa20

For a given key, the initial state vector for multiple blocks of Salsa20 encryption remains the same except for a counter value, that is incremented for each block. Hence it can be treated as a constant array, while the counter is taken care of by each thread kernel individually by its *threadID*. Keeping the initial state vector in fast read-only constant memory, instead of shared memory, is therefore useful as constant memory is optimized for broadcast due to data coalescing. Since each block of Salsa20 requires reading of initial state vector twice, once before the *DoubleRound* iterations and once after it (line no. 6 and 9 respectively in Algorithm 2), the use of constant memory is highly suited. CUDA specific function *cudaMemcpyToSymbol* writes the initial state vector in the constant memory. This strategy cannot be generalized to all ciphers. However, a prudent use of a faster memory, whenever applicable, always enhances performance for CUDA applications. This factor alone boosts the peak throughput for Salsa20/12 (for 1 GB of plaintext) by 4 GBps.

4   Autotuning Framework for Performance Optimizations

In context of a CUDA back-end application, our autotuning framework automatically chooses tunable parameters of application mapping with the aim of improving a designated Figure of Merit (FoM). The tunable parameters may be application specific, device dependent or compiler optimizations. Finding the optimal values of these parameters may require extensive experimentation on a case-by-case basis. Apart from promising a performance boost, another reason for developing an autotuning framework is the provision of portability across different devices belonging to the same architecture family. Some common figures of merit are listed below.

Occupancy. Device occupancy (or concurrency) is the ratio of active resident threads to the maximum number of resident threads whose resources can be stored on-chip simultaneously. Occupancy serves as a guideline for performance, but does not guarantee optimized throughput.

GFLOPs. Comparison of application's GFLOPs (Giga FLoating point Operations Per Second) against peak GFLOPs specified for a device. However, peak GFLOPs is quoted strictly for Floating point instructions.

Throughput. It is the measured output rate ($Bytes/sec$) of an application using timing functions on the device. Given the application in hand, we chose it to be our FoM.

The aim of an autotuning framework is to admit a large range of tunable parameters to CUDA application and select the one that makes the kernel run most efficiently. The range of these tunable parameters may be dependent on the constraints imposed by the device, application or both. The task of identifying *what* parameters should be subjected to tuning is critical, since they vary widely between algorithms. Keeping in mind the operation flow of the framework, we classify these parameters optimizations as *Compiler-specific* and *Device-specific*.

4.1   Device-Specific Optimizations

Device-specific optimizations are the ones that are tweakable at the runtime of the application, e.g., device grid dimensions. The given CUDA application is enhanced by the addition of the provision of the kernel variants being subjected to all possible combinations of these parameters after pruning by certain checks. Benchmarking for throughput is also added for later use. As shown in Fig. 3, enhancement of the application with the addition of device specific optimizations and benchmarking provision is the first step of the autotuning framework. However, execution of these enhancements does not manifest before various compiler optimizations have been done and multiple copies of the code executables are ready. Final runs of these programs result in sifting the fastest implementation with recommended parametrization choices.

Algorithm 3 gives the pseudo-code of the device specific optimizations setup. Out of the 4 different inputs, device properties (obtained by *cudaDeviceProp* function) and compute capability properties (obtained from a lookup table corresponding to major and minor compute capability) are device dependent. Kernel constraints are application dependent and are obtained after compilation of the

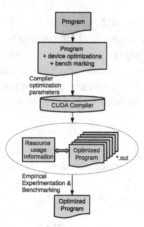

Fig. 3. Autotuning Framework Flowchart

program. *minOccupancy* is specified by user to filter out kernels with too low occupancy from experimentation. A higher value will prune the search space more but might miss the fastest kernels too; a lower value, on the other hand, will compromise on speed due to large search space for the fastest kernel.

All possible values of the two critical device parameters, *threadsPerBlock* and *blocksPerSM*, are considered for experimentation within their permitted range in nested loops as specified by line no. 1 and 2. Threadblock size should always be a multiple of *warpSize*, because kernels issue instructions in warps. The next four lines of code calculate the resource budget for the current configuration of the device parameters. Total resident *threadsPerSM* is a product of *blocksPerSM* and *threadsPerBlock*. The next two lines calculate the resource usage of register and shared memory per kernel from the application specific parameters.

A kernel is subjected to experimentation with a set of possible device parameters configuration after pruning by 4 checks as specified from line no. 7 to 10. Check 1 ensures that the maximum number of possible threads executable on an SM is not exceeded. Check 2 and 3 ensure that the register budget and the shared memory budget specific to one kernel is not exceeded. Check 4 makes sure that the current device configuration has an occupancy higher than the minimum specified by the user. Functions to calculate the time elapsed before and after the kernel call are used to carryout the time duration benchmarking.

4.2   Compiler-Specific Optimizations

Compiler-specific optimizations are the ones that are subjected to the *nvcc* compiler at the compile time, e.g., preprocessor directives. As shown in Fig. 3, this step generates a number of optimized programs, each pertaining to a possible permutation out of the range of all the compiler-specific optimization parameters. Other than getting these executables, compiler generates information regarding the resource usage of the application in question, i.e., global, constant memory usage per grid, register count, local memory and shared memory usage

---

Input: 4 types of inputs:
1. **Device:** $warpSize$, $maxRegsPerBlock$, $maxShMemPerBlock$, $maxThreadsPerBlock$, $maxSM$;
2. **Compute Capability:** $maxBlocksPerSM$, $maxWarpsPerSM$;
3. **Kernel constraints:** $regsPerThread$, $shMemPerThread$;
4. **User constraints:** $minOccupancy$.
Output: Valid parameter variants for benchmarking

1 **for** $threadsPerBlock = warpSize$ **to** $maxThreadsPerBlock$ **step** $warpSize$ **do**
2   **for** $blocksPerSM = 1$ **to** $maxBlocksPerSM$ **step** 1 **do**
3          $threadsPerSM = blocksPerSM \times threadsPerBlock$;
4          $regsPerSM = threadsPerSM \times regsPerThread$;
5          $ShMemPerSM = threadsPerSM \times shMemPerThread$;
6          $occupancy = \frac{threadsPerSM}{(maxWarpsPerSM \times warpSize)}$;

7   **Check1:** $threadsPerSM \leq (maxWarpsPerSM \times warpSize)$;
8   **Check2:** $regsPerSM \leq maxRegsPerBlock$;
9   **Check3:** $ShMemPerSM \leq maxShMemPerBlock$;
10   **Check4:** $occupancy \geq minOccupancy$;

11         $blocksPerGrid = maxSM \times blocksPerSM$;
12         $success = kernel\_launch \lll blocksPerGrid, threadsPerBlock \ggg$
    end
end

Algorithm 3. Device-specific optimizations: Search space generation and pruning

per kernel. These resources are used as constraints during the empirical experimentation before reaching the performance-optimized kernel. The two compiler-specific optimizations applicable for the current application are loop unrolling and restricting per kernel register budget. Both of these manifest as a compromise between parallelism and register pressure.

Unroll Factor. Loop Unrolling replaces the main body of a loop with multiple copies of itself, adjusting the control logic accordingly. *#pragma unroll n* is a preprocessing directive where n defines the unroll factor ($n = 1$ means no unrolling, $n = k$ means full unrolling, where the trip count of the loop is k). On the positive side, loop unrolling results in reduced dynamic instructions (compare and jump) count, boosting speedup. On the negative side, however, unrolling increases the total instruction count of the loop body and leads to an increased register pressure. Since registers are partitioned among threadblocks, an increased use of registers per threadblock reduces the device occupancy. This may or may not affect the throughput and requires experimentation for assurance. For Salsa20, the three flavors of the algorithm iterate for 4, 6 and 10 times for Salsa20/8, Salsa20/12 and Salsa20/20 respectively. For each of these, unroll factor from no unrolling to maximum unrolling is considered for experimentation.

Register Budget. A CUDA programmer can force a restricted number of registers by specifying cuda-nvcc-opts=-maxrregcount R, limiting the register use to R per kernel. Lowering register count allows increased occupancy which may result in increased throughput. On a negative note, it may cause spilling into the local memory when the register limit is exceeded. The local memory is as slow as the global memory and spilling into it can consequently cause severe performance degradation despite the higher occupancy. For all Salsa20 kernel variants, the register budget varies from 26 to 43 for no unrolling to maximum unrolling. For parametrization of register budget, all the multiples of 5 within this minimum and maximum register use are considered. Lowering the register budget any further than the minimum limit causes spilling and hence these cases are omitted from benchmarking.

5   Results and Discussion

In this section, we present detailed experimental results and compare them with the available state-of-the art benchmarks.

5.1   Experimental Setup

Throughput performances of Salsa20 stream cipher is reported for NVIDIA GeForce GTX 590, although the autotuning framework is generic enough to cater for any Fermi NVIDIA GPU device. To quantify the speedup against a general purpose computer, a single threaded application program written in C was run on an AMD Phenom 1055T Processor (clockspeed 2.8 GHz) with 8 GB of RAM and Linux operating system. For a good approximation, each experiment was run 100 times and the timing results were averaged.

5.2   Search Space Generation and Pruning

Table 1 gives the possible range of parameters for the Salsa20 application kernel for NVIDIA GeForce GTX 590. The register budget range was chosen within the minimum and the maximum register requirements with no unroll and full unroll respectively, in steps of 5. All possible values of the unroll factor are taken into consideration. Grid dimension's permitted range is dependent on the device. Minimum occupancy was chosen to be 0.16, i.e., 256 threads per SM (256/1536), since tests with selective lower occupancies gave inferior throughputs.

In order to give an idea of the magnitude of the possible kernel configurations on which the autotuning framework carried out experimentation, some numbers are presented. For Salsa20/20, 10 possible unroll factors generate 10

Table 1. Range of parameters for autotuning Salsa20 kernel on a GTX 590

	Parameter	Range
Compiler-specific optimizations	Register budget	$26, 30, 35, 40, 43$
	Unroll factor	$1, 2, ..., r/2$
Device-specific optimizations	Threads per block	$32, 64, ..., 1024$
	Blocks per SM	$1, 2, ...8$
	Minimum occupancy	0.16

optimized versions of the program, each a candidate for experimentation. Further, each of them is subjected to restricted register budget to generate multiple versions. Considering only the case of full unroll and unrestricted use of registers for Salsa20/20 kernel subjected to device specific constraints, the number of allowed grid size combinations comes out to be 55. Extensive experimentation of all possible combinations of parameters after pruning was carried out for benchmarking.

5.3   Compile Time Optimization of Register Pressure

To find the optimal trade-off between active concurrent threads and registers availability per thread, two parameters have been tweaked. These are the restricted use of register budget and register unrolling. Restricting the use of registers per thread was benchmarked to always have a deteriorating effect on the throughput, in spite of increased occupancy. Changing the unroll factor, however, gives improved performance results. Fig. 4(a) gives the effect of unrolling factor on the registers used per thread for Salsa20/8, Salsa20/12 and Salsa20/20. Since Salsa20/20 requires 10 loop iterations of *DoubleRound* function, unrolling factors range from 1 to 10. Unrolling an n-iteration loop more than n times makes no difference and is considered by the CUDA compiler as a full unroll. Consequently, the unrolling of Salsa20/8 and Salsa20/12 kernels show no change after unroll factor of 4 and 6 respectively.

5.4   Register Unroll vs. Throughput

The register unrolling positively effects the throughput in general as shown in Fig. 4(b). These results are obtained after benchmarking more than 2500 kernel variants considering the full range of unroll factors and all grid dimensions supported by the device. Constraint of *minOccupancy* is applied, but register use restriction at compile time is skipped since it does not boost the throughput. The size of plaintext is kept 32 KB for encryption by the kernel.

Interestingly, the highest throughput for a Salsa20 kernel variant is obtained when the inner loop is unrolled by a factor one *less* than the full unrolling. Considering the case of Salsa20/20, the registers used per kernel remain unchanged till the unroll factor is raised from 1 till half of the full unroll factor, i.e., 5 as given in Fig. 4(a). For an unroll factor of 6 to 9, the no. of registers per kernel increases from 39 and saturates to a maximum of 43 for the full unroll. By varying the grid dimension, we find that the best throughput figures are obtained when the unroll factor is 9. Although partially unrolled loops may require some cleanup code to account for loop iterations that are not an exact multiple of the unrolling factor, it may or may not decrease the performance in practice. Hence considering a range of unroll factors for experimentation proves beneficial in reaching the optimized performance.

Similarly, for the other flavors of Salsa20, i.e., for Salsa20/8 and Salsa20/12, the highest throughput is achieved when unroll factor is 3 and 5 respectively, as shown in Table 2. For these unroll factors, the register usage in the three kernels

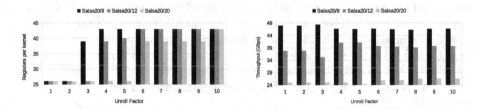

Fig. 4. (a) Register pressure and (b) kernel throughput against unroll factor

Table 2. Salsa20 optimized parameters for GTX 590 (32 KB plaintext)

Kernel variant	Unroll factor	Threads per block	Blocks per SM	Device occupancy	Throughput (GBps)	Throughput (GBps)	Improvement (GBps)
		Autotuned				Hand-tuned	
Salsa20/8	3	448	1	0.29	48.29	45.77	2.52
Salsa20/12	5	320	2	0.41	41.14	39.91	1.23
Salsa20/20	9	512	1	0.33	26.60	24.42	2.18

restricts the occupancy of the device. With 40 registers, no more than 25 warps can be launched on each SM for GTX 590 (register limit on the device being 32K) restricting the device occupancy to 0.52. Table 2 gives the throughput performance with hand-tuned parametrization for maximum device occupancy. The improvement in throughput obtained emphasizes the need of autotuning as a necessary requirement for performance enhancement of a CUDA application.

5.5 Workload vs. Performance

Fig. 5 shows the performance of Salsa20 variants on a GTX 590 for varying plaintext sizes. For throughput estimation, the plaintext blocksize is increased from 1 Byte till 1 GB. For GTX 590, we cannot go beyond 1.5 GB in one batch of plaintext encryption due to the size of the global memory (obtained from *cudaDeviceProp* function). It is easy to see that the performance of Salsa20 is highly dependent on the amount of parallel work it receives. We find the

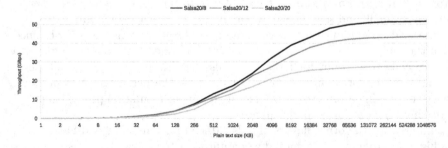

Fig. 5. Salsa20 throughput on GTX 590 for varying plaintext sizes (w/o mem trans.)

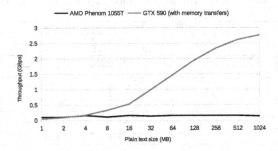

Fig. 6. Salsa20 throughput comparison on a CPU and GPU

peak throughput performance of Salsa20/8, Salsa20/12 and Salsa20/20 to reach 51.55, 43.44 and 27.65 GBps, respectively, outperforming the best reported GPU implementations so far.

We also took into account the overhead attributable to the plaintext data transfer from CPU to GPU and ciphertext data transfer from GPU to CPU to get the effective throughput, as given in Fig. 6. The peak performance for the GPU under consideration reaches around 2.8 GBps with memory transfer overheads. The severe drop in the throughput clearly indicates that the bottleneck in the system is the data transfer bandwidth: PCIe bandwidth. For the host CPU, i.e., for AMD Phenom 1055T, the peak performance reaches 157 MBps.

5.6   Comparison with Other Works

Table 3 gives a comparison of our work on Salsa20 acceleration on GPUs with the results presented by D. Stefan [13] and S. Neves [12]. We also compare the performance with the fastest reported AES implementation on GPUs [7]. For a fair comparison, we scale up the throughput figures of other devices (without memory transfers) in accordance with our newer GPU device by considering the number of processor cores per device. Although the processing frequency of our device is slower in comparison, we ignore this factor for scaling the throughput calculation. The throughputs (GBps) per core from   [13,12] is (5.3/480 and 9/192), which is multiplied with the number of cores of our device (512) to get 5.7 GBps and 24 Gbps respectively. These scaled throughputs are surpassed by our peak performance of 43.44 GBps. In [13], the maximum throughput of 5.3 GBps (without memory transfers) was achieved for Trivium and that is also far behind (even after scaling) the throughput of our implementation. Scaling on similar lines, the AES implementation by Iwai *et al.* [7] results in a throughput of 9.3 GBps which is about 4.6 times slower than our reported peak performance for Salsa20/12. This re-scaling formula would be invalid for throughput calculation with memory transfers, since, like most of the cryptographic algorithms, Salsa20 and AES are data intensive in nature and show performance dependence on external memory access speed. The main factor contributing to our performance gain is the use of constant memory instead of shared memory for keeping the copy

Table 3. Comparison of peak performance (Tp stands for Throughput)

	D. Stefan[13]	S. Neves[12]	This work	Iwai et al.[7]
Algorithm	Salsa20/12	Salsa20/12	Salsa20/12	AES
NVIDIA device	GTX 295	GTX 260	GTX 590 (one GF110)	GTX 285
Release (DD/MM/YYYY)	08/01/2009	16/06/2008	24/03/2011	15/01/2009
Compute Capability	1.3	1.2	2.0	1.3
Processor cores	480	192	512	240
Shader Frequency (MHz)	1242	1350	1215	1470
Threads / Block	256	256	320	512
Tp (GBps)(w/ m)	-	1.3	2.8	2.8
Tp (GBps)(w/o m)	5.3	9	43.44	4.4
Scaled Tp (GBps)(w/o m)	5.7	24	43.44	9.3

of the initial state vector. Moreover, our autotuning framework to sift out the choice of parameters maximizing throughput also helps in reaching the claimed performance. According to Table 3, our throughput with memory transfer is the same as the best result known for AES. However, the claimed speed of 2.8 GBps for AES with memory transfers is reported after being improved by 68% by optimization of overlapping GPU processing and data transfers [7]. Our current framework does not support this optimization. However, the search for an optimal transfer blocksize to hide the transfer latency is on our roadmap.

6   Conclusion

We present an autotuning framework for Salsa20 series of stream cipher. It not only guarantees fast portability for Fermi GPUs and optimized throughput performance, but it can be generalized and extended to other massive parallel cryptographic operations also. Moreover, our peak throughput figure of 43.44 GBps surpasses the fastest GPU based performance reported so far for all stream ciphers (both hardware and software) in the eSTREAM portfolio [12,13,8], as well as AES in CTR mode [7].

Regarding the future work, we plan to extend our efforts in different directions. Firstly, we intend to benchmark GPU implementation of other parallelizable stream ciphers (e.g., ChaCha [5], a variant of Salsa20). Secondly, we plan to extend our autotuning framework to handle plaintext data ordered as a 2-D array for multimedia applications. Generalization of our autotuning framework for optimizing any symmetric key cryptographic kernel is also intended.

References

1. eSTREAM: the ECRYPT Stream Cipher Project,
   http://www.ecrypt.eu.org/stream
2. Bernstein, D.J.: Hash functions and ciphers. In Notes on the ECRYPT Stream Cipher Project (eSTREAM), http://cr.yp.to/streamciphers/why.html
3. Bernstein, D.J.: eBACS: ECRYPT Benchmarking of Cryptographic Systems,
   http://bench.cr.yp.to/results-stream.html

4. Bernstein, D.J.: The salsa20 family of stream ciphers. In: Robshaw, M., Billet, O. (eds.) New Stream Cipher Designs. LNCS, vol. 4986, pp. 84–97. Springer, Heidelberg (2008)
5. Bernstein, D.J.: ChaCha, a variant of Salsa20. Workshop Record of SASC 2008: The State of the Art of Stream Ciphers, http://cr.yp.to/papers.html#chacha
6. Biagio, A., Barenghi, A., Agosta, G., Pelosi, G.: Design of a parallel AES for graphics hardware using the CUDA framework. In: International Symposium on Parallel & Distributed Processing (IPDPS), pp. 1–8. IEEE (2009)
7. Iwai, K., Nishikawa, N., Kurokawa, T.: Acceleration of AES encryption on CUDA GPU. International Journal of Networking and Computing 2(1), 131–145 (2012)
8. Khalid, A., Bagchi, D., Paul, G., Chattopadhyay, A.: Optimized GPU implementation and performance analysis of HC series of stream ciphers. In: Kwon, T., Lee, M.-K., Kwon, D. (eds.) ICISC 2012. LNCS, vol. 7839, pp. 293–308. Springer, Heidelberg (2013), http://eprint.iacr.org/2013/059
9. Kurzak, J., Tomov, S., Dongarra, J.: Autotuning GEMM kernels for the Fermi GPU. In: Transactions on Parallel and Distributed Systems, pp. 2045–2057. IEEE (2012)
10. Manavski, S.A.: CUDA compatible GPU as an efficient hardware accelerator for AES cryptography. In: International Signal Processing and Communications (ICSPC), pp. 65–68. IEEE (2007)
11. Murthy, G.S., Ravishankar, M., Baskaran, M.M., Sadayappan, P.: Optimal loop unrolling for GPGPU programs. In: International Symposium on Parallel & Distributed Processing (IPDPS), pp. 1–11. IEEE (2010)
12. Neves, S.: Cryptography in GPUs. Master's thesis (2009), http://eden.dei.uc.pt/~sneves/pubs
13. Stefan, D.: Analysis and Implementation of eSTREAM and SHA-3 Cryptographic Algorithms. Master's thesis (2011), https://github.com/deian/gSTREAM.
14. Wu, H.: The Stream Cipher HC-128, http://www.ecrypt.eu.org/stream/hcp3.html
15. NVIDIA's Next Generation CUDA Compute Architecture: Fermi, http://stanford-cs193g-sp2010.googlecode.com/svn/trunk/lectures/lecture_4/cuda_memories.pdf
16. CUDA C Programming Guide, http://docs.nvidia.com/cuda/cuda-c-programming-guide/index.html#ptx-compatibility
17. CUDA C Best Practices Guide, http://docs.nvidia.com/cuda/cuda-c-best-practices-guide/index.html
18. Basic Linear Algebra Subprograms Technical Forum Standard (August 2001), http://www.netlib.org/blas/blast-forum/blas-report.ps

Cryptanalysis of AES and Camellia
with Related S-boxes

Marco Macchetti

Kudelski Group
marco.macchetti@nagra.com

Abstract. Cryptanalysis mainly has public algorithms as target; however cryptanalytic effort has also been directed quite successfully to block ciphers that contain secret components, typically S-boxes. Known approaches can only attack reduced-round variants of the target algorithms, AES being a nice example. In this paper we present a novel cryptanalytic attack that can recover the specification of S-boxes from algorithms that resist to cryptanalysis, under the assumption that the attacker can work on a pair of such block ciphers that instantiate related S-boxes. These S-boxes satisfy the designer's requirements but are weakly diversified; the relationship between these unknown components is used in much the same way as relationship between secret keys is used in related-key attacks. This attack (called related S-box attack) can be used, under certain assumptions, to retrieve the content of the S-boxes in practical time. We apply our attack to two well known ciphers, AES and Camellia; these ciphers use 8-bit S-boxes but are structurally very different, and our attack adapts accordingly. This shows that most probably the same can be applied to other ciphers which can be customized to instantiate unknown 8-bit S-boxes.

1 Introduction

Block cipher design is a well developed research field; the AES contest has without doubt contributed to its growth. Today, not only we have a significant number of good algorithms, we also possess criteria that can be used to design ciphers that are robust against known cryptanalytic techniques, such as linear cryptanalysis [15], differential cryptanalysis [2], algebraic attacks [11].

Those algorithms that are fully public, and withstand all cryptanalytic attacks, are considered to be the best and therefore are used ubiquitously; this is, after all, the main motivation behind the AES contest (and the SHA-3 one). But in some cases there may be a need to keep at least part of an algorithm private. Although this is not commonly seen as good practice, being a contradiction of the famous (although perhaps overestimated) Kerckhoffs principle, it is not rarely done in practice as there may be a good justification.

Considering products such as RFIDs, smart-cards and conditional access tokens, adversaries may be able to compromise the security of part of the system with the ultimate goal of cloning the device (well-funded pirate organizations have the possibility and technical skills to pursue this goal). Basing the

A. Youssef, A. Nitaj, A.E. Hassanien (Eds.): AFRICACRYPT 2013, LNCS 7918, pp. 208–221, 2013.
© Springer-Verlag Berlin Heidelberg 2013

cryptographic constructions on completely standard algorithms thus gives the adversary an advantage because the cloning procedure is easier; assuming the adversary gains complete control over the block cipher under attack, he can choose a key value and by observing few encryptions he will be able to say which standard algorithm is used. On the other hand, an unknown algorithm forces the attacker to fully reverse-engineer the device, a thing which is definitely more difficult, and costly, than a partially invasive attack. Of course, the algorithms must still be based on solid constructions, well-analyzed and characterized by proofs of security.

In many algorithms, the S-box is a natural candidate for customization, for several reasons. As an example, the Rijndael SPN structure can be easily customized by replacing the standard S-box with a randomly picked 8-bit permutation; the resulting cipher still maintains all the structural properties of AES while it forces the adversary to reverse engineer an implementation to be able to clone the circuit.

Even if the particular S-box used in Rijndael has optimal differential and linear characteristics [12], these parameters can actually be relaxed, since a large margin of security exists with regards to classical attacks in the design of the cipher. For instance, the expected maximum entry in the Differential Distribution Table (DDT) of a random 8-bit S-box is 16 [17], whereas the maximum DDT value of the AES S-box is 4. This means that the probability of differential trails over 4 rounds is increased from 2^{-150} to 2^{-100}, a value that anyway render differential attacks over the full cipher impossible. Regarding algebraic properties, a random 8-bit bijection is likely to show up no monomial characterization, even if the algebraic degree will not be maximal. We also note that the recent biclique attacks which have been shown to work on the full AES [8] and are the most successful attack to date, combine the biclique concept with the use of meet-in-the-middle structures, for which known differentials must be used. These differentials are not known by the attacker if the S-box is unknown.

Thus a randomly-generated S-box (e.g. by means of a true random number generator and application of the Knuth shuffle [14]) is expected to behave well enough. The number of choices is extremely large; taking into account all permutations on 8 bits, we have a customization space of about $\log_2(256!) \approx 1684$ bits. Even giving the adversary the possibility to completely control the key, he cannot recover the content of the secret S-box and use it to clone the device. This is because in the known-key scenario the probability of differential and linear characteristics for the AES is low enough for them to be useless [19].

Even if it is intuitive that some security is added if part of a block cipher specification is kept private, there is little available quantitative analysis of the subject in the literature. We have to say that here we are not focusing on implementation-based attacks, such as fault injection or side channel analysis; it is today known that these techniques can be used to reverse engineer block ciphers, such as in SCARE attacks [10][16][18][13] and in FIRE attacks [3]. The primary goal of this paper is rather to consider the components of a block cipher (such as unknown S-boxes) as another design dimension, and to introduce a new class of quite powerful related-cipher attacks (that we call related-S-box attacks).

Related-key attacks are today accepted as a way to expose weaknesses of a block cipher, and are based on the fact that the adversary may know the relationship between a pair (or a bigger set) of otherwise secret keys [5]. The concept of related ciphers has been analyzed in [22][20], where the existence of ciphers parameterized by variable number of rounds was exploited to determine the value of the key. The relationship between different modes of operation has also be considered in related cipher attacks [21].

We present here a novel type of attack which follows in these footsteps and exploits the existence of a strong relationship between different, but unknown, S-boxes to break the cipher[1]. This is, to our knowledge, the first cryptanalytic attack that can obtain the specification of S-boxes instantiated in block ciphers with the strength of the AES; and under certain assumptions, we do it in practical time.

2   The Related S-box Attack

2.1   Overview and Assumptions

Let us examine the case of AES instantiating an unknown S-box, but with usual key-schedule, round function structure and number of rounds; let us limit our analysis to the 128-bit key variant. Consider the following definition:

Definition 1. *Two S-boxes $S_1, S_2 : F_{2^m} \to F_{2^n}$ are said to be δ-related if*

$$S_1(x) = S_2(x) \quad \Leftrightarrow \quad x \notin \Delta$$
$$|\Delta| = \delta \quad , \quad 2 \leq \delta \leq 2^m$$

This definition may seem a bit simplistic, in the sense that it considers the similarity between two S-boxes only in terms of the number of entries on which they agree; this is precisely the characteristic which is used by our attack, and we believe it is the most generic and agnostic notion of similarity. Of course one may think about linearly equivalent S-boxes [4], or even more complex relationships. These cases are also interesting, but the class of attacks that could stem from them is much more limited in the number of rounds that can be attacked[2].

Let us consider two identically structured block ciphers which instance two δ-related S-boxes S_1 and S_2 according to the definition above; we will by analogy call them δ-related block ciphers. Note that in our definition, low values of δ identify pairs of very similar S-boxes, and thus this parameter measures

---

[1] By *breaking* here we mean that the full specification of the algorithm is retrieved, since the goal of the attacker in our scenario is to clone the device. We assume that all block cipher inputs (including the key) are under the attacker's control.

[2] Our relationship definition has the advantage of capturing the case where physical attacks on memories or logic could result in few entries to be interchanged; more in general, the S-box generation algorithms could also be attacked or poorly implemented and give strongly related S-boxes as result.

the degree of *relationship* between two S-boxes; however, recall that δ is also a measure of the number of entries that *differ* between S_1 and S_2.

The attack starts from the simple but somewhat surprising consideration that these two ciphers behave in the same way in a relatively large amount of cases, depending on the value of δ and on the size and number of S-boxes. In the general case, if we key two block ciphers with the same key k, and we encrypt the same plaintext p, the chance of obtaining the same ciphertext is equal to the probability that no S-box receives as input one of the values in the set Δ (no S-box is Δ-active, in our terminology). If the block cipher contains s S-boxes, this probability is equal to:

$$P(c_1 = c_2 | p_1 = p_2, k_1 = k_2) = \left(\frac{2^m - \delta}{2^m} \right)^s \tag{1}$$

If we look at the case of AES, we have $s = 200$ and $m = 8$; if $\delta = 2$ (the minimum value possible for bijective S-boxes) then the probability becomes:

$$P(c_1 = c_2 | p_1 = p_2, k_1 = k_2) = \left(\frac{256 - 2}{256} \right)^{200} \approx 0.20833 \tag{2}$$

so we expect that in about 1 case out of 5 we observe a collision on the ciphertext values; in this computation we assume that all S-box inputs are uncorrelated and uniformly distributed, which is obviously not true in practice, however this probability is easily confirmed with experiments.

This fact seems somewhat in contradiction with the belief that a cipher like the AES has good randomization properties and such events should intuitively have a very low probability. If we consider the value of 2^{-64} as threshold for the collision probability, we have that δ can reach the value of 50, i.e. S_1 and S_2 are different for slightly less than one fifth of the values.

By looking at another well known cipher that contains 8-bit S-boxes, Camellia [1], we note that S-box s_1 is directly instantiated in the round function and key-schedule, and also used to derive the other three S-boxes s_2, s_3, s_4 in a way to preserve the value of δ. Since for Camellia we have $s = 192$ and $m = 8$, the collision probability for the different values of δ are even larger than those of AES.

In our attack, we use this collision probability as a tool to obtain the specification of the unknown components, i.e. the complete content of the S-boxes. The attack scenario is the following: we assume that the attacker is able to submit encryption queries to two δ-related block ciphers. We assume that the attacker can re-key the two ciphers as he likes; his goal is to recover the specification of the unknown component (the two δ-related S-boxes S_1, S_2). The attack we present here works on the full AES and Camellia block ciphers and is shown to work in *practical* time for δ up to 16.

Since in the attack we are mainly interested in verifying assumptions on the first round of encryption, we use ciphertext *almost* collisions, i.e. pairs of ciphertexts which differ in 8 or less byte positions. For both AES and Camellia, this means that a difference has most likely been originated within the last round of encryption, and this is sufficient for our goals. The probability of having such differentials springing from the first rounds should be around 2^{-64}, and therefore if the theoretical probability of collision is significantly greater than this, the approach will work well.

More precisely, by looking at a pair of ciphertexts obtained by encrypting the same plaintext with two δ-related AES ciphers we can make the following list of statements about the most probable explanation of the given observed difference pattern:

1. If the difference between the two ciphertexts is non null on only one byte position, then we most likely have a single Δ-active S-box in the final round of the cipher.
2. If the difference is a full row of the byte matrix, we most likely have a single Δ-active S-box in the key schedule computation of the last subkey (remember that the last round of AES has no MixColumn step).
3. If the difference pattern is (embedded by) a column of the byte matrix, the Δ-active S-box is in the round before the last.
4. If the difference is (embedded by) a double row of the byte matrix, the Δ-active S-box is in the key schedule computation of the second-to-last subkey.

These explanations implicitly consider that the event of having a single Δ-active S-box is much more probable than having two or more of them. Therefore, the probability of observing an almost-collision is equal to the probability of having zero Δ-active S-boxes among the first 160 S-boxes and at most 1 among the remaining 40 S-boxes. This means:

$$P \approx \left(\frac{256 - \delta}{256}\right)^{160} \left(\left(\frac{256 - \delta}{256}\right)^{40} + 40\left(\frac{256 - \delta}{256}\right)^{39} \frac{\delta}{256}\right) \approx 2^{-\delta} \quad \delta \leq 16 \quad (3)$$

Therefore to estimate attack workload in the rest of the paper we will use this approximated value of the almost-collision probability; the error for $\delta = 8$ is 2.4%.

The attack works in two phases, presented in the two sections below for both AES and Camellia.

2.2   First Phase

The aim of the first phase is to find the complete relationship between S_1 and S_2, i.e. a function $T : F_{2^8} \to F_{2^8}$ for which we have $S_2(T(x)) = S_1(x), \forall x$. Obviously T differs from the identity function in exactly δ values. Note that the knowledge of T says nothing about the values of S_1 and S_2, it is only characterizing their relationship.

AES — In AES the input of the 16 S-boxes of the first round is a XOR between plaintext bytes and key bytes (both controlled by the attacker); thus we can do the following: we initialize 2^m (256) counters, one for each possible S-box input. We then submit a certain number of random (p, k) queries to the two δ-related oracles; if the query results in a collision we increment the counters corresponding to the 16 inputs of the S-boxes of the first round. The counters corresponding to the values in Δ can never be incremented, except in the case where a difference propagating in the cipher is corrected at a later stage. Since the probability of this event is in general negligible compared to the collision probability, after having observed about 2^{10} collisions, the only counters which are left at 0 give us the S-box inputs in the Δ set. Note that this works even if we do not know in advance the value of δ as we will simply obtain it as $|\Delta|$. Once we know the number and positions of the differences in the two S-boxes S_1 and S_2 we proceed as follows: for all possible pair of values $d_i, d_j \in \Delta$ we generate a set of $2^{5+\delta}$ (p, k) pairs so that all S-boxes in the first round receive random inputs (not belonging to Δ) except one S-box, which will be fed with d_i in the first cipher, and with d_j in the second cipher. If $S_1(d_i) = S_2(d_j)$ we will observe almost collisions for the set of queries, otherwise not. Once all $\delta(\delta - 1)2^{5+\delta}$ queries are made we know T.

Camellia — For the Camellia cipher, we proceed in an analogous way, but since the subkey used for the first round is obtained with 4 applications of the round function, we cannot use them directly to obtain T. Instead, we will use the S-boxes in the first F function in the key schedule, whose inputs are completely controllable (key bytes are XORed with known constants). The targets are the two s_1 instances in the first F function of the key schedule, and we proceed with the same counter strategy we used for AES; since we have to compensate for the reduced number of S-boxes, we will need about $2^{14+\delta}$ encryptions. In the case of Camellia, the first phase stops here as we cannot use the same technique we used for AES to completely characterize T (this is due to the fact that a XOR difference pattern in a SPN network can be completely eliminated with one Δ-active S-box, while this is not possible in a Feistel structure). However, as we will see below, this has no impact on the attack.

The first phase requires at most 2^{30} encryptions if $\delta \leq 16$ for both algorithms.

2.3 Second Phase

The aim of the second phase is to use the knowledge we obtained on T in order to recover the full specification of the S-boxes S_1, S_2.

AES — The main tool is still the possibility to produce collisions between the two encryption oracles with non-negligible probability, and we use the subkey XORs and the interaction between the key schedule and the round function as a target for our attack. Since it is difficult to impose and verify conditions directly onto the S-box values, we will work on the XOR differences within the S-box entries; that is, we imagine to take an (unknown) entry of the S-box as reference, and we will try to determine the (XOR) difference between this reference value and all other S-box entries.

First, we choose the reference entry b; for simplicity we impose that $b \notin \Delta$. We then generate a set of (p, k) queries of a certain kind; the key value k is the following, where r stands for a random value (i.e. a byte value which changes for each pair and for each byte position), a and c are fixed byte values inside each set:

$$k = \begin{bmatrix} a \oplus (01) & r & r & b \\ a & r & r & b \\ a & r & r & b \\ a & r & r & b \end{bmatrix}$$

The associated plaintext value in the pair is the following:

$$p = \begin{bmatrix} c \oplus k(0,0) & r & r & r \\ r & c \oplus k(1,1) & r & r \\ r & r & c \oplus k(2,2) & r \\ r & r & r & c \oplus k(3,3) \end{bmatrix}$$

where $k(i,j)$ is the paired key byte at row i and column j; now, for each pair in the set the input of all S-boxes of the second round is:

$$\begin{bmatrix} S(b) \oplus S(c) \oplus a & r & r & r \\ S(b) \oplus S(c) \oplus a & r & r & r \\ S(b) \oplus S(c) \oplus a & r & r & r \\ S(b) \oplus S(c) \oplus a & r & r & r \end{bmatrix}$$

Pairs belonging to a set have fixed value for a and c and random values for bytes marked with r. Each set is made up by $2^{5+\delta}$ pairs. We have a total of 2^{16} sets which account for all possible combinations of values for a and c.

All the pairs of one set are submitted for encryption to the two oracles; if no almost-collision is observed, it means with high probability that the first column of S-boxes in the second round receive an input belonging to the set Δ, i.e.

$$S(b) \oplus S(c) \oplus a \in \Delta \tag{4}$$

For each value of c, this happens for δ values of a that we can denote as a_{δ_i}; let us call the set of these values A. By looking at equation 4 we easily realize that the set Δ and the set A are the same set of values, apart from an additive (XOR) constant, and this constant is precisely one entry of the S-box difference table at index c taking entry at b as base. Thus we can easily reconstruct the full XOR difference table of the S-box S_1; if $c \in \Delta$, we take the additional care of remapping all values of c in the query as submitted to the first oracle (instantiating S_1) with the value $T(c)$ in the query submitted to the second oracle (instantiating S_2). Once we have the complete XOR difference table of S_1, we just have to guess the value of $S_1(b)$ and with a mere 256 trial encryptions we will obtain the complete content of S_1; S_2 is then immediately obtained as we already know the remapping function T.

The computational cost of phase two is roughly equal to $2^{21+\delta}$ queries to the two ciphers. This algorithm has been implemented in C and tested to work; it takes few minutes on a ordinary PC to recover the complete specification of secret 8-related AES block ciphers; the search on 16-related S-boxes is still practical (2^{38} total encryptions). Note that we did not employ parallelization or dedicated HW for the search, two things which could make the algorithm practical for even bigger values of δ.

This algorithm works on the vast majority of cases, but there are some instances of the problem which are not tractable. The reason is that the set A is not ordered with respect to Δ, in other words we have no means to discriminate between a_{δ_i} and a_{δ_j} as all that we observe from the queries is that no almost collision could be observed, i.e. that we produce some value δ_i on the input of the S-boxes. Thus, when we look for the XOR constant that transforms the set A in Δ we may end up with multiple values. Let's try to define more precisely the problem.

Definition 2. *If we denote as $\sigma_{i,j}$ the XOR difference between δ_i and δ_j, we define the non-ordered set*

$$\Sigma_k = \sigma_{i,j} | i = k \qquad (5)$$

Now if $\Sigma_i = \Sigma_j, \forall i, j$ then the algorithm above is guaranteed not to work. Let's see a practical example.

Example 1. Let us take $\delta = 4$, and let's impose that $\sigma_{3,4} = \sigma_{1,2}$. Then we have that:

$$\delta_1 \oplus \delta_2 = \sigma_{1,2}$$
$$\delta_1 \oplus \delta_3 = \sigma_{1,3}$$
$$\delta_1 \oplus \delta_4 = \sigma_{1,4}$$

$$\delta_2 \oplus \delta_1 = \sigma_{1,2}$$
$$\delta_2 \oplus \delta_3 = \sigma_{1,2} \oplus \sigma_{1,3} = \sigma_{3,4} \oplus \sigma_{1,3} = \sigma_{1,4}$$
$$\delta_2 \oplus \delta_4 = \sigma_{1,2} \oplus \sigma_{1,4} = \sigma_{3,4} \oplus \sigma_{1,4} = \sigma_{1,3}$$

etc...

so that $\Sigma_1 = \Sigma_2 = \Sigma_3 = \Sigma_4$. Thus, when we look for the XOR constant that transforms A in Δ, we will get 4 such values, only one of the four being the true value of $S(b) \oplus S(c)$ for that given value of c. If the set Δ is randomly generated, the chance of falling into this case is the chance that $\delta_3 \oplus \delta_4 = \delta_1 \oplus \delta_2$, i.e. one out of 256. □

Note that the chance of getting such a hard instance is 2^{-32} for $\delta = 8$ and 2^{-64} for $\delta = 16$; thus for interesting cases, we will have only a small chance of not succeeding.

However, if we take $\delta = 2$, our search algorithm will never work; for this case we give here an additional step which can anyway retrieve S_1 and S_2, showing that with little more effort these difficult cases can be overcome. This is particularly interesting because additional properties of the AES algorithm are used and because the case of 2-related S-boxes can perhaps easily be produced on a single device by introducing faults in the S-box computation phase, targeting for instance a single S-box during the first round of computation. On the other hand, the generalized solution to these hard instances is left as an open problem.

For $\delta = 2$, we get two plausible values for each entry of the S-box XOR difference table; moreover all values come in pairs, so for instance, and depending on the S-box, if

$$S(b) \oplus S(c) \in \{x, y\} \quad , \quad x \oplus y = \delta_1 \oplus \delta_2 \tag{6}$$

then we also have another entry c' for each we also have that

$$S(b) \oplus S(c') \in \{x, y\} \tag{7}$$

and this is because for each real difference value $dS = S(b) \oplus S(c)$ there is always another one equal to $dS \oplus \delta_1 \oplus \delta_2$.

The problem is that we do not know which of the two options is valid for each entry, i.e. in the end if:

$$S(c) = S(b) \oplus x \tag{8}$$

$$S(c') = S(b) \oplus y \tag{9}$$

or vice versa; establishing the real difference table with this information would cost 2^{126} encryption trials, as one can compare the output of each trial with that of the two oracles (in other words we do not need to guess the XOR differences at δ_1 and δ_2).

To solve this problem, we will leverage on the properties of the MixColumn operation which is carried out in the first round. We will use this operation to produce the δ_i values at the input of the second round and to establish relationships between the possible values in the XOR difference table of the secret S-box.

Let us choose one index of the secret S-box which is different from those in the set $\{b, \tilde{b}', \delta_1, \delta_2\}$; let us call this index p_1, let us denote its two possible difference values determined before as $dS'(p_1)$ and $dS''(p_1)$ and let us call its conjugate index $\tilde{p_1}$ (the index with the same set of plausible difference values). Then, consider the index p_2 (or its conjugate, it does not change anything) such that the following condition is valid:

$$(02) \cdot dS'(p_1) \oplus (03) \cdot dS'(p_2) = \delta_1 \tag{10}$$

where multiplication is carried out in $GF(2^8)$ using the AES polynomial. Index p_2 is unique and well determined (up to its conjugate) as:

$$dS'(p_2) = (03)^{-1} \cdot p_1 \oplus (03)^{-1} \cdot (02) \cdot dS'(p_1) \tag{11}$$

is an affine relationship. Note that if Equation 10 holds, then:

$$(02) \cdot dS'(p_1) \oplus (03) \cdot dS''(p_2) =$$
$$= (02) \cdot dS'(p_1) \oplus (03) \cdot dS'(p_2) \oplus (03) \cdot (\delta_1 \oplus \delta_2) =$$
$$= \delta_1 \oplus (03) \cdot \delta_1 \oplus (03) \cdot \delta_2 =$$
$$= (02) \cdot \delta_1 \oplus (03) \cdot \delta_2 \tag{12}$$

which is always different from both δ_1 and δ_2 since $\delta_1 \neq \delta_2$.

On the other hand:

$$(02) \cdot dS''(p_1) \oplus (03) \cdot dS''(p_2) =$$
$$= (02) \cdot dS'(p_1) \oplus (02) \cdot (\delta_1 \oplus \delta_2) \oplus (03) \cdot dS'(p_2) \oplus (03) \cdot (\delta_1 \oplus \delta_2) =$$
$$= \delta_1 \oplus (02) \cdot \delta_1 \oplus (03) \cdot \delta_1 \oplus (02) \cdot \delta_2 \oplus (03) \cdot \delta_2 =$$
$$= \delta_2 \tag{13}$$

and again it is easy to see that:

$$(02) \cdot dS''(p_1) \oplus (03) \cdot dS'(p_2) \neq \{\delta_1, \delta_2\} \tag{14}$$

For the sake of clearness let us define the Boolean function

$$\mu(a, b) = \begin{cases} \text{True} & \text{if} \quad (02) \cdot a \oplus (03) \cdot b \in \{\delta_1, \delta_2\} \\ \text{False} & \text{if} \quad (02) \cdot a \oplus (03) \cdot b \notin \{\delta_1, \delta_2\} \end{cases}$$

then we can summarize the discussion above by saying that

$$\mu(dS'(p_1), dS'(p_2)) \iff \mu(dS''(p_1), dS''(p_2)) \tag{15}$$
$$\mu(dS'(p_1), dS''(p_2)) \iff \mu(dS''(p_1), dS'(p_2)) \tag{16}$$

Now consider the two real values of the difference at indexes p_1 and p_2, we write them as $dS(p_1)$ and $dS(p_2)$. If we could produce the value $\mu(dS(p_1), dS(p_2))$ at the input of one S-box, we would build a set of pairs with this characteristic and if no collision would be observed in the two oracles, then we would know that a δ_i was produced, i.e. that:

$$dS(p_1) = dS'(p_1) \quad \Rightarrow \quad dS(p_2) = dS'(p_2) \tag{17}$$
$$dS(p_1) = dS''(p_1) \quad \Rightarrow \quad dS(p_2) = dS''(p_2) \tag{18}$$

and if collisions *could* be observed, then we would know that:

$$dS(p_1) = dS'(p_1) \quad \Rightarrow \quad dS(p_2) = dS''(p_2) \tag{19}$$
$$dS(p_1) = dS''(p_1) \quad \Rightarrow \quad dS(p_2) = dS'(p_2) \tag{20}$$

In other words, we would establish a link between the real XOR difference value at index p_1 and that at index p_2 and we would decrease by one bit the search space needed to find the real S-box table. The shape of the plaintext and key values in every pair of such set is the following:

$$k = \begin{bmatrix} (01) & r & r & r \\ r & r & r & b \\ r & r & r & r \\ r & r & r & r \end{bmatrix}$$

such that the second subkey is:

$$\begin{bmatrix} S(b) & r & r & r \\ r & r & r & r \\ r & r & r & r \\ r & r & r & r \end{bmatrix}$$

and the associated plaintext value is:

$$p = \begin{bmatrix} p_1 \oplus k(0,0) & r & r & r \\ r & p_2 \oplus k(1,1) & r & r \\ r & r & p_1 \oplus k(2,2) & r \\ r & r & r & p_1 \oplus k(3,3) \end{bmatrix}$$

Therefore at the end of the first round, right after the XOR with the second subkey the leftmost and topmost byte in the state matrix is equal to:

$$S(b) \oplus (02) \cdot S(p_1) \oplus (03) \cdot S(p_2) \oplus S(p_1) \oplus S(p_1) =$$
$$= S(b) \oplus (02) \cdot dS(p_1) \oplus (02) \cdot S(b) \oplus (03) \cdot dS(p_2) \oplus (03) \cdot S(b) =$$
$$= (02) \cdot dS(p_1) \oplus (03) \cdot dS(p_2) =$$
$$= \mu(dS(p_1), dS(p_2)) \tag{21}$$

while all other bytes are random. This is exactly the value we need to obtain the one bit of information from the set.

Once the link between p_1 and p_2 is established, we can iterate this procedure taking p_2 as starting point and so on; in the end, we will have established links between all XOR differences in the table and the real difference value at index p_1. Now, to obtain the complete S-box we will have just to guess the value at the reference index, $S(b)$, and the XOR difference value at index p_1. With an effort of about 2^{17} encryptions, the search space has thus been reduced to 2^9, which is trivial to brute-force. The procedure has been implemented in C and tested to work.

Camellia — The second phase of attack for the Camellia cipher is rather different from that of AES; our target will not be the XOR difference table of the secret S-box, we will instead retrieve the S-box itself. First, let's take a (p, k) query which leads to an almost-collision; we have already produced a lot of them in the first phase of attack; in the following we will keep the value of the key fixed at k, so that we are sure that no S-box in the key schedule is Δ-active. Then, consider the Camellia F function. First, all input bytes are XORed with subkey bytes (which in our analysis will be considered unknown); then they are passed through an array of S-boxes and then through the mixing layer, known as P function. Let us concentrate our attention on byte 5 of the F output of the first round, which is obtained as the XOR of bytes 1,2,6,7 and 8 of the input (after key addition and S-boxes have been applied). Let us keep the values of the input bytes 1,2,6 and 7 to some values which lead to a ciphertext collision; then, let us prepare $2^{5+\delta}$ queries for each possible combination of values of input bytes 8 and 12; byte 12 is the byte which is XORed with byte 5 of the F output to form an input byte for the second round.

If, for a given value of input byte 8, we find that all values of byte 12 lead to no collision, it means that we are querying the s_1 S-box on byte 8 with a value in the Δ set; this happens for exactly δ values of byte 8 and from those we can easily obtain the value of the subkey byte which XORs with input byte 8.

On the other hand, if for a value of byte 8 we find that exactly δ values of byte 12 lead to no collision, it means we are producing the set Δ on the input of the S-box in the second round. By comparing the set of these values of byte 12 with Δ, we obtain an entry of S-box s_1 XORed with an unknown but constant byte which is the combination of all unknown constant contributions from other input bytes of round 1 and the subkey byte of round 2. So if we write down all these values, and by making a re-arrangement implied by the subkey byte 8 we have found, we obtain an S-box table which is equal to s_1 apart from an additive constant. So with a mere additional 256 encryptions, we will recover the complete content of s_1, actually the content apart from the indexes in the set Δ. For those, we will have to guess the correct arrangement, i.e. an additional effort of $\delta!$ encryption trials.

The effort of phase 2 is equal to $2^{22+\delta} + \delta!$ encryptions; for $\delta = 16$ the factorial dominates and we have an effort of about 2^{44} encryptions.

3   Discussion and Conclusions

Our attack is easily applicable only if the size of S-boxes is such that the collision probability is high enough to practically employed; 8-bit S-boxes are good candidate. Apart from this, we have seen successful reverse engineering of two quite different ciphers (AES and Camellia); we expect that the attack can be applied also to other ciphers based on large S-boxes (Clefia, Twofish and Kasumi among the others). However, if we try to apply our attack to ciphers which instance 4-bit S-boxes, we see that the collision probability is too small to be used, even for the smallest values of δ. For example, 2-related instances of PRESENT [7] would show a collision probability of only 2^{-100}. This is a point in favor of such ciphers, which are in general more compact for hardware implementations and seem to be more flexible under this aspect.

Previous work exist on the utilization of cryptanalysis to retrieve the content of unknown S-boxes, see for instance [9][6]. These papers present techniques which can obtain the S-boxes of reduced-round variants of SPN ciphers. In this paper, we take a different approach and we show that even full ciphers which are designed to be hermetic and resistant to related cipher attacks, can be attacked, provided that the adversary has access to at least two δ-related instances.

If instances of S-boxes are randomly chosen, the probability of success of the presented attack is negligible. In general, we can conclude that the probability of collision between different block cipher instances should be verified to be sufficiently low during the design phase, because it is a tool that can be used by attackers whose goal is to obtain the complete specification of the algorithm.

Also, care should be taken w.r.t to physical attacks, such as fault injection, because it is imaginable that this type of attack could make a single faulty circuit behave like a pair of δ-related ciphers. In this case an attacker may be able to attack a single instance of unknown AES-like cipher using the techniques presented in this paper. We think that this could be an interesting direction for future research, especially considering the fact that FIRE attacks on AES ciphers have not yet been developed.

References

[1] Aoki, K., Ichikawa, T., Kanda, M., Matsui, M., Moriai, S., Nakajima, J., Tokita, T.: *Camellia*: A 128-bit block cipher suitable for multiple platforms - design and analysis. In: Stinson, D.R., Tavares, S. (eds.) SAC 2000. LNCS, vol. 2012, pp. 39–56. Springer, Heidelberg (2001)

[2] Biham, E., Shamir, A.: Differential cryptanalysis of des-like cryptosystems. Journal of Cryptology 4, 3–72 (1991)

[3] Biham, E., Shamir, A.: Differential Fault Analysis of Secret Key Cryptosystems. In: Kaliski Jr., B.S. (ed.) CRYPTO 1997. LNCS, vol. 1294, pp. 513–525. Springer, Heidelberg (1997)

[4] Biryukov, A., De Cannire, C., Braeken, A., Preneel, B.: A toolbox for cryptanalysis: Linear and affine equivalence algorithms. In: Biham, E. (ed.) EUROCRYPT 2003. LNCS, vol. 2656, pp. 648–648. Springer, Heidelberg (2003)

[5] Biryukov, A., Khovratovich, D.: Related-Key Cryptanalysis of the Full AES-192 and AES-256. In: Matsui, M. (ed.) ASIACRYPT 2009. LNCS, vol. 5912, pp. 1–18. Springer, Heidelberg (2009)

[6] Biryukov, A., Shamir, A.: Structural cryptanalysis of SASAS. In: Pfitzmann, B. (ed.) EUROCRYPT 2001. LNCS, vol. 2045, pp. 395–405. Springer, Heidelberg (2001)

[7] Bogdanov, A.A., Knudsen, L.R., Leander, G., Paar, C., Poschmann, A., Robshaw, M., Seurin, Y., Vikkelsoe, C.: PRESENT: An Ultra-Lightweight Block Cipher. In: Paillier, P., Verbauwhede, I. (eds.) CHES 2007. LNCS, vol. 4727, pp. 450–466. Springer, Heidelberg (2007)

[8] Bogdanov, A., Khovratovich, D., Rechberger, C.: Biclique cryptanalysis of the full AES. In: Lee, D.H., Wang, X. (eds.) ASIACRYPT 2011. LNCS, vol. 7073, pp. 344–371. Springer, Heidelberg (2011)

[9] Borghoff, J., Knudsen, L.R., Leander, G., Thomsen, S.S.: Cryptanalysis of PRESENT-Like Ciphers with Secret S-Boxes. In: Joux, A. (ed.) FSE 2011. LNCS, vol. 6733, pp. 270–289. Springer, Heidelberg (2011)

[10] Clavier, C.: An improved SCARE cryptanalysis against a secret A3/A8 GSM algorithm. In: McDaniel, P., Gupta, S.K. (eds.) ICISS 2007. LNCS, vol. 4812, pp. 143–155. Springer, Heidelberg (2007)

[11] Courtois, N.T., Pieprzyk, J.: Cryptanalysis of Block Ciphers with Overdefined Systems of Equations. In: Zheng, Y. (ed.) ASIACRYPT 2002. LNCS, vol. 2501, pp. 267–287. Springer, Heidelberg (2002)

[12] Daemen, J., Rijmen, V.: The Design of Rijndael - AES - The Advanced Encryption Standard. Springer (2002)

[13] Guilley, S., Sauvage, L., Micolod, J., Réal, D., Valette, F.: Defeating Any Secret Cryptography with SCARE Attacks. In: Abdalla, M., Barreto, P.S.L.M. (eds.) LATINCRYPT 2010. LNCS, vol. 6212, pp. 273–293. Springer, Heidelberg (2010)

[14] Knuth, D.E.: The Art of Computer Programming, 3rd edn. Addison-Wesley (1997)

[15] Matsui, M.: Linear Cryptanalysis Method for DES Cipher. In: Helleseth, T. (ed.) EUROCRYPT 1993. LNCS, vol. 765, pp. 386–397. Springer, Heidelberg (1994)

[16] Novak, R.: Side-Channel Attack on Substitution Blocks. In: Zhou, J., Yung, M., Han, Y. (eds.) ACNS 2003. LNCS, vol. 2846, pp. 307–318. Springer, Heidelberg (2003)

[17] O'Connor, L.: On the Distribution of Characteristics in Bijective Mappings. In: Helleseth, T. (ed.) EUROCRYPT 1993. LNCS, vol. 765, pp. 360–370. Springer, Heidelberg (1994)

[18] Réal, D., Dubois, V., Guilloux, A.-M., Valette, F., Drissi, M.: SCARE of an Unknown Hardware Feistel Implementation. In: Grimaud, G., Standaert, F.-X. (eds.) CARDIS 2008. LNCS, vol. 5189, pp. 218–227. Springer, Heidelberg (2008)

[19] Sasaki, Y.: Known-Key Attacks on Rijndael with Large Blocks and Strengthening *ShiftRow* Parameter. In: Echizen, I., Kunihiro, N., Sasaki, R. (eds.) IWSEC 2010. LNCS, vol. 6434, pp. 301–315. Springer, Heidelberg (2010)

[20] Sung, J., Kim, J., Lee, C., Hong, S.: Related-Cipher Attacks on Block Ciphers with Flexible Number of Rounds. In: Research in Cryptology - 1st Western European Workshop, WEWoRC 2005, Leuven-Heverlee,be. LNCS, p. 10. Springer (2005)

[21] Wang, D., Lin, D., Wu, W.: Related-mode attacks on ctr encryption mode. I. J. Network Security 4(3), 282–287 (2007)

[22] Wu, H.: Related-Cipher Attacks. In: Deng, R.H., Qing, S., Bao, F., Zhou, J. (eds.) ICICS 2002. LNCS, vol. 2513, pp. 447–455. Springer, Heidelberg (2002)

New Results on Generalization of Roos-Type Biases and Related Keystreams of RC4

Subhamoy Maitra[1], Goutam Paul[2,*], Santanu Sarkar[3,**], Michael Lehmann[4], and Willi Meier[4]

[1] Applied Statistics Unit, Indian Statistical Institute,
Kolkata 700 108, India
subho@isical.ac.in
[2] Department of Computer Science and Engineering,
Jadavpur University, Kolkata 700 032, India
goutam.paul@ieee.org
[3] Chennai Mathematical Institute, Chennai 603 103, India
sarkar.santanu.bir@gmail.com
[4] FHNW, Windisch, Switzerland
willi.meier@fhnw.ch, michael.lehmann87@gmail.com

Abstract. The first known result on RC4 cryptanalysis (presented by Roos in 1995) points out that the most likely value of the y-th element of the permutation after the key scheduling algorithm (KSA) for the first few values of y is given by $S_N[y] = f_y$, some linear combinations of the secret keys. While it should have been quite natural to study the association $S_N[y] = f_y \pm t$ for small positive integers t (e.g., $t \leq 4$), surprisingly that had never been tried before. In this paper, we study that problem for the first time and show that though the event $S_N[y] = f_y + t$ occurs with random association, there is a significantly high probability for the event $S_N[y] = f_y - t$. We also present several related non-randomness behaviour for the event $S_N[S_N[y]] = f_y - t$ of RC4 KSA in this direction. Further, we investigate near-colliding keys that lead to related states after the KSA and related keystream bytes. Our investigation reveals that near-colliding states do not necessarily lead to near-colliding keystreams. From this motivation, we present a heuristic to find a related key pair with differences in two bytes, that lead to significant matches in the initial keystream. In the process, we discover a class of related key distinguishers for RC4. The best one of these shows that given a random key and a related one to that (the last two bytes increased and decreased by 1 respectively), the first pair of bytes corresponding to the related keys are same with very high probability (e.g., approximately 0.011 for 16-byte keys to 0.044 for 30-byte keys).

Keywords: Bias, Cryptanalysis, Near Collision, RC4, Related Key Distinguisher, Stream Cipher.

---

* This work was done in part while the second author was visiting RWTH Aachen, Germany as an Alexander von Humboldt Fellow.
** This work was done in part while the third author was visiting National Institute of Standards and Technology, USA as a Post-Doctoral Fellow.

A. Youssef, A. Nitaj, A.E. Hassanien (Eds.): AFRICACRYPT 2013, LNCS 7918, pp. 222–239, 2013.

1 Introduction

RC4 is perhaps the simplest of all commercial ciphers, requiring only a few lines of code for its implementation. The cipher has been around for almost twenty five years since Ron Rivest designed it (alleged RC4) for RSA Data Security in 1987. In this long journey, the elegant design of RC4 and its reasonable security have attracted the attention of many cryptologists around the world. However, none of the existing cryptanalytic attacks on RC4 proves to be a serious threat to the cipher and it continues to be of significant interest in the cryptology community. The works [10, 12] summarize the entire literature of RC4 till date. This is "yet another effort" to study certain important results in the area of RC4 cryptanalysis. However, given the serious attention of several eminent researchers to this cipher for quite a long time, it is very competitive to obtain improved results in this area. The Roos bias [11] and the Mantin-Shamir distinguisher [7] are among the most celebrated cryptanalytic results in RC4 research. We generalize the first one, while improve the second in this effort. It is surprising that such natural observations eluded the community for a long period.

Like all stream ciphers, RC4 has two components, namely, the Key Scheduling Algorithm (KSA) and the Pseudo-Random Generation Algorithm (PRGA). All operations are done modulo $N = 256$. The KSA uses an l-byte secret key $k[0, \ldots, (l-1)]$ to scramble an identity permutation S over $\mathbb{Z}_N$. The PRGA uses the scrambled permutation to generate a pseudo-random sequence of keystream bytes, $z_1, z_2, \ldots$, that are bitwise XOR-ed with the plaintext to generate the ciphertext at the sender end and bitwise XOR-ed with the ciphertext to get back the plaintext at the receiver end.

Given the state-of-the-art cryptanalytic results, the practical range of l should be in between 16 to 30 bytes. For ease of description, the l-byte key is expanded into an N-byte array $K[0 \ldots (N-1)]$, such that $K[y] = k[y \bmod l]$ for any y, $0 \le y \le N - 1$. Both the KSA and the PRGA uses two indices i and j to access the permutation entries and swap a pair at every round. A brief description is given below.

KSA(K)	PRGA(S)
Initialization:	*Initialization:*
For $i = 0, \ldots, N - 1$	$i = j = 0;$
$S[i] = i;$	
$j = 0;$	*Keystream Generation Loop:*
	$i = i + 1;$
Scrambling:	$j = j + S[i];$
For $i = 0, \ldots, N - 1$	Swap($S[i], S[j]$);
$j = (j + S[i] + K[i]);$	$t = S[i] + S[j];$
Swap($S[i], S[j]$);	Output $z = S[t];$

Let S_r, i_r, j_r denote the permutation and the two indices after round r of RC4 KSA. Thus, the initial identity permutation is given by S_0 and the final permutation after the KSA is given by S_N. For $0 \le y \le N - 1$, let

$$f_y = \frac{y(y+1)}{2} + \sum_{x=0}^{y} K[x].$$

In 1995, Roos [11] argued that the most likely value of the y-th element of the permutation after the KSA for the first few values of y is given by $S_N[y] = f_y$. The experimental values of the probabilities $P(S_N[y] = f_y)$ for y reported in [11] steadily decrease from 0.37 to 0.006 as y varies from 0 to 47 and then slowly settles down to the random association of $1/N \approx 0.003906$. Much later, the theoretical proof appeared in [9]. While it was quite natural to explore little deviations of events like $S_N[y] = f_y$, it was not attempted before. The way the Roos biases appear is through the addition of the secret key bytes and the value of the indices. The study considers swaps in the S array for once, but not more than once. We show that looking at the number of swaps more than once reveals that $S_N[y]$ does not only relate to f_y, but also relates to $f_y - t$, for small integer values t.

In 2008, it was shown in [6] that not only the permutation bytes $S_N[y]$, but the nested permutation bytes, e.g., the bytes $S_N[S_N[y]]$, $S_N[S_N[S_N[y]]]$, and so on, are also biased to f_y. When we observe the generalization of Roos bias as above, it is quite natural that similar biases will be observed for the nested cases and we also prove those results generalizing the work of [6].

Another important problem related to RC4 KSA (in fact for any stream cipher) is to obtain related keys. The most important question is to obtain two different keys that will provide the same permutation after the KSA and thus, will generate the exact same keystream. The question was first raised in [2, Section 5.1], but the actual collision could be demonstrated in [8] for keys of length 24 bytes. This was later improved in [3] where colliding pairs for 22 byte keys could be obtained. Note that these are only a very few examples and it is not possible to obtain many of them in reasonable amount of computation. It is also noted in [3,8] that finding exact collisions for key-length of 16 bytes (which is mostly used in practice) is computationally very hard. Thus, we look into the problem of finding near-colliding keys that lead to near-colliding final states after the KSA for which we get related keystreams where a large number of initial bytes match. In [4], related keys (k_1, k_2) of RC4 of length 256 bytes were reported that lead to two keystreams whose many initial bytes match. These keys were of the following form: $k_2[d] = k_1[d] + \delta$, $k_2[d+1] = k_1[d+1] - \delta$, where $d = N - 2$ and $\delta = 127$ or 255. However, this method does not work with significant efficiency for practical key-length of 16 to 30 bytes. The kind of strategies used in [3,8] provides near-colliding keys quite easily, but unfortunately, the amount of matching keystream in initial bytes is not significant. The pair of keys used in [3,8] differ only in one byte. Instead, we consider difference in two bytes for this purpose.

These results, in obtaining key-pairs such that the corresponding keystreams match significantly in initial bytes, can be used as related-key distinguishers in the simplest sense. Given the key-length l (indexed by 0 to $l-1$), we consider the keys k_1 and k_2 such that $k_2[i] = k_1[i]$ for $i \le l - 3$, $k_2[l-2] = k_1[l-2] + 1$, and $k_2[l-1] = k_1[l-1] - 1$ (additions and subtractions are modulo N). Note that the key k_1 can be chosen randomly, and then modifying k_1 as explained, we obtain k_2. Applying these two related keys k_1, k_2 in RC4, we get the initial keystream bytes that match significantly. In particular, to obtain the most efficient distinguisher, we consider only the first byte in each of the keystream output and they are equal with significantly high probability. Experiments show that the first two bytes are same with probability 0.011 for 16-byte keys and it increases to 0.044 for 30-byte keys. For an ideal stream cipher, this probability should have been $\frac{1}{N}$ due to random association, which is 0.003906 for $N = 256$. For 16-byte key length, the expected number of samples (each sample consists of the first byte from the two related keystreams) needed to reliably mount the distinguisher equals 138. Quite naturally, it is even better as the key length increases. For example, for 30-byte key-length, the required number of samples is only 12.

The 2nd byte distinguisher reported in [7] was of probability $\frac{2}{N}$, which is less efficient than ours. It is well known that the initial keystream bytes of RC4 should not be used and thus our distinguisher (similar to all the distinguishers of RC4 based on initial keystreams including [7]) can also be made ineffective if the initial keystream bytes are discarded. However, given the design of the cipher, this is the best known distinguisher against RC4.

2   Extension of Roos Biases

The Roos biases [11] relate the permutation entry $S_N[y]$ to the key combination f_y. They were first reported in [11] and theoretically derived in [9]. We recapitulate the main results below.

Proposition 1. *[9, Lemma 1] If index j is pseudo-random at each KSA round, we have for $0 \leq y \leq N - 1$,*

$$\Pr\left(j_{y+1} = f_y\right) \approx \frac{1}{N} + \left(1 - \frac{1}{N}\right)^{1 + \frac{y(y+1)}{2}}.$$

Proposition 2. *[9, Lemma 2] For $1 \leq r \leq N$, $0 \leq y \leq r - 1$, we have $\Pr\left(S_r[y] = j_{y+1}\right) \approx \left(1 - \frac{y}{N}\right)\left(1 - \frac{1}{N}\right)^{r-1}.$*

The Roos bias is formally stated in the following result.

Proposition 3. *[9, Theorem 1] On completion of RC4 KSA, we have for $0 \leq y \leq N - 1$,*

$$\Pr(S_N[y] = f_y) \approx \frac{1}{N} + \left(1 - \frac{y}{N}\right) \cdot \left(1 - \frac{1}{N}\right)^{\frac{y(y+1)}{2} + N}.$$

In this section, we state and prove a set of new key-correlations in RC4, which are analogous to the biases observed by Roos [11], but address biases to the value $(f_y - t)$ instead of f_y, where $1 \leq t < y$.

The first new result, analogous to Proposition 1, is as follows.

Theorem 1. *If j is pseudo-random at each KSA round, we have for $2 \le y \le N-1$ and $1 \le t < y$, $\Pr\left(j_{y+1} = f_y - t\right) \approx \frac{1}{N} + \left(1 - \frac{1}{N}\right) p_{y,t}$, where*
$p_{y,t} = \sum_{u=0}^{y-t} \left(1 - \frac{y}{N}\right) \cdots \left(1 - \frac{y-u+1}{N}\right) \cdot \frac{1}{N} \cdot \left(1 - \frac{y-u-1}{N}\right) \cdots \left(1 - \frac{1}{N}\right).$

Proof. Let us define the event $A_{y,t,u}$ for $1 \le t < y$ and $0 \le u \le y - t$ as follows: $A_{y,t,u} : j_1 \notin \{1, \ldots, y\}, \ldots, j_u \notin \{u, \ldots, y\}, j_{u+1} = u + t, j_{u+2} \notin \{u+2, \ldots, y\}, \ldots, j_y \ne y.$

Note that for each $u \in \{0, \ldots, y - t\}$, the event $A_{y,t,u}$ implies that $S_r[r] = r$ for all $0 \le r \le y$ except for $r = u + t$, where we have $S_{u+t}[u + t] = u$, instead of $u + t$. This results in $j_{y+1} = f_y - t$, and thus provides a special path for our desired event.

One may compute

$$\Pr(A_{y,t,u}) = \left(1 - \frac{y}{N}\right) \cdots \left(1 - \frac{y - u + 1}{N}\right) \frac{1}{N} \left(1 - \frac{y - u - 1}{N}\right) \cdots \left(1 - \frac{1}{N}\right),$$

and if we define $A_{y,t} : \bigcup_{u=0}^{y-t} A_{y,t,u}$, we have the probability $\Pr(A_{y,t})$ as

$$p_{y,t} = \sum_{u=0}^{y-t} \left(1 - \frac{y}{N}\right) \cdots \left(1 - \frac{y - u + 1}{N}\right) \cdot \frac{1}{N} \cdot \left(1 - \frac{y - u - 1}{N}\right) \cdots \left(1 - \frac{1}{N}\right).$$

As the event $A_{y,t}$ leads to a favorable path for $j_{y+1} = f_y - t$, we may write

$$\begin{aligned}
\Pr\left(j_{y+1} = f_y - t\right) &= \Pr(A_{y,t}) \cdot \Pr\left(j_{y+1} = f_y - t \mid A_{y,t}\right) \\
&\quad + \Pr(\overline{A_{y,t}}) \cdot \Pr\left(j_{y+1} = f_y - t \mid \overline{A_{y,t}}\right) \\
&\approx p_{y,t} \cdot 1 + (1 - p_{y,t}) \cdot \frac{1}{N} = \frac{1}{N} + \left(1 - \frac{1}{N}\right) p_{y,t},
\end{aligned}$$

where we assume that for $\overline{A_{y,t}}$, the desired condition $j_{y+1} = f_y - t$ is true by random association. □

Remark 1. Note that the result does not hold for $t = y$. For example, when $t = y = 1$, the event $j_2 = f_1 - 1$ occurs if and only if $K[0] = 1$, which holds with probability $\frac{1}{N}$. So in general, we take $t < y$.

The following result is an extension to Proposition 3, which was observed by Roos [11].

Theorem 2. *On completion of RC4 KSA, we have for $2 \le y \le N-1$, $1 \le t < y$,*
$\Pr(S_N[y] = f_y - t) \approx \frac{1}{N} + \left(1 - \frac{y}{N}\right) \left(1 - \frac{1}{N}\right)^N p_{y,t}$, *where*
$p_{y,t} = \sum_{u=0}^{y-t} \left(1 - \frac{y}{N}\right) \cdots \left(1 - \frac{y-u+1}{N}\right) \cdot \frac{1}{N} \cdot \left(1 - \frac{y-u-1}{N}\right) \cdots \left(1 - \frac{1}{N}\right).$

Proof. We start by proving a general result for $1 \le t < y < r \le N$, as follows:

$$\Pr\left(S_r[y] = f_y - t\right) \approx \frac{1}{N} + \left(1 - \frac{y}{N}\right) \left(1 - \frac{1}{N}\right)^r p_{y,t},$$

Along one path, the event $(S_r[y] = f_y - t)$ occurs if the following two conditions are satisfied.

1. The event $A_{y,t}$ defined in the proof of Theorem 1 occurs, with probability $p_{y,t}$, as before, and

2. The event $(S_r[y] = j_{y+1})$ occurs, with probability $\left(1 - \frac{y}{N}\right)\left(1 - \frac{1}{N}\right)^{r-1}$, as in Proposition 2.

Assuming the above two events are independent, the probability that both the conditions are satisfied is given by $(1 - \frac{y}{N})(1 - \frac{1}{N})^{r-1}p_{y,t}$. If one of the condition fails, still S_r can equal $f_y - t$ by random association, and the probability of this path is given by $(1 - (1 - \frac{y}{N})(1 - \frac{1}{N})^{r-1}p_{y,t}) \cdot \frac{1}{N}$. Adding the contributions from the two paths, and substituting $r = N$, we get the result. $\square$

Remark 2. From the proof of Theorem 1, it is clear that the main source of the bias towards $f_y - t$ (with $t > 0$) is the event $S_{u+t}[u+t] = u$, instead of $u+t$. This is a high probability event. To have biases towards $f_y + t$, $1 \le t \le N - 1 - f_y$, we would need $S_{u-t}[u - t] = u$, which is essentially random. Our experimental observations also support this fact.

In Table 1, we show theoretical values of $\Pr(S_N[y] = f_y - t)$ for some sample values of y and t. These values closely match with the empirical values averaged over 1 billion key schedulings, each with randomly generated 16-byte key. For a comparative study, we list the Roos biases of Proposition 3 in the second column.

Table 1. Theoretical values of $\Pr(S_N[y] = f_y - t)$. These are much more significant than the random association having probability $\frac{1}{256} = 0.003906$.

y	$t = 0$ (Roos bias)	$t = 1$	$t = 2$	$t = 3$	$t = 4$
10	0.288393	0.015191	0.014083	0.012970	0.011853
15	0.220008	0.016815	0.015978	0.015138	0.014295
20	0.152696	0.015733	0.015164	0.014593	0.014020
25	0.096759	0.013039	0.012692	0.012342	0.011992
30	0.056426	0.009979	0.009789	0.009597	0.009405

The idea of extending the correlation between the S array locations and the secret key to the correlation between the keystream bytes and secret key has been pointed out in [5] and later it has been studied in [6]. We like to summarize the main idea behind this. We use the notation S^G to denote the permutation during the PRGA.

Proposition 4. *[6, Lemma 4] Let ϕ be some function related to the secret key bytes of RC4. Then, for $r \ge 1$, we have $\Pr(z_r = r - \phi) = \frac{1}{N} \cdot (1 + \omega_r)$, where $\Pr(S^G_{r-1}[r] = \phi) = \omega_r$.*

Now let us present the result how the keystream bytes during the PRGA rounds provide some information about the secret keys.

Theorem 3. *During RC4 PRGA, for $y \ge 2$, $1 \le t < y$, we have $\Pr(z_y = y - f_y + t) = \frac{1}{N} \cdot (1 + \omega_y)$, where $\omega_y = p_{y,t}\left(1 - \frac{y}{N}\right)\left(1 - \frac{1}{N}\right)^{N-1}\left(1 - \frac{1}{N}\right)^{y-1}\left(1 - \frac{1}{N}\right) + \frac{1}{N}$.*

Proof. Consider the event $A_{y,t}$ occurs during KSA. So we have $j_{y+1} = f_y - t$. Now applying Proposition 2 and assuming that the y-th location during PRGA is not touched by the index j in $y - 1$ rounds, we get the probability that $S_{y-1}^G[y] = f_y - t$ in this path is $w_y' = p_{y,t} \left(1 - \frac{y}{N}\right) \left(1 - \frac{1}{N}\right)^{N-1} \left(1 - \frac{1}{N}\right)^{y-1}$. On the other hand, the event $S_{y-1}^G[y] = f_y - t$ holds randomly with probability $\frac{1}{N}$. Thus, total probability

$$w_y = \Pr(S_{y-1}^G[y] = f_y - t) = w_y' + (1 - w_y')\frac{1}{N}.$$

Following Proposition 4, $\Pr(z_y = y - f_y + t) = \frac{1}{N} \cdot (1 + w_y)$. $\qquad\square$

Table 2. Theoretical and Experimental values of $\Pr(z_y = y - f_y + t)$

y	Data	$t = 1$	$t = 2$	$t = 3$
10	Theory	0.00396406	0.00395988	0.00395569
	Exp	0.00395183	0.00394792	0.00393499
15	Theory	0.00396924	0.00396615	0.00396304
	Exp	0.00395311	0.00395563	0.00393300
20	Theory	0.00396440	0.00396233	0.00396026
	Exp	0.00395378	0.00394837	0.00394727
25	Theory	0.00395399	0.00395275	0.00395151
	Exp	0.00393417	0.00393210	0.00392483
30	Theory	0.00394268	0.00394202	0.00394135
	Exp	0.00393857	0.00393038	0.00392104

In Table 2, we compare the theoretical and experimental values of $\Pr(z_y = y - f_y + t)$ for some sample values of y and t. The experimental values are generated from 1 billion key schedulings, each with randomly generated 16-byte key.

2.1   Generalizing Nested Biases of [6]

In [6], Roos correlations were extended to nested permutation entries in the direction that not only the elements of S have non-random association with secret keys, but the nested elements also have significant correlations. The main existing result in this direction is as follows.

Proposition 5. *[6, Theorem 2] On completion of RC4 KSA, for $0 \leq y \leq 31$, $\Pr(S_N[S_N[y]] = f_y)$ is approximately given by*

$$\left(\frac{y}{N} + \frac{1}{N}\left(1 - \frac{1}{N}\right)^{2-y} + \left(1 - \frac{y}{N}\right)^2\left(1 - \frac{1}{N}\right)\right)\left(1 - \frac{1}{N}\right)^{\frac{y(y+1)}{2}+2N-4}.$$

We have a new result analogous to that of Roos along the same line. But to prove the main result, we first need the following two technical results for nested indices.

Lemma 1. *For $2 \leq y \leq N - 1$ and $1 \leq t < y$, we have*

$$\Pr\left((S_{y+1}[S_{y+1}[y]] = f_y - t) \wedge (S_{y+1}[y] \leq y)\right)$$

$$\approx \frac{y}{N} \cdot \left(1 - \frac{2}{N}\right)^{y-1} \cdot \phi_{y,t} + \frac{1}{N} \cdot \psi_{y+1,y,t},$$

where $\phi_{y,t} = \Pr\left(j_{y+1} = f_y - t\right)$, as in Theorem 1, and $\psi_{r,y,t} = \Pr(S_r[y] = f_y - t)$, as in Theorem 2.

Proof. Note that the condition $(S_{y+1}[y] \leq y)$ implies that $S_{y+1}[y]$ can only take the values $0, 1, \ldots, y$. First, suppose $S_{y+1}[y] = x$, where $0 \leq x \leq y - 1$. Then $S_{y+1}[S_{y+1}[y]] = S_{y+1}[x] = f_y - t$ can occur along one path in the following way.

1. From round 1 to x (i.e., for $i = 0$ to $x - 1$), index j does not touch the places x and $f_y - t$. Thus, after round x, $S_x[x] = x$ and $S_x[f_y - t] = f_y - t$. This happens with probability $\left(\frac{N-2}{N}\right)^x$.

2. In round $x + 1$ (i.e., for $i = x$), index j_{x+1} equals $f_y - t$, and after the swap, $S_{x+1}[x] = f_y - t$ and $S_{x+1}[f_y - t] = x$. The probability of this event is $\Pr(j_{x+1} = f_y - t) \approx 1/N$.

3. From round $x + 2$ to y (i.e., for $i = x + 1$ to $y - 1$), index j does not touch the places x and $f_y - t$ once again. Thus, after round y, we have $S_y[x] = f_y - t$ and $S_y[f_y - t] = x$. This occurs with probability $\left(\frac{N-2}{N}\right)^{y-x-1}$.

4. In round $y + 1$ (i.e., for $i = y$), index j_{y+1} equals $f_y - t$, and after the swap, $S_{y+1}[y] = S_y[f_y - t] = x$ and $S_{y+1}[S_{y+1}[y]] = S_{y+1}[x] = S_y[x] = f_y - t$. According to Theorem 1, this happens with probability $\phi_{y,t}$.

Considering the above events to be independent, the probability of the joint event

$$\left((S_{y+1}[S_{y+1}[y]] = f_y - t) \wedge (S_{y+1}[y] = x)\right)$$

can be computed as $(1 - \frac{2}{N})^x \cdot \frac{1}{N} \cdot (1 - \frac{2}{N})^{y-x-1} \cdot \phi_{y,t} = \frac{1}{N} \cdot (1 - \frac{2}{N})^{y-1} \cdot \phi_{y,t}$. Summing for all x in $[0, \ldots, y - 1]$, as considered in the path above, we get

$$\Pr\left((S_{y+1}[S_{y+1}[y]] = f_y - t) \wedge (S_{y+1}[y] \leq y - 1)\right) = \frac{y}{N} \cdot \left(1 - \frac{2}{N}\right)^{y-1} \cdot \phi_{y,t}.$$

For the case $S_{y+1}[y] = y$, we can have $S_{y+1}[S_{y+1}[y]] = S_{y+1}[y] = f_y - t$ only if $f_y - t = y$, which happens with probability $1/N$. We also require $S_{y+1}[y] = f_y - t$, which happens with probability $\psi_{y+1,y,t}$, obtained by substituting $r = y + 1$ in the expression of $\psi_{r,y,t} = \Pr(S_r[y] = f_y - t)$ as computed in the proof of Theorem 2. Hence, we have $P\left((S_{y+1}[S_{y+1}[y]] = f_y - t) \wedge (S_{y+1}[y] = y)\right) = \frac{1}{N} \cdot \psi_{y+1,y,t}$. Adding the contributions for $S_{y+1}[y] \leq y - 1$ and $S_{y+1}[y] = y$, we get the result. □

Lemma 2. *If we denote $q_r(y, t) = \Pr\left((S_r[S_r[y]] = f_y - t) \wedge (S_r[y] \leq r - 1)\right)$ for $1 \leq r \leq N$, $2 \leq y \leq N - 1$, and $1 \leq t < y$, then the following recurrence relation holds:*

$$q_r(y, t) = \left(1 - \frac{2}{N}\right) q_{r-1}(y, t) + \frac{1}{N}\left(1 - \frac{1}{N}\right)^{r-1} \psi_{r-1,y,t} + \frac{1}{N^2},$$

for $1 \leq t < y$, $2 \leq y \leq N - 1$, $y + 2 \leq r \leq N$, where $\psi_{r,y,t} = \Pr(S_r[y] = f_y - t)$, as in Theorem 2.

Proof. Then event $((S_r[S_r[y]] = f_y - t) \wedge (S_r[y] \leq r - 1))$ can occur in two mutually exclusive ways:

$$((S_r[S_r[y]] = f_y - t) \wedge (S_r[y] \leq r - 2)) \text{ and }$$

$$((S_r[S_r[y]] = f_y - t) \wedge (S_r[y] = r - 1)).$$

We compute the contribution of each separately.

In round r, we have $i_r = r - 1$, which does not touch the indices $0, \ldots, r - 2$. Thus, the event

$$((S_r[S_r[y]] = f_y - t) \wedge (S_r[y] \leq r - 2))$$

occurs if

$$((S_{r-1}[S_{r-1}[y]] = f_y - t) \wedge (S_{r-1}[y] \leq r - 2))$$

occurred in the previous round (i.e., for $i = r - 2$) and if simultaneously we have $j_r \notin \{y, S_{r-1}[y]\}$. Thus, the contribution of this path of the event is $q_{r-1}(y, t) \cdot (1 - \frac{2}{N})$. On the other hand, the event $((S_r[S_r[y]] = f_y - t) \wedge (S_r[y] = r - 1))$ may occur in one path as follows:

1. After round $r - 1$, we have $S_{r-1}[r - 1] = r - 1$. This happens if the location $r - 1$ is not touched during the rounds $i = 0, \ldots, r - 2$, thus occurs with probability $(1 - \frac{1}{N})^{r-1}$.
2. We have $S_{r-1}[y] = f_y - t$. This happens with probability $\psi_{r-1,y,t}$, as in Theorem 2.
3. In the r-th round (i.e., for $i = r - 1$), index j_r equals y causing a swap involving the indices y and $r - 1$. This happens with probability $1/N$.

The contribution of this path is $\frac{1}{N}(1 - \frac{1}{N})^{r-1}\psi_{r-1,y,t}$, and the other path owing to random association contributes $1/N^2$ towards the event. Adding all the contributions, we get the desired result. $\square$

The recurrence in Lemma 2 along with the base case proved in Lemma 1 completely specifies the probabilities $q_r(y, t)$ for all $1 \leq t < y < r \leq N$. On solving the recurrence for $r = N$, we get our desired key correlation analogous to Proposition 5, as follows.

Theorem 4. *On completion of RC4 KSA, for $2 \leq y \leq N - 1$ and $1 \leq t < y$, $\Pr(S_N[S_N[y]] = f_y - t)$ is approximately given by*

$$\frac{y\phi_{y,t}}{N}\left(1 - \frac{1}{N}\right)^{2(N-2)} + \frac{\psi_{y+1,y,t}}{N}\left(1 - \frac{1}{N}\right)^{2(N-y-1)} + \frac{1}{N}\left(1 - \frac{1}{N}\right)^{N-1}\left(1 - \frac{y+1}{N}\right)$$
$$+ p_{y,t}\left(1 - \frac{y+1}{N}\right)\left(1 - \frac{y}{N}\right)\left(1 - \frac{1}{N}\right)^{2(N-1)} + \frac{1}{2N}\left(1 - \left(1 - \frac{1}{N}\right)^{2(N-y-1)}\right),$$

where $\phi_{y,t} = \Pr(j_{y+1} = f_y - t)$, as in Theorem 1, $\psi_{r,y,t} = \Pr(S_r[y] = f_y - t)$, as in Theorem 2, and $p_{y,t}$ is as defined in Theorem 1.

Proof. The recurrence in Lemma 2 can be rewritten (with a few approximations) as

$$q_r(y,t) = b^2 q_{r-1}(y,t) + a^2 b^{r-1} + cb^{2(r-1)} + a^2,$$

where $a = \frac{1}{N}$, $b = 1 - a = 1 - \frac{1}{N}$ and $c = \frac{1}{N}\left(1 - \frac{y}{N}\right)p_{y,t}$. The solution of this for $r \geq y + 1$ is

$$q_r(y,t) = b^{2(r-y-1)}q_{y+1}(y,t) + ab^{r-1}\left(1 - b^{r-y-1}\right)$$
$$+ cb^{2(r-1)}(r - y - 1) + \frac{a}{1+b}\left(1 - b^{2(r-y-1)}\right).$$

We substitute the values of a, b, c as above, the base case $q_{y+1}(y,t)$ from Lemma 1, and perform a few natural approximations to obtain the result.   □

Table 3. Theoretical values of $\Pr(S_N[S_N[y]] = f_y - t)$

y	$t = 0$ (Nested Roos)	$t = 1$	$t = 2$	$t = 3$	$t = 4$
10	0.105889	0.007267	0.006855	0.006441	0.006025
15	0.080614	0.007854	0.007541	0.007228	0.006914
20	0.055678	0.007434	0.007221	0.007008	0.006793
25	0.034889	0.006410	0.006280	0.006148	0.006017
30	0.019835	0.005242	0.005170	0.005097	0.005025

We already discussed in Remark 2 that for $1 \leq t \leq N - 1 - f_y$, the event $(S_N[y] = f_y + t)$ occurs with probability $\frac{1}{N}$. The same is the case for the event $(S_N[S_N[y]] = f_y + t)$. In Table 3, we show theoretical values of $\Pr(S_N[S_N[y]] = f_y - t)$ for some sample values of y and t. These values closely match with the empirical values averaged over 1 billion key schedulings, each with randomly generated 16-byte key. For a comparative study, we list the nested Roos biases of Proposition 5 in the second column.

3 Near-Colliding States and Related Keystreams

The idea presented in [3,8] succeeded to provide a few examples of colliding key-pairs, but those are obtained with quite significant computational effort. The similar idea can be used to obtain near-collisions. However, it is not immediate that near-collisions will always provide good match in initial keystream bytes. In this section we explore how the ideas of [3,8] can be modified to obtain significant match in the initial keystream bytes of RC4 for related keys.

Consider a key-pair whose first byte-difference occurs in position d. Note that if one cannot access the internal state variables, then the best way to update the key would be to use the Roos biases. We know that $j_{d+1} = j_d + S_d[d] + k[d] \approx j_d + d + k[d]$. According to Proposition 1, the most likely value of j_d is $f_{d-1} = d(d - 1)/2 + \sum_{y=0}^{d-1} k[y]$. If we want $j_{d+1} = target$, then we must set $k[d] = target - \frac{d(d+1)}{2} - \sum_{y=0}^{d-1} k[y]$. We will see that the approach of [3,8] is to take $target = d$ (see Table 4) and our approach is to take $target = N - 1$ (see Table 5). However, if one can access the internal state variables, one can update $k[d]$ deterministically in both the approaches, as we will see shortly.

3.1 Analysis for Key-Pairs with One Key-Byte Difference

In [8], an algorithm is presented to find a colliding key-pair (k_1, k_2), such that all the bytes are the same except $k_2[d] = k_1[d] + 1$, for some d in $0 \le d \le l - 1$. The conditions required for such a key pair to collide is tabulated in [8, Table 3] which has been summarized in [3, Table 1] noting that the conditions can be expressed in terms of the index j corresponding to the key k_1 only. We reproduce these conditions in our notations in Table 4.

Table 4. Matsui's j conditions corresponding to key k_1 to achieve a collision

Period	Range of i	Class 1 conditions on j	Class 2 conditions on j
1	$[0, d+1]$	$j_{d+1} = d,\ j_{d+2} = d+l$	$j_r \ne d, d+1,\ r \in [1, d]$
2	$[d+2, d+l]$	$j_{d+l+1} = d+2l$	$j_r \ne d+l,\ r \in [d+3, d+l]$
$\cdots$		$\cdots$	$\cdots$
t	$[d+(t-2)l+1,$ $d+(t-1)l]$	$j_{d+(t-1)l+1} = d+tl$	$j_r \ne d+(t-1)l,\ r \in$ $[d+(t-2)l+2, d+(t-1)l]$
$\cdots$	$\cdots$	$\cdots$	$\cdots$
$n-1$	$[d+(n-3)l+1,$ $d+(n-2)l]$	$j_{d+(n-2)l+1} =$ $(d-1)+(n-1)l$	$j_r \ne d+(n-2)l,\ r \in$ $[d+(n-3)l+2, d+(n-2)l]$
n	$[d+(n-2)l+1,$ $d+(n-1)l-1]$	$j_{d+(n-1)l-1} = S_{d+(n-1)l-2}^{-1}[d],$ $j_{d+(n-1)l} = d+(n-1)l-1$	$j_r \ne d+(n-1)l-1,\ r \in [d+$ $(n-2)l+2, d+(n-1)l-2]$

The key is repeated $n = \lceil (N + l - 1 - d)/l \rceil$ times during the KSA, each of which we call a *period*.[1] Note that the Class 1 conditions are less probable and hence computationally more expensive than the Class 2 conditions. The collision-finding algorithm of Chen and Miyaji [3] is essentially the same algorithm as that of Matsui [8], albeit with certain tricks to reduce the search complexity. Even with the improvements, finding exact collision for 16 byte key-length remains practically infeasible.

We therefore shift our focus to finding near-colliding key-pairs that result in related states after the KSA with many bytes matching and related keystreams with many initial bytes matching. However, our detailed experimentation revealed that though the methods of [3,8] is effective in finding exact collision, it is not as effective as finding near-collisions with the above requirement. For example, the 20-byte near-colliding key reported by Matsui [8] lead to a match of only 20 bytes in the first 256 keystream bytes. We find that the key-pairs with two key-byte differences become more effective in this case which we analyze in next subsection.

---

[1] This is called a *round* in [3], but to avoid confusion with the N KSA rounds, we prefer to call each key repetition a *period*.

3.2   Analysis of Key-Pairs with Two Key-Byte Differences

In this section, first, we investigate the conditions required for collision of a key-pair differing in two bytes and use it to search for near-colliding key-pairs.

Along the same line as [2, Section 5.1], we consider two l-byte keys k_1, k_2 such that $k_1[i] = k_2[i]$ for all $i \in [0, l-1] \setminus \{d, d+1\}$ and $k_2[d] = k_1[d]+1$, $k_2[d+1] = k_1[d+1] - 1$ for some d in $0 \leq d < l - 1$.

A quick experiment with random key-pairs of the above form with $l = 16$ and $d = 14$ immediately shows that the average number of keystream bytes matching in initial 256 bytes is around 1.015, which is clearly more than 1. On the other hand, randomly generated 16-byte key-pairs of the form of [3, 8] with one key-byte difference give an average of 1.005 matches in the initial 256 keystream bytes.

We also define the number of periods as $n = \lfloor \frac{N+1-l}{l} \rfloor$. We list all the conditions required for collision in Table 5.

Table 5. Our j conditions corresponding to key k_1 to achieve a collision

Period	Range of i	Class 1 conditions on j	Class 2 conditions on j
1	$[0, d+1]$	$j_{d+1} = N - 1$	$j_r \neq 0, N-1, r \in [1, d]$
2	$[d+2, d+l+1]$	$j_{d+l+1} = N - 1$	$j_r \neq 0, N-1, r \in [d+3, d+l]$
...	...	...	...
n	$[d+(n-1)l+2,$ $d+nl+1]$	$j_{d+nl+1} = N - 1$	$j_r \neq 0, N-1, r \in$ $[d+(n-1)l+1, d+nl]$
$n+1$	$[d+nl+2,$ $N-1]$	$j_N = 0, S_{N-1}[j_{N-1}+$ $k[N-1]+N-1] = N-1$	$j_r \neq 0, N-1, r \in$ $[d+nl+2, N-1]$

By tracing the state evolution with the above conditions, the following result is immediately established.

Theorem 5. *For two secret keys k_1, k_2 of RC4 with $k_2[d] = k_1[d]+1$, $k_2[d+1] = k_1[d+1] - 1$, and otherwise same, if all the conditions of Table 5 are satisfied during the KSA, then the final states will be same and the intermediate states would differ in at most three bytes.*

Proof. Let $S_r^{(1)}$, $S_r^{(2)}$ be the states and $j_r^{(1)}$, $j_r^{(2)}$ be the j indices corresponding to the two keys respectively after the r-th round of the KSA. Note that at round $d+1$, when $i = d$, three differences are introduced at indices $j_{d+1}^{(1)} = N - 1, i = d$ and $j_{d+1}^{(1)} + 1 = 0$. Because of the Class 2 conditions of the first period, we have the following values.

$$S_{d+1}^{(1)}[0] = 0, \qquad S_{d+1}^{(1)}[d] = N - 1, \; S_{d+1}^{(1)}[N-1] = S_d^{(1)}[d],$$
$$S_{d+1}^{(2)}[0] = S_d^{(1)}[d], \; S_{d+1}^{(2)}[d] = 0, \qquad S_{d+1}^{(2)}[N-1] = N - 1.$$

The values at indices 0 and $N-1$ remain the same up to $i = N-1$ and the third difference is at an index $y \in [1, N-2]$. In the last round, the conditions we need to get a collision are [2] $j_N^{(1)} = 0$ and $j_N^{(2)} = j_{N-1}^{(2)} + S_{N-1}^{(2)}[N-1] + k_2[N-1] = j_{N-1}^{(1)} + k_1[N-1] + N - 1 = y$. We know that $S_{d+1}^{(1)}[y] = S_{N-1}^{(1)}[y] = N - 1$, and therefore the second condition becomes $S_{N-1}^{(1)}[y] = S_{N-1}^{(1)}[j_{N-1}^{(1)} + k_1[N-1] + N - 1] = N - 1$. □

As an example, consider $N = 32$, $l = 10$ and two keys as given below.
$k_1 = [24, 5, 21, 27, 9, 8, 27, 18, 16, 14]$, $k_2 = [24, 5, 21, 27, 9, 8, 27, 18, 17, 13]$.
The two state arrays do not differ till $i = 7$. During $i = 8, 9, 10$, they differ in positions 0, 8, 31. From $i = 11$ to 30 they differ in positions 0, 11, 31. When $i = 31$, they become identical once again.
We adopt the following strategies to reduce the search complexity.

Deterministically Passing Period 1. We run KSA up to $i = d - 1$. Now setting $k_1[d] = (N - 1 - j_d - S_d[d])$, we get $j_{d+1} = j_d + S_d[d] + k_1[d] = N - 1$.

Passing Period 2 with High Probability. We require $j_{d+l+1} = N - 1$. We first run KSA up to $i = l - 2$. We can write j_{d+l+1}
$$= j_{l-1} + S_{l-1}[l-1] + \ldots + S_{d+l}[d+l] + k_1[l-1] + \ldots + k_1[d+l]$$
$$\approx j_{l-1} + S_{l-2}[l-1] + \ldots S_{l-2}[d+l] + k_1[l-1] + \ldots + k_1[d+l].$$
The above approximation will be true, if the index $j_{l+x} \notin \{l+x, \ldots, d+l\}$ for $0 \leq x \leq d$ which happens with probability $(\frac{N-d-2}{N}) \cdots (\frac{N-1}{N})$.
Hence we choose $k_1[l-1]$ such that
$$j_{l-1} + S_{l-2}[l-1] + \ldots S_{l-2}[d+l] + k_1[l-1] + \ldots + k_1[d+l] = N - 1.$$

Choice of d. In general, if there exists a d such that $\lfloor \frac{N-1-d}{l} \rfloor = n$ and $N - 1 - (d + nl) = d$, then one can choose that d. But at the same time, we should have d as close to $l - 1$ as possible. Hence, we take $d = l - 2$.

This is because when $i = N - 1$, $S_{N-1}[d] = N - 1$ and $S_{N-1}[N-1] = d + nl$ holds with higher probability. Assume that all the previous periods are satisfied, i.e., $j_{d+1} = N - 1, \ldots, j_{d+nl+1} = N - 1$. So, if j does not touch the positions d and $d + nl$ in the first $N - 1$ rounds of the KSA, which happens with probability $(\frac{N-2}{N})^{N-3}$, then $S_{N-1}[d] = N - 1$ and $S_{N-1}[N-1] = d + nl$. Note that in the power, we have $N - 3$ instead of $N - 1$, because we already know that $j_{d+1} = N - 1$ and $j_{d+nl+1} = N - 1$. In such a situation, if the two arrays are $S^{(1)}$ and $S^{(2)}$, then in the last round with $i = N - 1$, they would differ in three places as follows:
$$S^{(1)}[0] = X, \ S^{(1)}[d] = N - 1, \ S^{(1)}[N-1] = d + nl,$$
$$S^{(2)}[0] = d + nl, \ S^{(2)}[d] = X, \ S^{(2)}[N-1] = N - 1.$$
Hence, we require $j_N^{(1)} = 0$ and $j_N^{(2)} = d$ for a collision to occur. Thus, if $N - 1 - (d + nl) = d$ and $j_N^{(1)} = 0$, $j_N^{(2)}$ will always be d. So, in the last period, the success probability would be much higher than $\frac{1}{N^2}$.

Multi-Key Modification. We can use similar kind of multi-key modification as [3]. For each periods to follow, two other key bytes need to be set at prescribed

---

[2] Note that in $k_1[r]$ or $k_2[r]$, index r should be reduced modulo l.

positions. Let t denoting the first period not passed. For $(t-1)l + d \geq r \geq (t-2)l + d - 1$, suppose we want to decrease $j_{r+1} = j_r + S_r[r] + k_1[r]$ by $D = j_{r+1} - target$. We require the new value of $S_r[r]$ to be

$$\Delta_r = S_r[r] - D = S_r[r] - j_{r+1} + target.$$

This can be achieved by setting new j_{Δ_r+1} as $j'_{\Delta_r+1} = j_{\Delta_r} + S_{\Delta_r}[\Delta_r] + k'_1[\Delta_r] = r$. Noting that $j'_{\Delta_r+1} - j_{\Delta_r+1} = k'_1[\Delta_r] - k_1[\Delta_r] = r - j_{\Delta_r+1}$, we find that the key has to be modified as follows.

$$k'_1[\Delta_r] = k_1[\Delta_r] + r - j_{\Delta_r+1}, \quad k'_1[\Delta_r + 1] = k_1[\Delta_r + 1] - r + j_{\Delta_r+1}.$$

Using the above strategy, for the same key-length, we can have the same number of Class 1 conditions for our case as that of [3], and hence the overall probability for holding Class 1 conditions would be of the same order. However, since we have two Class 2 conditions in each KSA round instead of one, the probability that all our Class 2 conditions are satisfied is of the order of $(\frac{N-2}{N}) \approx (\frac{N-1}{N})^2$, which is the square of the success probability of the Class 2 conditions of [3] (refer to Table 4).

3.3   Near-Collision Search Algorithm

Though finding exact collision using the two key-byte difference strategy is not as effective as [3,8], it provides significant improvement in finding near-colliding states that lead to a match of significant number of bytes in the initial portion of the keystream.

   When designing a near-collision search algorithm, it is quite natural to have the following goals in mind.

P1. *The position of the first mismatch in the state comes as late as possible.*
P2. *The total number of mismatches in the state is as low as possible.*

Considering the update and the output function of RC4, both the properties are in favor of colliding keystream bytes. As a consequence of the first property, with significant probability the colliding state bytes are carried in the output function, whereas the second property makes the j-values coincide in both the states, so that the updated state bytes continue to remain the same with good probability.

   Our search strategy is presented in Algorithm 1. The parameter $target$ denotes the admissible value(s) of j_{d+ml+1}. According to Table 5, $target = N - 1$. However, in practice, better results are obtained by a $target$ varying over a specific range of values. The parameter $ncond$ denotes the maximum number of periods up to which j_{d+ml+1} must equal to $target$; the algorithm proceeds even if the condition fails after $ncond$ periods. For a key-pair, the number of bytes that match in the corresponding final states after the KSA is denoted by $nstate$ and the number of matches in the first 256 keystream bytes is denoted by $nstream$.

   The larger the value of the parameter $ncond$, the higher is the probability of P1 being satisfied. P1 indirectly favours P2. However, for explicitly meeting

P2, one may specify a threshold $minState$ and restart the search with a new random key if $nstate < minState$. For better near-collision in keystream, one may also specify a parameter $minStream$ and repeat the search if $nstream < minStream$.

Input: Key length l, $d = l - 2$, $target$, $ncond$, $minState$, $minStream$.
Output: A near-colliding key-pair (k_1, k_2).

1 Generate a random key k_1;
2 Run the KSA until $i = d - 1$ after the swap. Modify $k_1[d] = target - j_d - S_d[d]$;
3 Keep running the KSA until $i = d$ after the swap. Modify

$$k_1[d+1] = -\sum_{y=d+1}^{d+l} S_{d+1}[y] - \sum_{y=d+2}^{d+l} k_1[y];$$

4 Rerun the KSA with checkpoints at $i \bmod l = d$, until $i = d + (ncond - 1)l$;
    If at these points, $j \neq target$, then call $NewSearch(k_1, i, target)$;
5 Set $k_2 = k_1$, $k_2[d] = k_1[d] + 1$, $k_2[d+1] = k_1[d+1] - 1$;
6 Run the KSA with k_1 and k_2 separately to compute $nstate$ and $nstream$;
7 **if** $nstate < minState$ or $nstream < minStream$ **then**
8 | go to Step 1;
  end
9 **else**
10 | return (k_1, k_2);
  end

Subroutine $NewSearch(k_1, i, target)$:
Set $m = (i - d)/l$;
for $r = d + (m - 1)l + 1$ to $d + ml$ in step of 1 **do**
    Set $\Delta_r = S_r[r] - j_{r+1} + target$;
    if $\Delta_r < d + (m - 1)l$ **then**
        $k_1[\Delta_r] = k_1[\Delta_r] + r - j_{\Delta_r+1}$, $k_1[\Delta_r + 1] = k_1[\Delta_r + 1] - r + j_{\Delta_r+1}$;
        return;
    end
end

Algorithm 1. Our Algorithm for Finding Near-Colliding Key Pairs.

In Table 6, we compare the average and the maximum values of $nstate$ and $nstream$ for different near-collision search strategies for 16-byte keys. The averages were computed over 10000 runs of each algorithm.

For the strategy in the last row, we consider $target = d$ for the following reason. Note that for any value of $target \neq d$, at period $i = d$ we generally get three differences in the two permutations (corresponding to the two keys), namely at indices d, j_{d+1} and j'_{d+1}, where j' denotes the index corresponding to the second key k_2. On the other hand, if $target = d$, we get only two differences in period $i = d$. Thus, towards favouring P2, we take $target = d$. We have also relaxed the check at Step 4 as $j_{d+ml+1} \leq d + ml$, which includes the checkpoint $j_{d+ml+1} = target = d$. This increases the probability of crossing the m-th checkpoint from $\frac{1}{N}$ to $\frac{ml+1}{N}$ and hence we consider a higher value of $ncond$, namely,

9. We give examples of two key-pairs corresponding to this strategy. The best results in terms of collisions in the state is obtained by the key-pair

$k_1 = [66, 59, 221, 167, 250, 203, 16, 105, 88, 24, 201, 249, 175, 1, 136, 239]$,
$k_2 = [66, 59, 221, 167, 250, 203, 16, 105, 88, 24, 201, 249, 175, 1, 137, 238]$,

resulting in $nstate = 229$, but with $nstream = 6$ only.

On the other hand, the best result in terms of collisions in the keystream is achieved by the key-pair

$k_1 = [134, 185, 224, 14, 229, 137, 109, 10, 210, 196, 84, 204, 124, 238, 114, 35]$,
$k_2 = [134, 185, 224, 14, 229, 137, 109, 10, 210, 196, 84, 204, 124, 238, 115, 34]$,

giving $nstate = 222$ and $nstream = 94$.

Table 6. Experimental results related to near-collisions

Search strategy	target	ncond	nstate		nstream	
			avg	max	avg	max
Random diff. of $+1$ at d	-	-	4.94	42	1.002	9
Random diff. of ± 1 at $d, d+1$	-	-	24.27	221	1.015	58
Chen-Miyaji's Method	Table 4, Class 1	5	4.78	18	0.980	6
Our Algorithm 1	$N - 1$	5	56.24	217	1.140	40
Our Algorithm 1	$\in [N - 13, N + 1]$	5	91.14	221	1.600	35
Our Algorithm 1	d	9	150.59	229	2.740	94

As evident from the third row of Table 6, Chen-Miyaji's method is not suitable for finding near-collisions either in state or in keystream. Sometimes a higher maximum for $nstate$ or $nstream$ is obtained by a modified version of Chen-Miyaji's method having the same conditions as in Table 4 but key-pairs having two-byte differences like Algorithm 1. However, the average values of $nstate$ and $nstream$ are always higher in case of Algorithm 1.

3.4 Related Keystream Distinguisher

From Table 6, it is evident that one can mount a distinguisher on RC4 using related keys as follows. Assume that the attacker can only modify the secret key and rerun the keystream generation algorithm as a black box as many time as he wants. For uniform random stream, the probability that the first keystream byte will match is $\frac{1}{N}$, which is 0.003906 for $N = 256$. However, if the key-pairs are chosen with two-byte differences ($k_2[d] = k_1[d] + 1$, $k_2[d+1] = k_1[d+1] - 1$, $d = l - 2$), we get that the first pair of keystream bytes in the related keystreams match with significantly higher probability. This can distinguish the keystream of RC4 from random keystream reliably.

The exact theoretical derivation of this distinguisher seems extremely hard due to dependence in key-length and non-randomness in pairs of j values. However, empirical evidences support that for two secret keys k_1, k_2 of RC4 with $k_2[d] = k_1[d] + 1$, $k_2[d+1] = k_1[d+1] - 1$, and otherwise same, the corresponding pseudo-random indices $j^{(1)}$ and $j^{(2)}$ differ for the first time after $3l + d$ rounds of the KSA (except when $i^{(1)} = i^{(2)} = d \bmod l$) with very high probability. If this holds, then for the remaining $N - 3l - d$ rounds, $j^{(1)}$ does not touch the indices 1, $S^{(1)}[1]$ and $S^{(1)}[1] + S^{(1)}[S^{(1)}[1]]$ in $S^{(1)}$ and $j^{(2)}$ does not touch the indices 1, $S^{(2)}[1]$ and $S^{(2)}[1] + S^{(2)}[S^{(2)}[1]]$ in $S^{(2)}$ with a probability $p = \left(\frac{N-3}{N}\right)^{2(N-3l-d)}$. Assuming a complimentary path of random association, an approximate estimate of the probability of the first two keystream bytes being equal is given by $p + (1-p)/N = \frac{1}{N} + \left(1 - \frac{1}{N}\right)\left(\frac{N-3}{N}\right)^{2(N-3l-d)}$.

Table 7. Experimental (A) and estimated (B) values of probabilities of the first pair of keystream output bytes matching that are generated from related keys. (C) The average number of key pairs that provide the first match in the first byte. (D) The probability that we get at least one match in certain number of trials given in parenthesis.

l	16	18	20	22	24	26	28	30
Emp. Prob. (A)	0.011	0.014	0.017	0.021	0.026	0.031	0.037	0.044
Est. Prob. (B)	0.014	0.016	0.019	0.022	0.026	0.030	0.036	0.042
Data Comp. (C)	89.16	70.70	58.07	47.34	38.73	32.61	27.48	22.88
Data Comp. (D)	0.676 (100)	0.679 (80)	0.702 (70)	0.722 (60)	0.730 (50)	0.712 (40)	0.727 (35)	0.737 (30)

Table 7 shows the experimental (averaged over one million runs) and estimated values of the above probability for different values of key-length l with $d = l-2$. It is clear that our results are improved than the observation on second byte in [7]. The probability, that the second output byte of RC4 keystream is zero, is $\frac{2}{N}$ [7]. In our case, for 16 (respectively 30) byte key-length, the probability is 0.011 (respectively 0.044) which is approximately $\frac{3}{N}$ (respectively $\frac{11}{N}$) for $N = 256$. This is significantly greater than the observation of [7] and thus provides the most efficient distinguisher for RC4.

According to [1, Section 4.1], if an event occurs in random stream with probability p and in RC4 (or in any algorithm) with probability $p(1+q)$, the number of samples required for mounting the distinguishing attack with success probability 69.15% is given by $\frac{\left(\sqrt{1-p}+\sqrt{(1+q)(1-p(1+q))}\right)^2}{4pq^2}$. This general formula (instead of $\frac{1}{pq^2}$) needs to be used in this case as q is not small here. Plugging in the appropriate values of our biases from Table 7, one can easily see that our distinguisher requires only 138 samples (i.e., pair of 1st bytes of keystreams from 138 key-pairs) for 16-byte key and 12 samples (i.e., pair of 1st bytes of keystreams from 12 key-pairs) for 30-byte key.

Experimentally, we also note in how many trials the first byte is matching. The results show the average data required in Table 7 corresponding to each key-length. Similarly, we also consider what is the probability of at least one match in x many trials. The probability and the value of x is also reported in Table 7. These are also easily theoretically justified (using the mean of Geometric distribution and the tail of Binomial distribution to estimate the rows C and D respectively) given our basic observation of probabilities as described in row A (the second row) of Table 7. One may note that this is significantly distinguishable from the cases of random associations.

4   Conclusion

In this paper, we revisit Roos biases, i.e., biases of the permutation bytes $S[y]$ after the KSA towards $f_y = y(y+1)/2 + \sum_{y=0}^{l-1} k[y]$. We show that these biases can be extended towards $f_y - t$ for small positive integers t. The same extension is also possible for the nested permutation entries $S[S[y]]$. We also look into the problem searching colliding key pairs in RC4. The existing algorithms are not efficient enough for finding near-collisions that will provide significant matches in initial keystream bytes. We propose an algorithm for finding near-colliding RC4 keys that can yield related states after the KSA for a good amount of match in initial keystream bytes. This result, provides a class of related key distinguishers for RC4. We particularly explain the one that is the best ever known distinguisher for RC4.

References

1. Basu, R., Ganguly, S., Maitra, S., Paul, G.: A Complete Characterization of the Evolution of RC4 Pseudo Random Generation Algorithm. Journal of Mathematical Cryptology 2(3), 257–289 (2008)
2. Biham, E., Dunkelman, O.: Differential Cryptanalysis in Stream Ciphers. IACR Eprint Server, eprint.iacr.org, number 2007/218 (June 6, 2007)
3. Chen, J., Miyaji, A.: How to Find Short RC4 Colliding Key Pairs. In: Lai, X., Zhou, J., Li, H. (eds.) ISC 2011. LNCS, vol. 7001, pp. 32–46. Springer, Heidelberg (2011)
4. Grosul, A.L., Wallach, D.S.: A Related Key Cryptanalysis of RC4 (July 6, 2000), http://www.wisdom.weizmann.ac.il/~itsik/RC4/Papers/GrosulWallach.ps
5. Klein, A.: Attacks on the RC4 stream cipher. Later appeared in Designs, Codes and Cryptography 48(3), 269–286 (2006, 2008)
6. Maitra, S., Paul, G.: New Form of Permutation Bias and Secret Key Leakage in Keystream Bytes of RC4. In: Nyberg, K. (ed.) FSE 2008. LNCS, vol. 5086, pp. 253–269. Springer, Heidelberg (2008)
7. Mantin, I., Shamir, A.: A Practical Attack on Broadcast RC4. In: Matsui, M. (ed.) FSE 2001. LNCS, vol. 2355, pp. 152–164. Springer, Heidelberg (2002)
8. Matsui, M.: Key Collisions of the RC4 Stream Cipher. In: Dunkelman, O. (ed.) FSE 2009. LNCS, vol. 5665, pp. 38–50. Springer, Heidelberg (2009)
9. Paul, G., Maitra, S.: Permutation After RC4 Key Scheduling Reveals the Secret Key. In: Adams, C., Miri, A., Wiener, M. (eds.) SAC 2007. LNCS, vol. 4876, pp. 360–377. Springer, Heidelberg (2007)
10. Paul, G., Maitra, S.: RC4 Stream Cipher and Its Variants. CRC Press (2011)
11. Roos, A.: A class of weak keys in the RC4 stream cipher. Posts in sci.crypt, message-id 43u1eh$1j3@hermes.is.co.za and 44ebge$11f@hermes.is.co.za (1995)
12. Sen Gupta, S., Maitra, S., Paul, G., Sarkar, S. (Non-)Random Sequences from (Non-)Random Permutations - Analysis of RC4 stream cipher. To appear in Journal of Cryptology (November 3, 2012) (accepted), http://eprint.iacr.org/2011/448

Impact of Sboxes Size upon Side Channel Resistance and Block Cipher Design

Louis Goubin[1], Ange Martinelli[1,2], and Matthieu Walle[2]

[1] Versailles Saint-Quentin-en-Yvelines University
Louis.Goubin@prism.uvsq.fr
[2] Thales Communications
{Pjean.martinelli,matthieu.walle}@thalesgroup.com

Abstract. Designing a cryptographic algorithm requires to take into account various cryptanalytic threats. Since the 90's, Side Channel Analysis (SCA) has become a major threat against cryptographic algorithms embedded on physical devices. Protecting implementation of ciphers against such attacks is a very dynamic topic of research and many countermeasures have been proposed to thwart these attacks. The most common countermeasure for block cipher implementations is masking, which randomizes the variables by combining them with one or several random values. In this paper, we propose to investigate the impact of the size of the words processed by an algorithm on the security against SCA. For this matter we describe two AES-like algorithms operating respectively on 4 and 16-bit words. We then compare them with the regular AES (8 bits) both in terms of complexity and security with respect to various masking schemes. Our results show that SCA is a determinant criterion for algorithms design and that cryptographers may have various possibilities depending on their security and complexity requirements.

Keywords: Side Channel Analysis (SCA), S-boxes, Word size, Masking Countermeasure, Higher-Order SCA, AES Implementation, FPGA.

1 Introduction

When designing a block cipher, cryptographers take into account various cryptanalytic threats in order to prevent flaws in their algorithm. The most common methods are linear [19] and differential [5] cryptanalysis, interpolation [17] or related key attacks [4]. All these attacks target the mathematical primitive independently of its implementation.

In the 90's, a new kind of cryptanalysis was developed: *Side Channel Analysis* (SCA). SCA is a cryptanalytic method in which an attacker does not attack the algorithm itself, but rather its implementation. Namely, the attacker analyzes the *side channel leakage* (*e.g.* the power consumption, the electromagnetic emanations, ...) produced during the execution of a cryptographic algorithm embedded on a physical device. SCA exploits the fact that this leakage is statistically dependent on the intermediate variables that are involved in the computation. Some of these variables are called *sensitive* in that they are related to a

A. Youssef, A. Nitaj, A.E. Hassanien (Eds.): AFRICACRYPT 2013, LNCS 7918, pp. 240–259, 2013.

secret data (*e.g.* the key) and a known data (*e.g.* the plain text), and recovering information on them therefore enables efficient key recovery attacks [18,6,14].

However it is a hard task to improve the intrinsic security of a cryptographic algorithm against SCA. Designers can nonetheless foresee the implementation of countermeasures, and design their algorithm in order to help these implementations. As many countermeasures [1,22,15] use the arithmetic structure of the AES S-box, it seems a good option for designers to keep such type of structure.

To evaluate the efficiency of a countermeasure, Prouff *et al.* introduce in [23] a methodology to compute the optimal correlation between a leakage measure and a (multivariate) known variable. They give the optimal correlation for boolean masking. We can observe that this correlation depends on the noise standard variation, the order of the masking, but also on the size of the words targeted by the attack: the longer the words, the better the security. The goal of this paper is to study the impact of the word size on both the complexity and the security of the scheme.

Related work. This paper is mainly related to three kinds of previous works. In [12,24,15], the authors propose countermeasures that provide a good security/complexity compromise for some security level, and propose practical results implementing their countermeasures to the AES. In [8], small scale variants of AES are designed in order to compare different cryptanalytic methods. Eventually, various optimized hardware implementations of the AES S-box can be found in [7,20].

Our contribution. In this paper, we propose an evaluation of the impact of the word size on the security of an algorithm with respect to various masking schemes, namely boolean [22], affine [12] and polynomial masking [24,15]. We thus define two AES-like algorithms operating respectively on 4 and 16 bits words, and discuss their implementation. Then we compare the security and the complexity of each algorithm depending on the countermeasure scheme. We finally give practical implementation results on a hardware device for equivalent level of security.

Organization of the paper. The remainder of this paper is organized as follows. In the second section we pursue a theoretical analysis about the impact of the size of the words manipulated by an algorithm upon its resistance to SCA. In section 3, we recall the AES algorithm and detail the two AES-like algorithms we have implemented. In section 4 we compare implementation costs of the three algorithms first theoretically, then on practical hardware implementations. Then section 5 details simulations results on the SCA resistance of these algorithms and the AES with respect to various masking schemes. We conclude our work in section 6.

2 Impact of Sboxes Size upon Side Channel Resistance

S-boxes are the most sensitive layer with respect to the resistance of a block cipher against higher order side channel attacks. For a matter of implementation, an S-box must have a short dimension n and therefore, the input block is shared in n-bit words for the independent internal computations of the algorithm. The size of these words is determined by designers with respect to the needed properties of the algorithm and to its resistance against known cryptanalysis.

In the state of the art, we can find block ciphers manipulating 8 bits words (*e.g.* the AES [9]), 4 bits words (*e.g.* *Serpent* [2]) or non-square S-boxes such as those of the DES [11]. In the following of this section we investigate, for various masking schemes, the impact of this dimensions upon the resistance of a block cipher algorithm against SCA.

In what follows, we shall consider that an intermediate variable U_i is associated with a leakage variable L_i representing the information leaking about U_i through side channel. We will assume that the leakage can be expressed as a deterministic *leakage function* φ of the intermediate variable U_i with an independent additive noise B_i. Namely, we will assume that the leakage variable L_i satisfies:

$$L_i = \varphi(U_i) + B_i \ . \tag{1}$$

In the following, we call d^{th}-*order leakage* a tuple $\mathbf{L}$ of d leakage variables L_i corresponding to d different intermediate variables U_i that jointly depend on some sensitive variable. Moreover we place ourselves in the Hamming weight model, *i.e.* $\varphi = HW$.

2.1   Security Against HO-DPA

In order to compare various scales of implementation with respect to various masking schemes, we compute, for each case, the optimal correlation value following the methodology described in [23] for decreasing signal-to-noise ratio (SNR). Namely we considered the value of

$$\rho_{\text{opt}} = \sqrt{\frac{\text{Var}\left[\text{E}\left[\mathcal{C}(\mathbf{L})|Z=z\right]\right]}{\text{Var}\left[\mathcal{C}(\mathbf{L})\right]}} \tag{2}$$

where Z is a sensitive variable and $\mathcal{C}$ is a combining function that converts the multivariate leakage $\mathbf{L}$ into a univariate signal. In our evaluations we have chosen $\mathcal{C}$ to be the normalized product. In [25,12,15], authors give equations for evaluating the value of ρ_{opt} respectively $\rho_{\text{bool-d}}$ for d-th order boolean masking, ρ_{aff} for affine masking and $\rho_{\text{polynomial}}$ for first order polynomial masking. Let us consider an gaussian noise B_i with 0 mean and standart deviation σ. We have:

$$\rho_{\text{bool-d}} = (-1)^d \frac{\sqrt{n}}{(n + 4\sigma^2)^{\frac{d+1}{2}}}, \tag{3}$$

$$\rho_{\text{aff}} = \frac{n}{(4\sigma^2 + n)\sqrt{2^n - 1}} \tag{4}$$

and

$$\rho_{\text{polynomial}} = \sqrt{\frac{n^3 \cdot (2^{n+1} - 4^n - 1)}{\alpha_2 \cdot \sigma^4 + \alpha_1 \cdot \sigma^2 + \alpha_0}}, \tag{5}$$

where

$$\alpha_2 = 192 \cdot 2^n - 2^{4n+4} - 64 - 208 \cdot 4^n + 96 \cdot 8^n$$
$$\alpha_1 = (40 \cdot 8^n - 64 \cdot 4^n - 8 \cdot 16^n + 32 \cdot 2^n)n^2$$
$$+(88 \cdot 8^n + 128 \cdot 2^n - 2^{4n+4} - 168 \cdot 4^n - 32)n$$
$$\alpha_0 = (8^n - 3 \cdot 4^n + 6 \cdot 2^n - 4)n^4 + (-4 \cdot 16^n + 14 \cdot 8^n - 16 \cdot 4^n + 2 \cdot 2^n + 4)n^3$$
$$+(-4 \cdot 16^n + 23 \cdot 8^n - 44 \cdot 4^n + 34 \cdot 2^n - 8)n^2$$
$$+(-3 \cdot 8^n + 10 \cdot 4^n - 9 \cdot 2^n + 2)n.$$

$$(6)$$

As a matter of comparaison, the optimal correlation ρ_{unmasked} for a non-masked implementation is:

$$\rho_{\text{unmasked}} = \frac{\sqrt{n}}{\sqrt{(n + 4\sigma^2)}}, \qquad (7)$$

We then evaluate these values for any bit size $n \in \{4, 8, 16\}$, any SNR $\in \{1, 1/2, 1/5, 1/10\}$, and for the variables targeted respectively in [25,12,24]:

- 1O-boolean masking, with targeted variables $(x \oplus r_1 ; r_1)$
- 2O-boolean masking, with targeted variables $(x \oplus r_1 \oplus r_2 ; r_1 ; r_2)$
- 3O-boolean masking, with targeted variables $(x \oplus r_1 \oplus r_2 \oplus r_3 ; r_1 ; r_2 ; r_3)$
- Affine masking, with targeted variables $(r2 \cdot x \oplus r_1 ; r_1)$

Table 1. Theoretical correlation values

Word length \SNR	$+\infty$	1	1/2	1/5	1/10
2O-DPA against 1O-boolean masking					
4-bits	0.5	0.25	0.1	0.083333	0.045455
8-bits	0.353553	0.176777	0.117851	0.058926	0.032141
16-bits	0.25	0.125	0.083333	0.041667	0.022727
3O-DPA against 2O-boolean masking					
4-bits	0.25	0.088388	0.022361	0.017010	0.006853
8-bits	0.125	0.044194	0.024056	0.008505	0.003426
16-bits	0.0625	0.022097	0.012028	0.004253	0.001713
4O-DPA against 3O-boolean masking					
4-bits	0.125	0.031250	0.005	0.003472	0.001033
8-bits	0.044194	0.011049	0.004910	0.001228	0.000365
16-bits	0.015625	0.003906	0.001736	0.000434	0.000129
2O-DPA against Affine masking [12]					
4-bits	0.258199	0.129099	0.015188	0.009931	0.002556
8-bits	0.062622	0.020874	0.006958	0.001228	0.000312
16-bits	0.003906	0.000781	0.000230	0.000039	0.000010
2O-DPA against 1O-polynomial masking [15]					
4-bits	0.030589	0.023984	0.015187	0.013612	0.009063
8-bits	0.001854	0.001473	0.001243	0.000876	0.000607
16-bits	0.0000074	0.0000060	0.0000051	0.0000037	0.0000027

- 1O-polynomial masking, with targeted variables $((r_1, r3 \cdot r_1 \oplus x)$; $(r_2, r3 \cdot r_2 \oplus x))$ where $r_1 \neq r_2 \neq 0$.

Note that we only consider the best available attacks against these masking. Table 1 summarizes the theoretical correlations ρ_{opt}.

We can state that the security of each scheme evolves in different ways when the word size increases. Indeed the value of the optimal correlation for boolean masking decreases polynomialy in n, whereas it decreases exponentially for both affine and polynomial masking. Intuitively, this can be explained seeing that, when using boolean masking, every bit of the mask operates on a single bit of the sensitive variable. Thus the security overhead of a larger bit size is roughly linear in the word length. In the case of affine and polynomial maskings, the relation between the targeted values and the sensitive one is much more complex, which highly improves the security when increasing the word length.

Moreover, in the case of boolean masking, the optimal correlation coefficient decrease exponentially in the order d. In order to compare the different implementations of (higher order) boolean masking we represent in Figure 1 the correlation value for various amount of noise. As expected, we can state that a higher order scheme provides a better security in (almost) all cases. Notably

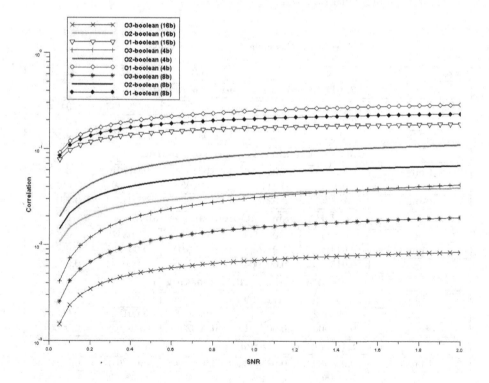

Fig. 1. Correlation value of boolean maskings with respect to SNR

$(d + 1)$-th order boolean masking applied to the 4-bits AES variation provides better security than d-th order masking applied to the regular 8-bits AES.

2.2   Information Theoretic Analysis

The analysis of the optimal correlation allows us to evaluate precisely the security of a countermeasure against CPA but does not give any general informations independently of the chosen distinguisher. In [26] the authors propose to evaluate the mutual information between the leakage vector $\mathbf{L}$ and the sensitive variable Z in order to compute the theoretic leakage induced by the computation. Nevertheless, this metric does not give any direct information of the complexity of an attack but only gives the amount of information leaked during the computation. We can then efficiently compare two countermeasures implemented on the same algorithm in terms of security against SCA, but the comparison between two distinct algorithms does not seems to be so relevant, especially when the word sizes are distinct.

A third security analysis can be the practical attack simulation but it needs the definition of a complete algorithm. Such an analysis is the topic of section 5.

In this section we have shown that non-linear masking techniques applied to large S-boxes (typically 16 bits) provides the best theoratical security among the considered countermeasures. We may wonder what is the practical complexity of the implementation of such countermeasures. In the following, we evaluate the complexity of some implementations in order to emphasize the most interesting one in terms of complexity for a given security level.

3   Design of the Sboxes

The main goal of this article is to evaluate the optimal word size to implement countermeasures against SCA. In order to achieve this goal, we define variants of the AES using different word sizes. For matters of simplicity, we focus on powers of 2. After recalling the processing of the AES, we propose in this section two AES-like algorithms working on respectively 4 and 16 bit words. Both are operating on 128-bit blocks.

3.1   The Advanced Encryption Standard

The Advanced Encryption Standard (AES) is a Substitution-Permutation Network (SPN) introduced by J. Daemen and V. Rijmen in [9]. It iterates 10 times (for the 128-bit version) a transformation involving four steps : AddRoundKey, ShiftRows, MixColumn, and SubByte. Details about these steps can be founded in appendix A. In particular the AES S-box is designed as

$$Sb_8[x] = Q(x) + a(x) \cdot P(x) \bmod [X^8 + 1]$$

where $a(x)$ is the inverse of x in the field $\mathbb{F}_{2^8}$, and P and Q are polynomials chosen such that it ensure a complicated algebraic expression when combined with the inverse mapping, and that there is no fixed points ($Sb_8[x] = x$) and no 'opposite fixed points' ($Sb_8[x] = \bar{x}$). This construction ensures a good resistance against linear and differential cryptanalysis. Following the formalism introduced in [10], the AES sbox achieve a prop-ratio and an input-output correlation respectively equal to 2^{-6} and 2^{-3}.

Remark 1. The inversion of $a \in \mathbb{F}_{2^8}$ as described in [22] can be implemented using 4 multiplications, 7 squares and 1 refreshMask operation.

3.2 4-Bit Variation

Let $\mathbb{F}_{2^4} = \mathbb{F}_2[x]/(x^4 + x + 1)$.
We define the 4-bit AES-like Sbox as follows :

$$Sb_4[x] = Q(x) + a(x) \cdot P(x) \bmod [X^4 + 1]$$

where $a(x)$ is the inverse of x in the field $\mathbb{F}_{2^4}$, and P and Q are polynomials chosen according to [9] such that : $P(x) = x^3 + x + 1$ and $Q(x) = x^3 + x^2 + 1$.
A look-up table for the Sbox Sb_4 is given (notation in hexadecimal) in table 2.

Table 2. Sbox over $\mathbb{F}_{2^4}$

Input	0	1	2	3	4	5	6	7	8	9	A	B	C	D	E	F
Output	D	6	B	9	5	C	F	4	2	A	E	8	7	3	1	0

Remark 2. The inversion of $a \in \mathbb{F}_{2^4}$ can be computed as $a^{-1} = a^{14} = (a \cdot a^2)^4 \cdot a^2$. It can then be implemented using only two multiplications and 3 squares.

Resistance Against Known Attacks : In order to evaluate the security of the S-box against differential and linear cryptanalysis, we have to compute respectively the prop-ratio and the input-output correlation (see appendix D). We can then evaluate the length of an efficient linear or differential trail and adapt the number of round adequately.

- prop-ratio : 2^{-2}
- input-output correlation : 2^{-1}

Keeping the original ShiftRows and MixColumn operating on 8-bits words, we obtain no 12-round differential trail with a predicted prop ratio above 2^{-150} (which is sufficient for the 128-bit block length), and no 12-round linear trail with a correlation above 2^{-75}. In this case, in order to keep an equivalent security, we have to extend the round number to 30. Moreover this construction is not directly compatible with every masking schemes operating on 4 bits.

In order to bypass this incompatibility and to optimize the complexity of the overall cipher, we define a diffusion layer composed of a MixColumns operation designed using a 8×8 MDS matrix over $\mathbb{F}_{2^4}$ (see appendix C), combined with a ShiftRows operation as designed for the AES. We then consider the internal state as a 8×4 matrix over $\mathbb{F}_{2^4}$. The branch number is thus 9. For example the matrix involved in the MixColumns operation can be chosen as a circulant matrix with first line equal to:

$$[1 \ 1 \ 2 \ 1 \ 3 \ 4 \ 2 \ 3].$$

We can deduce that there is no 4-round differential trail with a predicted prop ratio above 2^{-98} and no 4-round linear trail with a correlation above 2^{-49}. The round number can thus be fixed to 15 without any security loss.

However, in [8] the authors design simplified version of AES in order to try their security against algebraic attacks. For the 4-bit version the succed only with a small number of round and using a sub-optimal linear layer. Our construction clearly ensure a much better security against such attacks.

3.3   16-Bit Variation

Let $\mathbb{F}_{2^{16}} = \mathbb{F}_2[x]/(x^{16} + x^{13} + x^{10} + x^9 + x^2 + x + 1)$.
   We define the 16 bit AES-like S-box as following :

$$Sb_{16}[x] = Q(x) + a(x) \cdot P(x) \bmod [X^{16} + 1]$$

where $a(x)$ is the inverse of x in the field $\mathbb{F}_{2^{16}}$, and P and Q are polynomials chosen according to [9] such that : $P(x) = x^{15} + x^8 + x^3 + x + 1$ and $Q(x) = x^{15} + x^9 + x^8 + x^7 + x^2 + x + 1$.

Remark 3. The inversion of $a \in \mathbb{F}_{2^{16}}$ can be computed following:

$$\begin{aligned} b &= (a^2.a)^2.a = a^7 \\ c &= b^8.b \quad\ \ = a^{63} \\ c &= c^{64}.c \quad\ = a^{4095} \\ c &= c^{16}.b^2 \ = a^{65534} \end{aligned}$$

The inversion can then be implemented using only 5 multiplications and 16 squares.

Resistance Against Known Attacks : As previously, we compute respectively the prop-ratio and the input-output correlation and evaluate the length of an efficient linear or differential trail and adapt the number of rounds adequately.

– Sb_{16} prop-ratio : 2^{-14}.
– Sb_{16} input-output correlation : 2^{-7}.

As for the 4 bit case, the original linear layer of the AES is not compatible with every masking schemes operating on 16 bits. A good alternative is to use a 8×8 circulant MDS matrix instead of the ShiftRows and MixColumn operations. Such a matrix can even be optimized allowing the Hamming weight of each of its component to be 1 and thus leads to the most optimized overall design for this size of words. For example such a matrix can be chosen as a circulant matrix with first line equal to (in hexadecimal):

$$[0001\ 0001\ 0020\ 0001\ 0100\ 0400\ 0200\ 0040].$$

The branch number of this linear layer is maximal (*i.e.* equal to 9), ensuring that there is no 1-round differential trail with a predicted prop ratio above 2^{-125}, and no 1-round linear trail with a correlation above 2^{-63}. As for the 4 bit case, the round number can thus be lower to 6 rounds without any security loss.

4   Complexity

Previously, we have seen that increasing the word size improves the security of a device against SCA. However, this security improvement should not lead to an unreasonable cost. In this section, we compare several implementations to study the impact of S-boxes sizes on the complexity. Firstly in a theoretical manner, then by doing a comparative analysis of hardware implementations.

4.1   Overall Complexity

For each masking scheme, we consider the AES variations described in Section 3 (see appendix B for details about multiplication implementations) :

- − 4-bit words (using look-up table),
- − 8-bit words (using log/alog tables),
- − 16-bit words (using log/alog tables),
- − 16-bit words (using tower fields method)

The evolution of the theoretical complexity of each masking scheme according to the word size is given in Table 3.

Remark 4. Each implementation of the affine masking is optimised using the most appropriate variation of the scheme : that is the reference implementation for the original AES and the 4-bit variation, and the least memory expensive variation for the 16-bit (see [12] for details about each variation). Similarly the implementation of the polynomial masking is made using the straightforward adaptation of [3] as explained in [24,16].

   The 8-bit affine masking appears to be a very good option both in terms of security and complexity. The complexity of the 4-bit variation is not as low as it can be expected because of its heavy linear layer. Using a similar S-box in a Feistel scheme could solve this probleme though, but such a construction is not in the scope of this paper. With respect to boolean masking, we can state that, for a high amount of noise, the 4-bit variation provides a very low complexity for a good security level. For instance, with a SNR near 1, the third order boolean masking implemented on a 4-bit algorithm provides a better complexity and a better security than a second order boolean masking implemented on a 8-bit algorithm. The 16-bit variation does not seem an interesting choice because of its huge memory requirements. However the very high security provided by polynomial masking on this variation may justify its implementation on very low restricted devices.

Table 3. Theoretical Complexity of cipher implementations

Implementation	XORs/ANDs/shifts	Table look-ups	Random bits	RAM (bits)	ROM (bits)
1-st order Boolean Masking					
4-bit	8580	7920	5760	284	192
8-bit	17640	16144	16896	312	6128
16-bit (log/alog)	7704	9273	13056	368	2097120
16-bit (tower field)	40269	50385	13056	368	1022
2-nd order Boolean Masking					
4-bit	14340	14760	15360	328	192
8-bit	37800	32272	46080	352	6128
16-bit (log/alog)	16200	16257	36086	448	2097120
16-bit (tower field)	72549	90273	36086	448	1022
3-rd order Boolean Masking					
4-bit	26820	23520	28800	328	192
8-bit	65640	54160	87552	400	6128
16-bit (log/alog)	25656	25257	69120	544	2097120
16-bit (tower field)	114429	141969	69120	544	1022
Affine Masking					
4-bit	2176	1224	2400	448	1088
8 bits	3424	1840	1552	4392	8176
16 bits (log/alog)	526560	394456	800	1048912	3145696
16 bits (tower field)	2500080	1971288	800	1048912	1022
1st order polynomial Masking					
4 bits	9480	19440	3840	328	192
8 bits	58560	65824	27792	400	6128
16 bits (log/alog)	39840	57568	18592	544	2097120
16 bits (tower field)	321360	409856	18592	544	1022

4.2   Complexity of Chosen Hardware Implementations

The theoretical complexities given in the previous section provide a good overview of the implementations' security. However we want to evaluate the practical feasability of some chosen implementations on hardware devices. As boolean masking is the most widely implemented scheme, we limited ourselves to implement Rivain and Prouff's scheme from [22] at orders 1, 2 and 3.

We developed in VHDL a small system on chip (SoC) embedding a simple serial interface and a 128-bits masked AES implementation running in ECB mode. The implementations are fully parallelized, notably all Sboxes are processed simultaneously. As proposed in [22], the multiplicative inverse is computed using a d-order secure square-and-multiply algorithm. To do so, each Sbox encompasses a secure multiplier as well as a square operator, both working on d shares. Then, alternating the square and the multiply module in a sequential way ensure to use the minimal area.

The SoC is built on an Altera Cyclone III EP3C25 (24,624 Logic Elements, Speed grade -7, -8) with no particular optimization technique and an automated place-and-route stage. The resulting maximal clock frequency is 125 Mhz for all implementations. Following this fully parallelized design, no protected version of the 16-bits scheme can be realistically implemented on the SoC.

We implemented multiplier and square blocks for $GF(2^8)$ and $GF(2^4)$ in the same way, that is fully combinatorial with input/output register. In that case, a secure multiplication takes 3 cycles since some variables have to be processed

Table 4. Implementations areas (in logic elements) and performances

Word Size	Global System (LEs)	S-box (LEs)	SecMult (LEs)	Clock cycles	Throughput (MB/s)
1-st order Boolean Masking					
4 bits	4350	112	54	197	18.87
8 bits	8089	380	212	282	7.09
2-nd order Boolean Masking					
4 bits	7500	207	117	197	18.87
8 bits	13435	690	487	282	7.09
3-rd order Boolean Masking					
4 bits	11300	350	212	197	18.87
8 bits	21299	1170	870	282	7.09

at different time as explained in [22], squaring is done in 2 clock cycles and only one cycle is needed to refresh the masks. Eventually, the linear layer is also combinatorial so the total number of cycles to process the whole round is equal to the the number of cycles required for the inversion.

For the 8-bit version, we have 4 secMult + 7 Square + 2 refreshMask = $4 \times 3 + 7 \times 2 + 2 = 28$ cycles per round. Since there are 10 rounds, we obtain 280 cycles for the encryption + 2 cycles to handle the I/O, hence 282 cycles in total.

For the 4-bit version, we have 2 secMult + 3 Square + 1 refreshMask = $2 \times 3 + 3 \times 2 + 1 = 13$ cycles per round. Since there are 15 rounds, we obtain 195 cycles for the encryption + 2 cycles to handle the I/O, hence 197 cycles in total.

As expected, the resulting 8-bit S-box is at least 3 times bigger than the 4-bit version for a given order of masking. The interesting fact is that this inequality still hold for different order of masking : a d-order 8-bit S-box is *bigger* than a $(d+1)$-order 4-bit Sbox. Now if we look at the theoretical correlation of each of this implementations (see Table 1), we observe that any $(d+1)$-order 4-bit AES is *more* secure the d-order 8-bit version. Notably the 4-bit 2nd-order AES is more secure *and* smaller than the regular 8-bit AES using only one mask.

As a matter of fact, we can observe that between the two implementations, the difference of size of the global circuit is not so important for 1-st order masking but increases with the order. It can be explained by noticing that the expensive layer for the 8-bit scheme is clearly the S-box (and in particular the SecMult operation), while it is the permutation layer in the case of the 4 bit variation. Indeed, in this case, the cost of the S-box is roughly 4 times lower than the one operating on 8 bits. Moreover the cost of the S-box transformation is quadratic in d while the linear layer is only linear in d, and so the difference increases with the order.

By taking into account the Sbox size during the design of an implementation, it is possible to improve the security of the device without increasing the size of the circuit excessively. Actually the linear layer may take a non-negligible place. Indeed, in order to avoid to increase excessively the number of round, this layer has to be improved. This leads to a bigger linear layer and the global system size increases. Anyway, the $(d+1)$-order 4-bit variation is wholly more secure and smaller and faster than the d-order 8-bit variation.

5   Attack Simulations

To confirm the theoretical analysis conducted in section 2, we performed several attacks simulations. Formally we applied several side-channel distinguishers to simulated leakages. The leakage measurements have been simulated as samples of the random variables L_i defined according to equation (1) with $\varphi = \mathrm{HW}$ and $B_i \sim \mathcal{N}(0, \sigma^2)$ where the different B_i's are assumed mutually independent. For all the attacks, the sensitive variable Z was chosen to be an S-box output of the targeted algorithm of the form $\mathsf{S}(M \oplus k^\star)$ where M represents a varying plaintext byte and $k^\star$ represents the key byte to recover.

Side-channel distinguishers. We applied two kind of side-channel distinguishers: higher-order DPA such as described in section 2.1 and higher-order MIA [21,13]. In a HO-MIA, the distinguisher is the mutual information: the guess k is tested by estimating $\mathrm{I}(\hat{\varphi}(Z(k)); \mathbf{L})$. As mutual information is a multivariate operator, this approach does not involve a combining function.

Targeted variables. Each attack was applied against leakage values associated to boolean masking, affine masking and polynomial masking. The target variables are those listed in section 2.1 where $x = \mathsf{S}(X \oplus k^\star)$:

Prediction functions. For each DPA, we choose $\hat{\varphi}$ to be the optimal prediction function :

$$\hat{\varphi} : z \mapsto \mathrm{E}\left[\mathcal{C}(\mathbf{L})|Z = z\right]. \tag{8}$$

This leads us to select the Hamming weight function in the attacks against both $1O$-polynomial masking and dO-boolean masking and the Dirac function δ_0 for the affine masking.

For the MIA attacks, we choose $\hat{\varphi}$ such that it maximizes the mutual information $\mathrm{I}(\hat{\varphi}(Z(k)); \mathbf{L})$ for $k = k^\star$ while ensuring that the mutual information is lower for $k \neq k^\star$. In our case, every HO-MIA against both polynomial and Boolean masking is performed with $\hat{\varphi} = \mathrm{HW}$ since the distribution of $(\mathrm{HW}(Z \oplus m_0), \mathrm{HW}(m_0))$ (resp. $(\mathrm{HW}(Z \oplus a_0 \cdot x_0, x_0), \mathrm{HW}(Z \oplus a_0 \cdot x_1, x_1))$) only depends on $\mathrm{HW}(Z)$. Therefore

$$\mathrm{I}\big(Z; (\mathrm{HW}(Z \oplus m_0), \mathrm{HW}(m_0))\big) = \mathrm{I}\big(\mathrm{HW}(Z); (\mathrm{HW}(Z \oplus m_0), \mathrm{HW}(m_0))\big).$$

Note that the same relation holds at every masking order. Every HO-MIA against affine masking is performed using $\hat{\varphi} = \delta_0$ since the distribution of the leakage functions is identically distributed for any $Z \neq 0$, and is only remarkable for $Z = 0$ [12].

Pdf Estimation Method. For the (HO-)MIA attacks, we use the histogram estimation method with rule of [14] for the *bin-widths* selection.

Table 5. Number of leakage measurements for a 90% success rate against 4, 8 and 16-bits algorithms

Word size \ SNR	$+\infty$	1	1/2	1/5	1/10
2O-CPA against 1O-boolean masking					
4-bit	70	160	400	1 800	13 000
8-bit	150	500	1 500	6 000	20 000
16-bit	400	1 400	2 000	10 000	25 000
2O-MIA against 1O-boolean masking					
4-bit	80	1 000	5 000	10 000	33 000
8-bit	100	5 000	15 000	50 000	160 000
3O-CPA against 2O-boolean masking					
4-bit	370	1 700	20 000	50 000	300 000
8-bit	1 500	9000	35 000	280 000	$> 10^6$
16-bit	6 500	20 000	85 000	900 000	$> 10^6$
3O-MIA against 2O-boolean masking					
4-bit	120	10 000	200 000	800 000	$> 10^6$
8-bit	160	160 000	650 000	$> 10^6$	$> 10^6$
2O-CPA against affine masking					
4-bit	300	1400	20 000	100 000	400 000
8-bit	6500	20 000	45 000	170 000	650 000
16-bit	55 000	200 000	800 000	$> 10^6$	$> 10^6$
2O-MIA against affine masking					
4-bit	270	10 000	100 000	800 000	$> 10^6$
8-bit	5500	100 000	600 000	$> 10^6$	$> 10^6$
2O-CPA against 1O-polynomial masking					
4-bit	15 000	40 000	100 000	150 000	250 000
8-bit	$> 10^6$	$> 10^6$	$> 10^6$	$> 10^6$	$> 10^6$
16-bit	$> 10^6$	$> 10^6$	$> 10^6$	$> 10^6$	$> 10^6$
2O-MIA against 1O-polynomial masking					
4-bit	100 000	300 000	600 000	$> 10^6$	$> 10^6$
8-bit	500 000	$> 10^6$	$> 10^6$	$> 10^6$	$> 10^6$

Attack simulation results. Each attack simulation is performed 100 times for various SNR values ($+\infty$, 1, 1/2, 1/5 and 1/10). Table 5 summarizes the number of leakage measurements required to observe a success rate of 90% in retrieving $k^\star$ for the different attacks.

Remark 5. No MIA processed against an implementation of the 16-bits algorithm had succeeded. This can be explained by the complexity of estimation of the probability density functions needed by the attack.

The simulation results confirm the security intuition introduced in section 2 that the security of an algorithm is highly dependant of its word size. We can indeed state that the number of measurements needed for a 90% success rate increase with the word size. In particular these results show that the security improvement induced by boolean masking on longer words increase more slowly than that induced by non-linear masking scheme. Moreover we are able to give practical results for the efficiency of MIA upon the considred implementations. These results shows that the security improvement of a longer word size has the same kind of impact on both CPA and MIA.

6 Conclusion

In this paper, we investigated the influence of the size of the words in an cryptographic algorithm on the efficiency and the security of the scheme against side channel analysis. We designed for this matter two algorithms operating respectively on 4 and 16-bit words, and compared them to the original 8-bits AES both in terms of complexity and SCA resistance.

The 16-bit variation provides a very good security, particularly assiciated with a non-linear masking, but the complexity overhead is consequent. On the contrary, we have shown that in some situations, using smaller Sboxes associated with higher order masking technique improves the security of a device with almost no extra cost. Our results show that indeed, a 2nd order boolean masking applied on the 4-bits AES provides both a better resistance as well as better performances than 1st order boolean masking applied on the 8-bit AES.

The S-boxes size and the masking order can be viewed as two complementary parameters. By choosing these parameters, one can adapt the performances (area, thoughput, security) of a device to match a specific need. Table 6 recall implementations complexities and the corresponding CPA simulation results for a realistic amount of noise (SNR= 1/2).

Table 6. Comparison of two distinct implementations

CPA (traces)	Global System (LEs)	S-box (LEs)	Clock cycles	Throughput (MB/s)
8-bits AES secured by 1-st order Boolean Masking				
1 500	8089	380	282	7.09
4-bits AES secured by 2-nd order Boolean Masking				
20 000	7500	207	197	18.87

References

1. Akkar, M.-L., Giraud, C.: An Implementation of DES and AES, Secure against Some Attacks. In: Koç, Ç.K., Naccache, D., Paar, C. (eds.) CHES 2001. LNCS, vol. 2162, pp. 309–318. Springer, Heidelberg (2001)
2. Anderson, R., Biham, E., Knudsen, L.: Serpent: A Proposal for the Advanced Encryption Standard (1998)
3. Ben-Or, M., Goldwasser, S., Wigderson, A.: Completeness theorems for non-cryptographic fault-tolerant distributed computation (extended abstract). In: STOC, pp. 1–10. ACM (1988)
4. Biham, E.: New types of cryptanalytic attacks using related keys. In: Helleseth, T. (ed.) EUROCRYPT 1993. LNCS, vol. 765, pp. 398–409. Springer, Heidelberg (1994)

5. Biham, E., Shamir, A.: Differential Cryptanalysis of DES-like Cryptosystems. Journal of Cryptology 4(1), 3–72 (1991)
6. Brier, E., Clavier, C., Olivier, F.: Correlation Power Analysis with a Leakage Model. In: Joye, M., Quisquater, J.-J. (eds.) CHES 2004. LNCS, vol. 3156, pp. 16–29. Springer, Heidelberg (2004)
7. Canright, D.: A Very Compact S-Box for AES. In: Rao, J.R., Sunar, B. (eds.) CHES 2005. LNCS, vol. 3659, pp. 441–455. Springer, Heidelberg (2005)
8. Cid, C., Murphy, S., Robshaw, M.: Small scale variants of the AES. In: Gilbert, H., Handschuh, H. (eds.) FSE 2005. LNCS, vol. 3557, pp. 145–162. Springer, Heidelberg (2005)
9. Daemen, J., Rijmen, V.: AES Proposal: Rijndael (September 1999)
10. Daemen, J.: Cipher and hash function design strategies based on linear and differential cryptanalysis. PhD thesis, K.U.Leuven (1995)
11. FIPS PUB 46-3. Data Encryption Standard (DES). National Institute of Standards and Technology (October 1999)
12. Fumaroli, G., Martinelli, A., Prouff, E., Rivain, M.: Affine masking against higher-order side channel analysis. In: Biryukov, A., Gong, G., Stinson, D.R. (eds.) SAC 2010. LNCS, vol. 6544, pp. 262–280. Springer, Heidelberg (2011)
13. Gierlichs, B., Batina, L., Preneel, B., Verbauwhede, I.: Revisiting Higher-Order DPA Attacks: Multivariate Mutual Information Analysis. Cryptology ePrint Archive, Report 2009/228 (2009), http://eprint.iacr.org/
14. Gierlichs, B., Batina, L., Tuyls, P., Preneel, B.: Mutual Information Analysis. In: Oswald, E., Rohatgi, P. (eds.) CHES 2008. LNCS, vol. 5154, pp. 426–442. Springer, Heidelberg (2008)
15. Goubin, L., Martinelli, A.: Protecting AES with shamir's secret sharing scheme. In: Preneel, B., Takagi, T. (eds.) CHES 2011. LNCS, vol. 6917, pp. 79–94. Springer, Heidelberg (2011)
16. Goubin, L., Martinelli, A.: Protecting aes with shamir's secret sharing scheme - extended version. IACR Cryptology ePrint Archive 2011, 516 (2011)
17. Jakobsen, T., Knudsen, L.R.: The interpolation attack on block ciphers. In: Biham, E. (ed.) FSE 1997. LNCS, vol. 1267, pp. 28–40. Springer, Heidelberg (1997)
18. Kocher, P., Jaffe, J., Jun, B.: Differential Power Analysis. In: Wiener, M. (ed.) CRYPTO 1999. LNCS, vol. 1666, pp. 388–397. Springer, Heidelberg (1999)
19. Matsui, M.: Linear cryptanalysis method for DES cipher. In: Helleseth, T. (ed.) EUROCRYPT 1993. LNCS, vol. 765, pp. 386–397. Springer, Heidelberg (1994)
20. Morioka, S., Satoh, A.: An optimized s-box circuit architecture for low power AES design. In: Kaliski Jr., B.S., Koç, Ç.K., Paar, C. (eds.) CHES 2002. LNCS, vol. 2523, pp. 172–186. Springer, Heidelberg (2003)
21. Prouff, E., Rivain, M.: Theoretical and Practical Aspects of Mutual Information Based Side Channel Analysis. In: Abdalla, M., Pointcheval, D., Fouque, P.-A., Vergnaud, D. (eds.) ACNS 2009. LNCS, vol. 5536, pp. 499–518. Springer, Heidelberg (2009)
22. Rivain, M., Prouff, E.: Provably Secure Higher-Order Masking of AES. In: Mangard, S., Standaert, F.-X. (eds.) CHES 2010. LNCS, vol. 6225, pp. 413–427. Springer, Heidelberg (2010)
23. Prouff, E., Rivain, M., Bévan, R.: Statistical Analysis of Second Order Differential Power Analysis. IEEE Trans. Comput. 58(6), 799–811 (2009)
24. Prouff, E., Roche, T.: Higher-order glitches free implementation of the AES using secure multi-party computation protocols. In: Preneel, B., Takagi, T. (eds.) CHES 2011. LNCS, vol. 6917, pp. 63–78. Springer, Heidelberg (2011)

25. Rivain, M., Prouff, E., Doget, J.: Higher-Order Masking and Shuffling for Software Implementations of Block Ciphers. In: Clavier, C., Gaj, K. (eds.) CHES 2009. LNCS, vol. 5747, pp. 171–188. Springer, Heidelberg (2009)
26. Standaert, F.-X., Malkin, T.G., Yung, M.: A Unified Framework for the Analysis of Side-Channel Key Recovery Attacks. In: Joux, A. (ed.) EUROCRYPT 2009. LNCS, vol. 5479, pp. 443–461. Springer, Heidelberg (2009)
27. Wolkerstorfer, J., Oswald, E., Lamberger, M.: An ASIC Implementation of the AES SBoxes. In: Preneel, B. (ed.) CT-RSA 2002. LNCS, vol. 2271, pp. 67–78. Springer, Heidelberg (2002)

A   Original AES Steps

In this section, we recall the four main operations involved in each round of the AES encryption Algorithm. For each of them, we denote by $\mathbf{s} = (s_{i,j})_{0 \leq i,j \leq 3}$ the 16-byte state at the input of the transformation, and by $\mathbf{s}' = (s'_{i,j})_{0 \leq i,j \leq 3}$ the state at the output of the transformation.

1. AddRoundKey: Let $\mathbf{k} = (k_{i,j})_{0 \leq i,j \leq 3}$ denote the round key. Each byte of the state is XOR-ed with the corresponding round key byte:

$$(s'_{i,j}) \leftarrow (s_{i,j}) \oplus (k_{i,j}).$$

2. SubBytes: each byte of the state passes through the 8-bit S-box S:

$$s'_{i,j} \leftarrow S(s_{i,j}).$$

For all x in $\mathrm{GF}(2^8)$, the AES S-box is defined as follows :

$$S[x] = Q(x) + a(x) \cdot P(x) \bmod [X^8 + 1]$$

where $a(x)$ is the inversion function in the field $\mathrm{GF}(2^8)$, $P(x) = x^7 + x^6 + x^5 + x^4 + 1$ coprime to the modulus, and $Q(x) = x^7 + x^6 + x^2 + x$ chosen such that the S-box has no fixed points $(S(x) = x)$ and no "opposite fixed point" $(S(x) = \bar{x})$.

3. ShiftRows: each row of the state is cyclically shifted by a certain offset:

$$s'_{i,j} \leftarrow s_{i,j-i \bmod 4}.$$

4. MixColumns: each column of the state is modified as follows:

$$(s'_{0,c}, s'_{1,c}, s'_{2,c}, s'_{3,c}) \leftarrow \mathsf{MixColumns}_c(s_{0,c}, s_{1,c}, s_{2,c}, s_{3,c})$$

where $\mathsf{MixColumns}_c$ implements the following operations:

$$\begin{cases} s'_{0,c} \leftarrow (02 \cdot s_{0,c}) \oplus (03 \cdot s_{1,c}) \oplus s_{2,c} \oplus s_{3,c} \\ s'_{1,c} \leftarrow s_{0,c} \oplus (02 \cdot s_{1,c}) \oplus (03 \cdot s_{2,c}) \oplus s_{3,c} \\ s'_{2,c} \leftarrow s_{0,c} \oplus s_{1,c} \oplus (02 \cdot s_{2,c}) \oplus (03 \cdot s_{3,c}) \\ s'_{3,c} \leftarrow (03 \cdot s_{0,c}) \oplus s_{1,c} \oplus s_{2,c} \oplus (02 \cdot s_{3,c}), \end{cases}$$

where $\cdot$ and $\oplus$ respectively denote the multiplication and the addition in the field $\mathrm{GF}(2)[X]/p(X)$ with $p(X) = X^8 + X^4 + X^3 + X + 1$, and where 02 and 03 respectively denote the elements X and $X + 1$.

B   Implementations of the Field Multiplication

The main problem encountered when implementing AES is the implementation of the field multiplication. In this section we will view several possibilities of implementations with respect to the words bit-size.

B.1   4-Bit Multiplication :

In the case of a 4-bit implementation, a natural idea is to pre-compute the field multiplications and store them in a 16×16 entries table. The multiplication is then resumed to 1 table look-up. Such table can be stored on 128 bytes.

B.2   8-Bit Multiplication :

A classical method to implement the multiplication over GF(256) in software is to use *log/alog* tables. These tables are constructed using the fact that all non-zero elements in a finite field $GF(2^n)$ can be obtained by exponentiation of a generator in this field. For a generator α of GF(256)* we define $\log(\alpha^i) = i$ and $\mathrm{alog}(i) = \alpha^i$. This results are stored in two tables of $2^n - 1$ words of n bits.

If a, b are non-zero, then the product $a \cdot b$ can be computed using log/alog tables as

$$a \cdot b = \mathrm{alog}[(\log(a) + \log(b)) \bmod (2^n - 1)]. \qquad (9)$$

With this representation, computing a product over GF(256) requires 3 table look-ups, and two additions modulo 256. Both tables can be stored in ROM on 510 bytes.

On hardware systems, the multiplication can easily be implemented using composite field method using the methodology given in [27], or simple combinatorial multipliers.

B.3   16-Bits Multiplication :

In order to compute multiplication over $GF(2^{16})$, two tools can be used: log/alog tables or the tower field method (see [27]).

Using log/alog table requires 3 table look-ups, and two additions $\bmod(2^{16})$. Both tables can be stored in ROM on 262140 bytes.

For more memory-restricted implementations, the tower field methodology can be applied. It consists in considering $GF(2^{16})$ as $GF(2^8) \times GF(2^8)$, thus making product in the smaller field $GF(2^8)$.

In [27], Wolkerstorfer *et al.* give an efficient hardware implementation of multiplications and inversions in $GF(2^8)$. They represent $GF(2^8)$ as a quadratic extension of the field $GF(2^4)$ then exhibit an isomorphism between $GF(2^8)$ and $GF(2^4) \times GF(2^4)$. The multiplication can thus be implemented in

$\mathrm{GF}(2^4) \times \mathrm{GF}(2^4)$, instead of $\mathrm{GF}(2^8)$. We want to develop here the same methodology in order to implement the multiplication in $\mathrm{GF}(2^{16})$ using multiplication in $\mathrm{GF}(2^8)$.

Let α be the class of X in $\mathrm{GF}(2^{16})$. Let $Q(X) = X^2 + X + \alpha^7$ be an irreducible polynomial over $\mathrm{GF}(2^8)$. Let us consider the field $\mathrm{GF}(2^8) \times \mathrm{GF}(2^8) = \mathrm{GF}(2^8)[X]/Q(X)$. If β is the class of X in $\mathrm{GF}(2^8)^2$, then every element of $\mathrm{GF}(2^8)^2$ can be written as $a \cdot \beta + b$ with a and b in $\mathrm{GF}(2^8)$.

Let η and ν be two elements of $\mathrm{GF}(2^8)^2$ such that $\eta = u_1\beta + u_0$ and $\nu = v_1\beta + v_0$. Then we have :

$$\begin{aligned} \eta \cdot \nu &= (u_1\beta + u_0)(v_1\beta + v_0) \\ &= u_1v_1\beta^2 + (u_1v_0 + u_0v_1)\beta + u_0v_0 \\ &= (u_1v_0 + u_0v_1 + u_1v_1)\beta + (u_1v_1\alpha^7 + u_0v_0) \end{aligned} \qquad (10)$$

Hence the product in $\mathrm{GF}(2^8)^2$ can be performed using 5 multiplications in $\mathrm{GF}(2^8)$ and 3 XORs. In order to compute the isomorphism $I : \mathrm{GF}(2^{16}) \longrightarrow \mathrm{GF}(2^8) \times \mathrm{GF}(2^8)$ and its inverse, we simply have to define base changing equations from the relation $I(\alpha) = 2A\beta + 1C$. Base changing can then be computed following algorithm 1.

Algorithm 1. Base changing

INPUT: An element a in the input field F, $(\mu_0, \ldots, \mu_{15})$ the base changing value
OUTPUT: The corresponding element a' in the ouput field G

---

1. $a' \leftarrow 0$
2. **for** $i = 0$ **to** 15 **do**
3. $a' \leftarrow a' \oplus (a_i \cdot \mu_i)$
4. **return** a'

---

where a_i is the ith bit of a.

Remark 6. As both words in $\mathrm{GF}(2^8)$ depend on every 16 bits of the input, there is no security loss in this implementation.

Using this method, each multiplication in $\mathrm{GF}(2^8) \times \mathrm{GF}(2^8)$ can be performed using 5 multiplications in $\mathrm{GF}(2^8)$ (using log/alog tables) and 3 XORs. Both isomorphisms I and I^{-1} can be computed using 16 XORs and 16 ANDs (and 16 shifts in software) knowing both 32-bytes tables of base changing.

C   Linear Layer and MDS Matrix

We have seen that the linear layer of the AES is composed of two operations: ShiftRows and MixColumn. This linear layer allows a reasonable diffusion entwined with a very low complexity. However we can define optimal diffusion function using MDS matrices as follows.

Let C be an (m, k, d)-error correcting code over $\mathbb{F}_{2^n}$. Then m is the word size of C, k is its dimension, and d is the minimal distance between two words of the code (or the minimal weight of a non-zero word of the code). Let us have the following definition:

Definition 1 (MDS code). $\mathcal{C}$ *is said MDS (Maximum Distance Separable) if it is a linear code that reaches the Singleton bound, i.e. if and only if* $d = m - k + 1$.

An important parameter in the security of a block cipher against linear and differential cryptanalysis is its *branch number* B.

Let b be the linear (respectively differential) bias (see section D) associated to a transformation $S : \mathrm{GF}(q) \to \mathrm{GF}(q)$ of the substitution layer, then the global resistance provided by N rounds of the algorithm can be evaluated by

$$b^{B \cdot \frac{N}{2}}.$$

Let θ be a diffusion layer with input k elements of $\mathrm{GF}(q)$ and with output m elements of $\mathrm{GF}(q)$ then

$$\theta : \left| \begin{array}{l} \mathrm{GF}(q)^k \to \mathrm{GF}(q)^m \\ \quad x \quad \mapsto \quad \theta(x). \end{array} \right.$$

Then θ's branch number is given by

$$\mathcal{B}(\theta) = \min_{\xi \in (\mathrm{GF}(q)^k)^\star} \{\omega(\xi) + \omega(\theta(\xi))\}.$$

Proposition 1. *We have* $\mathcal{B}(\theta) \leq m + 1$.

Let now $\mathcal{C}$ be a $(2k, k, k + 1)$-MDS code over $\mathbb{F}_{2^n}$ with generator matrix $G = (I \parallel M)$ with I the identity and $I, M \in \mathcal{M}_{k \times k}(\mathbb{F}_{2^n})$. M is then called an MDS matrix. Let us have the following proposition:

Proposition 2. *Let* M *be an MDS matrix over* $\mathbb{F}_{2^n}$. *We can then define an optimal, i.e having the maximal branch number, invertible SPN-diffusion layer* $\theta_\mathcal{C}$ *as*

$$\theta_\mathcal{C} : \left| \begin{array}{l} \mathbb{F}_{2^n}^k \to \mathbb{F}_{2^n}^k \\ \quad x \quad \mapsto M x. \end{array} \right.$$

In this case, the branch number of the linear layer is maximal, and equal to $k + 1$.

D   Linear and Differential Cryptanalysis

Differential and Linear cryptanalysis were first described respectively by Eli Biham and Adi Shamir [5] in 1991 and by Mitsuru Matsui [19] in 1993. Both attacks aims to recover the last round's subkey of the algorithm by exploiting statistical bias in the propagation of the message through the algorithm called linear or differential trails. The efficiency of these attacks depends of the length of the trails, *i.e.* the round number. Basically, the round number can be derived from the branch number of the linear layer and both the prop ratio and the input-output correlation of the S-boxes [10].

In practice we evaluate the security of an S-box S against differential cryptanalysis by computing the prop-ratio R_S. Let (a', b') be a pair where a' is a difference of input values and b' the difference of the corresponding outputs. The prop-ratio $R_S(a', b')$ of S associated to (a', b') is:

$$R_S(a', b') = 2^{-n} \sum_a \delta(b' \oplus S(a \oplus a') \oplus S(a)) \tag{11}$$

where δ is the Dirac function.

Similarly, we can evaluate the security of S against linear cryptanalysis by computing its input-output correlation. Let (a', b') be an input-output pair, then the correlation $c_S(a', b')$ of S associated to (a', b') is:

$$c_S(a', b') = 2 \cdot p_X [a' \cdot S(x) = b' \cdot x] - 1 \tag{12}$$

Formally, for a cipher operating on n bits blocks to be resistant against Differential Cryptanalysis, it is a necessary condition that there is no differential trail with a predicted prop ratio higher than 2^{1-n}.

Similarly, to be resistant against Linear Cryptanalysis, it is a necessary condition that there is no linear trail with a input-output correlation coefficient higher than $2^{n/2}$.

Efficient Multiparty Computation for Arithmetic Circuits against a Covert Majority

Isheeta Nargis, Payman Mohassel, and Wayne Eberly

Department of Computer Science,
University Of Calgary, Calgary, AB, Canada T2N 1N4
{inargis,eberly}@ucalgary.ca, pmohasse@cpsc.ucalgary.ca

Abstract. We design a secure multiparty protocol for arithmetic circuits against covert adversaries in the dishonest majority setting. Our protocol achieves a deterrence factor of $\left(1 - \frac{1}{t}\right)$ with $O(Mn^2t^2s)$ communication complexity and $O(Mn^3t^2)$ exponentiations where s is the security parameter, n is the number of parties and M is the number of multiplication gates. Our protocol builds on the techniques introduced in (Mohassel and Weinreb, CRYPTO'08), extending them to work in the multiparty case, working with higher deterrence factors, and providing simulation-based security proofs. Our main underlying primitive is a lossy additive homomorphic public key encryption scheme where the lossiness is critical for the simulation-based proof of security to go through. Our concrete efficiency measurements show that our protocol performs better than previous solutions for a range of deterrence factors, for functions such as AES and matrix multiplication.

Keywords: Multiparty Computation, Covert Adversary, Dishonest Majority, Arithmetic Circuit, Homomorphic Encryption, Lossy Encryption.

1 Introduction

In a secure multiparty computation (MPC) problem, a group of parties compute a possibly randomized function of their inputs in such a way that the privacy of their inputs are maintained and the computed output follows the distribution of the function definition. MPC is a very strong primitive in cryptography since almost all cryptographic problems can be solved, in principle, by a general secure MPC protocol. In many applications, arithmetic circuits offer a more efficient representation than boolean circuits. This includes applications such as secure auction [6], secure linear algebra [19], distributed key generation [7], financial data analysis [5] and privacy preserving data mining [18].

Much of the literature on secure MPC concerns security against either *passive* or *active* adversaries. To achieve a trade-off between higher security in the active model and better efficiency in the passive model, a third type of adversary called covert adversary was introduced by Aumann and Lindell [1]. In the covert adversary model, a corrupted party may cheat in an arbitrary way (similar to an active adversary model) and if a party cheats, then it is guaranteed to get caught

A. Youssef, A. Nitaj, A.E. Hassanien (Eds.): AFRICACRYPT 2013, LNCS 7918, pp. 260–278, 2013.

with a reasonable (though not all but negligible) probability. This probability is called the *deterrence factor* of the model and denoted by ϵ. In many real world settings, the loss of reputation associated with getting caught is such that the parties involved in a computation will possibly avoid cheating when a realistic probability of getting caught is ensured. Protocols with covert security tend to be noticeably more efficient than their active counterparts.

For efficiency of an MPC protocol, we focus on the communication and computational complexity. The communication complexity of a protocol is defined as the total number of bits transferred among the parties during the execution of the protocol. As for the computation complexity, we mainly focus on the number of exponentiations since they are the dominant factor in practical efficiency.

MPC against Covert Adversaries for Boolean Circuits. Aumann and Lindell [1] designed a two-party protocol secure against covert adversaries, with deterrence $\epsilon = (1 - \ell^{-1})(1 - 2^{-m+1})$ in constant rounds and $O(\ell s|C_b| + ms)$ communication complexity where s is the security parameter and $|C_b|$ is the size of the boolean circuit representing the function. Goyal et al. [14] designed a two-party protocol secure against covert adversaries, that achieves deterrence $\epsilon = \left(1 - \frac{1}{t}\right)$ in constant rounds and $O(|C_b| + sq + t)$ communication complexity where q is the input size.

Goyal et al. [14] devised a multiparty protocol secure against covert adversaries with a dishonest majority, with deterrence $\epsilon = \left(1 - \frac{1}{t}\right)$ requiring $O(\log n)$ rounds and $O(n^3 ts|C_b|)$ communication complexity where n is the number of parties.

For multiparty dishonest majority setting, Lindell et al. [17] designed a compiler that converts a protocol secure against passive adversaries into a protocol secure against covert adversaries with deterrence $\epsilon = 1 - e^{-0.25} = 0.2212$.

MPC for Arithmetic Circuits. Starting with the work of [3] and [9], a long line of research on MPC considers secure computation of arithmetic circuits with an honest majority. But our paper is focused on the dishonest majority setting which includes the important two-party setting as a special case. In this setting, there are only a handful of protocols for arithmetic circuits with active/covert security [11,12] that provide practical efficiency. The other constructions such as [17] are mostly of theoretical interest.

Damgård and Orlandi [11] designed a multiparty protocol that is secure with probability $(1 - 2^{-\kappa})$ against active adversaries with a dishonest majority. Their protocol needs $O(Mn^2 s + \kappa n^2 s)$ communication complexity where κ is the statistical security parameter, M is the number of multiplication gates in the circuit and s is the (computational) security parameter. The constant factors of [11] are quite large – it needs $(126Mn^2 s + 457n^2 \kappa s + 564Mns + 2030n\kappa s + O(Mn^2 + n^2\kappa))$ communication complexity and $(6Mn^3 + 211Mn^2 + 211Mn + 20n^3\kappa + 7n^3 + 769n^2\kappa + O(n\kappa))$ exponentiations. This motivates the design of more efficient protocols in this setting.

Damgård et al. [10] designed a protocol secure against covert adversaries with a dishonest majority, with deterrence $\epsilon = min\{1 - \frac{1}{p}, \frac{1}{2(n-1)}\}$ where p is the size of the finite field over which the computation is performed. This protocol uses lattice-based cryptosystems. The lattice-based constructions tend to have better computation complexity but significantly larger communication costs than protocols based on number-theoretic complexity assumptions. However, a more accurate comparison is not possible as the underlying primitives are quite different. Hence, for the purpose of this paper, we focus on comparison with [11] which uses similar underlying operations.

Our Contribution. We designed a secure multiparty protocol for arithmetic circuits in the presence of covert adversaries with a *dishonest majority*. Our protocol achieves a deterrence factor $\epsilon = \left(1 - \frac{1}{t}\right)$ with $O(Mn^2t^2s)$ communication complexity and $O(Mn^3t^2)$ exponentiations.

The protocol of Goyal et al. [14] works for boolean circuits, in the presence of covert adversaries with a dishonest majority. Their protocol requires $O(n^3ts|C_b|)$ communication complexity to achieve deterrence $\epsilon = \left(1 - \frac{1}{t}\right)$ where $|C_b|$ represents the size of the boolean circuit representing the function. For functions involving many addition and multiplication operations, arithmetic circuits give a more efficient representation than boolean circuits. As an example, the secure multiplication of two shared matrices of size $\ell \times \ell$ needs $|C_b| \in O(\ell^{2.38})$. The set of $\ell \times \ell$ matrices over a finite field $\mathbb{F}$ forms a matrix ring, denoted $\mathbb{M}_\ell(\mathbb{F})$, under matrix addition and matrix multiplication. Our protocol can be generalized to work over matrix rings $\mathbb{M}_\ell(\mathbb{F})$. [1] So the number of multiplication in the matrix ring $\mathbb{M}_\ell(\mathbb{F})$, $M = 1$. The communication complexity is multiplied by ℓ^2 (The communication of an $\ell \times \ell$ matrix requires the communication of ℓ^2 field elements.). For this problem, the protocol of [14] needs $O(n^3t\ell^{2.38}s)$ communication complexity and our protocol needs $O(n^2t^2\ell^2s)$ communication complexity.

The security of our protocol is comparable to the security of protocol π_{AMPC} of Damgård and Orlandi [11] if we set $\kappa = \log_2 t$. For deterrence factor $\frac{1}{2}$, the asymptotic communication complexity of our protocol and π_{AMPC} is similar, but the constant factor for our leading coefficient is four times smaller than that of π_{AMPC}. Asymptotically, the number of exponentiations needed by the two protocols are similar (i.e. both $O(Mn^3)$). For evaluating an AES cipher with deterrence $\frac{1}{2}$, our protocol is up to **11** times more communication efficient than π_{AMPC} and needs up to **5** times less exponentiations than π_{AMPC}. For performing matrix multiplication of size 128×128 with deterrence $\frac{1}{2}$, our protocol is up to **33** times more communication efficient than π_{AMPC} and needs up to **17** times fewer exponentiations than π_{AMPC}.

Techniques. Our protocol builds on the techniques introduced by Mohassel and Weinreb [19], extending them to work in the multiparty case, working with higher deterrence factors, and providing simulation-based security proofs. Our protocol is based on the application of cut-and-choose techniques to additive sharing of the inputs and intermediate values, and a *lossy* additive homomorphic public key encryption scheme. In all stages of the computation it is maintained that each

---

[1] In this generalization, the functionality is defined over inputs and outputs from $\mathbb{M}_\ell(\mathbb{F})$.

party holds an additive share, an associated randomness and an encrypted share of other parties for each wire that has been evaluated so far. Parties evaluate the multiplication gates by splitting their shares into subshares, broadcasting encryptions of subshares and performing homomorphic multiplication by their own subshares to these encryptions while a randomly selected portion of the computations are opened for verification. After evaluating all the gates in the circuit, each party sends its share and randomness calculated for the output wire of each other party. The receiving party holds the encryption of the shares of the remaining parties for this wire and uses this encryption to check the consistency of the received share and randomness. In this way, the encryption acts as a commitment to ensure that a party trying to send an invalid share gets caught. *Due to the binding property of a traditional encryption scheme, a simulation-based proof of the above idea is not possible.* At the end, the simulator has to generate shares and randomness on behalf of the honest parties, in a way that is consistent with the actual outputs of the corrupted parties (based on the actual inputs of the honest parties) and the messages transmitted so far (based on dummy inputs of the honest parties). This task is not possible given a traditional encryption scheme. But we show that this goal is achievable if a lossy encryption scheme is used. In a lossy encryption scheme, a ciphertext generated using a lossy key can be opened as an encryption of any message of choice. But this raises another security issue. A corrupted party can try to cheat by using a lossy key in the protocol. To prevent such an attack, a cut-and-choose verification of the key generation is also incorporated in the protocol.

2 Background

2.1 Covert Adversary Model

Aumann and Lindell [1] introduced the covert adversary model. Let ϵ denote the deterrence factor. The security is defined in an ideal/real world simulation-based paradigm [8]. In the ideal world, there are two additional types of inputs that a party can send to the trusted party. If a party P_i sends an input $corrupted_i$, then the trusted party sends $corrupted_i$ to all honest parties and halts. If a party P_i sends an input $cheat_i$, then the trusted party performs the following step according to the outcome of a random coin toss. With probability ϵ, the trusted party sends $corrupted_i$ to all honest parties and halts. With probability $(1 - \epsilon)$, the trusted party sends $undetected$ and the inputs of the honest parties to the adversary. Then the adversary sends a set of outputs of its choice for the honest parties to the trusted party. The trusted party sends these outputs to the honest parties as their outputs. For full security definition of this model, see [1].

2.2 Homomorphic Encryption Scheme

An encryption scheme is called *additive homomorphic* if it is possible to compute an encryption of $(m_1 + m_2)$ from the encryptions of m_1 and m_2. This operation is called homomorphic addition, denoted by $+_h : E_k(m_1+m_2) = E_k(m_1)+_h E_k(m_2)$. We can similarly define homomorphic subtraction, we denote this by $-_h$. For an additive homomorphic scheme, for an unknown plaintext m and a constant c in the plaintext space, it is possible to compute an encryption of $c \cdot m$ from the encryption of m. This operation is called homomorphic multiplication by constant, denoted by $\times_h : E_k(c \cdot m) = c \times_h E_k(m)$. The Paillier encryption scheme [20] is an additive homomorphic encryption scheme.

2.3   Lossy Encryption

Bellare et al. defined *Lossy Encryption* in [2], extending the definition of *Dual-Mode Encryption* of [21] and *Meaningful/Meaningless Encryption* of [16]. In a lossy encryption scheme, there are two modes of operations. In the injective mode, encryption is an injective function of the plaintext. In the lossy mode, the ciphertexts generated are independent of the plaintext.

For a probabilistic polynomial time Turing machine A, let $a \xleftarrow{\$} A(x)$ denote that a is obtained by running A on input x where a is distributed according to the internal randomness of A. Let $coins(A)$ denote the distribution of the internal randomness of A. For a set R, let $r \xleftarrow{\$} R$ denote that r is obtained by sampling uniformly from R. Let $E_{pk}(m, r)$ denote the result of encryption of plaintext m using encryption key pk and randomness r. Let $D_{sk}(c)$ denote the result of decryption of ciphertext c using decryption key sk.

Definition 1. *(**Lossy Public Key Encryption Scheme [2]**) A lossy public-key encryption scheme is a tuple (G, E, D) of probabilistic polynomial time algorithms such that*
  − *keys generated by $G(1^s, inj)$ are called injective keys.*
  − *keys generated by $G(1^s, lossy)$ are called lossy keys.*
The algorithms must satisfy the following properties.
  1. ***Correctness on injective keys.*** *For all plaintexts $x \in X$,*
$$Pr[(pk, sk) \xleftarrow{\$} G(1^s, inj); r \xleftarrow{\$} coins(E) : D_{sk}(E_{pk}(x, r)) = x] = 1.$$
  2. ***Indistinguishability of keys.*** *The public keys in lossy mode are computationally indistinguishable from the public keys in the injective mode.*
  3. ***Lossiness of lossy keys.*** *If $(pk_{lossy}, sk_{lossy}) \xleftarrow{\$} G(1^s, lossy)$, then for all $x_0, x_1 \in X$, the distributions $E_{pk_{lossy}}(x_0, R)$ and $E_{pk_{lossy}}(x_1, R)$ are statistically indistinguishable.*
  4. ***Openability.*** *If $(pk_{lossy}, sk_{lossy}) \xleftarrow{\$} G(1^s, lossy)$ and $r \xleftarrow{\$} coins(E)$, then for all $x_0, x_1 \in X$ with overwhelming probability, theres exists $r' \in coins(E)$ such that*
$$E_{pk_{lossy}}(x_0, r) = E_{pk_{lossy}}(x_1, r').$$
That is, there exists a (possibly inefficient) algorithm opener that can open a lossy ciphertext to any arbitrary plaintext with all but negligible probability.

The semantic security of a lossy encryption scheme is implied by definition, as follows. For any $x_0, x_1 \in X$, $E_{proj(G(1^s, inj))}(x_0, R) \overset{c}{\equiv} E_{proj(G(1^s, lossy))}(x_0, R) \overset{s}{\equiv} E_{proj(G(1^s, lossy))}(x_1, R) \overset{c}{\equiv} E_{proj(G(1^s, inj))}(x_1, R)$.

Definition 2. *(Key Pair Detection) A lossy encryption scheme (G, E, D) is said to satisfy key pair detection, if it holds that it can be decided in polynomial time whether a given pair (PK_i, SK_i) of keys generated by invoking G is a lossy pair or an injective pair.*

In our protocol, we use a public key encryption scheme that satisfies the following properties.

1. additive homomorphic,
2. lossy encryption with an efficient (polynomial time) *Opener* algorithm, and
3. key pair detection.

Hemenway et al. [15] designed a lossy encryption scheme based on Paillier's encryption scheme. This scheme satisfies all these required properties.

3   Problem Description and Useful Subprotocols

All the computation will be performed over a finite field $\mathbb{F}$ of size p. Here p is either a prime or a prime power. For simplicity we assume that the security parameter $s = \log p$. Let n denote the number of parties. Let $P_1, \ldots, P_n$ denote the parties. Let $f : \mathbb{F}^n \to \mathbb{F}^n$ denote an n-party functionality where $f(x_1, \ldots, x_n) = \{f_1(x_1, \ldots, x_n), \ldots, f_n(x_1, \ldots, x_n)\}$. P_i has input $x_i \in \mathbb{F}$ and output $y_i = f_i(x_1, \ldots, x_n) \in \mathbb{F}$.

Let C denote an arithmetic circuit representing f.[2] C consists of three types of gates – addition gates, multiplication-by-constant gates and multiplication gates. No output wire of C is connected as an input to any gate of C. Then it is possible to define a topological ordering of the gates of C. Let θ denote the number of gates in C. Let w_{z_δ} denote the output wire of gate g_δ of C. If g_δ is an addition or multiplication gate, then g_δ has two input wires – let w_{u_δ} and w_{v_δ} denote the input wires of g_δ. Otherwise, g_δ has one input wire – let w_{u_δ} denote the input wire of g_δ. Let ρ be the number of wires in C. The wires of C are numbered $(w_1, \ldots, w_\rho)$ in such a way that the input wires of each gate in C has smaller indices than the index of its output wire. The input wires of C are numbered $w_1, \ldots, w_n$ where w_i denotes the wire holding the input x_i of P_i. Let $\gamma = (\rho - n)$. The output wires of C are numbered $w_{\gamma+1}, \ldots, w_{\gamma+n}$ where the wire $w_{\gamma+i}$ holds the output y_i of P_i.

The parties communicate through an authenticated broadcast channel. The communication model is synchronous. The adversary is a probabilistic polynomial time Turing machine and can corrupt at most $(n-1)$ parties. The adversary is assumed to be static, meaning that the adversary fixes the set of parties to corrupt before the start of the protocol. Let $[n]$ denote the set $\{1, \ldots, n\}$.

We assume the existence of the following multiparty protocols, which are secure against active adversaries.

---

[2] For simplicity of presentation, we describe the protocol for arithmetic circuits such that each gate has one output wire (this type of arithmetic circuits are called arithmetic formula). The protocol can be generalized for any arithmetic circuit simply by repeating the step of the protocol that is described for the single output wire of a gate, for each of the output wires of the gates with more than one output wires.

1. Simulatable Coin Flipping from Scratch, $CoinFlipPublic$: This protocol generates a shared random string σ. A simulator controlling a single party can control the result. This string σ is used as the common reference string in the commitment and opening protocols described below.

   The commitment and opening protocols described below are designed in the CRS model. [3].

2. Committed Coin Flipping, $CommittedCoinFlipPublic_\sigma$: This protocol generates a commitment to a random string in such a way that each party is committed to an additive share of the string.

3. Open Commitment, $OpenCom_\sigma$: In this protocol, the parties open their commitments in the broadcast channel. A simulator controlling a subset of the parties can generate a valid simulation where all the honest parties lie about their committed values but all the corrupted parties are bound to tell the truth or otherwise they will get caught.

There exists secure multiparty protocols against active adversaries for these tasks, based on the Decisional Diffie-Hellman assumption. More details about these protocols can be found in [14,13].

4   Protocol for Secure Computation

Let w_k be a wire in C. Let $S_{k,i}$ denote the share of P_i for the wire w_k. Let $rs_{k,i}$ denote the randomness of P_i for the wire w_k. For each $j \in [n] \setminus \{i\}$, let $ES_{k,i,j}$ denote the encrypted share of P_j that P_i holds for the wire w_k.

We designed a protocol $Circuit$ for computing functionality f in the presence of covert adversaries. The main stages of protocol $Circuit$ are presented in Fig. 1. Each stage is presented in a separate figure later. Unless otherwise specified, we describe the action of each party $P_i, i \in [n]$, in the protocol.

In the *CRS generation stage* (see Fig. 2), parties generate a common reference string σ. σ is used as the common reference string in commitment and opening subprotocols used during the rest of the protocol.

In the *key generation stage* (see Fig. 3), each party generates two pair of keys and broadcasts the public keys of each pair. One pair is randomly selected to verify that it is an injective key of the lossy encryption scheme being used.[4] The unopened keys are set as the keys of the corresponding parties to be used during the rest of the protocol.

In the *input sharing stage* (see Fig. 4), each party distributes two sets of additive shares of its input and their encryptions to other parties. One set is randomly selected for verification and the unopened shares are used in the computation. [5] Then we say that the input wires $w_1, \ldots, w_n$ of C have been *evaluated*. By saying

---

[3] In the common reference string (CRS) model, it is assumed that all parties have access to a common string that is selected from some specified distribution [4].

[4] For verification, the generating party has to broadcast the private key of the selected pair.

[5] This verification is only making sure that the encryptions of the shares are done correctly.

Protocol Circuit.
  Common Inputs:
  1. An arithmetic circuit C describing $f : \mathbb{F}^n \to \mathbb{F}^n = \{f_1, \ldots, f_n\}$,
  2. A topological ordering $(g_1, \ldots, g_\theta)$ of gates in C, and
  3. An ordering $(w_1, \ldots, w_\rho)$ of wires in C such that the input wires of each gate in C has smaller indices than the index of its output wire.
  Input of $P_i : x_i \in \mathbb{F}$.
  Output of $P_i : y_i = f_i(x_1, \ldots, x_n) \in \mathbb{F}$.
  In some steps of the protocol, each party is supposed to broadcast some message. If some party P_j does not broadcast any message in one of these steps, then P_i aborts.
  1. **CRS Generation Stage.** Parties generate a common reference string.
  2. **Key Generation Stage.** Parties generate keys for the encryption scheme.
  3. **Input Sharing Stage.** Each party distributes additive shares of its input and the encryptions of these shares to other parties.
  4. **Computation Stage.** Parties evaluate the circuit gate-by-gate.
  5. **Output Generation Stage.** Parties evaluate their outputs.

Fig. 1. Protocol $Circuit$

Parties generate a common reference string σ of length $p_1(s)$ using the protocol $CoinFlipPublic$. Here $p_1(s)$ is a polynomial of s.

Fig. 2. The CRS generation stage

---

that a wire w_k has been evaluated we mean that the share $S_{k,i}$, randomness $rs_{k,i}$ of each party P_i for w_k and the encrypted share $ES_{k,i,j}$ that each party P_i holds for each other party P_j for w_k, have been fixed.[6]

In the *computation stage* (see Fig. 5), parties evaluate the circuit gate-by-gate, in the order $(g_1, \ldots, g_\theta)$. When the evaluation of gate g_δ is finished, we say that the output wire w_{z_δ} of g_δ has been evaluated. At each point of computation the following holds for each wire w_k of C that has been evaluated so far: each party P_i holds an additive share $S_{k,i}$, an associated randomness $rs_{k,i}$ and an encrypted share $ES_{k,i,j}$ of each other party P_j for the wire w_k.

If g_δ is an addition gate, each party P_i sets its share $S_{z_\delta,i}$ for the output wire of g_δ to the sum of the shares of P_i for the input wires of g_δ. Each party P_i computes its randomness $rs_{z_\delta,i}$ for the output wire of g_δ such that $E_{PK_i}(S_{z_\delta,i}, rs_{z_\delta,i})$ equals the result of homomorphic addition of the ciphertexts ($ES_{u_\delta,j,i}$ and $ES_{v_\delta,j,i}$, $j \neq i$) that other parties hold for the input wires of g_δ for P_i. Each party P_i computes the encrypted share $ES_{z_\delta,i,j}$ of each other party P_j for the output wire of g_δ locally, by performing homomorphic addition of the encrypted shares of P_j that P_i holds for the input wires of g_δ.

If g_δ is a multiplication-by-constant gate, each party P_i sets its share $S_{z_\delta,i}$ for the output wire of g_δ to the product of the share of P_i for the input wire of g_δ and q_δ where q_δ is the known constant multiplicand for gate g_δ. This constant q_δ is

---

[6] $ES_{k,i,j}$ is supposed to be $E_{PK_j}(S_{k,j}, rs_{k,j})$.

1. **Parties generate challenge.** Parties generate a challenge $m_{key} \in \{1,2\}$ using the protocol $CommittedCoinFlipPublic_\sigma$ in the CRS model.
2. **Parties broadcast injective public keys.** P_i generates two pairs of keys in injective mode, that is, P_i generates $(U_{i,j}, V_{i,j}) = G(1^s, inj)$, for each $j \in \{1,2\}$. P_i broadcasts the public keys $U_{i,1}$ and $U_{i,2}$.
3. **Parties open the challenge.** Parties open the challenge m_{key}, using the protocol $OpenCom_\sigma$ in the CRS model.
4. **Parties respond to challenge.** P_i broadcasts the private key $V_{i,m_{key}}$.
5. **Parties verify the responses.** For each $j \in [n] \setminus \{i\}$, P_i verifies that the key pair $(U_{j,m_{key}}, V_{j,m_{key}})$ is a valid injective pair of keys for the lossy encryption scheme being used. If this is not the case, then P_i broadcasts $corrupted_j$ and aborts.
6. **Parties fix their keys.** P_i performs the following steps.
   (a) P_i sets (PK_i, SK_i) to $(U_{i,3-m_{key}}, V_{i,3-m_{key}})$,
   (b) For each $j \in [n] \setminus \{i\}$, P_i sets PK_j to $U_{j,3-m_{key}}$.

Fig. 3. The key generation stage

1. **Parties generate challenge.** Parties generate a challenge $m_{in} \in \{1,2\}$ using the protocol $CommittedCoinFlipPublic_\sigma$ in the CRS model.
2. **Parties broadcast encrypted shares.** P_i randomly selects two sets of shares $\{B_{1,i,j}\}_{j \in [n]}$ and $\{B_{2,i,j}\}_{j \in [n]}$ such that $\sum_{j \in [n]} B_{1,i,j} = \sum_{j \in [n]} B_{2,i,j} = x_i$. P_i randomly selects two sets of strings $\{b_{1,i,j}\}_{j \in [n]}$ and $\{b_{2,i,j}\}_{j \in [n]}$. For each $\ell \in \{1,2\}$ and each $j \in [n]$, P_i broadcasts $Y_{\ell,i,j} = E_{PK_j}(B_{\ell,i,j}, b_{\ell,i,j})$.
3. **Parties send share and randomness to the designated parties.** P_i sends $\{B_{1,i,j}, B_{2,i,j}, b_{1,i,j}, b_{2,i,j}\}$ to P_j, for each $j \in [n] \setminus \{i\}$.
4. **Parties open the challenge.** Parties open the challenge m_{in} using the protocol $OpenCom_\sigma$ in the CRS model.
5. **Parties respond to challenge.** P_i broadcasts the sets $\{B_{m_{in},i,j}\}_{j \in [n] \setminus \{i\}}$ and $\{b_{m_{in},i,j}\}_{j \in [n] \setminus \{i\}}$.
6. **Parties verify the responses.** For each $j \in [n] \setminus \{i\}$ and each $k \in [n] \setminus \{j\}$, P_i verifies that $Y_{m_{in},j,k} = E_{PK_k}(B_{m_{in},j,k}, b_{m_{in},j,k})$.
   If any of the equalities does not hold, then P_i broadcasts $corrupted_j$ and aborts.
7. **Parties fix their shares, randomness and encrypted shares of other parties.** For each $k \in [n]$, P_i sets the followings for the input wire w_k of C.
   (a) P_i sets $S_{k,i}$ to $B_{3-m_{in},k,i}$,
   (b) P_i sets $rs_{k,i}$ to $b_{3-m_{in},k,i}$, and
   (c) For each $j \in [n] \setminus \{i\}$, P_i sets $ES_{k,i,j}$ to $Y_{3-m_{in},k,j}$.

Fig. 4. The input sharing stage

part of the description of C and known to all parties. Each party P_i computes its randomness $rs_{z_{\delta,i}}$ for the output wire of g_δ such that $E_{PK_i}(S_{z_{\delta,i}}, rs_{z_{\delta,i}})$ equals the result of homomorphic addition of the ciphertexts ($ES_{u_\delta,j,i}$ and $ES_{v_\delta,j,i}$, $j \neq i$) that other parties hold for the input wires of g_δ for P_i. Each party P_i computes the encrypted share $ES_{z_{\delta,i,j}}$ of each other party P_j for the output wire of g_δ locally, by performing homomorphic multiplication by q_δ to the encrypted share of P_j that P_i holds for the input wire of g_δ.

If g_δ is a multiplication gate, then each party P_i first generates two sets of random shares and broadcasts their encryptions. One set is randomly selected for verification and the unopened set will be used as the set $\{C_{i,j}\}_{j \in [n]\setminus\{i\}}$ during the evaluation of g_δ. Each party P_i splits its shares $A_i = S_{u_\delta,i}$ and $B_i = S_{v_\delta,i}$ (for the input wires of g_δ) into two additive subshares ($A_{i,1} + A_{i,2} = A_i$ and $B_{i,1} + B_{i,2} = B_i$), then broadcasts the encryptions of these subshares. Each party P_i splits its random share $C_{i,j}$ into four additive subshares ($H_{i,j,1,1}, H_{i,j,1,2}, H_{i,j,2,1}$, and $H_{i,j,2,2}$) and broadcasts their encryptions, for each $j \in [n] \setminus \{i\}$. Each party P_i performs homomorphic multiplication by its own subshare $A_{i,k}$ to the encryption $Y_{i,\ell}$ of its own subshare $B_{i,\ell}$, then adds a random encryption of zero and broadcasts the resulting ciphertext $L_{i,k,\ell}$, for each $k, \ell \in \{1,2\}^2$. Each party P_i performs homomorphic multiplication by its own subshare $A_{i,k}$ to the encryption $Y_{j,\ell}$ of the subshare $B_{j,\ell}$ of each other party P_j, then adds a random ciphertext of $H_{i,j,k,\ell}$ and broadcasts the resulting ciphertext $K_{i,j,k,\ell}$, for each $k, \ell \in \{1,2\}^2$. After receiving the results of these calculations from other parties, each party P_i decrypts the ciphertexts $\{K_{j,i,k,\ell}\}_{j \in [n]\setminus\{i\}, k,\ell \in \{1,2\}^2}$ under its own key PK_i, sums the results of decryptions up, then subtracts its own randomness ($\{C_{i,j}\}_{j \in [n]\setminus\{i\}}$) to get its share $S_{z_\delta,i}$ of the product. Each party P_i sets its randomness $rs_{z_\delta,i}$ for the output wire of g_δ to a string such that encrypting $S_{z_\delta,i}$ under PK_i using this string as randomness would result in the ciphertext ($ES_{z_\delta,j,i}, j \neq i$) that the other parties would hold as the encryption of the share of P_i for the output wire of g_δ. Each party P_i computes the encryption $ES_{z_\delta,i,j}$ of the share of each other party P_j for the output wire of g_δ, by performing the corresponding homomorphic additions to the corresponding ciphertexts as all the ciphertexts (including the results after calculations) are available to all parties. Exactly half of all the calculations (splitting into subshares, encryption of subshares, homomorphic multiplication by own subshares to own encrypted subshares, and homomorphic multiplication by own subshares to other parties' encrypted subshares) are randomly selected for verification, ensuring that a party attempting to cheat gets caught with probability at least $\frac{1}{2}$. It is also verified that the homomorphic encryptions of the additive subshares (e.g. the encryptions $X_{i,1}$ and $X_{i,2}$ of subshares $A_{i,1}$ and $A_{i,2}$) and an encryption of zero (e.g. $E_{PK_i}(0, a_{i,0})$) results in the encryption of the original share (e.g. $EA_{j,i} = ES_{u_\delta,j,i}, j \neq i$, the encryption of A_i that P_j holds, that is, the encryption of the share of P_i that P_j holds for the input wire u_δ of g_δ) – the splitting party P_i has to broadcast the string to be used as randomness to encrypt zero (e.g. $a_{i,0}$) to prove this equality. A party attempting to cheat gets caught with probability 1 in this case.

In the *output generation stage* (see Fig. 6), each party P_i sends its share $S_{\gamma+k,i}$ and randomness $rs_{\gamma+k,i}$ for the output wire $w_{\gamma+k}$ to each other party P_k. The receiving party P_k holds the encryption $ES_{\gamma+k,k,i}$ of each other party P_i for its output wire $w_{\gamma+k,i}$. The receiving party P_k checks the consistency of the received input and randomness with the corresponding ciphertexts (P_k checks whether $ES_{\gamma+k,k,i} = E_{PK_i}(S_{\gamma+k,i}, rs_{\gamma+k,i})$ or not).

5   Security of Protocol *Circuit*

We have the following Theorem on the security of protocol *Circuit*.

Theorem 1. *Assuming the existence of lossy additive homomorphic public key encryption schemes with efficient Opener algorithm and secure coin tossing protocols in the presence of active adversaries, protocol Circuit securely computes functionality f with deterrence $\frac{1}{2}$ in the presence of static covert adversaries that can corrupt up to $(n-1)$ parties.*

Let M denote the number of multiplication gates in the arithmetic circuit representing f. Protocol Circuit runs in $O(M \log n)$ rounds and needs $O(Mn^2 s)$ communication among the parties where s is the security parameter.

The parties generate two challenges for evaluating each multiplication gate. The challenge generation step requires $O(\log n)$ rounds, so the overall round complexity is $O(M \log n)$. Parties do not need any interaction for evaluating addition gates or multiplication-by-constant gates. So the number of these two types of gates in the circuit does not affect the communication complexity or round complexity. The evaluation of one multiplication gate requires $O(n^2 s)$ communication complexity, so the total communication complexity is $O(Mn^2 s)$.

Now we describe the main intuition behind the security of protocol *Circuit*. A full proof is deferred to the full version due to lack of space. The security is proved by constructing a simulator S that acts as the ideal-world adversary. Let $\mathcal{A}$ denote the adversary in the real-world.

In each stage of protocol *Circuit*, S checks the responses of $\mathcal{A}$ for both values of the challenge, by rewinding the challenge generation step and using the simulator for the coin flip protocol to set the challenge to the other possible value. If $\mathcal{A}$ cheats on behalf of a corrupted party P_i for both values of the challenge, then S simulates catching P_i. If $\mathcal{A}$ cheats on behalf of a corrupted party P_i for exactly one value of the challenge and that value is selected for verification, then S simulates catching P_i. If $\mathcal{A}$ cheats on behalf of a corrupted party P_i for exactly one value of the challenge and that value is not selected for verification, then the simulation continues to the next stage. If $\mathcal{A}$ does not cheat on behalf of any corrupted party for both values of the challenge, then S proceeds to the next stage.

In the *key generation stage*, at step 2, S randomly selects $d_{key} \in \{1, 2\}$. For each honest party P_i, S generates one injective pair $(u_{i,d_{key}}, v_{i,d_{key}})$ and one lossy pair $(u_{i,3-d_{key}}, v_{i,3-d_{key}})$. S rewinds to step 2 until m_{key} equals d_{key}. [7] By the "key indistinguishability" property of the lossy encryption scheme, public key of a lossy pair and public key of an injective pair are computationally indistinguishable. No honest party gets caught as the opened keys of the honest

---

[7] The expected number of rewinds until $d_{key} = m_{key}$ is 2. That means the execution of steps 2–4 of this stage of S needs expected constant time. To bound the running time of these steps within a polynomial of the security parameter s, we can continue rewinding S at most s^ℓ times where $\ell \in \mathbb{N}$. If $d_{key} \neq m_{key}$ after s^ℓ rewinds, then S fails. The probability of failure of S is negligible.

For each $\delta \in [\theta]$, parties perform the following actions, depending on the type of gate g_δ.

Case 1: g_δ is an addition gate.

1. P_i sets $S_{z_\delta,i} = S_{u_\delta,i} + S_{v_\delta,i}$.
2. P_i computes $rs_{z_\delta,i}$ such that the following equality holds. $E_{PK_i}(S_{z_\delta,i}, rs_{z_\delta,i}) = E_{PK_i}(S_{u_\delta,i}, ra_{u_\delta,i}) +_h E_{PK_i}(S_{v_\delta,i}, ra_{v_\delta,i})$.
3. For each $j \in [n] \setminus \{i\}$, P_i sets $ES_{z_\delta,i,j} = ES_{u_\delta,i,j} +_h ES_{v_\delta,i,j}$.

Case 2: g_δ is a multiplication-by-constant gate.

Let $q_\delta \in \mathbb{F}$ be the constant with which the multiplication will be done.

1. P_i sets $S_{z_\delta,i} = q_\delta \cdot S_{u_\delta,i}$.
2. P_i computes $rs_{z_\delta,i}$ such that the following equality holds. $E_{PK_i}(S_{z_\delta,i}, rs_{z_\delta,i}) = q_\delta \times_h E_{PK_i}(S_{u_\delta,i}, rs_{u_\delta,i})$.
3. For each $j \in [n] \setminus \{i\}$, P_i sets $ES_{z_\delta,i,j} = q_\delta \times_h ES_{u_\delta,i,j}$.

Case 3: g_δ is a multiplication gate.

For each $i \in [n]$, let A_i, B_i, ra_i and rb_i denote $S_{u_\delta,i}, S_{v_\delta,i}, rs_{u_\delta,i}$, and $rs_{v_\delta,i}$, respectively. For each $i \in [n]$ and each $j \in [n] \setminus \{i\}$, let $EA_{i,j}$ and $EB_{i,j}$ denote $ES_{u_\delta,i,j}$ and $ES_{v_\delta,i,j}$, respectively.

1. **Parties generate random shares.**
   (a) **Parties generate challenge.** Parties generate challenge $m_r \in \{1,2\}$ using the protocol $CommittedCoinFlipPublic_\sigma$ in the CRS model.
   (b) **Parties generate random shares.** P_i randomly selects two sets of shares $\{Q_{1,i,j}\}_{j \in [n] \setminus \{i\}}$ and $\{Q_{2,i,j}\}_{j \in [n] \setminus \{i\}}$ and two sets of strings $\{q_{1,i,j}\}_{j \in [n] \setminus \{i\}}$ and $\{q_{2,i,j}\}_{j \in [n] \setminus \{i\}}$.
   (c) **Parties broadcast encrypted shares.** For each $\ell \in \{1,2\}$ and each $j \in [n] \setminus \{i\}$, P_i broadcasts $Y_{\ell,i,j} = E_{PK_i}(Q_{\ell,i,j}, q_{\ell,i,j})$.
   (d) **Parties open the challenge.** Parties open the challenge m_r using the protocol $OpenCom_\sigma$ in the CRS model.
   (e) **Parties respond to challenge.** P_i broadcasts $Q_{m_r,i,j}$ and $q_{m_r,i,j}$ for each $j \in [n] \setminus \{i\}$.
   (f) **Parties verify the responses.** For each $j \in [n] \setminus \{i\}$ and each $k \in [n] \setminus \{j\}$, P_i verifies that $Y_{m_r,j,k} = E_{PK_j}(Q_{m_r,j,k}, q_{m,j,k})$. If any of these equalities does not hold for party P_j, then P_i broadcasts $corrupted_j$ and aborts.
   (g) **Parties fix their randomness.** P_i performs the following steps.
      i. For each $j \in [n] \setminus \{i\}$, P_i sets $C_{i,j}$ to $Q_{3-m_r,i,j}$ and $rc_{i,j}$ to $q_{3-m_r,i,j}$.
      ii. For each $j \in [n] \setminus \{i\}$ and each $k \in [n] \setminus \{j\}$, P_i sets $EC_{j,k}$ to $Y_{3-m_r,j,k}$.
2. **Parties generate challenge.** Parties generate a challenge $m \in \{1,2\}$ using the protocol $CommittedCoinFlipPublic_\sigma$ in the CRS model.
3. **Parties split their shares into subshares.**
   (a) P_i chooses $A_{i,1}$ and $B_{i,1}$ uniformly at random from $\mathbb{F}$.
   (b) P_i sets $A_{i,2} = A_i - A_{i,1}$, and $B_{i,2} = B_i - B_{i,1}$.
   (c) P_i generates two random strings $r_{i,1}$ and $r_{i,2}$.
   (d) For each $j \in \{1,2\}$, P_i broadcasts $X_{i,j} = E_{PK_i}(A_{i,j}, r_{i,j})$, and $Y_{i,j} = E_{PK_i}(B_{i,j}, r_{i,j})$.
   (e) For each $j \in [n] \setminus \{i\}$, and each $k, \ell \in \{1,2\}^2$, P_i chooses $H_{i,j,k,\ell}$ uniformly at random from $\mathbb{F}$ such that $\sum_{k,\ell \in \{1,2\}^2} H_{i,j,k,\ell} = C_{i,j}$.
   (f) For each $j \in [n] \setminus \{i\}$, and each $k, \ell \in \{1,2\}^2$, P_i chooses a random string $h_{i,j,k,\ell}$ and broadcasts $G_{i,j,k,\ell} = E_{PK_i}(H_{i,j,k,\ell}, h_{i,j,k,\ell})$.
4. **Parties prove their sums.**
   (a) P_i computes two strings $a_{i,0}$ and $b_{i,0}$ such that
   $E_{PK_i}(0, a_{i,0}) = E_{PK_i}(A_i, ra_i) -_h X_{i,1} -_h X_{i,2}$, and
   $E_{PK_i}(0, b_{i,0}) = E_{PK_i}(B_i, rb_i) -_h Y_{i,1} -_h Y_{i,2}$.
   (b) P_i broadcasts $a_{i,0}$ and $b_{i,0}$.

Fig. 5. The computation stage

Case 3: g_δ is a multiplication gate.

4. **Parties prove their sums.**
   (c) For each $j \in [n] \setminus \{i\}$, P_i performs the following two actions.
      i. P_i computes a string $c_{i,j,0}$ such that $E_{PK_i}(0, c_{i,j,0}) = E_{PK_i}(C_{i,j}, rc_{i,j}) -_h$
      $G_{i,j,1,1} -_h G_{i,j,1,2} -_h G_{i,j,2,1} -_h G_{i,j,2,2}$.
      ii. P_i broadcasts $c_{i,j,0}$.
5. **Parties send output parts that depend only on their own subshares.** For
   each $k, \ell \in \{1,2\}^2$, P_i selects a random string $vv_{i,k,\ell}$, then broadcasts $L_{i,k,\ell} = A_{i,k} \times_h Y_{i,\ell} +_h E_{PK_i}(0, vv_{i,k,\ell})$.
6. **Parties perform computations on other parties' encrypted subshares.** For
   each $k, \ell \in \{1,2\}^2$ and each $j \in [n] \setminus \{i\}$, P_i selects a random string $hh_{i,j,k,\ell}$,
   performs the following computation on the encrypted inputs of P_j, then broadcasts
   $K_{i,j,k,\ell} = A_{i,k} \times_h Y_{j,\ell} +_h E_{PK_j}(H_{i,j,k,\ell}, hh_{i,j,k,\ell})$.
7. **Parties open the challenge.** Parties open the challenge m using the protocol
   $OpenCom_\sigma$ in the CRS model.
8. **Parties respond to challenge.**
   (a) P_i broadcasts $A_{i,m}, B_{i,m}, r_{i,m}, vv_{i,m,1}$ and $vv_{i,m,2}$.
   (b) For each $j \in [n] \setminus \{i\}$, and each $\ell \in \{1,2\}$, P_i broadcasts $H_{i,j,m,\ell}, h_{i,j,m,\ell}$ and
   $hh_{i,j,m,\ell}$.
9. **Parties verify the responses.** P_i verifies the following for each party $P_j, j \in [n] \setminus \{i\}$:
   (a) P_j**'s encryption sums.**
      i. $E_{PK_j}(0, a_{j,0}) = EA_{i,j} -_h X_{j,1} -_h X_{j,2}$.
      ii. $E_{PK_j}(0, b_{j,0}) = EB_{i,j} -_h Y_{j,1} -_h Y_{j,2}$.
      iii. For each $k \in [n] \setminus \{j\}$,
      $E_{PK_j}(0, c_{j,k,0}) = EC_{j,k} -_h G_{j,k,1,1} -_h G_{j,k,1,2} -_h G_{j,k,2,1} -_h G_{j,k,2,2}$.
   (b) P_j **knows its encrypted data.**
      i. $X_{j,m} = E_{PK_j}(A_{j,m}, r_{j,m})$.
      ii. $Y_{j,m} = E_{PK_j}(B_{j,m}, r_{j,m})$.
      iii. For each $k \in [n] \setminus \{j\}$ and each $\ell \in \{1,2\}$,
      $G_{j,k,m,\ell} = E_{PK_j}(H_{j,k,m,\ell}, h_{j,k,m,\ell})$.
   (c) **The computations performed by P_j on its own subshares are correct.**
   For each $\ell \in \{1,2\}$, $L_{j,m,\ell} = A_{j,m} \times_h Y_{j,\ell} +_h E_{PK_j}(0, vv_{j,m,\ell})$.
   (d) **The computations performed by P_j on other parties' subshares are
   correct.** For each $k \in [n] \setminus \{j\}$, and for each $\ell \in \{1,2\}$, $K_{j,k,m,\ell} = A_{j,m} \times_h Y_{k,\ell} +_h E_{PK_k}(H_{j,k,m,\ell}, hh_{j,k,m,\ell})$.
   If P_j fails in any of the verifications, then P_i broadcasts $corrupted_j$ and aborts.
10. **Parties compute their shares of product.**
   (a) For each $j \in [n] \setminus \{i\}$, P_i performs the following two actions.
      i. P_i computes $V_{j,i} = K_{j,i,1,1} +_h K_{j,i,1,2} +_h K_{j,i,2,1} +_h K_{j,i,2,2}$.
      ii. P_i computes $W_{j,i} = D_{SK_i}(V_{j,i}) = A_j B_i + C_{j,i}$.
   (b) P_i computes its share $S_{z_\delta,i}$ as follows.
   $S_{z_\delta,i} = A_i B_i + \sum_{j \in [n] \setminus \{i\}} W_{j,i} - \sum_{j \in [n] \setminus \{i\}} C_{i,j}$
   $= (\sum_{j \in [n]} S_{u_\delta,j}) S_{v_\delta,i} + \sum_{j \in [n] \setminus \{i\}} C_{j,i} - \sum_{j \in [n] \setminus \{i\}} C_{i,j}$.
   (c) P_i computes its randomness $rs_{z_\delta,i}$ such that the following equality holds.
   $E_{PK_i}(S_{z_\delta,i}, rs_{z_\delta,i}) = \sum_{k,\ell} L_{i,k,\ell} + \sum_{k \in [n] \setminus \{i\}} \sum_{\ell,q \in \{1,2\}^2} K_{k,i,\ell,q} -_h$
   $\sum_{k \in [n] \setminus \{i\}} E_{PK_i}(C_{i,k}, rc_{i,k})$.
11. **Parties compute encryption of the shares of other parties.** For each $j \in [n] \setminus \{i\}$, P_i computes the encrypted share $ES_{z_\delta,i,j}$ of P_j as follows.
$ES_{z_\delta,i,j} = \sum_{k,\ell} L_{j,k,\ell} +_h \sum_{k \in [n] \setminus \{j\}} \sum_{\ell,q \in \{1,2\}^2} K_{k,j,\ell,q} -_h \sum_{k \in [n] \setminus \{j\}} U_{j,k}$.

Fig. 5. (*continued*)

Note that the wire $w_{\gamma+k}$ is supposed to carry the output y_k of party P_k.

1. For each $k \in [n]$, the parties perform the following steps.
   (a) **Parties send their shares and randomness for the wire $w_{\gamma+k}$ to P_k.**
       $P_i, i \in [n] \setminus \{k\}$, sends $S_{\gamma+k,i}$ and $rs_{\gamma+k,i}$ to P_k. If some P_j does not send these to P_k in this step, then P_k aborts.
   (b) P_k **verifies the responses.** For each $i \in [n] \setminus \{k\}$, P_k compares $E_{PK_i}(S_{\gamma+k,i}, rs_{\gamma+k,i})$ with the ciphertext $ES_{\gamma+k,k,i}$ that P_k holds as the encryption of the share of P_i for the wire $w_{\gamma+k}$. If the ciphertexts do not match, then P_k broadcasts $corrupted_i$ and aborts.
   (c) P_k **computes its output.** P_k computes $L_k = \sum_{i \in [n]} S_{\gamma+k,i}$.
   (d) If P_k broadcasts $corrupted_j$ for some j during step 1(b) of this stage, then $P_i, i \in [n] \setminus \{k\}$, aborts.
2. $P_i, i \in [n]$, outputs L_i.

Fig. 6. The output generation stage

parties are injective keys. At the end of the *key generation stage* of the simulation, the key pair of each honest party P_i is set to a lossy pair $(u_{i,3-d_{key}}, v_{i,3-d_{key}})$ of keys. From the responses of $\mathcal{A}$ for both values of the challenge, $\mathcal{S}$ learns the decryption keys $(\{v_{i,3-m_{key}}\}_{i \in I})$ of the corrupted parties that will be used during the rest of the protocol. By the "key detection" property of the lossy encryption scheme, a corrupted party attempting to cheat by using lossy keys gets caught with probability at least $\frac{1}{2}$. A corrupted party P_i attempting to cheat by using lossy keys may not get caught with probability at most $\frac{1}{2}$.[8] In that case, the key pair of P_i is set to a lossy pair for the rest of the protocol.

In the *input sharing stage*, $\mathcal{S}$ uses zero as the inputs of the honest parties and sends shares of zero to the adversary. By the semantic security of the lossy encryption scheme, the ciphertexts are indistinguishable in two worlds. Since there is at least one honest party, the set of shares of the corrupted parties for the input wire w_i in two worlds are identically distributed, for each honest party P_i. At step 5, for each honest party P_i, $\mathcal{S}$ sends $(n-1)$ out of n additive shares of zero $(\{B_{m_{in},i,j}\}_{j \in [n] \setminus \{i\}})$ to $\mathcal{A}$, the set of these $(n-1)$ shares in both worlds are identically distributed. A corrupted party attempting to cheat by sending invalid encryption to other parties gets caught with probability at least $\frac{1}{2}$. A corrupted party P_i may modify its input by modifying its own shares $B_{1,i,i}$ and $B_{2,i,i}$ – such a behavior can not be prohibited in the ideal world as well. By rewinding the challenge generation step, $\mathcal{S}$ learns $Y_{1,i,i}$ and $Y_{2,i,i}$ for each corrupted party P_i. $\mathcal{S}$ computes the shares $B_{1,i,i}$ and $B_{2,i,i}$ by decrypting $Y_{1,i,i}$ and $Y_{2,i,i}$ using the decryption key of P_i, and thereby learns the replaced input x_i' of P_i for f, for each corrupted party P_i.

In the *computation stage*, for addition gates and multiplication-by-constant gates, for each honest party P_i, $\mathcal{S}$ performs the same steps as an honest P_i would. For evaluating a multiplication gate g_δ, $\mathcal{S}$ behaves honestly during the random

---

[8] This happens only if P_i selects one lossy pair and one injective pair, and the injective pair of P_i is opened for verification.

share generation step 1. At step 3, $\mathcal{S}$ randomly selects $d \in \{1,2\}$. For each honest party P_i, $\mathcal{S}$ honestly generates $A_{i,d}, r_{i,d}$, computes $X_{i,d} = E_i(A_{i,d}, ra_{i,d})$ honestly and sets $X_{i,3-d}$ to a ciphertext such that the homomorphic addition of $X_{i,1}, X_{i,2}$ and a random encryption of zero under PK_i ($E_{PK_i}(0, a_{i,0})$) equals $EA_{q,i}$ where $q \in I$.[9] $\mathcal{S}$ performs similarly for the shares of B_i for each honest party P_i. Similarly, for each honest party P_i and each $j \in [n] \setminus \{i\}$, $\mathcal{S}$ honestly generates three shares $G_{i,j,d,1}, G_{i,j,d,2}, G_{i,j,3-d,1}$, corresponding randomness and performs their encryptions honestly, and sets $H_{i,j,3-d,2}$ to a ciphertext such that the homomorphic addition of $H_{i,j,1,1}, H_{i,j,1,2}, H_{i,j,2,1}, H_{i,j,2,2}$ and a random encryption of zero under PK_i equals $EC_{q,i,j}$ where $q \in I$. $\mathcal{S}$ rewinds to step 2 until the challenge m equals d. So no honest party gets caught during the verification. At steps 5 and 6, on behalf of each honest party P_i, $\mathcal{S}$ honestly performs the homomorphic multiplications by generated subshares of P_i. By the semantic security of the lossy encryption scheme, the ciphertexts are indistinguishable in two worlds. For each honest party P_i, exactly one out of two subshares of A_i and B_i are opened for verification – the distribution of the opened subshares are identical in two worlds. The same holds for $C_{i,j}$ as exactly two out of four subshares are opened for verification. Since exactly half of all the calculations are checked for verification, a party attempting to cheat in any calculation gets caught with probability at least $\frac{1}{2}$. Let $q \in I$. For each honest party P_i, $\mathcal{S}$ computes $ES_{z_\delta,q,i}$ by performing homomorphic addition of the ciphertexts $\{L_{i,k,\ell}\}_{k,\ell \in \{1,2\}^2}$ and $\{K_{j,i,k,\ell}\}_{j \in [n] \setminus \{i\}, k,\ell \in \{1,2\}^2}$, then homomorphically subtracting the ciphertexts $\{EC_{q,i,j}\}_{j \in [n] \setminus \{i\}}$ from the result.[10] For each honest party P_i, $\mathcal{S}$ randomly selects its share $S_{z_\delta,i}$ of the output wire of g_δ and computes its randomness $rs_{z_\delta,i}$ by running the *Opener* algorithm on inputs PK_i, $S_{z_\delta,i}$ and $ES_{z_\delta,q,i}$.[11]

In the *output generation stage*, $\mathcal{S}$ sends the replaced inputs $\{x'_i\}_{i \in I}$ of the corrupted parties to the trusted party and receives the actual outputs $\{yO_i\}_{i \in I}$ of the corrupted parties. For each corrupted party P_k, $\mathcal{S}$ selects a set $\{S'_{\gamma+k,i}\}_{i \in [n] \setminus I}$ of shares of the honest parties such that the sum of these shares and the set $\{S_{\gamma+k,i}\}_{i \in I}$ of the shares of the corrupted parties for the wire $w_{\gamma+k}$ equals the actual output yO_k of P_k. Note that the key pair of each honest party P_i was set to a lossy pair in the *key generation stage* in the simulation. For each corrupted party P_k and each honest party P_i, $\mathcal{S}$ computes fake randomness $rs'_{\gamma+k,i}$ by running the *Opener* algorithm on inputs PK_i, $S'_{\gamma+k,i}$ and $ES_{\gamma+k,k,i}$ so that it satisfies $E_{PK_i}(S'_{\gamma+k,i}, rs'_{\gamma+k,i}) = ES_{\gamma+k,k,i}$ ($ES_{\gamma+k,k,i}$ is the ciphertext held by $\mathcal{A}$ for P_i for the wire $w_{\gamma+k}$). For each honest party P_k, $\mathcal{S}$ simulates the consistency checking as an honest P_k would. If a corrupted party P_i sends inconsistent share and randomness to an honest party P_k, then one of the following two situations may happen. If the key pair of P_i is an injective pair of keys, then P_i gets caught in both worlds. If the key pair of P_i is a lossy pair of keys, then P_i does not get

---

[9] $EA_{q,i} = ES_{u_\delta,q,i}$ is the encryption of the share $A_i = S_{u_\delta,i}$ of P_i that $\mathcal{A}$ holds for the input wire w_{u_δ} of g_δ.

[10] $ES_{z_\delta,q,i}$ is the encrypted share of P_i that $\mathcal{A}$ will hold for the output wire of g_δ.

[11] Since the key of honest P_i is a lossy key, this will satisfy that $E_{PK_i}(S_{z_\delta,i}, rs_{z_\delta,i}) = ES_{z_\delta,q,i}$.

caught in both worlds and the output of honest P_k is set to an output desired by the adversary. [12] By definition of security in the covert adversary model, such an event can happen with probability at most $(1 - \epsilon)$ in the ideal world as well.[13] In our protocol too, such an event can happen with probability at most $(1 - \epsilon) = 1 - \frac{1}{2} = \frac{1}{2}$.

5.1 Achieving a Higher Deterrence Factor

Here we outline what modifications are performed to protocol $Circuit$, to achieve a deterrence factor $\epsilon = 1 - \frac{1}{t}$ where t is a positive integer. In the *key generation stage* and the *input sharing stage*, each party P_i generates t key pairs and t sets of shares where $(t - 1)$ pair of keys and $(t - 1)$ sets of shares are opened for verification. For evaluating a multiplication gate, each party splits each of its two shares into t subshares. $\frac{t-1}{t}$ fraction of all the calculations are opened for verification. We have the following Corollary.

Corollary 1. *Assuming the existence of lossy additive homomorphic public key encryption schemes with efficient Opener algorithm and secure coin tossing protocols in the presence of active adversaries, protocol $Circuit$ securely computes functionality f with deterrence $\left(1 - \frac{1}{t}\right)$ in the presence of static covert adversaries that can corrupt up to $(n - 1)$ parties.*

Let M denote the number of multiplication gates in the arithmetic circuit representing f. Then the aforementioned runs in $O(M \log n)$ rounds and needs $O(M n^2 t^2 s)$ communication among the parties where s is the security parameter.

6 Efficiency and Applications

The security of protocol $Circuit$ is comparable to the security of protocol π_{AMPC} of [11] if we set $\kappa = \log_2 t$. Both protocols work in the arithmetic circuit representation and in the dishonest majority setting. In both protocols, the computation is performed over a finite field of size p, so the security parameter $s = \log p$.

In our protocol $Circuit$, we use the lossy encryption scheme based on Paillier cryptosystem [15] and a multiparty secure commitment scheme against active adversaries, based on ElGamal encryption scheme [14,13]. A commitment performs two exponentiations. The size of a commitment is s. In our protocol, the public key (N, g) of the Paillier cryptosystem must satisfy $N > p$, so we can assume that $\log N = \log p + 1 = s + 1$. Then the size of a ciphertext is $(2s + 2)$. Each encryption performs two exponentiations and each decryption performs one exponentiation. Homomorphic multiplication of a ciphertext by a constant performs one exponentiation.

Table 1 compares the communication and computation complexity of protocols π_{AMPC} and $Circuit$. Moreover, protocol π_{AMPC} needs $(30Mn^2s + 108n^2\kappa s +$

---

[12] This can happen with probability at most $\frac{1}{2}$, as described in the *key generation stage*.
[13] This happens when the trusted party replies with *undetected*.

Table 1. Comparison of [11] and *Circuit* protocol, $\kappa = \log_2 t$. Here CC and Exp denotes the communication complexity and the number of exponentiations required by the protocols, respectively.

Criterion	Protocol	Performance
CC	[11]	$(126Mn^2 + 564Mn + 457n^2\kappa - 12n^2 + 2030n\kappa - 230n)s + 82Mn^2 - 82Mn + 297n^2\kappa - 39n^2 - 298n\kappa + 39n$
	Circuit	$(7Mn^2t^2 + Mn^2t - Mn^2 - 5Mnt^2 + 8Mnt - 2Mn + 6n^2t + 3nt - 2n)s + 6Mn^2t^2 + Mn^2t - 4Mnt^2 - Mn + 5n^2t + 4nt - 2n$
Exp	[11]	$6Mn^3 + 211Mn^2 + 211Mn + 20n^3\kappa + 7n^3 + 769n^2\kappa - 75n^2 + 653n\kappa - 82n$
	Circuit	$5Mn^3t^2 - 3Mn^3t - 2Mn^2t^2 + 9Mn^2t + 6Mn^2 - 2Mnt + 2n^3t - 2n^3 - 2n^2t + 10n^2 + 2nt - 4n$

$163Mns + 599n\kappa s + O(n^2))$ storage complexity for the preprocessing phase that our protocol does not need.

Next we present the concrete efficiency comparison of these two protocols for two applications.

6.1 Application 1: AES Cipher

The AES cipher has become a standard benchmark for evaluating the performance of secure multiparty protocols [10].

Table 2. Comparison of [11] and *Circuit* protocol for $t = 2, s = 40$ for AES

n	Communication Complexity			Number of exponentiations		
	[11]	Circuit	Ratio	[11]	Circuit	Ratio
3	328,056,840	28,672,320	11.44	7,767,408	1,468,908	5.29
5	694,453,000	81,948,800	8.47	20,413,980	6,134,800	3.33
7	1,465,584,640	162,554,560	6.88	40,004,384	16,008,020	2.50
10	2,127,364,000	334,705,600	6.36	84,226,910	44,815,400	1.88

In Table 2, we compare the communication complexity and the number of exponentiations of both protocols for evaluating AES cipher on a shared input with $t = 2, s = 40$ for various number of parties. Here the column 'Ratio' shows the factor by which our protocol improves the previous work (in both communication and number of exponentiations). In this case, our protocol is up to 11 times more communication efficient than π_{AMPC}. Our protocol needs up to 5 times less exponentiations than π_{AMPC}.

6.2 Application 2: Matrix Multiplication

We consider the problem of performing secure multiplication of two shared matrices of size $\ell \times \ell$ over a finite field $\mathbb{F}$ of size p. Our protocol can be generalized to work over matrix rings $\mathbb{M}_\ell(\mathbb{F})$. The number of multiplication over the ring $\mathbb{M}_\ell(\mathbb{F})$, $M = 1$. This application is in high contrast to our first application AES in the sense that AES needs a huge number of multiplications while matrix multiplication problem performs a single secure multiplication.

Table 3. Comparison of [11] and *Circuit* protocol for $t = 2, s = 40$ for matrix multiplication of size 128×128 for n parties

	Communication Complexity			Number of exponentiations		
n	[11]	Circuit	Ratio	[11]	Circuit	Ratio
3	8,049,082,368	242,663,424	33.17	2,741,403,648	159,940,608	17.14
5	17,212,948,480	679,034,880	25.35	8,313,733,120	665,681,920	12.49
7	29,415,063,552	1,335,083,008	22.03	17,494,736,896	1,738,440,704	10.06
10	53,414,952,960	2,731,048,960	19.56	39,181,680,640	4,875,223,040	8.04

Table 3 compares the communication complexity and the number of exponentiations needed by both protocol for performing matrix multiplication of size 128×128 for $t = 2, s = 40$ for various number of parties. Our protocol is up to 33 times more communication efficient than π_{AMPC}. Our protocol requires up to 17 times less exponentiations than π_{AMPC}. For performing matrix multiplication of this size with $t = 4$ and deterrence $\epsilon = (1 - \frac{1}{4}) = \frac{3}{4}$, our protocol requires up to 19 times less communication than π_{AMPC} and needs up to 14 times fewer exponentiations than π_{AMPC}.

From these analysis we see that asymptotic analysis does not tell us everything – in order to analyze the performance, it is sometimes necessary to perform the exact calculation of runtime and communication complexity. The exact analysis shows that our protocol performs much better than protocol π_{AMPC}, both in terms of communication complexity and the time complexity, measured by the number of exponentiations.

References

1. Aumann, Y., Lindell, Y.: Security Against Covert Adversaries: Efficient Protocols for Realistic Adversaries. J. Cryptology 23(2), 281–343 (2010)
2. Bellare, M., Hofheinz, D., Yilek, S.: Possibility and Impossibility Results for Encryption and Commitment Secure under Selective Opening. In: Joux, A. (ed.) EUROCRYPT 2009. LNCS, vol. 5479, pp. 1–35. Springer, Heidelberg (2009)
3. Ben-Or, M., Goldwasser, S., Wigderson, A.: Completeness Theorems for Non-Cryptographic Fault-Tolerant Distributed Computation. In: STOC, pp. 1–10 (1988)
4. Blum, M., Feldman, P., Micali, S.: Non-Interactive Zero-Knowledge and Its Applications (Extended Abstract). In: STOC, pp. 103–112 (1988)
5. Bogdanov, D., Talviste, R., Willemson, J.: Deploying Secure Multi-Party Computation for Financial Data Analysis. In: Keromytis, A.D. (ed.) FC 2012. LNCS, vol. 7397, pp. 57–64. Springer, Heidelberg (2012)

6. Bogetoft, P., et al.: Secure Multiparty Computation Goes Live. In: Dingledine, R., Golle, P. (eds.) FC 2009. LNCS, vol. 5628, pp. 325–343. Springer, Heidelberg (2009)
7. Boneh, D., Franklin, M.K.: Efficient Generation of Shared RSA Keys. In: Kaliski Jr., B.S. (ed.) CRYPTO 1997. LNCS, vol. 1294, pp. 425–439. Springer, Heidelberg (1997)
8. Canetti, R.: Security and Composition of Multiparty Cryptographic Protocols. J. Cryptology 13(1), 143–202 (2000)
9. Chaum, D., Crépeau, C., Damgård, I.: Multiparty Unconditionally Secure Protocols. In: STOC, pp. 11–19 (1988)
10. Damgård, I., Keller, M., Larraia, E., Miles, C., Smart, N.P.: Implementing AES via an Actively/Covertly Secure Dishonest-Majority MPC Protocol. In: Visconti, I., De Prisco, R. (eds.) SCN 2012. LNCS, vol. 7485, pp. 241–263. Springer, Heidelberg (2012)
11. Damgård, I., Orlandi, C.: Multiparty Computation for Dishonest Majority: From Passive to Active Security at Low Cost. In: Rabin, T. (ed.) CRYPTO 2010. LNCS, vol. 6223, pp. 558–576. Springer, Heidelberg (2010)
12. Damgård, I., Pastro, V., Smart, N., Zakarias, S.: Multiparty Computation from Somewhat Homomorphic Encryption. In: Safavi-Naini, R. (ed.) CRYPTO 2012. LNCS, vol. 7417, pp. 643–662. Springer, Heidelberg (2012)
13. Goyal, V., Mohassel, P., Smith, A.: Efficient Two Party and Multi Party Computation against Covert Adversaries, http://research.microsoft.com/en-us/um/people/vipul/eff-mpc.pdf
14. Goyal, V., Mohassel, P., Smith, A.: Efficient Two Party and Multi Party Computation Against Covert Adversaries. In: Smart, N.P. (ed.) EUROCRYPT 2008. LNCS, vol. 4965, pp. 289–306. Springer, Heidelberg (2008)
15. Hemenway, B., Libert, B., Ostrovsky, R., Vergnaud, D.: Lossy Encryption: Constructions from General Assumptions and Efficient Selective Opening Chosen Ciphertext Security. In: Lee, D.H., Wang, X. (eds.) ASIACRYPT 2011. LNCS, vol. 7073, pp. 70–88. Springer, Heidelberg (2011)
16. Kol, G., Naor, M.: Cryptography and Game Theory: Designing Protocols for Exchanging Information. In: Canetti, R. (ed.) TCC 2008. LNCS, vol. 4948, pp. 320–339. Springer, Heidelberg (2008)
17. Lindell, Y., Oxman, E., Pinkas, B.: The IPS Compiler: Optimizations, Variants and Concrete Efficiency. In: Rogaway, P. (ed.) CRYPTO 2011. LNCS, vol. 6841, pp. 259–276. Springer, Heidelberg (2011)
18. Lindell, Y., Pinkas, B.: Privacy Preserving Data Mining. In: Bellare, M. (ed.) CRYPTO 2000. LNCS, vol. 1880, pp. 36–54. Springer, Heidelberg (2000)
19. Mohassel, P., Weinreb, E.: Efficient Secure Linear Algebra in the Presence of Covert or Computationally Unbounded Adversaries. In: Wagner, D. (ed.) CRYPTO 2008. LNCS, vol. 5157, pp. 481–496. Springer, Heidelberg (2008)
20. Paillier, P.: Public-Key Cryptosystems Based on Composite Degree Residuosity Classes. In: Stern, J. (ed.) EUROCRYPT 1999. LNCS, vol. 1592, pp. 223–238. Springer, Heidelberg (1999)
21. Peikert, C., Vaikuntanathan, V., Waters, B.: A Framework for Efficient and Composable Oblivious Transfer. In: Wagner, D. (ed.) CRYPTO 2008. LNCS, vol. 5157, pp. 554–571. Springer, Heidelberg (2008)

Impact of Optimized Field Operations AB, AC and $AB + CD$ in Scalar Multiplication over Binary Elliptic Curve

Christophe Negre[1,2] and Jean-Marc Robert[1,2]

[1] Team DALI, Université de Perpignan, France
[2] LIRMM, UMR 5506, Université Montpellier 2 and CNRS, France

Abstract. A scalar multiplication over a binary elliptic curve consists in a sequence of hundreds of multiplications, squarings and additions. This sequence of field operations often involves a large amount of operations of type AB, AC and $AB + CD$. In this paper, we modify classical polynomial multiplication algorithms to obtain optimized algorithms which perform these particular operations AB, AC and $AB + CD$. We then present software implementation results of scalar multiplication over binary elliptic curve over two platforms: Intel Core 2 and Intel Core i5. These experimental results show some significant improvements in the timing of scalar multiplication due to the proposed optimizations.

Keywords: Optimized field operations AB, AC and $AB + CD$, double-and-add, halve-and-add, parallel, scalar multiplication, software implementation, carry-less multiplication.

1 Introduction

Finite field arithmetic is widely used in elliptic curve cryptography (ECC) [13,11] and coding theory [4]. The main operation in ECC is the scalar multiplication which is computed as a sequence of multiplications and additions in the underlying field [6,8]. Efficient implementations of these sequences of finite field operations are thus crucial to get efficient cryptographic protocols.

We focus here on the special case of software implementation of scalar multiplication on elliptic curve defined over an extended binary field $\mathbb{F}_{2^m}$. An element in $\mathbb{F}_{2^m}$ is a binary polynomial of degree at most $m - 1$. In practice m is a prime integer in the interval $[160, 600]$. An addition and a multiplication of field elements consist in a regular binary polynomial addition and multiplication performed modulo the irreducible polynomial defining $\mathbb{F}_{2^m}$. An addition and a reduction are in practice faster than a multiplication of size m polynomials. Specifically, an addition is a simple bitwise XOR of the coefficients: in software, this consists in computing several independent word bitwise XORs (WXOR). Concerning the reduction, when the irreducible polynomial which defines the field $\mathbb{F}_{2^m}$ is sparse, reducing a polynomial can be expressed as a number of word shifts and word XORs.

A. Youssef, A. Nitaj, A.E. Hassanien (Eds.): AFRICACRYPT 2013, LNCS 7918, pp. 279–296, 2013.

Until the end of 2009 the fastest algorithm for software implementation of polynomial multiplication was the Comb method of Lopez and Dahab [12]. This method essentially uses look-up tables, word shifts (Wshift), ANDs and XORs. One of the most recent implementation based on this method was done by Aranha *et al.* in [1] on an Intel Core 2. But, since the introduction by Intel of a new carry-less multiplication instruction on the new processors i3, i5 and i7, the authors in [16] have shown that the polynomial multiplication based on Karatsuba method [15] outperforms the former approaches based on Lopez-Dahab multiplication. In the sequel, we consider implementations on two platforms: processor without carry-less multiplication (Intel Core 2) and processor i5 which has such instruction.

Our Contributions. In this paper, we investigate some optimizations of the operations AB, AC and $AB+CD$. The fact that we can optimize two multiplications AB, AC which have a common input A, is well known, it was for example noticed in [2]. Indeed, since there is a common input A, the computations depending only on A in AB and AC can be shared.

We also investigate a new optimization based on $AB + CD$. In this situation, we show that we can save in Lopez-Dahab polynomial multiplication algorithm $60N$ WShifts and $30N$ WXORs if the inputs are stored on N computer words. We also show that this approach can be adapted to the case of Karatsuba multiplication and we evaluate the resulting complexity.

We present implementation results of scalar multiplication which involve the previously mentioned optimizations. The reported results on an Intel Core 2 were obtained using Lopez-Dahab polynomial multiplication for field multiplication, and the reported results on an Intel Core i5 were obtained with Karatsuba multiplication.

Organization of the Paper. In Section 2, we review the best known algorithms for software implementation of polynomial multiplication of size $m \in [160, 600]$. In Section 3, we then present optimized versions of these algorithms for the operations AB, AC and $AB+CD$. In Section 4, we describe how to use the proposed optimizations in a scalar multiplication and give implementation results obtained on an Intel Core 2 and on an Intel Core i5. Finally, in Section 5, we give some concluding remarks.

2   Review of Multiplication Algorithms

The problem considered in this section is to compute efficiently a multiplication in a binary field $\mathbb{F}_{2^m}$. A field $\mathbb{F}_{2^m}$ can be defined as the set of binary polynomials modulo an irreducible polynomial $f(x) \in \mathbb{F}_2[x]$ of degree m. Consequently, a multiplication in $\mathbb{F}_{2^m}$ consists in multiplying two polynomials of degree at most $m - 1$ and reducing the product modulo $f(x)$. The fields considered here are described in Table 1 and are suitable for elliptic curve cryptography. The irreducible polynomials in Table 1 have a sparse form. This implies that the reduction can be expressed as a number of shifts and additions (the reader may refer for example to [8] for further details).

We then focus on efficient software implementation of binary polynomial multiplication: we review the best known algorithms for polynomial of cryptographic size. An element $A = \sum_{i=0}^{m-1} a_i x^i \in \mathbb{F}_2[x]$ is coded over $N = \lceil m/64 \rceil$ computer words of size 64 bits $A[0], \ldots, A[N-1]$. In the sequel, we will often use a nibble decomposition of A: $A = \sum_{i=0}^{n-1} A_i x^{4i}$ where $\deg A_i < 4$ and $n = \lceil m/4 \rceil$ is the nibble size of A. In Table 1 we give the value of N and n for the field sizes $m = 233$ and $m = 409$ considered in this paper.

Table 1. Irreducible polynomials and word/nibble sizes of field elements

m the field degree	Irreducible polynomial	N (64-bit word size)	n (nibble size)
233	$x^{233} + x^{74} + 1$	4	59
409	$x^{409} + x^{87} + 1$	7	103

2.1 Comb Multiplication

One of the best known methods for software implementation of the multiplication of two polynomials A and B was proposed by Lopez and Dahab in [12]. This algorithm is generally referred as the left-to-right comb method with window size w. We present this method for the window size $w = 4$ since, based on our experiments and several other experimental results in the literature [1,8], this seems to be the best case for the platform considered here (Intel Core 2). This method first computes a table T containing all products $u \cdot A$ for $u(x)$ of degree < 4. The second input B is decomposed into 64-bit words and nibbles as follows

$$B = \sum_{j=0}^{N-2} \sum_{k=0}^{15} B_{16j+k} x^{64j+4k} + \sum_{k=0}^{n-16(N-1)-1} B_{16(N-1)+k} x^{64(N-1)+4k}$$

where $\deg B_{16j+k} < 4$. Then the product $R = A \times B$ is expressed by expanding the above expression of B as follows

$$R = A \cdot \left(\sum_{j=0}^{N-2} \sum_{k=0}^{15} B_{16j+k} x^{64j+4k} + \sum_{k=0}^{n-16(N-1)-1} B_{16(N-1)+4k} x^{64(N-1)+4k} \right)$$

$$= \sum_{j=0}^{N-2} \sum_{k=0}^{15} (A \cdot B_{16j+k} x^{64j+4k}) + \sum_{k=0}^{n-16(N-1)-1} (A \cdot B_{16(N-1)+k}) x^{64(N-1)+4k}$$

$$= \sum_{k=0}^{n-16(N-1)-1} x^{4k} \left(\sum_{j=0}^{N-1} A \cdot B_{16j+k} x^{64j} \right)$$

$$+ \sum_{k=n-16(N-1)}^{15} x^{4k} \left(\sum_{j=0}^{N-2} (A \cdot B_{16j+k} x^{64j}) \right).$$

The above expression can be computed through a sequence of accumulations $R \leftarrow R + T[B_{16j+k}] x^{64j}$, corresponding to the terms $A \cdot B_{16j+k} x^{64i}$, followed by multiplications by x^4. This leads to Algorithm 1 for a pseudo-code formulation and Algorithm 6 in the appendix for a C-like code formulation.

Complexity. We evaluate the complexity of the corresponding C-like code (Algorithm 6; see p. 294) of the CombMul algorithm in terms of the number

---

Algorithm 1. CombMul(A,B)

Require: Two binary polynomials $A(x)$ and $B(x)$ of degree $< 64N - 4$, and $B(x) = \sum_{j=0}^{N-1} \sum_{k=0}^{15} B_{16j+k} x^{4k+64j}$ is decomposed in 64-bit words and nibbles.

Ensure: $R(x) = A(x) \cdot B(x)$

// Computation of the table T containing $T[u] = u(x) \cdot A(x)$ for all u such that $\deg u(x) < 4$

$T[0] \leftarrow 0;$

$T[1] \leftarrow A;$

for k **from** 1 **to** 7 **do**

   $T[2k] \leftarrow T[k] \cdot x;$

   $T[2k+1] \leftarrow T[2k] + A;$

end for

// right-to-left shifts and accumulations

$R \leftarrow 0$

for k **from** 15 **downto** 0 **do**

   $R \leftarrow R \cdot x^4$

   for j **from** $N-1$ **downto** 0 **do**

      $R \leftarrow R + T[B_{16j+k}]x^{64j}$

   end for

end for

---

of 64-bit word operations (WXOR, WAND and WShift). We do not count the operations performed for the loop variables $k, j, \ldots$. Indeed, when all the loops are unrolled, these operations can be precomputed. We have separated the complexity evaluation of the CombMul algorithm into three parts: the computation of the table T, the accumulations $R \leftarrow R + T[B_{16j+k}]x^{64j}$ and the shifts $R \leftarrow R \cdot x^4$ of R.

- *Table computation.* The loop on k is of length 7, and performs one WXOR and one WShift plus $2(N-1)$ WXORs and $2(N-1)$ WShifts in the inner loop on i.
- *Shifts by* 4. There are two nested loops: the one on k is of length 15 and the loop on i is of length $2N$. The loop operations consist in two WShifts and one WXOR.
- *Accumulations.* The number of accumulations $R \leftarrow R + T[B_{16j+k}]x^{64j}$ is equal to n, the nibble length of B. This results in nN WXOR, n WAND and $n - N$ WShift operations, since a single accumulation $R \leftarrow R + T[B_{16j+k}]x^{64j}$ requires N WXOR, one WAND and one WShift (except for $k = 0$).

As stated in Table 2, the total number of operations is equal to $nN + 44N - 7$ WXORs, $n + 73N - 7$ WShifts and n WANDs.

2.2   Karatsuba Multiplication

We review the Karatsuba approach for binary polynomial multiplication. Let A and B be two binary polynomials of size $64N$ and assume that N is even. Then, we first split A and B in two halves $A = A_0 + x^{64N/2}A_1$ and $B = B_0 +$

Table 2. Complexity of the C code of the Comb multiplication

Operation	#WXOR	#WShift	#WAND
Table T	$14N - 7$	$14N - 7$	0
$R \leftarrow R + T[B_{16j+k}]x^{64j}$	nN	$n - N$	n
Shift $R \leftarrow R << 4$	$30N$	$60N$	0
Total	$nN + 44N - 7$	$n + 73N - 7$	n

$x^{64N/2}B_1$ and then we re-express the product $A \times B$ in terms of three polynomial multiplications of half size:

$$R_0 = A_0 B_0, \quad R_1 = A_1 B_1, \quad R_2 = (A_0 + A_1)(B_0 + B_1),$$
$$C = R_0 + x^{64N/2}(R_0 + R_1 + R_2) + x^{64N} R_1. \tag{1}$$

The resulting recursive approach is given in `KaratRec` algorithm (Algorithm 2). In this case the inputs A and B are supposed to be of size $64N$ bits where $N = 2^s$ and packed in an array of N computer words. The three products R_0, R_1 and R_2 are computed recursively until we reach inputs of size one computer word. Then the word products are computed with a `Mult64` operation. We further assume that this `Mult64` operation is performed using a single processor instruction: this is the case of the Intel Cores i3, i5 and i7.

Algorithm 2. `KaratRec(A,B,N)`

Require: A and B on $N = 2^s$ computer words.
Ensure: $R = A \times B$
  if $N = 1$ **then**
    return ($Mult64(A, B)$)
  else
    // Split in two halves of word size $N/2$.
    $A = A_0 + x^{64N/2}A_1$
    $B = B_0 + x^{64N/2}B_1$
    // Recursive multiplication
    $R_0 \leftarrow$ `KaratRec`$(A_0, B_0, N/2)$
    $R_1 \leftarrow$ `KaratRec`$(A_1, B_1, N/2)$
    $R_2 \leftarrow$ `KaratRec`$(A_0 + A_1, B_0 + B_1, N/2)$
    // Reconstruction
    $R \leftarrow R_0 + (R_0 + R_1 + R_2)X^{64N/2} + R_1 X^{64N}$
    return (R)
  end if

Complexity of KaratRec Approach. We briefly compute the complexity of the `KaratRec` algorithm in terms of the number of WXOR and `Mult64` operations. One single recursion of the Karatsuba formula with inputs of word size N requires N WXORs for the additions $A_0 + A_1$ and $B_0 + B_1$, and $5N/2$ WXORs for the reconstruction of R. We obtain the recursive complexity given in the left side of (2). We rewrite the complexity in the non-recursive form given in the right side of (2).

$$\begin{cases} \#WXOR(N){=}4N + 3\#WXOR(N/2), \\ \#WXOR(1){=}0. \end{cases} \implies \#WXOR(N) = 8N^{\log_2(3)} - 8N$$

$$\begin{cases} \#Mult64(N){=}3\#Mult64(N/2), \\ \#Mult64(1){=}1. \end{cases} \implies \#Mult64(N) = N^{\log_2(3)}.$$

$$(2)$$

3 Optimization of the Operations $AB + CD$ and AB, AC

In this section, we present our main building blocks for the optimization of software implementation of elliptic curve scalar multiplication. The main idea is that the scalar multiplication involves operations of type $AB + CD$ or AB, AC. In such operations $AB + CD$ and AB, AC some computations can be saved resulting in a more efficient software implementation. This idea was previously mentionned for example in [2] for AB, AC for the CombMul algorithm. We extend this idea to the variants based on Karatsuba multiplication. We also study the optimization based on the operation $AB + CD$ in the case of CombMul algorithm and in the case of the variants of Karatsuba multiplication.

3.1 Optimizations of $AB + CD$ and AB, AC in the CombMul Approach

Optimization AB, AC in the CombMul Algorithm. The fact that we have to compute two multiplications with the same operand A, implies that the table T in the CombMul algorithm, which contains the products $T[u] = u \cdot A$, can be computed only once for the two multiplications AB and AC. This saves $14N - 7$ WXORS and $14N - 7$ Shifts operations in the computation of AC. The resulting complexity of the CombMul_ABAC algorithm is shown in Table 3.

Optimization $AB + CD$ in the CombMul Algorithm. We optimize the operation $AB + CD$ by performing the final addition $(AB) + (CD)$ during the accumulation step of the CombMul algorithm. Specifically, we keep the table computation stage $T[u] = u \cdot A$ and $S[u] = u \cdot C$ for u of degree < 4 unchanged. But we accumulate $T[B_{16j+k}]$ and $S[D_{16j+k}]$ in the same variable $R \leftarrow R + (T[B_{16j+k}] + S[B_{16j+k}])x^{64j}$. The shifts by 4 are then performed only on R.

The complexity of Algorithm 3 can be easily deduced from the complexity of the CombMul algorithm (Table 2):

- We have in the CombMul_ABplusCD algorithm two table computations which contribute to twice the complexity of the table computation in Table 2.
- The accumulations $R \leftarrow R + (T[B_{16j+k}] + S[D_{16j+k}])x^{64j}$ also contribute to twice the complexity of the accumulation step in Table 2.
- We have the same amount of shifts $R \leftarrow R \cdot x^4$ as in the CombMul algorithm.

The resulting complexity is given in Table 3.

Algorithm 3. `CombMul_ABplusCD(A,B)`

Require: Four binary polynomials A, B, C and D of degree $< 64N - 4$, and $B(x) = \sum_{j=0}^{N-1} \sum_{k=0}^{15} B_{16j+k} x^{4k+64j}$ with $\deg B_{16j+k} < 4$ and $D(x) = \sum_{j=0}^{N-1} \sum_{k=0}^{15} D_{16j+k} x^{4k+64j}$ with $\deg D_{16j+k} < 4$

Ensure: $R(x) = A(x) \cdot B(x) + C(x) \cdot D(x)$

// Computation of the table T and S such that $T[u] = u(x) \cdot A(x)$ and $S[u] = u(x) \cdot B(x)$ for all $\deg u(x) < 4$

$T[0] \leftarrow 0; S[0] \leftarrow 0;$

$T[1] \leftarrow A; S[1] \leftarrow C;$

for k **from** 1 **to** 7 **do**

  $T[2k] \leftarrow T[k] \cdot x; S[2k] \leftarrow S[k] \cdot x;$

  $T[2k + 1] \leftarrow T[2k] + A; S[2k + 1] \leftarrow S[2k] + C;$

end for

// right-to-left shift Comb multiplication

$R \leftarrow 0$

for k **from** 15 **downto** 0 **do**

  $R \leftarrow R \cdot x^4$

  for j **from** $N - 1$ **downto** 0 **do**

    $R \leftarrow R + (T[B_{16j+k}] + S[D_{16j+k}])x^{64j}$

  end for

  return (R)

end for

Table 3. Complexity of the optimizations AB, AC and $AB + CD$ on `CombMul`

Algorithm	#WXOR	#WShift	#WAND
CombMul_ABAC	$2nN + 74N - 7$	$2n + 132N - 7$	$2n$
CombMul_ABplusCD	$2nN + 58N - 14$	$2n + 86N - 14$	$2n$

3.2 Optimizations $AB + CD$ and AB, AC in the `KaratRec` Approach

The optimization based on AB, AC can be extended to the `KaratRec` algorithm. Indeed the recursive splitting and the addition of the two halves $A_0 + A_1$ can be performed only once for the polynomial A. This approach is described in Algorithm 5.

We also adapt the optimization $AB+CD$ as follows: the addition is performed before the reconstruction of the two products AB and AC, this means that we have only one recursive reconstruction instead of two. This approach is specified in Algorithm 4.

Complexity of `KaratRec_ABAC`. In the first recursion we have $3N/2$ WXORs for $A_0 + A_1$, $B_0 + B_1$ and $C_0 + C_1$ plus $5N$ WXORs for the reconstructions of R and S. This leads to the following complexity:

Algorithm 4.
KaratRec_ABpCD(A,B, C,D,N)

require: A, B, C and D are polynomials of word size $N = 2^s$ each.
ensure: $R = AB + CD$
if $N = 1$ **then**
return($Mul64(A, B) + Mul64(C, D)$)
else
// Splitting in two halves of $N/2$ 64-bit words.
$A = A_0 + x^{64N/2}A_1$, $B = B_0 + x^{64N/2}B_1$,
$C = C_0 + x^{64N/2}C_1$, $D = D_0 + x^{64N/2}D_1$
// Additions of the halves
$A_2 = A_0 + A_1$, $B_2 = B_0 + B_1$
$C_2 = C_0 + C_1$, $D_2 = D_0 + D_1$
// Recursive multiplications/additions
$R_0 \leftarrow$ KaratRec_ABpCD($A_0, B_0, C_0, D_0, N/2$)
$R_1 \leftarrow$ KaratRec_ABpCD($A_1, B_1, C_1, D_1, N/2$)
$R_2 \leftarrow$ KaratRec_ABpCD($A_2, B_2, C_2, D_2, N/2$)
// Reconstruction
$R \leftarrow R_0 + (R_0 + R_1 + R_2)x^{64N/2} + R_1 x^{64N}$
return(R)
end if

Algorithm 5. KaratRec_ABAC(A,B,C,N)

require: A, B and C are polynomials of word size $N = 2^s$ each.
ensure: $R = A \cdot B$ and $S = A \cdot C$
if $N = 1$ **then**
return($Mul64(A, B), Mul64(A, C)$)
else
// Splitting in two halves of $N/2$ 64-bit words.
$A = A_0 + x^{64N/2}A_1$, $B = B_0 + x^{64N/2}B_1$,
$C = C_0 + x^{64N/2}C_1$
// Additions of the halves
$A_2 = A_0 + A_1$, $B_2 = B_0 + B_1$, $C_2 = C_0 + C_1$
// Recursive multiplications
$R_0, S_0 \leftarrow$ KaratRec_ABAC($A_0, B_0, C_0, N/2$)
$R_1, S_1 \leftarrow$ KaratRec_ABAC($A_1, B_1, C_1, N/2$)
$R_2, S_2 \leftarrow$ KaratRec_ABAC($A_2, B_2, C_2, N/2$)
// Reconstruction
$R \leftarrow R_0 + (R_0 + R_1 + R_2)x^{64N/2} + R_1 x^{64N}$
$S \leftarrow S_0 + (S_0 + S_1 + S_2)x^{64N/2} + S_1 x^{64N}$
return(R, S)
end if

$$\begin{cases} \#WXOR(N)=13N/2 + 3 \ \#\text{WXOR}(N/2), \\ \#WXOR(1)=0. \end{cases} \implies \begin{array}{c} \#WXOR(N)=13N^{\log_2(3)} \\ -13N \end{array}$$

$$\begin{cases} \#Mult64(N)=3\#Mult64(N/2), \\ \#Mult64(1)=2. \end{cases} \implies \#Mult64(N) = 2N^{\log_2(3)}.$$

Complexity of KaratRec_ABpCD. In the first recursion we have $2N$ WXORs for the computations $A_0 + A_1$, $B_0 + B_1$, $C_0 + C_1$ and $D_0 + D_1$ plus $5N/2$ WXORs for the reconstruction of R. The complexity for $N = 1$ is equal to $2Mult64$ plus one WXOR. Based on this, we derive the complexity for the KaratRec_ABpCD algorithm:

$$\#WXOR(N) = 10N^{\log_2(3)} - 9N, \qquad \#Mult64(N) = 2N^{\log_2(3)}.$$

3.3   Complexity Comparison and Implementation Results

Using the complexity results determined in the former subsections, we can compute the complexities of the multiplication algorithms and their optimized AB, AC and $AB + CD$ counter parts for the polynomial sizes $m = 233$ and $m = 409$. We implemented these algorithms on the platforms Intel Core 2 and Intel Core i5. Our implementation uses 128-bit registers and vector instructions available on these two processors. On the Core 2 we used the modified CombMul algorithm of [5,1] which uses mostly shifts by multiple of 8; cheaper than an arbitrary shift for 128-bit data. On the Core i5 we implemented the KaratRec multiplication method with the PCLMUL instruction which performs carry-less multiplication of two 64 bit inputs contained in 128-bit registers.

The resulting complexities and timings are reported in Table 4 and Table 5.

Table 4. Complexity/timing results of the CombMul variants on a Core 2 (2.5 GHz)

Algorithm	Overall complexity in terms of word operations	233		409	
		#W.Op.	#CC	#W.Op.	#CC
CombMul	$nN + 2n + 117N - 14$	808	336	1732	795
CombMul_ABAC	$2nN + 4n + 206N - 14$	1511	555	3282	1597
CombMul_ABplusCD	$2nN + 4n + 144N - 28$	1256	564	2834	1737

#W. Op. = number of word operations (WXOR, WAND, WShift).
#CC = number of clock cycles.

Table 5. Complexity/timing results of the KaratRec variants on a Core i5 (2.5 GHz)

Algorithm	Complexity for $N = 2^s$		233			409		
	#WXOR	#Mul64	#WXOR	#Mul64	#CC	#WXOR	#Mul64	#CC
KaratRec	$8N^{\log_2(3)} - 8N$	$N^{\log_2(3)}$	40	9	107	152	27	286
KaratRec_ABAC	$13N^{\log_2(3)} - 13N$	$2N^{\log_2(3)}$	65	18	189	247	54	566
KaratRec_ABpCD	$10N^{\log_2(3)} - 9N$	$2N^{\log_2(3)}$	54	18	182	198	54	541

Based on the results presented above, we notice that the optimization $AB + CD$ has always a better complexity than the optimization AB, AC and better than two independent multiplications. Concerning the timings we note that:

– On the Core 2 the optimization $ABplusCD$ is always slower than the optimization AB, AC. Moreover, the optimizations $ABplusCD$ and AB, AC are effective only for $m = 233$, since in this case they are faster than two independent multiplications. This seems to contradict the corresponding complexity results since the complexity differences appear quite large.

– On the Core i5 the timing results are more related to the complexity values: for the two considered degrees $ABplusCD$ and AB, AC are faster than two independent multiplications and $ABplusCD$ is always faster than AB, AC.

In the literature we can find some timing of the CombMul algorithm over a Core 2 in [1]. The authors in [1] report implementation timings in the range of $[241, 276]$ clock-cycles for a polynomial multiplication of size $m = 233$ and in the range of $[690, 751]$ for $m = 409$, which are both better than the results reported in Table 4. Our results on the Core i5 compares favorably with the results reported in [16]: 128 clock-cycles for $m = 233$ and 345 clock-cycles for $m = 409$. These reported timings may include the reduction operation (this is not clearly specified in [16]). The same authors reported later in [17] better timings on the same processor and compiler: 100 clock-cycles for $m = 233$ and 270 clock-cycles for $m = 409$.

4 Implementations of Scalar Multiplication Based on the Optimizations AB, AC and $AB + CD$

In this section, we present our experimental results for scalar multiplication based on the optimizations AB, AC and $AB + CD$ presented in the previous section. We first review best known elliptic curve point operation formulas, and describe how we use the optimizations AB, AC and $AB + CD$ in these formulas. Then we describe the strategies we used for our implementations: scalar multiplication algorithms and implementations of field operations (squaring, inversion, ...). Finally, we present the implementation results on an Intel Core 2 and an Intel Core i5.

4.1   Elliptic Curve Arithmetic

The considered curves are ordinary binary elliptic curve defined by the following Weierstrass equation

$$y^2 + xy = x^3 + x^2 + b \text{ where } b \in \mathbb{F}_{2^m}.$$

We will more specifically focused on the two NIST [14] curves $B233$ and $B409$.

Optimization AB, AC and $ABplusCD$ in Curve Operation. We review Kim-Kim elliptic curve operations [10] in order to describe how the optimized operations AB, AC and $AB + CD$ can be used in the curve operations. Kim and Kim in [10] use a specific projective coordinates $P = (X : Y : Z : T)$ which corresponds to the affine point $(X/Z, Y/T)$ where $T = Z^2$. In the following formulas we use the following notations: $A \cdot B$ is a non reduced polynomial multiplication, and $[R]$ represents the reduction of the polynomial R modulo the irreducible polynomial defining the field $\mathbb{F}_{2^m}$.

• *Point doubling in Kim-Kim coordinates.* We compute the doubling $P_1 = (X_1 : Y_1 : Z_1 : T_1) = 2 \cdot (X : Y : Z : T)$ of a point $P = (X : Y : Z : T)$ by performing the following sequence of operations

$$A=X^2, \ B=[Y]^2.$$

and then:

$$Z_1=[T \cdot A], \ T_1=[Z_1^2], \ X_1=[A^2 + \underbrace{b \cdot T^2}_{AB,AC}], \ Y_1=\overbrace{B \cdot (B + X_1 + Z_1) + \underbrace{b \cdot T_1}_{AB,AC}}^{ABplusCD} +T_1.$$

• *Point addition in Kim-Kim coordinates.* We review the Kim-Kim formula for mixed point addition: we add one point $P_1 = (X_1 : Y_1 : Z_1 : T_1)$ which has a regular Kim-Kim projective coordinates with a point $P_2 = (X_2 : Y_2 : 1 : 1)$ which is in affine coordinates, i.e., $Z_2 = T_2 = 1$. The coordinates of $P_3 = (X_3 : Y_3 : Z_3 : T_3)$ is then computed with the following sequence of operations:

$$A=X_1 + [X_2 \cdot Z_1], \ B=[Y_1 + Y_2 \cdot T_1], \ C=[A \cdot Z_1], \ D=\underbrace{[C \cdot (B + C)]}_{AB,AC}.$$

and then deduce $Z_3 = [C^2]$, $T_3 = [Z_3^2]$, and

$$X_3 = [B^2 + \underbrace{C \cdot [A^2]}_{AB,AC}] + D, \quad Y_3 = \overbrace{[(X_3 + [X_2 \cdot Z_3]) \cdot D + (X_2 + Y_2) \cdot T_3]}^{ABplusCD}.$$

In the above formulas, we indicated the operations which can be performed with the optimization $AB + CD$ and the operations which can be performed with the optimization AB, AC.

- **Optimization AB, AC and $ABplusCD$ in other curve operation formulas.** We consider the following two cases: Lopez-Dahab formulas, which are variants of the Kim-Kim formulas, and Montgomery laddering. For the Lopez-Dahab formulas the optimizations AB, AC and $ABplusCD$ can be applied in both doubling and mixed addition. For the Montgomery laddering we can just apply one optimization $ABplusCD$ in the inner loop operation.

Another interesting operation is the point halving, but, unfortunately, we could not apply any of the optimizations AB, AC or $ABplusCD$ in the halving formula of [8] (Algorithm 3.81 [8], p. 131). Indeed, this point halving consists in one half-trace operation, followed by one multiplication, one trace computation and one square root, so no optimization based on combined multiplications can be applied.

Scalar Multiplication Algorithm. The scalar multiplication on the curve $E(\mathbb{F}_{2^m})$ consists in the computation of $r \cdot P$ for a given point $P \in E(\mathbb{F}_{2^m})$ and an ℓ-bit integer r where ℓ is the bit length of the order of P. We implemented the following methods for scalar multiplication:

- *Double-and-add.* This approach consists in a sequence of doublings and additions on the curve. The integer r is generally recoded with the NAF_w algorithm [8] with window size $w = 4$ in order to reduce the number of additions performed during the double-and-add algorithm. The scalar multiplication then requires a table precomputation $T[i] = i \cdot P$ for the odd integers $0 < i < 2^{w-1}$. In our implementations we used the Kim-Kim (cf. Subsection 4.1) and the Lopez-Dahab [8] doubling and addition formulas.
- *Halve-and-add.* This approach consists in a sequence of halvings and additions on the curve. The integer r is first recoded in $r' = r \cdot 2^{\ell-1} \mod \#<P>$ since in this case we have:

$$r = r'2^{-(\ell-1)} = (\sum_{i=0}^{\ell-1} r_i'2^i)2^{-(\ell-1)} = (\sum_{i=0}^{\ell-1} r_i'2^{i-(\ell-1)})$$

and we can then compute $r \cdot P$ as a sequence of halvings and additions. We use again the NAF_w algorithm for $w = 4$ to recode r' and the variant of the halve-and-add approach to perform the scalar multiplication. The reader may refer to Section 3.6 in [8] for further details on point halving approaches.

- *Parallel (Double-and-add, Halve-and-add).* This approach, proposed in [16,17], splits the computation of the scalar multiplication in two parts: one uses double-and-add approach and the other uses halve-and-add approach. This requires some recoding of the scalar r similar to the one used in halve-and-add approach.
- *Montgomery.* The last approach we considered is the Montgomery laddering (cf. Algorithm 3.40, p.103 in [8]): it is a variant of the double-and-add approach. The main difference is that two points are computed in the inner for loop of the algorithm: P_1 and P_2 which have a constant difference $P_1 - P_2 = P$. This approach has some nice properties as counter measure against side channel attacks.

4.2   Implementation Aspects

We use the following strategies to implement the field operations required in scalar multiplication algorithms:

- *Multiplication.* The considered multiplication strategies have already been described in Subsection 3.3. Specifically, on the Intel Core 2 platform, we use the version of the CombMul algorithm of [5,1] which uses 128-bit instruction sets. On the Intel Core i5 platform we use the Karatsuba algorithm along with vector instructions and more precisely the carry-less instruction which performs binary polynomial multiplication of size 64 bits.
- *Squaring.* For the squaring we use the strategy described in [1]. Specifically, we use a 128-bit word Sq which stores in each byte the squaring of a 4-bit polynomial. Then for each 128-bit word $A[i]$ of A we separate odd and even nibbles with a masking and a shift and then apply _mm_shuffle_epi8 intrinsinc function with left input value Sq and right input value the word containing even or odd nibbles of $A[i]$. The result is a 128-bit word containing the squaring of each nibble. The bytes are then reordered and repacked into two 128-bit words. The reader may refer to Algorithm 1 in [1] for further details.
- *Square-root.* The square root is based on the expression $\sqrt{A} = (\sum_{i=0}^{\lceil m/2 \rceil} a_{2i} X^i) + \sqrt{x}(\sum_{i=0}^{\lceil m/2 \rceil} a_{2i+1} X^i)$. Following [1], we separate odd and even coefficients of A using the intrinsinc function named _mm_shuffle_epi8 and by reordering the resulting bytes. Then the multiplication by $\sqrt{x}$ is done through a number of shifts and additions since for $m = 233$ and $m = 409$, $\sqrt{x}$ has a sparse expression.
- *Reduction.* The reduction follows the strategy of [8]: the considered irreducible polynomials are sparse (cf. Table 1), this makes possible to perform a reduction with a short sequence of shifts and WXORs.
- *Inversion.* The inversion is computed using the Itoh-Tsujii algorithm [9]. This algorithm consists in a sequence of multiplications and multi-squarings. This sequence of multiplication and squaring reconstructs step by step the exponent of $A^{-1} = A^{2^m - 2}$ following an addition chain in the exponent. For example, for $m = 233$, the inverse of A is given by $(A^{2^{232}-1})^2$, and is obtained

with the addition chain $1 \to 2 \to 3 \to 6 \to 7 \to 14 \to 28 \to 29 \to 58 \to 116 \to 232$ in the exponent. For multi-squaring consisting in long sequence of squaring we use a look-up table approach.

- *Half-trace.* In the halving curve operation, we have to compute half-trace (HT) of an element: $HT(A) = \sum_{i=0}^{(m-1)/2} A^{2^{2i}}$. Our implementation is again inspired from [16] and [7] and uses the intrinsic function _mm_shuffle_epi8 to compute the half-trace of the even bits of A and look-up table to compute the half-trace of the odd bits of A. For further details on this the reader may refer to [16,17].

Lazy Reductions. An optimization called *lazy-reduction* can be used to optimize curve operations (cf. [2,3]). This consists in removing unnecessary reduction operations performed during the sequence of multiplications and squarings in the curve operation formulas. Here we considered the following two lazy reduction optimizations:

- *Lazy-reduction 1 (LR1).* This optimization regroups reduction operations corresponding to distinct squarings or multiplications. For example in the sequence of operations $A^2 + C \cdot D$ we can perform the addition (addition of polynomial of degree $2m - 2$) before performing the reduction. This reduces the total number of WXORs and WShifts. In the considered elliptic curve operation formulas (Kim-Kim, Lopez-Dahab and Montgomery) the bracket [·] specifies the reduction operations corresponding to this LR1 optimization (cf. Subsection 4.1).
- *Lazy-reduction 2 (LR2).* In this case the reduction modulo the irreducible polynomial is partially done, this results in a polynomial with a degree larger than $m - 1$. We have applied this approach for $m = 233$: the polynomial is reduced to a degree 255 instead of 232. Since the KaratRec algorithm multiplies polynomials of size 256, we don't have to reduce the coefficients in the range [233, 256], so we can use a *lazy reduction* of this kind. Figure 4.2 illustrates this strategy: we can see in this figure that the LR2 approach saves the computations involved in the reduction of the word containing coefficients $c_{255}, \ldots, c_{233}$.

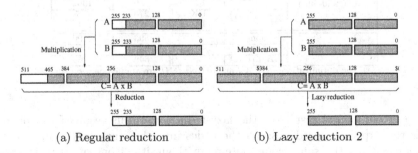

(a) Regular reduction   (b) Lazy reduction 2

Fig. 1. Regular reduction vs lazy reduction 2

We did not apply this strategy in the case of Intel Core 2 since the `CombMul` approach multiplies polynomials of degree 232 and not 256. For the case of degree 409, the LR2 approach does not provide any saving in the number of words which have to be reduced, so, again, we did not implement such LR2 optimization.

4.3   Implementation Results on an Intel Core 2

The timings of our implementation are reported in Table 6. These values were obtained on a Linux Ubuntu 11.10 platform with GCC 4.6.1. The reported clock-cycles were obtained with the following strategy: we used the cycle counter `rdtsc` attached to each core in the Intel Core 2 to get the number of clock cycles. The reported values are average timings for randomly generated input datas.

Table 6. Timings in terms of 10^3 clock-cycles of scalar multiplication on an Intel Core 2 (2.50GHz)

Optimization	Formulas	$m = 233$ $(\#CC)/10^3$	$m = 409$ $(\#CC)/10^3$
Double-and-add	none KK	592	2125
	none LD	613	2192
	LR1 KK	1249	2207
	LR1 LD	1179	2832
	AB, AC KK	558	6217
	AB, AC LD	928	2917
	$ABplusCD$ KK	542	2187
	$ABplusCD$ LD	553	2296
Halve-and-add	none KK	387	1504
	none LD	403	1575
	LR1 KK	651	1706
	LR1 LD	855	1837
	AB, AC KK	858	2277
	AB, AC LD	887	2359
	$ABplusCD$ KK	375	1504
	$ABplusCD$ LD	386	1640
Parallel[*] (Double-and-add + Halve-and-add)	none KK	280	965
	none LD	295	999
	LR1 KK	335	1042
	LR1 LD	315	1104
	AB, AC KK	270	2311
	AB, AC LD	289	1362
	$ABplusCD$ KK	273	977
	$ABplusCD$ LD	277	1014
Montgomery	none -	593	2190
	LR1 -	637	2482
	$ABplusCD$ -	549	2289

(∗) The optimizations AB, AC and $ABplusCD$ are applied only on the double-and-add part.

The experimental results of the lazy-reduction optimization (LR1) do not show the expected speed-ups: all the codes involving such lazy-reduction are all slower than the same code running without it. Consequently, we have not combined this optimization with the two other optimizations AB, AC and $ABplusCD$.

Table 7. Timings in terms of 10^3 clock-cycles of scalar multiplication on an Intel Core i5 (2.5 GHz)

	Optimizations	Curve Formulas	$m = 233$ $\#CC/10^3$	$m = 409$ $\#CC/10^3$
Double-and-add	none	KK	246	917
	none	LD	252	940
	LR1 and LR2$^{(**)}$	KK	220	906
	LR1 and LR2$^{(**)}$	LD	228	959
	AB, AC and LR1 and LR2$^{(**)}$	KK	219	903
		LD	226	961
	$ABplusCD$ and LR1 and LR2$^{(**)}$	KK	214	877
		LD	222	903
Halve-and-add	none	KK	165	667
	none	LD	169	719
	LR1 and LR2$^{(**)}$	KK	150	723
	LR1 and LR2$^{(**)}$	LD	155	708
	AB, AC and LR1 and LR2$^{(**)}$	KK	149	733
		LD	155	720
	$ABplusCD$ and LR1 and LR2$^{(**)}$	KK	150	696
		LD	154	689
Parallel$^{(*)}$	none	KK	131	466
	none	LD	133	478
	LR1 and LR2$^{(**)}$	KK	116	458
	LR1 and LR2$^{(**)}$	LD	122	474
	AB, AC and LR1 and LR2$^{(**)}$	KK	117	457
		LD	123	476
	$ABplusCD$ and LR1 and LR2$^{(**)}$	KK	117	452
		LD	118	467
Montgomery	none	-	244	924
	LR1 and LR2$^{(**)}$	-	229	886
	$ABplusCD$ and LR1 and LR2$^{(**)}$	-	220	883

$(*)$ The optimizations AB, AC and $ABplusCD$ are applied only on the double-and-add part.
$(**)$ The optimizations LR2 is applied only for $m = 233$

Based on the results reported in Table 6, we remark that the proposed optimization $AB + CD$ provides some significant speed-up for the field sizes 233 only. The optimization AB, AC does also provide some speed-up compared to non-optimized results in the case of $m = 233$, but in some cases we obtain some sudden loose of performance like in halve-and-add or double-and-add/LD cases. In the case $m = 409$, none of the optimizations provide any improvement, this confirms the timings we get in Table 4.

We could not find in the literature any timing on a Core 2 for the same curves and same fields. We just mention that Aranha *et al.* in [1] report in the range [785000,858000] clock-cycles over the curve NIST-B283 and [4310000,4754000] clock-cycles over the curve NIST-B571 for double-and-add scalar multiplication on an Intel Core 2. This means that our timings seem to be in the expected range of values.

4.4 Implementation Results on an Intel Core i5

In Table 7 we report our timings obtained on an Intel Core i5 using implementation strategies discussed in Subsections 4.1 and 4.2. The codes were compiled

Algorithm 6. CombMul_C_Code

Require: A and B two N 64-bit words polynomials of nibble length n
Ensure: $R = A \times B$

```
for(i = 0; i < N; i + +){
    T[0][i] = 0;
    T[1][i] = A[i]; }
for(k = 2; k < 16; k+= 2){
    T[k][0] = (T[k >> 1][0] << 1);
    T[k + 1][0] = T[k][0] ∧ A[0];
    for(i = 1; i < N; i + +){
        T[k][i] = (T[k >> 1][i] << 1)
                ∧(T[k >> 1][i − 1] >> 63);
        T[k + 1][i] = T[k][i] ∧ A[i]; }}
for(k = 15; k >= n − 16(N − 1); k − −){
    for(j = 0; j < N − 1; j + +){
        u = (B[j] >> (4 ∗ k)) & 0xf
        for(i = 0; i < N; i + +){
            R[i + j] = R[i + j] ∧ T[u][i];
        }
    }
    carry = 0
    for(i = 0; i < 2 ∗ N; i + +){
        temp = R[i];
        R[i] = (R[i] << 4) ∧ carry;
        carry = temp >> 60; }
}
for(k = n − 16(N − 1) − 1; k > 0; k − −){
    for(j = 0; j < N − 1; j + +){
        u = (B[j] >> (4 ∗ k)) & 0xf
        for(i = 0; i < N; i + +){
            R[i + j] = R[i + j] ∧ T[u][i];
        }
    }
    carry = 0
    for(i = 0; i < 2 ∗ N; i + +){
        temp = R[i];
        R[i] = (R[i] << 4) ∧ carry;
        carry = temp >> 60; }
}
for(j = 0; j < N; j + +){
    u = B[j] & 0xf;
    for (i = 0; i < N − 1; i + +){
        R[i + j] = (R[i + j] << 4) ∧ T[u]; }
}
```

Right-side annotations:

Table: $T[u] = u \cdot A$ with $\deg u < 4$
$\#WXOR = 7(2(N − 1) + 1) = 14N − 7$
$\#WShift = 7(2(N − 1) + 1) = 14N − 7$

Accumulation $R \leftarrow R + x^{64j} B_{k+16j} A$
$\#WXOR = N$
$\#WShift = 1$
$\#WAND = 1$

Shift $R \leftarrow R << 4$
$\#WXOR = 2N$
$\#WShift = 4N$

Accumulation $R \leftarrow R + x^{64j} B_{k+16j} A$
$\#WXOR = N$
$\#WShift = 1$
$\#WAND = 1$

Shift $R \leftarrow R << 4$
$\#WXOR = 2N$
$\#WShift = 4N$

Accumulation $R \leftarrow R + x^{64j} B_{16j+k} A$
$\#WXOR = N$
$\#WShift = 0$
$\#WAND = 1$

with GCC 4.7.2 on a Linux Ubuntu 12.10. We also disabled the turbo mode of the Core i5 in order to avoid miss-evaluations on the timings.

We note that, the lazy reduction optimizations provide a significant speed-up compared to regular implementations. We also remark that, except in some rare cases, the optimizations $AB + CD$ and AB, AC provide a speed-up compared to non-optimized or LR-optimized implementations. In the case of halve-and-add, the speed-up is less than in the case of double-and-add, but this can be explained by the fact that, in halve-and-add approach, the optimizations are only used in the curve additions which are less frequent than the point halvings. Moreover, the optimization $AB + CD$ is generally more efficient than AB, AC. The only cases in which neither $AB + CD$ nor AB, AC provide the best timing result is the parallel implementation for $m = 233$ and halve-and-add implementation for $m = 409$.

Let us briefly compare our results with the ones obtained by Aranha *et al.* over an Intel Core i5 with a GCC compiler in [17]. We remark that, except for parallel implementation when $m = 409$, our results are competitive with the timings of [17]. This means that our implementations reach the level of performance of [17] and that the proposed optimized operations are efficient when included in the best known implementation strategies for Intel Core i5.

5 Conclusion

The goal of this paper was to study software optimizations of binary field operations AB, AC and $AB + CD$ for scalar multiplication on binary elliptic curves. We have established several algorithms for these optimizations and have evaluated the complexity of the corresponding C-like codes of these algorithms. We have then presented implementation results for scalar multiplication on an Intel Core 2 and on an Intel Core i5. In our implementations of scalar multiplication we have used best known algorithms. We have also tested lazy reduction optimizations. The experimental results have shown that the proposed $AB + CD$ optimization improves the timing of scalar multiplication on an Intel Core 2 only for the small field $\mathbb{F}_{2^{233}}$. On an Intel Core i5, the optimization provides the best results for scalar multiplication over the two considered fields $\mathbb{F}_{2^{233}}$ and $\mathbb{F}_{2^{409}}$. For the case of Intel Core i5, we have reached the level of performance of the best known results found in the literature [16].

Acknowledgment. We are greatful to the members of the team DALI (University of Perpignan) for their helpful comments on the preliminary versions of this work. This work was supported by the ANR Pavois.

References

1. Aranha, D.F., López, J., Hankerson, D.: Efficient Software Implementation of Binary Field Arithmetic Using Vector Instruction Sets. In: Abdalla, M., Barreto, P.S.L.M. (eds.) LATINCRYPT 2010. LNCS, vol. 6212, pp. 144–161. Springer, Heidelberg (2010)
2. Avanzi, R.M., Thériault, N.: Effects of Optimizations for Software Implementations of Small Binary Field Arithmetic. In: Carlet, C., Sunar, B. (eds.) WAIFI 2007. LNCS, vol. 4547, pp. 69–84. Springer, Heidelberg (2007)
3. Avanzi, R.M., Thériault, N., Wang, Z.: Rethinking low genus hyperelliptic Jacobian arithmetic over binary fields: interplay of field arithmetic and explicit formulæ. J. Mathematical Cryptology 2(3), 227–255 (2008)
4. Berlekamp, E.R.: Bit-serial Reed-Solomon encoder. IEEE Trans. on Inform. Theory IT-28 (1982)
5. Beuchat, J.-L., López-Trejo, E., Martínez-Ramos, L., Mitsunari, S., Rodríguez-Henríquez, F.: Multi-core Implementation of the Tate Pairing over Supersingular Elliptic Curves. In: Garay, J.A., Miyaji, A., Otsuka, A. (eds.) CANS 2009. LNCS, vol. 5888, pp. 413–432. Springer, Heidelberg (2009)

6. Cohen, H., Miyaji, A., Ono, T.: Efficient Elliptic Curve Exponentiation Using Mixed Coordinates. In: Ohta, K., Pei, D. (eds.) ASIACRYPT 1998. LNCS, vol. 1514, pp. 51–65. Springer, Heidelberg (1998)
7. Fong, K., Hankerson, D., López, J., Menezes, A.: Field Inversion and Point Halving Revisited. IEEE Trans. Computers 53(8), 1047–1059 (2004)
8. Hankerson, D., Menezes, A., Vanstone, S.: Guide to Elliptic Curve Cryptography. Springer-Verlag New York, Inc., Secaucus (2003)
9. Itoh, T., Tsujii, S.: A Fast Algorithm for Computing Multiplicative Inverses in $GF(2^m)$ Using Normal Bases. Information and Computation 78, 171–177 (1988)
10. Kim, K.H., Kim, S.I.: A New Method for Speeding Up Arithmetic on Elliptic Curves over Binary Fields. Technical report, National Academy of Science, Pyongyang, D.P.R. of Korea (2007)
11. Koblitz, N.: Elliptic curve cryptosystems. Mathematics of Computation 48, 203–209 (1987)
12. López, J., Dahab, R.: High-Speed Software Multiplication in $\mathbb{F}_{2^m}$. In: Roy, B., Okamoto, E. (eds.) INDOCRYPT 2000. LNCS, vol. 1977, pp. 203–212. Springer, Heidelberg (2000)
13. Miller, V.: Use of elliptic curves in cryptography. In: Williams, H.C. (ed.) CRYPTO 1985. LNCS, vol. 218, pp. 417–426. Springer, Heidelberg (1986)
14. National Institute of Standards and Technology (NIST). Recommended elliptic curves for federal government use. NIST Special Publication (July 1999)
15. Paar, C.: A New Architecture for a Parallel Finite Field Multiplier with Low Complexity Based on Composite Fields. IEEE Trans. on Comp. 45, 856 (1996)
16. Taverne, J., Faz-Hernández, A., Aranha, D.F., Rodríguez-Henríquez, F., Hankerson, D., López, J.: Software Implementation of Binary Elliptic Curves: Impact of the Carry-Less Multiplier on Scalar Multiplication. In: Preneel, B., Takagi, T. (eds.) CHES 2011. LNCS, vol. 6917, pp. 108–123. Springer, Heidelberg (2011)
17. Taverne, J., Faz-Hernández, A., Aranha, D.F., Rodríguez-Henríquez, F., Hankerson, D., López, J.: Speeding scalar multiplication over binary elliptic curves using the new carry-less multiplication instruction. J. Cryptographic Engineering 1(3), 187–199 (2011)

An Attack on RSA Using LSBs
of Multiples of the Prime Factors

Abderrahmane Nitaj

Laboratoire de Mathématiques Nicolas Oresme
Université de Caen, Basse Normandie, France
`abderrahmane.nitaj@unicaen.fr`

Abstract. Let $N = pq$ be an RSA modulus with a public exponent e and a private exponent d. Wiener's famous attack on RSA with $d < N^{0.25}$ and its extension by Boneh and Durfee to $d < N^{0.292}$ show that using a small d makes RSA completely insecure. However, for larger d, it is known that RSA can be broken in polynomial time under special conditions. For example, various partial key exposure attacks on RSA and some attacks using additional information encoded in the public exponent e are efficient to factor the RSA modulus. These attacks were later improved and extended in various ways. In this paper, we present a new attack on RSA with a public exponent e satisfying an equation $ed - k(N+1-ap-bq) = 1$ where $\frac{a}{b}$ is an unknown approximation of $\frac{q}{p}$. We show that RSA is insecure when certain amount of the Least Significant Bits (LSBs) of ap and bq are known. Further, we show that the existence of good approximations $\frac{a}{b}$ of $\frac{q}{p}$ with small a and b substantially reduces the requirement of LSBs of ap and bq.

Keywords: RSA, Cryptanalysis, Factorization, Lattice, LLL algorithm, Coppersmith's method.

1 Introduction

The RSA cryptosystem was invented by Rivest, Shamir and Adleman [16] in 1977 and is today's most important public-key cryptosystem. The standard notations in RSA are as follows:

- p and q are two large primes of the same bit size.
- $N = pq$ is the RSA modulus and $\phi(N) = (p-1)(q-1)$ is Euler's totient function.
- e and d are respectively the public and the private exponents and satisfy $ed - k\phi(N) = 1$ for some positive integer k.

There have been a large number of attacks on RSA. Some attacks, called small private key attacks can break RSA in polynomial time when the private key is small. For example, Wiener [17] showed that if the private key satisfies $d < \frac{1}{3}N^{\frac{1}{4}}$, then N can be factored and Boneh and Durfee [4] showed that RSA is insecure if

A. Youssef, A. Nitaj, A.E. Hassanien (Eds.): AFRICACRYPT 2013, LNCS 7918, pp. 297–310, 2013.

$d < N^{0.292}$. Some attacks, called partial key exposure attacks exploit the knowledge of a portion of the private exponent or of one of the prime factors. Partial key exposure attacks are mainly motivated by using side-channel attacks, such as fault attacks, power analysis and timing attacks ([10], [11]). Using a side-channel, an attacker can expose a part of one of the modulus prime factors p or q or of the private key d. In 1998, Boneh, Durfee and Frankel [5] presented several partial key exposure attacks on RSA with a public key $e < N^{1/2}$ where the attacker requires knowledge of most significant bits (MSBs) or least significant bits (LSBs) of the private exponent d. In [2], Ernest et al. [7] proposed several partial key exposure attacks that work for $e > N^{1/2}$. Notice that Wiener's attack[17] and the attack of Boneh and Durfee[4] can be seen as partial key exposure attacks because the most significant bits of the private exponent are known and are equal to zero. Sometimes, it is possible to factor the RSA modulus even if the private key is large and no bits are exposed. Such attacks exploit the knowledge of special conditions verified by the modulus prime factors or by the exponents. In 2004, Blömer and May [3] showed that RSA can be broken if the public exponent e satisfies an equation $ex = y + k\phi(N)$ with $x < \frac{1}{3}N^{\frac{1}{4}}$ and $|y| < N^{-\frac{3}{4}}ex$. At Africacrypt 2009, Nitaj [15] presented an attack when the exponent e satisfies an equation $eX - (N - (ap + bq))Y = Z$ with the constraints that $\frac{a}{b}$ is an unknown convergent of the continued fraction expansion of $\frac{q}{p}$, $1 \le Y \le X < \frac{1}{2}\frac{N^{\frac{1}{4}}}{\sqrt{a}}$, $\gcd(X, Y) = 1$, and Z depends on the size of $|ap - bq|$. Nitaj's attack combines techniques from the theory of continued fractions, Coppersmith's method [6] for finding small roots of bivariate polynomial equations and the Elliptic Curve Method [12] for integer factorization.

In this paper we revisit Nitaj's attack by studying the generalized RSA equation $ed - k(N + 1 - ap - bq) = 1$ with different constraints using Coppersmith's method [6] only. We consider the situation when an amount of LSBs of ap and bq are exposed where $\frac{a}{b}$ is an unknown approximation of $\frac{q}{p}$, that is when $a = \left\lceil \frac{bq}{p} \right\rceil$. More precisely, assume that $ap = 2^{m_0}p_1 + p_0$ and $bq = 2^{m_0}q_1 + q_0$ where m_0, p_0 and q_0 are known to the attacker. We show that one can factor the RSA modulus if the public key e satisfies an equation $ed_1 - k_1(N + 1 - ap - bq) = 1$ where $e = N^\gamma$, $d_1 < N^\delta$, $2^{m_0} = N^\beta$ and $a < b < N^\alpha$ satisfy

$$\delta \le \begin{cases} \delta_1 & \text{if} \quad \gamma \ge \frac{1}{2}(1 + 2\alpha - 2\beta), \\ \delta_2 & \text{if} \quad \gamma < \frac{1}{2}(1 + 2\alpha - 2\beta). \end{cases}$$

with

$$\delta_1 = \frac{7}{6} + \frac{1}{3}(\alpha - \beta) - \frac{1}{3}\sqrt{4(\alpha - \beta)^2 + 4(3\gamma + 1)(\alpha - \beta) + 6\gamma + 1},$$
$$\delta_2 = \frac{1}{4}(3 - 2(\alpha - \beta) - 2\gamma).$$

We notice the following facts

- When $a = b = 1$, the equation becomes $ed_1 - k_1(N + 1 - p - q) = 1$ as in standard RSA.
- When $\gamma = 1$ and $\alpha = \beta$, the RSA instance is insecure if $d < \frac{7}{6} - \frac{\sqrt{7}}{3} \approx 0.284$. This is a well known boundary in the cryptanalysis of RSA (see e.g. [4]).
- When $\gamma = 1$ and $\beta = 0$, that is no LSBs of ap nor of bq are known, the RSA instance is insecure if $\delta < \frac{7}{6} + \frac{1}{3}\alpha - \frac{1}{3}\sqrt{\alpha^2 + 16\alpha + 7}$. This considerably improve the bound $\delta < \frac{1}{4}(1 - 2\alpha)$ of [15].
- The ANSI X9.31 standard [1] requires that the prime factors p and q shall not be near the ratio of two small integers. Our new attack shows that this requirement is necessary and can be easily checked once one has generated two primes simply by computing the convergents of the continued fraction expansion of $\frac{q}{p}$.

The rest of the paper is organized as follows. In Section 2 we review some basic results from lattice theory and their application to solve modular equations as well as two useful lemmas. In Section 3 we describe the new attack on RSA. In Section 4, we present various numerical experiments. Finally, we conclude in Section 5.

2 Preliminaries

2.1 Lattices

Let ω and n be two positive integers with $\omega \leq n$. Let $b_1, \cdots, b_\omega \in \mathbb{R}^n$ be ω linearly independent vectors. A lattice $\mathcal{L}$ spanned by $\{b_1, \cdots, b_\omega\}$ is the set of all integer linear combinations of $b_1, \cdots, b_\omega$, that is

$$\mathcal{L} = \left\{ \sum_{i=1}^{\omega} x_i b_i \mid x_i \in \mathbb{Z} \right\}.$$

The set $\langle b_1 \ldots, b_\omega \rangle$ is called a lattice basis for $\mathcal{L}$. The lattice dimension is $\dim(\mathcal{L}) = \omega$. We say that the lattice is full rank if $\omega = n$. If the lattice is full rank, then the determinant of $\mathcal{L}$ is equal to the absolute value of the determinant of the matrix whose rows are the basis vectors $b_1, \cdots, b_\omega$. In 1982, Lenstra, Lenstra and Lovász [13] invented the so-called LLL algorithm to reduce a basis and to find a short lattice vector in time polynomial in the bit-length of the entries of the basis matrix and in the dimension of the lattice. The following lemma, gives bounds on LLL-reduced basis vectors.

Theorem 1 (Lenstra, Lenstra, Lovász). *Let $\mathcal{L}$ be a lattice of dimension ω. In polynomial time, the LLL- algorithm outputs two reduced basis vectors v_1 and v_2 that satisfy*

$$v_1 \leq 2^{\frac{\omega}{2}} \det(\mathcal{L})^{\frac{1}{\omega}}, \quad v_2 \leq 2^{\frac{\omega}{2}} \det(\mathcal{L})^{\frac{1}{\omega-1}}.$$

Using the LLL algorithm, Coppersmith [6] proposed a method to efficiently compute small roots of bivariate polynomials over the integers or univariate modular polynomials. Howgrave-Graham [8] gave a simple reformulation of Coppersmith's method in terms of the norm of the polynomial $f(x, y) = \sum a_{ij}x^i y^j$ which is defined by

$$\|f(x, y)\| = \sqrt{\sum a_{ij}^2}.$$

Theorem 2 (Howgrave-Graham). *Let $f(x, y) \in \mathbb{Z}[x, y]$ be a polynomial which is a sum of at most ω monomials. Suppose that $f(x_0, y_0) \equiv 0 \pmod{e^m}$ where $|x_0| < X$ and $|y_0| < Y$ and $\|f(xX, yY)\| < \frac{e^m}{\sqrt{\omega}}$. Then $f(x_0, y_0) = 0$ holds over the integers.*

2.2 Useful Lemmas

Let $N = pq$ be an RSA modulus. The following lemma is useful to find a value of $ap - bq$ using a known value of $ap + bq$.

Lemma 1. *Let $N = pq$ be an RSA modulus with $q < p < 2q$ and S be a positive integer. Suppose that $ap + bq = S$ where $\frac{a}{b}$ is an unknown approximation of $\frac{q}{p}$. Then*

$$ab = \left\lfloor \frac{S^2}{4N} \right\rfloor \quad and \quad |ap - bq| = \sqrt{S^2 - 4 \left\lfloor \frac{S^2}{4N} \right\rfloor N}.$$

Proof. Observe that multiplying $q < p < 2q$ by p gives $N < p^2 < 2N$ and consequently $\sqrt{N} < p < \sqrt{2}\sqrt{N}$. Suppose that $\frac{a}{b}$ is an approximation of $\frac{q}{p}$, that is $a = \left\lceil \frac{bq}{p} \right\rceil$. Hence $\left| a - \frac{bq}{p} \right| \le \frac{1}{2}$, which gives

$$|ap - bq| \le \frac{p}{2} \le \frac{\sqrt{2}\sqrt{N}}{2} < 2\sqrt{N}.$$

Next, suppose that $ap + bq = S$. We have $S^2 = (ap + bq)^2 = (ap - bq)^2 + 4abN$. Since $|ap - bq| < 2\sqrt{N}$, then the quotient and the remainder in the Euclidean division of S^2 by $4N$ are respectively ab and $(ap - bq)^2$. Hence

$$ab = \left\lfloor \frac{S^2}{4N} \right\rfloor \quad and \quad |ap - bq| = \sqrt{S^2 - 4abN},$$

which terminates the proof.   □

The following lemma shows how to factor $N = pq$ using a known value of $ap + bq$.

Lemma 2. *Let $N = pq$ be an RSA modulus with $q < p < 2q$ and S be a positive integer. Suppose that $ap + bq = S$ where $\frac{a}{b}$ is an unknown approximation of $\frac{q}{p}$. Then N can be factored.*

Proof. Suppose that $\frac{a}{b}$ is an approximation of $\frac{q}{p}$ and that $ap + bq = S$. By Lemma 1, we get $ab = \left\lfloor \frac{S^2}{4N} \right\rfloor$ and $|ap - bq| = D$ where

$$D = \sqrt{S^2 - 4abN}.$$

Hence $ap - bq = \pm D$. Combining with $ap + bq = S$, we get $2ap = S \pm D$. Since $a < q$, then $\gcd(N, S \pm D) = \gcd(N, 2ap) = p$. This gives the factorization of N.   □

3   The New Attack

Let e, d_1, k_1 be positive integers such that $ed_1 - k_1(N + 1 - ap - bq) = 1$. In this section, we consider the following parameters.

- $2^{m_0} = N^\beta$ where m_0 is a known integer.
- $a < b < N^\alpha$ with $\alpha < \frac{1}{2}$ where $\frac{a}{b}$ is an unknown approximation of $\frac{q}{p}$.
- $ap = 2^{m_0} p_1 + p_0$ where p_0 is a known integer.
- $bq = 2^{m_0} q_1 + q_0$ where q_0 is a known integer.
- $e = N^\gamma$.
- $d_1 = N^\delta$.

The aim in this section is to prove the following result.

Theorem 3. *Suppose that* $ap = 2^{m_0} p_1 + p_0$ *and* $bq = 2^{m_0} q_1 + q_0$ *where* m_0, p_0 *and* q_0 *are known with* $2^{m_0} = N^\beta$ *and* $\frac{a}{b}$ *is an unknown approximation of* $\frac{q}{p}$ *satisfying* $a, b < N^\alpha$. *Let* $e = N^\gamma$, $d_1 = N^\delta$ *and* k_1 *be positive integers satisfying an equation* $ed_1 - k_1(N + 1 - ap - bq) = 1$. *Then one can factor* N *in polynomial time when*

$$\delta \leq \begin{cases} \delta_1 & \text{if} \quad \gamma \geq \frac{1}{2}(1 + 2\alpha - 2\beta), \\ \delta_2 & \text{if} \quad \gamma \leq \frac{1}{2}(1 + 2\alpha - 2\beta), \end{cases}$$

where

$$\delta_1 = \frac{7}{6} + \frac{1}{3}(\alpha - \beta) - \frac{1}{3}\sqrt{4(\alpha - \beta)^2 + 4(3\gamma + 1)(\alpha - \beta) + 6\gamma + 1},$$

$$\delta_2 = \frac{1}{4}(3 - 2(\alpha - \beta) - 2\gamma).$$

Proof. Suppose that $ap = 2^{m_0} p_1 + p_0$ and $bq = 2^{m_0} q_1 + q_0$ with known m_0, p_0 and q_0. Then $ap + bq = 2^{m_0}(p_1 + q_1) + p_0 + q_0$. Starting with the variant RSA equation $ed_1 - k_1(N + 1 - ap - bq) = 1$, we get

$$ed_1 - k_1 \left(N + 1 - p_0 - q_0 - 2^{m_0}(p_1 + q_1) \right) = 1.$$

Reducing modulo e, we get

$$-2^{m_0} k_1 (p_1 + q_1) + (N + 1 - p_0 - q_0)k_1 + 1 \equiv 0 \pmod{e}.$$

Observe that $\gcd(2^{m_0}, e) = 1$. Then multiplying by -2^{-m_0} (mod e), we get

$$k_1(p_1 + q_1) + a_1 k_1 + a_2 \equiv 0 \pmod{e},$$

where

$$a_1 \equiv -(N + 1 - p_0 - q_0)2^{-m_0} \pmod{e},$$
$$a_2 \equiv -2^{-m_0} \pmod{e}.$$

Consider the polynomial

$$f(x, y) = xy + a_1 x + a_2.$$

Then $(x, y) = (k_1, p_1 + q_1)$ is a modular root of the equation $f(x, y) \equiv 0 \pmod{e}$. Assuming that $\alpha \ll \frac{1}{2}$, we get

$$k_1 = \frac{ed_1 - 1}{N + 1 - ap - bq} \sim N^{\gamma + \delta - 1}.$$

On the other hand, we have

$$p_1 + q_1 < \frac{ap + bq}{2^{m_0}} < N^{\frac{1}{2} + \alpha - \beta}.$$

Define the bounds X and Y as

$$X = N^{\gamma + \delta - 1}, \quad Y = N^{\frac{1}{2} + \alpha - \beta}.$$

To find the small modular roots of the equation $f(x, y) \equiv 0 \pmod{e}$, we apply the extended strategy of Jochemsz and May [9]. Let m and t be positive integers to be specified later. For $0 \le k \le m$, define the set

$$M_k = \bigcup_{0 \le j \le t} \left\{ x^{i_1} y^{i_2 + j} \;\middle|\; x^{i_1} y^{i_2} \text{ monomial of } f^m(x, y) \right.$$

$$\left. \text{and} \quad \frac{x^{i_1} y^{i_2}}{(xy)^k} \text{ monomial of } f^{m-k} \right\}.$$

Observe that $f^m(x, y)$ satisfies

$$f^m(x, y) = \sum_{i_1 = 0}^{m} \binom{m}{i_1} x^{i_1} (y + a_1)^{i_1} a_2^{m - i_1}$$

$$= \sum_{i_1 = 0}^{m} \binom{m}{i_1} x^{i_1} \left(\sum_{i_2 = 0}^{i_1} \binom{i_1}{i_2} y^{i_2} a_1^{i_1 - i_2} a_2^{m - i_1} \right)$$

$$= \sum_{i_1 = 0}^{m} \sum_{i_2 = 0}^{i_1} \binom{m}{i_1} \binom{i_1}{i_2} x^{i_1} y^{i_2} a_1^{i_1 - i_2} a_2^{m - i_1}.$$

Hence, $x^{i_1} y^{i_2}$ is a monomial of $f^m(x, y)$ if

$$i_1 = 0, \ldots, m, \quad i_2 = 0, \ldots, i_1.$$

Consequently, for $0 \leq k \leq m$, when $x^{i_1} y^{i_2}$ is a monomial of $f^m(x, y)$, then $\frac{x^{i_1} y^{i_2}}{(xy)^k}$ is a monomial of $f^{m-k}(x, y)$ if

$$i_1 = k, \ldots, m, \quad i_2 = k, \ldots, i_1.$$

Hence, for $0 \leq k \leq m$, we obtain

$$x^{i_1} y^{i_2} \in M_k \quad \text{if} \quad i_1 = k, \ldots, m, \quad i_2 = k, \ldots, i_1 + t.$$

Similarly,

$$x^{i_1} y^{i_2} \in M_{k+1} \quad \text{if} \quad i_1 = k+1, \ldots, m, \quad i_2 = k+1, \ldots, i_1 + t.$$

For $0 \leq k \leq m$, define the polynomials

$$g_{k, i_1, i_2}(x, y) = \frac{x^{i_1} y^{i_2}}{(xy)^k} f(x, y)^k e^{m-k} \quad \text{with} \quad x^{i_1} y^{i_2} \in M_k \backslash M_{k+1}.$$

For $0 \leq k \leq m$, these polynomials reduce to the following sets

$$\begin{cases} k = 0, \ldots, m, \\ i_1 = k, \ldots, m, \\ i_2 = k, \end{cases} \quad \text{or} \quad \begin{cases} k = 0, \ldots, m, \\ i_1 = k, \\ i_2 = k+1, \ldots, i_1 + t. \end{cases}$$

This gives rise to the polynomials

$$G_{k, i_1}(x, y) = x^{i_1 - k} f(x, y)^k e^{m-k}, \quad \text{for} \quad k = 0, \ldots m, \quad i_1 = k, \ldots m,$$
$$H_{k, i_2}(x, y) = y^{i_2 - k} f(x, y)^k e^{m-k}, \quad \text{for} \quad k = 0, \ldots m, \quad i_2 = k+1, \ldots, k+t.$$

Let $\mathcal{L}$ denote the lattice spanned by the coefficient vectors of the polynomials $G_{k, i_1}(xX, yY)$ and $H_{k, i_2}(xX, yY)$. The ordering of two monomials $x^{i_1} y^{i_2}$, $x^{i'_1} y^{i'_2}$ is as in the following rule: if $i_1 < i'_1$, then $x^{i_1} y^{i_2} < x^{i'_1} y^{i'_2}$ and if $i_1 = i'_1$ and $i_2 < i'_2$, then $x^{i_1} y^{i_2} < x^{i'_1} y^{i'_2}$. Notice that the matrix is left triangular. For $m = 2$ and $t = 1$, the coefficient matrix for $\mathcal{L}$ is presented in Table 1. The non-zero elements are marked with an '$\circledast$'.

From the triangular form of the matrix, the $\circledast$ marked values do not contribute in the calculation of the determinant. Hence, the determinant of $\mathcal{L}$ is

$$\det(\mathcal{L}) = e^{n_e} X^{n_X} Y^{n_Y}. \tag{1}$$

From the construction of the polynomials $G_{k, i_1}(x, y)$ and $H_{k, i_2}(x, y)$, we get

$$n_e = \sum_{k=0}^{m} \sum_{i_1=k}^{m} (m - k) + \sum_{k=0}^{m} \sum_{i_2=k+1}^{k+t} (m - k) = \frac{1}{6} m(m+1)(2m + 3t + 4).$$

Table 1. The coefficient matrix for the case $m = 3$, $t = 1$

	1	x	x^2	x^3	y	xy	x^2y	x^3y	xy^2	x^2y^2	x^3y^2	x^2y^3	x^3y^3	x^3y^4
$G_{0,0}$	e^3	0	0	0	0	0	0	0	0	0	0	0	0	0
$G_{0,1}$	0	Xe^3	0	0	0	0	0	0	0	0	0	0	0	0
$G_{0,2}$	0	0	X^2e^3	0	0	0	0	0	0	0	0	0	0	0
$G_{0,3}$	0	0	0	X^3e^3	0	0	0	0	0	0	0	0	0	0
$H_{0,1}$	0	0	0	0	Ye^3	0	0	0	0	0	0	0	0	0
$G_{1,1}$	⊛	⊛	0	0	0	XYe^2	0	0	0	0	0	0	0	0
$G_{1,2}$	0	⊛	0	0	0	0	X^2Ye^2	0	0	0	0	0	0	0
$G_{1,3}$	0	0	⊛	⊛	0	0	0	X^3Ye^2	0	0	0	0	0	0
$H_{1,2}$	0	0	0	0	⊛	⊛	0	0	XY^2e^2	0	0	0	0	0
$G_{2,2}$	⊛	⊛	⊛	0	0	⊛	⊛	0	0	X^2Y^2	0	0	0	0
$G_{2,3}$	0	⊛	⊛	⊛	0	0	⊛	⊛	0	0	X^3Y^2e	0	0	0
$H_{2,3}$	0	0	0	0	⊛	⊛	⊛	⊛	⊛	⊛	0	X^2Y^3e	0	0
$G_{3,3}$	⊛	⊛	⊛	⊛	0	⊛	⊛	⊛	⊛	0	⊛	0	X^3Y^3	0
$H_{3,4}$	0	0	0	0	⊛	⊛	⊛	⊛	⊛	⊛	⊛	⊛	⊛	X^3Y^4

Similarly, we have

$$n_X = \sum_{k=0}^{m}\sum_{i_1=k}^{m} i_1 + \sum_{k=0}^{m}\sum_{i_2=k+1}^{k+t} k = \frac{1}{6}m(m+1)(2m+3t+4),$$

and

$$n_Y = \sum_{k=0}^{m}\sum_{i_1=k}^{m} k + \sum_{k=0}^{m}\sum_{i_2=k+1}^{k+t} i_2 = \frac{1}{6}(m+1)(m^2+3mt+3t^2+2m+3t).$$

Finally, we can calculate the dimension of $\mathcal{L}$ as

$$\omega = \sum_{k=0}^{m}\sum_{i_1=k}^{m} 1 + \sum_{k=0}^{m}\sum_{i_2=k+1}^{k+t} 1 = \frac{1}{2}(m+1)(m+2t+2).$$

For the following asymptotic analysis we let $t = \tau m$. For sufficiently large m, the exponents n_e, n_X, n_Y and the dimension ω reduce to

$$n_e = \frac{1}{6}(3\tau + 2)m^3 + o(m^3),$$

$$n_X = \frac{1}{6}(3\tau + 2)m^3 + o(m^3),$$

$$n_Y = \frac{1}{6}(3\tau^2 + 3\tau + 1)^2 m^3 + o(m^3),$$

$$\omega = \frac{1}{2}(2\tau + 1)m^2 + o(m^2).$$

To apply Theorem 2 to the shortest vector in the LLL-reduced basis of $\mathcal{L}$, we have to set

$$2^{\frac{\omega}{2}} \det(\mathcal{L})^{\frac{1}{\omega-1}} < \frac{e^m}{\sqrt{\omega}}.$$

This transforms to

$$\det(\mathcal{L}) < \frac{1}{\left(2^{\frac{\omega}{2}}\sqrt{\omega}\right)^\omega} e^{m(\omega-1)} < e^{m\omega}.$$

Using (1), we get

$$e^{n_e} X^{n_X} Y^{n_Y} < e^{m\omega}.$$

Plugging n_e, n_X, n_Y, ω as well as the values $e = N^\gamma$, $X = N^{\gamma+\delta-1}$, and $Y = N^{\frac{1}{2}+\alpha-\beta}$, we get

$$\frac{1}{6}(3\tau + 2)m^3\gamma + \frac{1}{6}(3\tau + 2)m^3(\gamma + \delta - 1) + \frac{1}{6}(3\tau^2 + 3\tau + 1)^2 m^3 (\frac{1}{2} + \alpha - \beta)$$
$$< \frac{1}{2}(2\tau + 1)m^3\gamma,$$

which transforms to

$$6(\alpha - \beta + 1)\tau^2 + 3(2\alpha + 2\delta - 2\beta - 1)\tau + (2\gamma + 2\alpha + 4\delta - 2\beta - 3) < 0. \quad (2)$$

Next, we consider the cases $\tau \neq 0$ and $\tau = 0$ separately. First, we consider the case $\tau > 0$. The optimal value for τ in the left side of (2) is

$$\tau = \frac{1 + 2\beta - 2\alpha - 2\delta}{2(1 + 2\alpha - 2\beta)}. \quad (3)$$

Observe that for $\alpha < \frac{1}{2}$ and $\beta < \frac{1}{2}$, we have $1 + 2\alpha - 2\beta > 0$. To ensure $\tau > 0$, δ should satisfy $\delta < \delta_0$ where

$$\delta_0 = \frac{1}{2}(1 - 2(\alpha - \beta)). \quad (4)$$

Replacing τ by the optimal value (3) in the inequation (2), we get

$$-12\delta^2 + 4(7 + 2\alpha - 2\beta)\delta + 4(\alpha - \beta)^2 + 4(4\gamma - 1)(\alpha - \beta) + 8\gamma - 15 < 0,$$

which will be true if $\delta < \delta_1$ where

$$\delta_1 = \frac{1}{3}(\alpha - \beta) + \frac{7}{6} - \frac{1}{3}\sqrt{4(\alpha - \beta)^2 + 4(3\gamma + 1)(\alpha - \beta) + 6\gamma + 1}. \quad (5)$$

Since δ has to satisfy both $\delta < \delta_0$ and $\delta < \delta_1$ according to (4) and (5), let us find the minimum $\min(\delta_0, \delta_1)$. A straightforward calculation shows that

$$\min(\delta_0, \delta_1) = \begin{cases} \delta_0 & \text{if } \gamma \leq \frac{1}{2}(1 + 2\alpha - 2\beta), \\ \delta_1 & \text{if } \gamma \geq \frac{1}{2}(1 + 2\alpha - 2\beta). \end{cases}$$

Now, consider the case $\tau = 0$, that is $t = 0$. Then the inequation (2) becomes

$$2\gamma + 2\alpha + 4\delta - 2\beta - 3 < 0,$$

which leads to $\delta < \delta_2$ where

$$\delta_2 = \frac{1}{4}(2\beta + 3 - 2\gamma - 2\alpha). \tag{6}$$

To obtain an optimal value for δ, we compare δ_2 as in (6) to $\min(\delta_0, \delta_1)$, obtained respectively with $\tau > 0$ and $\tau = 0$. First suppose that $\gamma \le \frac{1}{2}(1 + 2\alpha - 2\beta)$. Then

$$\min(\delta_0, \delta_1) - \delta_2 = \delta_0 - \delta_2 = \frac{1}{2}\left(g - \frac{1}{2}(1 + 2\alpha - 2\beta)\right) \le 0.$$

Hence $\min(\delta_0, \delta_1) \le \delta_2$. Next suppose that $\gamma \ge \frac{1}{2}(1 + 2(\alpha - \beta))$. Then

$$\min(\delta_0, \delta_1) - \delta_2 = \delta_1 - \delta_2$$
$$= \frac{5}{6}(\alpha - \beta) + \frac{1}{2}\gamma + \frac{5}{12}$$
$$- \frac{1}{3}\sqrt{4(\alpha - \beta)^2 + 4(3\gamma + 1)(\alpha - \beta) + 6\gamma + 1}.$$

On the other hand, we have

$$\left(\frac{5}{6}(\alpha - \beta) + \frac{1}{2}\gamma + \frac{5}{12}\right)^2 - \left(\frac{1}{3}\sqrt{4(\alpha - \beta)^2 + 4(3\gamma + 1)(\alpha - \beta) + 6\gamma + 1}\right)^2$$
$$= \frac{1}{16}(1 + 2(\alpha - \beta) - 2\gamma)^2,$$

which implies that $\min(\delta_0, \delta_1) \ge \delta_2$.
Summarizing, the attack will succeed to find k_1, $p_1 + q_1$ and $d_1 = N^\delta$ when $\delta < \delta'$ with

$$\delta' = \begin{cases} \delta_1 & \text{if} \quad \gamma \ge \frac{1}{2}(1 + 2\alpha - 2\beta), \\ \delta_2 & \text{if} \quad \gamma \le \frac{1}{2}(1 + 2\alpha - 2\beta), \end{cases}$$

where δ_1 and δ_2 are given by (5) and (6).
Next, using the known value of $p_1 + q_1$, we can precisely calculate the value $ap + bq = 2^{m_0}(p_1 + q_1) + p_0 + q_0 = S$. Then using Lemma 1 and Lemma 2, we can find p and q. Since every step in the method can be done in polynomial time, then N can be factored in polynomial time. This terminates the proof.   □

For example, consider the standard instance with the following parameters

- $2^{m_0} = N^\beta$ with $\beta = 0$.
- $a \le b \le N^\alpha$ with $\alpha = 0$, that is $ap + bq = p + q$.
- $ap = 2^{m_0}p_1 + p_0 = p_1$, that is $p_0 = 0$.
- $bq = 2^{m_0}q_1 + q_0 = q_1$, that is $q_0 = 0$.
- $e = N^\gamma$ with $\gamma = 1$.
- $d_1 = N^\delta$.

Then $\gamma \ge \frac{1}{2}(1 + 2\alpha - 2\beta) > \frac{1}{2}$ and the instance is insecure if $\delta < \delta_1$, that is if $\delta < \frac{7}{6} - \frac{\sqrt{7}}{3} \approx 0.284$ which is the same boundary as in various cryptanalytic approaches to RSA (see e.g. [4]).

Now suppose that $\gamma = 1$ and that a, b are small. Then $\alpha \approx 0$ and the boundary (5) becomes

$$\delta_1 < \frac{7}{6} - \frac{1}{3}\beta - \frac{1}{3}\sqrt{4\beta^2 - 16\beta + 7},$$

where the right side increases from 0.284 to 1 when $\beta \in \left[0, \frac{1}{2}\right[$. This implies that the existence of good approximation $\frac{a}{b}$ of $\frac{q}{p}$ substantially reduces the requirement of LSBs of ap and bq for the new attack. This confirms the recommendation of the X9.31-1997 standard for public key cryptography [1] regarding the generation of primes, namely that $\frac{q}{p}$ shall not be near the ratio of two small integers.

4   Experimental Results

We have implemented the new attack for various parameters. The machine was with Windows 7 and Intel(R) Core(TM)2 Duo CPU, 2GHz and the algebra system was Maple 12 [14]. For each set of parameters, we solved the modular equation $f(x, y) \equiv 0 \pmod{e}$ using the method described in Section 3. We obtained two polynomials $f_1(x, y)$ and $f_2(x, y)$ with the expected root $(k_1, p_1 + q_1)$. We then solved the equation obtained using the resultant of $f_1(x, y)$ and $f_2(x, y)$ in one of the variables. For every instance, we could recover k_1 and $p_1 + q_1$ and hence factor N. The experimental results are shown in Table 2

Table 2. Experimental results

N	γ	β	α	δ	lattice parameters	LLL-time (sec)
2048	0.999	0.219	0.008	0.340	$m = 2$, $t = 1$, dim=9	54
2048	0.999	0.230	0.018	0.340	$m = 3$, $t = 2$, dim=18	2818
2048	0.999	0.172	0.114	0.273	$m = 2$, $t = 1$, dim=9	22
2048	0.999	0.150	0.096	0.272	$m = 2$, $t = 1$, dim=9	20
2048	0.999	0.091	0.019	0.280	$m = 2$, $t = 1$, dim=9	16
1024	0.999	0.326	0.123	0.368	$m = 3$, $t = 2$, dim=18	429
1024	0.999	0.326	0.123	0.339	$m = 2$, $t = 1$, dim=9	7
1024	0.998	0.229	0.050	0.326	$m = 2$, $t = 1$, dim=9	7
1024	0.995	0.102	0.008	0.297	$m = 2$, $t = 1$, dim=9	4
1024	0.999	0.131	0.123	0.239	$m = 2$, $t = 1$, dim=9	4

In the rest of this section, we present a detailed numerical example. Consider an instance of a 200-bit RSA public key with the following parameters.

- $N = 2463200821438139415679553190953343235761287240746891883363309$.
- $e = 2666252898014064620417496175410895131584066512832204161816153$.
  Hence $e = N^\gamma$ with $\gamma = 0.984$.

- $m_0 = 35$. Hence $2^{m_0} = N^\beta$ with $\beta = 0.174$.
- $a < b < N^{0.080}$. Hence $\alpha = 0.080$.
- $m = 4$, $t = 2$.

Now suppose we know $p_0 = 28297245379$ and $q_0 = 28341074839$ such that $ap = 2^{m_0}p_1 + p_0$ and $bq = 2^{m_0}q_1 + q_0$. The modular equation to solve is then $f(x, y) = xy + a_1 x + a_2 \equiv 0 \pmod{e}$, where

$$a_1 = 3964784709534486659618115970154533670674093676299708171329 7,$$
$$a_2 = 2308706621061057850011169368805615354669031076933179855381 02.$$

Working with $m = 4$ and $t = 2$, we get a lattice with dimension $\omega = 25$. Using the parameters $\gamma = 0.984$, $\alpha = 0.080$, and $\beta = 0.174$, the method will succeed with the bounds X and Y satisfying

$$p_1 + q_1 < X = N^{\gamma + \delta - 1} \approx 2^{52},$$
$$k_1 < Y = N^{\frac{1}{2} + \alpha - \beta} \approx 2^{81},$$

if $\delta < 0.356$. Applying the LLL algorithm, we find two polynomials $f_1(x, y)$ and $f_2(x, y)$ sharing the same integer solution. Then solving the resultant equation in y, we get $x = 4535179907267444$ and solving the resultant equation in x, we get $y = 3609045068101717298446784$. Hence

$$p_1 + q_1 = 4535179907267444,$$
$$k_1 = 3609045068101717298446784.$$

Next, define

$$S = 2^{m_0}(p_1 + q_1) + p_0 + q_0 = 124005844298295748786131327649328730.$$

Then S is a candidate for $ap + bq$, and using Lemma 1, we get

$$ab = \left\lfloor \frac{S^2}{4N} \right\rfloor = 1560718201,$$

$$|ap - bq| = D = \sqrt{S^2 - 4abN} = 108928763058542141383405605909 2.$$

Using S for $ap + bq$ and D for $|ap - bq|$, we get $2ap = S - D$, and finally

$$p = \gcd(N, S - D) = 2973592513804257910045501261169.$$

Hence $q = \frac{N}{p} = 828358562917839001533347328061$. This terminates the factorization of the modulus N. Using the equation $ed_1 = k_1(N + 1 - ap - bq) + 1$, we get $d_1 = 41897971798817657 \approx N^{0.275}$. We notice that, with the standard RSA equation $ed - k\phi(N) = 1$, we have $d \equiv e^{-1} \pmod{\phi(N)} \approx N^{0.994}$ which is out of reach of the attack of Boneh and Durfee as well as the attack of Blömer and May. Also, using $2ap = S - D$, we get $a = \frac{S-D}{2p} = 20851$. Similarly, using $2bq = S + D$, we get $b = \frac{S+D}{2q} = 74851$. We notice that $\gcd(a, b) = 1$ and $\frac{a}{b}$ is not among the convergents of $\frac{q}{p}$. This shows that Nitaj's attack as presented in [15] can not succeed to factor the RSA modulus in this example.

5   Conclusion

In this paper, we propose a new polynomial time attack on RSA with a public exponent satisfying an equation $ed_1 - k_1(N + 1 - ap - bq) = 1$ where $\frac{a}{b}$ is an unknown approximation of $\frac{q}{p}$ and where certain amount of the Least Significant Bits of ap and aq are known to the attacker. The attack is based on the method of Coppersmith for solving modular polynomial equations. This attack can be seen as an extension of the well known partial key attack on RSA when $a = b = 1$ and certain amount of the Least Significant Bits of one of the modulus prime factors is known.

References

1. ANSI Standard X9.31-1998, Digital Signatures Using Reversible Public Key Cryptography For The Financial Services Industry (rDSA)
2. Blömer, J., May, A.: New partial key exposure attacks on RSA. In: Boneh, D. (ed.) CRYPTO 2003. LNCS, vol. 2729, pp. 27–43. Springer, Heidelberg (2003)
3. Blömer, J., May, A.: A generalized Wiener attack on RSA. In: Bao, F., Deng, R., Zhou, J. (eds.) PKC 2004. LNCS, vol. 2947, pp. 1–13. Springer, Heidelberg (2004)
4. Boneh, D., Durfee, G.: Cryptanalysis of RSA with private key d less than $N^{0.292}$. In: Stern, J. (ed.) EUROCRYPT 1999. LNCS, vol. 1592, pp. 1–11. Springer, Heidelberg (1999)
5. Boneh, D., Durfee, G., Frankel, Y.: An attack on RSA given a small fraction of the private key bits. In: Ohta, K., Pei, D. (eds.) ASIACRYPT 1998. LNCS, vol. 1514, pp. 25–34. Springer, Heidelberg (1998)
6. Coppersmith, D.: Small solutions to polynomial equations, and low exponent RSA vulnerabilities. Journal of Cryptology 10(4), 233–260 (1997)
7. Ernst, M., Jochemsz, E., May, A., de Weger, B.: Partial key exposure attacks on RSA up to full size exponents. In: Cramer, R. (ed.) EUROCRYPT 2005. LNCS, vol. 3494, pp. 371–386. Springer, Heidelberg (2005)
8. Howgrave-Graham, N.: Finding small roots of univariate modular equations revisited. In: Darnell, M.J. (ed.) Cryptography and Coding 1997. LNCS, vol. 1355, pp. 131–142. Springer, Heidelberg (1997)
9. Jochemsz, E., May, A.: A strategy for finding roots of multivariate polynomials with new applications in attacking RSA variants. In: Lai, X., Chen, K. (eds.) ASIACRYPT 2006. LNCS, vol. 4284, pp. 267–282. Springer, Heidelberg (2006)
10. Kocher, P.C.: Timing attacks on implementations of Diffie-Hellman. In: Koblitz, N. (ed.) CRYPTO 1996. LNCS, vol. 1109, pp. 104–113. Springer, Heidelberg (1996)
11. Kocher, P.C., Jaffe, J., Jun, B.: Differential power analysis. In: Wiener, M. (ed.) CRYPTO 1999. LNCS, vol. 1666, pp. 388–397. Springer, Heidelberg (1999)
12. Lenstra, H.W.: Factoring integers with elliptic curves. Annals of Mathematics 126, 649–673 (1987)
13. Lenstra, A.K., Lenstra, H.W., Lovász, L.: Factoring polynomials with rational coefficients. Mathematische Annalen 261, 513–534 (1982)

14. Maple, http://www.maplesoft.com/products/maple/
15. Nitaj, A.: Cryptanalysis of RSA using the ratio of the primes. In: Preneel, B. (ed.) AFRICACRYPT 2009. LNCS, vol. 5580, pp. 98–115. Springer, Heidelberg (2009)
16. Rivest, R., Shamir, A., Adleman, L.: A Method for Obtaining digital signatures and public-key cryptosystems. Communications of the ACM 21(2), 120–126 (1978)
17. Wiener, M.: Cryptanalysis of short RSA secret exponents. IEEE Transactions on Information Theory 36, 553–558 (1990)

Modification and Optimisation
of an ElGamal-Based PVSS Scheme

Kun Peng

Institute for Infocomm Research, Singapore
dr.kun.peng@gmail.com

Abstract. Among the existing PVSS schemes, a proposal by Shoemakers is a very special one. It avoids a common problem in PVSS design and costly operations by generating the secret to share in a certain way. Although its special secret generation brings some limitations to its application, its improvement in simplicity and efficiency is significant. However, its computational cost is still linear in the square of the number of share holders. Moreover, appropriate measures need to be taken to extend its application. In this paper, the PVSS scheme is modified to improve its efficiency and applicability. Firstly, a more efficient proof technique is designed to reduce the computational cost of the PVSS scheme to be linear in the number of share holders. Secondly, its secret generation procedure is extended to achieve better flexibility and applicability.

1 Introduction

In many secure information systems, some secret information is distributed among multiple parties to enhance security and robustness of systems. The first secret sharing technique is Shamir's t-out-of-n secret sharing [22]. A dealer has a secret s and wants to share it among n share holders. The dealer builds a polynomial $f(x) = \sum_{j=0}^{t-1} \alpha_j x^j$ and sends $s_i = f(i) \bmod q$ to the ith share holder for $i = 1, 2, \ldots, n$ through a secure communication channel where $\alpha_0 = s$ and q is an integer larger than any possible secret to share. Any t shares can be used to reconstruct the secret $s = \sum_{i \in S} s_i u_i \bmod q$ where $u_i = \prod_{j \in S, j \neq i} j/(j-i) \bmod q$ and S contains the indices of the t shares. Moreover, no information about the secret is obtained if the number of available shares is less than t. Secret sharing is widely employed in various secure information systems like e-auction, e-voting and multiparty computation. As most of the applications require public verifiability, very often secret sharing must be publicly verifiable. Namely, it must be publicly verified that all the n shares are consistently generated from a unique share generating polynomial and any t of them reconstruct the same secret. As the verification is public in those distributed systems, any wrong-doing can be publicly detected by any one and thus is undeniable.

Publicly verifiable secret sharing is usually called PVSS. It is widely employed in various applications like mix network [1], threshold access control [20], e-voting [9,12,14,13], distributed encryption algorithm [6], zero knowledge proof [7], anonymous token [11] and fair exchange [15]. In PVSS the dealer is not

A. Youssef, A. Nitaj, A.E. Hassanien (Eds.): AFRICACRYPT 2013, LNCS 7918, pp. 311–327, 2013.
© Springer-Verlag Berlin Heidelberg 2013

trusted and may deviate from the secret sharing protocol. He may distribute inconsistent shares to the share holders such that some shares reconstruct a secret while some other shares reconstruct a different secret. This cheating behaviour compromises security of the applications of secret sharing. For example, in secret-sharing-based multiple-tallier e-voting [9,12], it allows the talliers to tamper with the votes. Famous PVSS schemes include [23,4,10,21], where one of the two PVSS protocols in [23] is a development of the proposal in [8] and [17] is a revisit to [4] with a different encryption algorithm and stronger security. Usually, PVSS is a combination of secret sharing and publicly verifiable encryption of an encryption algorithm, which is employed to implement the secure communication channel for share distribution. The dealer encrypts shares for the share holders using their public keys and publishes the ciphertexts. Each share holder can decrypt the ciphertext for him and obtain his share, while the dealer can publicly prove that the messages encrypted in the ciphertexts are shares for a unique secret without revealing the secret or its shares. The following three important security properties are desired in PVSS.

- Correctness: if the dealer is honest and does not deviate from the PVSS protocol, he can always successfully prove validity of the shares.
- Soundness: only with an exception of a negligible probability, the shares are guaranteed to be generated by the same secret-generating polynomial such that any t of them reconstruct the same secret.
- Privacy: no information about the secret or any of its shares is revealed in the proof of validity of the shares. More precisely, a private PVSS scheme should employ zero knowledge proof techniques, which do not reveal any secret information as their proof transcripts can be simulated without any difference by a party without any knowledge of the secret or any of its shares.

Those security properties can be defined in a fornal way as follows.

Definition 1 *There is a proof function $Val(s_1, s_2, \ldots, s_n)$ for a dealer to prove validity of the shares. If his proof returns $Val(s_1, s_2, \ldots, s_n) = TRUE$, validity of $s_1, s_2, \ldots, s_n$ are accepted. If it returns $FALSE$, the shares are regarded as invalid.*

- *Correctness: an honest dealer can always achieve $Val(s_1, s_2, \ldots, s_n) = TRUE$ and $Pr(Val(s_1, s_2, \ldots, s_n) = FALSE$ and $\exists f(x) = \sum_{j=0}^{t-1} \alpha_j x^j$ such that $s_i = f(i) \bmod q) = 0$ where $Pr()$ stands for the probability of an event.*
- *Soundness: $Pr(Val(s_1, s_2, \ldots, s_n) = TRUE$ and $\nexists f(x) = \sum_{j=0}^{t-1} \alpha_j x^j$ such that $s_i = f(i) \bmod q)$ is negligible.*
- *Privacy: suppose the transcript of $Val(s_1, s_2, \ldots, s_n)$ is $TRANS$, and then anyone without knowledge of s or any share can generate a simulating transcript $TRAN'$ such that distribution of the two transcripts is indistinguishable.*

Although public verification of some operations is discussed in a VSS (verifiable secret sharing) scheme [18], it is not a PVSS proposal as it does not implement publicly verifiable encryption.

There are two basic and compulsory requirements in PVSS, security of encryption algorithm and publicly verifiable encryption, which are explained as follows.

- A secure encryption algorithm must be employed for the dealer to encrypt the shares before distributing them to the share holders. The employed encryption algorithm protect privacy of the encrypted messages such that no polynomial adversary can obtain any information about the secret or any of its shares from the ciphertexts of all the shares. This assumption is called *basic* assumption as it is inevitable in any PVSS scheme.
- The employed encryption algorithm must support publicly verifiable encryption such that an encrypted share can be publicly proved and verified to be generated by a secret share-generating polynomial. The existing PVSS schemes specify the public proof and verification as follows.
  1. The dealer publishes $c_i = E_i(s_i)$ for $i = 1, 2, \ldots, n$ where $E_i()$ is the encryption function of the ith share holder's encryption algorithm.
  2. An integer g with multiplicative order q is chosen.
  3. $A_j = g^{\alpha_j}$ for $j = 0, 1, \ldots, t-1$ are published by the dealer as a public commitment to the share-generating polynomial. To enhance privacy, α_j can be committed to in $g^{\alpha_j} h^{r_j}$ where h is in the cyclic group generated by g, $\log_g h$ is secret and r_j is a random integer smaller than the order of g. For simplicity of description, we only discuss the simpler commitment algorithm, while replacing it with the more complex commitment algorithm does not change the specification of PVSS in essence.
  4. A commitment to every share s_i is publicly available: $C_i = \prod_{j=0}^{t-1} A_j^{i^j}$.
  5. The dealer has to publicly prove that the same integer is committed to in C_i and encrypted in c_i.

1.1   A Dilemma in Choosing Encryption Algorithm

Choice of encryption algorithm is a subtle question in PVSS. Most existing PVSS schemes [4,10,17] choose RSA or Paillier encryption [16] for the share holders. Those two encryption algorithms have a common property: suppose the decryption function of the ith share holder's encryption algorithm is $D_i()$ and the modulus used in the calculations in $D_i()$ is q_i and then $q_1, q_2, \ldots, q_n$ must be different from each other and so cannot be equal to q. More precisely, $D_i(E_i(m))$ is the remainder of m modulo q_i and is not necessary to be $m \bmod q$ as at least $n-1$ of the q_is cannot be equal to q. So if a dealer encrypts a message larger than q_i in c_i what the ith share holder obtains through $D_i(c_i)$ is not equal to what the dealer encrypts in c_i modulo q. Therefore, a malicious dealer can cheat as follows.

1. He publishes $c_i = E_i(f(i) + kq)$ where k is an integer to satisfy $f(i) + kq > q_i$.
2. The ith share holder obtains $D_i(c_i) = f(i) + kq \bmod q_i$, which is not equal to $f(i)$ modulo q.
3. The dealer can still prove that the same integer (namely $f(i) + kq$) is committed to in C_i and encrypted in c_i. But $D_i(c_i)$ cannot be used as a share to reconstruct s, the secret committed to in C_0.

The simplest way to prevent this attack is to set $q_1 = q_2 = \ldots = q_n = q$. However this setting is impossible with RSA or Paillier encryption. So in the PVSS schemes employing RSA or Paillier encryption [4,10,17], the dealer has to prove that the message committed to in C_i and encrypted in c_i is smaller than q_i for $i = 1, 2, \ldots, n$. Therefore, n instances of range proof is needed where a range proof proves that a secret integer is within a certain range without revealing any other information about it. The most efficient general-purpose range proof is proposed by Boudot in Section 3 of [5]. Although the range proof by Boudot costs only a constant number of exponentiations, it is still not efficient enough. His range proof consists of two proof operations to handle the upper bound and lower bound of the range respectively, while each of them costs about 20 exponentiations. n instances of such range proof in PVSS is a high cost. Range proof can be more efficient in terms of the number of exponentiations when a secret integer is chosen from a range and then proved to be in a much larger range. This condition is called expansion rate and explained in details in Section 1.2.3 of [5]. For range proof in a range R, a much smaller range R' is chosen in the middle of R. The prover chooses a message v in R' and publishes its monotone function $z = w + cv$ in Z where w is randomly chosen from a set S_1 and c is randomly chosen from a set S_2. Then z is verified to be in a range R''. Of course, both v and w must be sealed in some appropriate commitments (or ciphertexts) and satisfaction of $z = w + cv$ is proved by appropriately processing the commitments without revealing v or w. More details of this efficient special range proof (e.g. how to set the sizes of the ranges and sets) can be found in Section 1.2.3 of [5], which calls it CFT proof and shows that when the parameters are properly chosen v can be guaranteed to be in R with an overwhelmingly large probability. Obviously, this range proof is a special solution instead of a general-purpose range proof technique. Usage of this method is not easy and liable to many limitations. So its application should be cautious. For example, R must be at least billions of times larger than R'. Moreover, to minimize the information about v revealed from its monotone function[1] z, w must be at least billions of times larger than cv. In addition, for soundness of range proof, the relation between R'' and the other parameters must be delicately designed. As a result, R'' is usually very large and contains extremely large integers. Extra large integers and computation of them in Z without any modulus bring additional cost and should be taken into account in efficiency analysis.

It is easy to notice that ElGamal encryption supports the simple solution $q_1 = q_2 = \ldots = q_n = q$. However, it is not easy to prove that the message in an ElGamal ciphertext is committed to in a commitment as the exponent to a public base. The first PVSS scheme employing ElGamal encryption [23] uses the cut-and-choose strategy and needs to repeat its proof quite a few times to guarantee that the same integer is committed to in C_i and encrypted in c_i with a large enough probability. So it is inefficient. This dilemma in choice of encryption algorithm is partially solved in [21]. On one hand it employs ElGamal encryption

---

[1] Revealing of secret information is inevitable as z is calculated in Z as a monotone function of v.

and sets $q_1 = q_2 = \ldots = q_n = q$ and so avoids range proof of the shares. On the other hand, it generates the secret to share in a special way to avoid the cut-and-choose proof mechanism. However, in the PVSS scheme in [21], the dealer must know the discrete logarithm of the secret to share to a public base and generate the secret from the discrete logarithm. As explained in section 2.1, this special secret generation procedure limits application of the PVSS scheme (e.g. in circumstances where the discrete logarithm of the secret to share is unknown).

1.2 Our Contribution

The analysis above has shown that the RSA and Paillier based PVSS schemes [4,10,17] depend on n instances of range proof and ElGamal-based PVSS schemes [23,21] have their own drawbacks like complex and costly proof mechanism and limited application area. Another point we need to mention is that even the PVSS scheme in [21] is not efficient enough and cost $O(tn)$ in computation. Our task is to overcome all those drawbacks.

In this paper, the PVSS scheme in [21] is modified and optimised to achieve higher efficiency and better applicability. The new PVSS scheme sets $q_1 = q_2 = \ldots = q_n = q$ and thus is inherently free of any range proof. It not only avoids cut-and-choose in proof of validity of secret sharing but also achieves much higher efficiency than any existing PVSS scheme. It only costs $O(n)$ in computation. Moreover, it addresses the problem of limited application caused by the special secret generation mechanism in [21] and provides an alternative solution. Another advantage of our new PVSS scheme over the existing PVSS schemes is that it only needs the basic assumption and a verifier in zero knowledge proof (who generates random challenges and can be replaced by a (pseudo)random function in the random oracle model), both of which are inevitable in any PVSS scheme. In comparison, the existing PVSS schemes need to publish a public commitment to the share-generating polynomial as recalled earlier in this section. As no commitment algorithm is completely assumption-free and any commitment algorithm needs some computational assumption to guarantee that the secret message in the commitment is private and cannot be changed, they may need some additional computational assumption(s)[2].

2 Background and Preliminaries

Background knowledge and security model are given in this section.

2.1 The PVSS Scheme by Schoenmakers

In the PVSS scheme in [21], a dealer employs a special sharing function to share a specially-generated secret, while the other PVSS schemes employs the normal

---

[2] Actually, most of the existing PVSS schemes do need some additional computational assumptions.

sharing function by Shamir [22] and can share any secret in general. Moreover, a corresponding special secret reconstruction function is employed in [21] to reconstruct the secret accordingly. The special sharing function reconstruction function in [21] are described as follows while its public proof of validity of secret sharing follows the principle recalled in Section 1, which is commitment-based and applies to all the existing PVSS.

– Setting

p and q are large primes and q is a factor of $p - 1$. G_q is the cyclic group with order q in Z_p^* and g is a generator of G_q.

– Sharing

1. The dealer firstly chooses δ and then calculates the secret $s = g^\delta$.
2. He builds a polynomial $f(x) = \sum_{j=0}^{t-1} a_j x^j$ with $a_0 = \delta$ and a_j for $j = 1, 2, \ldots, t - 1$ are random integers.
3. He publishes ElGamal ciphertext $c_i = (g^{r_i}, g^{\delta_i} y_i^{r_i})$ for $i = 1, 2, \ldots, n$ where $\delta_i = f(i)$, y_i is P_i's ElGamal public key.

– Reconstruction

1. Each P_i decrypt c_i and obtains $s_i' = g^{\delta_i}$.
2. A set with at least t sharers are put together: $s = \prod_{i \in I} s_i'^{u_i}$ where $u_i = \prod_{j \in I, j \neq i} j/(j - i)$ and I is the set containing the indices of the t shares.

In the PVSS scheme in [21], $a_0 = \delta$, so actually discrete logarithm of the secret s is shared using the share-generating polynomial. Its reconstruction function is accordingly changed to reconstruct s using the shares of δ. So knowledge of discrete logarithm of the secret is compulsory to the dealer and due to hardness of DL problem its discrete logarithm must be fixed before the secret is generated in the PVSS scheme in [21]. Therefore, it is not suitable for some applications. One of the most common applications of PVSS is key sharing (or called distributed key generation in some cryptographic schemes). A secret key is usually chosen from a consecutive interval range instead of a cyclic group in many encryption algorithms (e.g. AES and normal ElGamal) and many other encryption algorithms (e.g. RSA and Paillier) do not first choose a discrete logarithm of the secret key and then calculate the secret key in a cyclic group either. Another important application of PVSS is sharing of password or accessing code in distributed access control. In most applications, a password or accessing code is usually randomly chosen by users and very often the users would like to choose some special password or accessing code like their birthdays or phone number. So it is very probable that discrete logarithm of the password or accessing code is unknown or even does not exist at all. In general, the secret to share cannot be generated in the discrete-logarithm-fixed-first manner in many applications. To guarantee that the secret generated in the discrete-logarithm-fixed-first manner is distributed in a distribution in a set R required by an application, the set $R' = \{x \mid g^x \in R\}$ must be calculated and thus discrete logarithm of the secret can be chosen from it first. However, due to hardness of DL problem, it is often hard to calculate R'.

In summary, although the PVSS scheme in [21] employs ElGamal encryption to avoid range proof and other complex proof operations for the first time and partially solved the dilemma explained in Section 1.1, it still has some drawbacks. Firstly, it still needs $O(tn)$ in computation and is thus not efficient enough. Secondly, its has some limitations in practical application.

2.2 Security Model

Soundness of PVSS is defined in Definition 2, while privacy of PVSS follows the simulation-based general definition of privacy of any proof protocol in Definition 3.

Definition 2 *(Soundness of PVSS). If the dealer's public proof of validity of the shares passes the public verification of a sound PVSS with a non-negligible probability, there exist integers $\alpha_0, \alpha_1, \ldots, \alpha_{t-1}$ such that $s_i = \sum_{j=0}^{t-1} \alpha_j i^j$ for $i = 1, 2, \ldots, n$ where s_i is the ith share.*

Definition 3 *(Privacy of a proof protocol). A proof protocol is private if its transcript can be simulated by a party without any knowledge of any secret such that the simulating transcript is indistinguishable from the real protocol transcript as defined in Definition 4.*

Definition 4 *(Indistinguishability of distributions). Suppose a set of variables have two transcripts respectively with two distributions T_1 and T_2. A random secret bit i is generated. If $i = 0$, two instances of T_2 are published; If $i = 1$, an instance of T_1 and an instance of T_2 are published. An algorithm can distinguish T_1 and T_2 if given the two published instances of transcripts he can calculate i with a probability non-negligibly larger than 0.5.*

3   New PVSS Based on ElGamal Encryption

Our new PVSS protocol employs ElGamal encryption algorithm to maintain consistency of modulus and avoid range proof like in [21] but uses a completely different proof technique. It does not employ any commitment algorithm to avoid additional computational assumption. Of course it still needs the basic assumption, namely semantic security of ElGamal encryption defined in the following.

Definition 5 *(Semantic security of encryption algorithm) A polynomial adversary chooses two messages m_0 and m_1 in the message space of the encryption algorithm in any way he likes and submits them to an encryption oracle, who randomly chooses a bit i and returns the adversary $E(m_i)$ where $E()$ denotes the encryption algorithm. The encryption algorithm is semantically secure if the probability that the adversary obtains i is not non-negligibly larger than 0.5.*

The main idea of the new PVSS technique is simple: given two sets of shares $s_1, s_2, \ldots, s_n$ and $k_1, k_2, \ldots, k_n$ and a random challenge R, if $k_1 + Rs_1, k_2 + Rs_2, \ldots, k_n + Rs_n$ are a set of consistent shares of the same secret, $s_1, s_2, \ldots, s_n$ are a set of consistent shares of the same secret with an overwhelmingly large probability. It is described as follows where the denotations in [21] are inherited and the share holders are $P_1, P_2, \ldots, P_n$.

1. The dealer firstly chooses δ in Z_q and then calculates $s = g^\delta \bmod p$. In this way, he generates an s in G_q such that he knows $\delta = \log_g s$.

2. He builds a polynomial $f_1(x) = \sum_{j=0}^{t-1} \alpha_j x^j$ with $\alpha_0 = \delta$ and α_j for $j = 1, 2, \ldots, t-1$ are random integers chosen from Z_q.

3. He randomly chooses ϵ in Z_q and builds a polynomial $f_2(x) = \sum_{j=0}^{t-1} \beta_j x^j$ with $\beta_0 = \epsilon$ where β_j for $j = 1, 2, \ldots, t-1$ are random integers chosen from Z_q.

4. He publishes $c_i = (g^{r_i} \bmod p, g^{\delta_i} y_i^{r_i} \bmod p)$ and $c_i' = (g^{r_i'} \bmod p, g^{\epsilon_i} y_i^{r_i'} \bmod p)$ for $i = 1, 2, \ldots, n$ where $\delta_i = f_1(i) \bmod q$, $\epsilon_i = f_2(i) \bmod q$ and r_i, r_i' are randomly chosen from Z_q.

5. A random challenge R in Z_q is publicly generated by one or more verifier(s) or a (pseudo)random function like in any publicly verifiable zero knowledge proof protocols (e.g. those zero knowledge proof primitives used in all the existing PVSS schemes). A simple and non-interactive method to publicly generate R is $R = H(c_1, c_1', c_2, c_2', \ldots, c_n, c_n')$ where $H()$ is a (pseudo)random function (e.g. a hash function). As mentioned in Section 1.2, this randomness generation procedure is the same as the randomness generation procedure necessary in any existing PVSS scheme and needs no additional assumption or technique.

6. He publishes $\gamma_j = \beta_j + R\alpha_j \bmod q$ for $j = 0, 1, \ldots, t-1$ and $R_i = Rr_i + r_i' \bmod q$ for $i = 1, 2, \ldots, n$ and any one can publicly verify

$$a_i^R a_i' = g^{R_i} \bmod p \qquad (1)$$
$$b_i^R b_i' = g^{S_i} y_i^{R_i} \bmod p \qquad (2)$$

for $i = 1, 2, \ldots, n$ where $c_i = (a_i, b_i)$, $c_i' = (a_i', b_i')$ and

$$S_i = \sum_{j=0}^{t-1} \gamma_j i^j \bmod q. \qquad (3)$$

7. Share decryption and secret reconstruction are as follows.
   (a) Each P_i decrypts c_i and obtains $s_i = b_i/a_i^{x_i} \bmod p$.
   (b) A set with at least t sharers are put together: $s = \prod_{i \in S} s_i^{u_i} \bmod p$ where $u_i = \prod_{j \in S, j \neq i} j/(j-i) \bmod q$ and S is the set containing the indices of the t shares.

Soundness and privacy of the new ElGamal-based PVSS protocol are proved in Theorem 1 and Theorem 2 respectively, following Definition 2 and Definition 3 respectively.

Theorem 1. *Proof: If Equations (1), (2) and (3) hold for $i = 1, 2, \ldots, n$ with a probability larger than $1/q$, then there exists a t-out-of-n share-generating polynomial $f()$ such that $\log_g D_i(c_i) = f(i) \bmod q$ for $i = 1, 2, \ldots, n$ where $D_i()$ denotes the ElGamal decryption function of P_i.*

Equations (1), (2) and (3) for $i = 1, 2, \ldots, n$ with a probability larger than $1/q$ imply

$$D_i(c_i^R c_i') = g^{S_i} = g^{\sum_{j=0}^{t-1} \gamma_j i^j} \bmod p \text{ for } i = 1, 2, \ldots, n \qquad (4)$$

with a probability larger than $1/q$. So, there must exist two different integers in Z_q, R and R', such that the dealer can produce $\gamma_0, \gamma_1, \ldots, \gamma_{t-1}$ and $\gamma_0', \gamma_1', \ldots, \gamma_{t-1}'$ respectively to satisfy

$$D_i(c_i^R c_i') = g^{\sum_{j=0}^{t-1} \gamma_j i^j} \bmod p \text{ for } i = 1, 2, \ldots, n \qquad (5)$$

$$D_i(c_i^{R'} c_i') = g^{\sum_{j=0}^{t-1} \gamma_j' i^j} \bmod p \text{ for } i = 1, 2, \ldots, n. \qquad (6)$$

Otherwise, there is at most one R in Z_q for the dealer to produce a set of integers $\gamma_0, \gamma_1, \ldots, \gamma_{t-1}$ to satisfy (4) and the probability that (4) is satisfied is no larger than $1/q$, which is a contradiction.

(5) divided by (6) yields

$$D_i(c_i)^{R-R'} = g^{\sum_{j=0}^{t-1} (\gamma_j - \gamma_j') i^j} \bmod p \text{ for } i = 1, 2, \ldots, n.$$

Namely,

$$\log_g D_i(c_i)^{(R-R')} = \sum_{j=0}^{t-1} (\gamma_j - \gamma_j') i^j \bmod q \text{ for } i = 1, 2, \ldots, n$$

and thus

$$\log_g D_i(c_i) = \sum_{j=0}^{t-1} ((\gamma_j - \gamma_j')/(R - R')) i^j \bmod q \text{ for } i = 1, 2, \ldots, n. \qquad \square$$

Theorem 2. *Privacy of the new ElGamal-based PVSS protocol is completely achieved, only dependent on the basic assumption and randomness of the challenge R. More precisely, their privacy is formally provable on the assumption that the employed ElGamal encryption algorithm is semantically secure and R is random.*

Proof: Both protocols have the same PVSS transcript $R, c_1, c_2, \ldots, c_n, c_1', c_2', \ldots, c_n', R_1, R_2, \ldots, R_n, \gamma_0, \gamma_1,$ $\ldots, \gamma_{t-1}$. So their privacy can be universally proved in one simulation. A party without any access to the secret or any of its shares can simulate the PVSS transcript and generate a simulated transcript as follows.

1. $R, \alpha_0, \alpha_1, \ldots, \alpha_{t-1}, \beta_0, \beta_1, \ldots, \beta_{t-1}, r_1, r_2, \ldots, r_n, r_1', r_2', \ldots, r_n'$ are randomly chosen from Z_q.

2. $R_i = r'_i + Rr_i \bmod q$ for $i = 1, 2, \ldots, n$.
3. $\gamma_j = R\alpha_j + \beta_j \bmod q$ for $j = 0, 1, \ldots, t - 1$.
4. $s_i = \sum_{j=0}^{t-1} \alpha_j i^j \bmod q$ for $i = 1, 2, \ldots, n$.
5. $k_i = \sum_{j=0}^{t-1} \beta_j i^j \bmod q$ for $i = 1, 2, \ldots, n$.
6. $c_i = (g^{r'_i} \bmod p, g^{s_i} y_i^{r'_i} \bmod p)$ for $i = 1, 2, \ldots, n$.
7. $c'_i = (g^{r_i} \bmod p, g^{k_i} y_i^{r'_i} \bmod p)$ for $i = 1, 2, \ldots, n$.

In this simulated transcript of c_i, c'_i, R_i for $i = 1, 2, \ldots, n$, $R, \gamma_0, \gamma_1, \ldots, \gamma_{t-1}$,

- each of c_i and c'_i alone is uniformly distributed in the ciphertext space of the employed ElGamal encryption algorithm for $i = 1, 2, \ldots, n$;
- each of $R, \gamma_0, \gamma_1, \ldots, \gamma_{t-1}$ is uniformly distributed in Z_q;
- each R_i is uniformly distributed in Z_q for $i = 1, 2, \ldots, n$;
- $D_1(c_1), D_2(c_2), \ldots, D_n(c_n)$ are shares of an integer uniformly distributed in G_q;
- $D_1(c'_1), D_2(c'_2), \ldots, D_n(c'_n)$ are shares of an integer uniformly distributed in G_q.

In comparison, in the real transcript of c_i, c'_i, R_i for $i = 1, 2, \ldots, n$, $R, \gamma_0, \gamma_1, \ldots, \gamma_{t-1}$ published by the dealer,

- each of c_i and c'_i alone is uniformly distributed in the ciphertext space of the employed ElGamal encryption algorithm for $i = 1, 2, \ldots, n$;
- each of $R, \gamma_0, \gamma_1, \ldots, \gamma_{t-1}$ is uniformly distributed in Z_q;
- each R_i is uniformly distributed in Z_q for $i = 1, 2, \ldots, n$;
- $D_1(c_1), D_2(c_2), \ldots, D_n(c_n)$ are shares of the shared secret s or $g^s \bmod p$, depending on which of the two protocols is referred to;
- $D_1(c'_1), D_2(c'_2), \ldots, D_n(c'_n)$ are shares of an integer uniformly distributed in G_q.

The only difference between the two transcripts lies in distribution of $c_1, c_2, \ldots, c_n$, which are encrypted shares of a random integer in the simulated transcript and encrypted shares of s or $g^s \bmod p$ in the real transcript. The transcript of $c_1, c_2, \ldots, c_n$ in the real ElGamal-based new PVSS protocols is denoted as T_1; while the simulated transcript of $c_1, c_2, \ldots, c_n$ is denoted as T_2. If an algorithm can compromise privacy of the ElGamal-based new PVSS protocols, according to Definition 3, it can distinguish T_1 and T_2 as defined in Definition 4. This algorithm, denoted as A, can be employed to win a game as follows.

1. An adversary sets $m_0 = s$ and randomly chooses m_1 from G_q and submits them to a dealer. The dealer randomly chooses a bit i and shares m_i among the share holders using an ElGamal-based new PVSS protocol. The party needs to calculate i to win the game using the encrypted shares, which is denoted as T'_1. Note that T'_1 and T_1 have the same distribution if $i = 0$; while T'_1 and T_2 have the same distribution if $i = 1$.
2. The adversary randomly chooses an integer from G_q and shares it among the share holders using the ElGamal-based new PVSS protocol. He generates the encrypted shares of this integer, which is denoted as T'_2. T'_2 and T_2 have the same distribution.

3. He inputs T_1' and T_2' to A, which outputs i'. As A can distinguish T_1 and T_2 as defined in Definition 4, with a probability non-negligibly larger than 0.5, it returns $i' = 0$ if $i = 1$ and returns $i' = 1$ if $i = 0$. So $1 - i'$ is a correct solution for i with a probability non-negligibly larger than 0.5.

However, to win the game with a probability non-negligibly larger than 0.5 will leads to a contradiction as follows. For simplicity of proof, suppose $2t \geq n + 1$ and like many existing PVSS descriptions the addition modulus is not explicitly specified. If there is an polynomial algorithm A to win the game above, it is illustrated in the following that this algorithm can be employed to break semantic security of the employed encryption algorithm.

1. The adversary in Definition 5 needs to obtain i where the encryption algorithm is $E()$.
2. He calculates integers $\alpha_0, \alpha_1, \ldots, \alpha_{t-1}, \alpha_0', \alpha_1', \ldots, \alpha_{t-1}'$ and $s_2, s_3, \ldots, s_n$ such that

$$m_0 = \sum_{j=0}^{t-1} \alpha_j$$
$$m_1 = \sum_{j=0}^{t-1} \alpha_j'$$
$$s_i = \sum_{j=0}^{t-1} \alpha_j i^j \text{ for } i = 2, 3, \ldots, n$$
$$s_i = \sum_{j=0}^{t-1} \alpha_j' i^j \text{ for } i = 2, 3, \ldots, n.$$

As $2t \geq n + 1$, he can always find such integers using efficient linear algebra computations.
3. He inputs (α_0, α_0') to A as the two possible secrets to share. He inputs $(c, c_2, c_3, \ldots, c_n)$ to A as the encrypted shares where $c_i = E_i(s_i)$ for $i = 2, 3, \ldots, n$ and each $E_i()$ denotes the same type of encryption algorithm as $E()$ but with a different key.
4. A guesses which secret is shared in the encrypted shares. As A can break semantic security of CCSD of the PVSS protocol, the probability that it gets a correct guess is P, which is non-negligibly larger than 0.5.
5. If A returns 0, the adversary outputs $i = 0$; if A returns 1, he outputs $i = 1$. Note that
   - if $c = E(m_0)$, then $(c, c_2, c_3, \ldots, c_n)$ are encrypted shares of α_0;
   - if $c = E(m_1)$, then $(c, c_2, c_3, \ldots, c_n)$ are encrypted shares of α_0'.
   So the probability that the adversary obtains i is P and non-negligibly larger than 0.5.

Therefore, semantic security of $E()$ is broken. When $2t < n + 1$ the proof can work as well but in a more complex way as sometimes neither of the two possible secrets is shared in the encrypted shares and A has to return a random output. Due to space limit, this extension is left to interested readers.

Since a contradiction is found, T_1 and T_2 cannot be can distinguished in polynomial time. Therefore, if ElGamal encryption is semantical secure, the ElGamal-based new PVSS protocol achieve privacy. $\square$

4 Broader Range of Applications and Batch Verification

Application of our new PVSS technique to a broader range and its efficiency improvement through batch verification are discussed in this section.

4.1 When the Logarithm-Known Secret Generation Mechanism Cannot Work

If s is not a random string to be freely appointed, but a secret with real meaning (e.g. a vote in an e-voting application or a bid in an e-auction system), the PVSS protocol with the logarithm-known secret generation mechanism in last section cannot work. However, a secret with real meaning is usually distributed in a relatively small space. For example, a vote is usually the name of one of the candidates and a bid is usually a biddable price. In this case, searching for discrete logarithm in the small space is not too costly (e.g. using Pollard's Lambda method) and thus our PVSS design can be slightly modified as follows to solve the problem.

1. To share a secret s, the dealer builds a polynomial $f_1(x) = \sum_{j=0}^{t-1} \alpha_j x^j$ with $\alpha_0 = s$ and α_j for $j = 1, 2, \ldots, t-1$ are random integers chosen from Z_q.
2. He randomly builds another polynomial $f_2(x) = \sum_{j=0}^{t-1} \beta_j x^j$ where β_j for $j = 0, 1, \ldots, t-1$ are random integers chosen from Z_q.
3. He publishes $c_i = (g^{r_i} \bmod p, \ g^{s_i} y_i^{r_i} \bmod p)$ and $c'_i = (g^{r'_i} \bmod p, \ g^{k_i} y_i^{r'_i} \bmod p)$ where $s_i = f_1(i) \bmod q$, $k_i = f_2(i) \bmod q$ and r_i, r'_i are randomly chosen from Z_q.
4. A random challenge R in Z_q is generated in the same way as in Section 3.
5. He publishes $\gamma_j = \beta_j + R\alpha_j \bmod q$ for $j = 0, 1, \ldots, t-1$ and $R_i = Rr_i + r'_i \bmod q$ for $i = 1, 2, \ldots, n$ and any one can check Equations (1), (2) and (3) for $i = 1, 2, \ldots, n$.
6. Share decryption and secret reconstruction are as follows.
   (a) Each P_i decrypt c_i and obtains $s'_i = b_i/a_i^{x_i} \bmod p$.
   (b) A set with at least t shares are put together: $s' = \prod_{i \in S} s'^{u_i}_i \bmod p$ where $u_i = \prod_{j \in S, j \neq i} j/(j - i) \bmod q$ and S is the set containing the indices of the t shares.
   (c) $\log_g s'$ is searched for in the space of the secret and the found discrete logarithm is the reconstruction result. As stated before, a space containing practical secret with real meaning is often not too large, so the search is often affordable.

If the secret to share is chosen from a large space and its discrete logarithm is unknown, its discrete logarithm can be calculated in advance to save real-time cost and then the first ElGamal-based PVSS protocol is employed to share the secret.

4.2 PVSS with Explicit Commitment

Although we have shown that in PVSS explicit commitments can be avoided to improve efficiency, some applications of PVSS may need an explicit commitment to the secret, especially when the shared secret has to be processed in the applications without being revealed. In such applications, our new PVSS scheme can be extended to support explicit commitment to the shared secret without compromising high efficiency. It is not necessary to commit to all the coefficients of the share-generating polynomial like in the existing PVSS schemes. Instead, we only commit to the shared secret using a simple commitment mechanism as follows.

1. The secret s is committed to in a public commitment $C = g^s h^\sigma \bmod p$ where σ is randomly chosen from Z_q and h is an integer in G_q such that $\log_g h$ is unknown.
2. When β_0 is employed in PVSS, it is committed to in the same way, namely in a public commitment $C' = g^{\beta_0} h^{\sigma'} \bmod p$ where σ' is randomly chosen in Z_q.
3. When R and γ_0 are published in PVSS, the dealer publishes $\tau = R\sigma + \sigma' \bmod q$ as well. Anyone can publicly verify $C^R C' = g^{\gamma_0} h^\tau \bmod p$ to be ensured that the secret committed in C is shared among the share holders with an overwhelmingly large probability as illustrated in Theorem 3.

Theorem 3. *If Equations (1), (2) and (3) hold for $i = 1, 2, \ldots, n$ and the dealer can calculate $\log_h(C^R C'/g^{\gamma_0})$ with a probability larger than $1/q$, he must have committed to the shared secret in C.*

Proof: Equations (1), (2) and (3) for $i = 1, 2, \ldots, n$ with a probability larger than $1/q$ imply

$$D_i(c_i^R c_i') = g^{S_i} = g^{\sum_{j=0}^{t-1} \gamma_j i^j} \bmod p \text{ for } i = 1, 2, \ldots, n$$

with a probability larger than $1/q$. So, there must exist two different integers in Z_q, R and R', such that the dealer can produce $\gamma_0, \gamma_1, \ldots, \gamma_{t-1}, \tau$ and $\gamma_0', \gamma_1', \ldots, \gamma_{t-1}', \tau'$ respectively to satisfy

$$D_i(c_i^R c_i') = g^{\sum_{j=0}^{t-1} \gamma_j i^j} \bmod p \text{ for } i = 1, 2, \ldots, n \tag{7}$$

$$D_i(c_i^{R'} c_i') = g^{\sum_{j=0}^{t-1} \gamma_j' i^j} \bmod p \text{ for } i = 1, 2, \ldots, n \tag{8}$$

$$C^R C' = g^{\gamma_0} h^\tau \bmod p \tag{9}$$

$$C^{R'} C' = g^{\gamma_0'} h^{\tau'} \bmod p. \tag{10}$$

Otherwise, there is at most one R in Z_q for the dealer to produce $\gamma_0, \gamma_1, \ldots, \gamma_{t-1}, \tau$ to satisfy $D_i(c_i^R c_i') = g^{\sum_{j=0}^{t-1} \gamma_j i^j} \bmod p$ for $i = 1, 2, \ldots, n$ and $C^R C' = g^{\gamma_0} h^\tau \bmod p$ and the probability that he can produce correct responses to satisfy the two equations is no larger than $1/q$, which is a contradiction.

(7)/(8) yields

$$D_i(c_i)^{R-R'} = g^{\sum_{j=0}^{t-1}(\gamma_j - \gamma_j')i^j} \mod p \text{ for } i = 1, 2, \ldots, n$$

and (9)/(10) yields

$$C^{R-R'} = g^{\gamma_0 - \gamma_0'} h^{\tau - \tau'} \mod p.$$

Note that R and R' are different integers in Z_q and q is a prime and so $(R - R')^{-1} \mod q$ exists. Therefore,

$$D_i(c_i) = g^{\sum_{j=0}^{t-1}(\gamma_j - \gamma_j')/(R-R')i^j} \mod p \text{ for } i = 1, 2, \ldots, n$$
$$C = g^{(\gamma_0 - \gamma_0')/(R-R')} h^{(\tau - \tau')/(R-R')} \mod p$$

and thus each $\log_g D_i(c_i)$ is a share generated by polynomial $f(x) = \sum_{j=0}^{t-1}((\gamma_j - \gamma_j')/(R - R'))x^j$ where discrete logarithm of the shared secret, $(\gamma_0 - \gamma_0')/(R - R') \mod N$, is committed to in C. □

4.3   Further Efficiency Improvement by Batch Verification

High efficiency for the dealer is very obvious in our new PVSS scheme. His only exponentiation computation in his proof of validity of his secret sharing includes encryption of k_is, namely n instances of ElGamal encryption, which costs $2n$ exponentiations. It is a great advantage over the existing PVSS schemes as to be detailed in Section 5. However, the computational cost for a verifier is not so efficient. Verification of Equations (1) and (2) for $i = 1, 2, \ldots, n$ costs $5n$ exponentiations, each with an exponent in Z_q. Verification of the two equations can be batched by a verifier using the idea in [3] as follows.

1. n integers $t_1, t_2, \ldots, t_n$ are randomly chosen by the verifier from Z_{2^L} where L is a security parameter such that $2^L < q$.
2. He verifies

$$(\prod_{i=1}^n a_i^{t_i})^R \prod_{i=1}^n a_i'^{t_i} = g^{\sum_{i=1}^n R_i t_i} \mod p \tag{11}$$
$$(\prod_{i=1}^n b_i^{t_i})^R \prod_{i=1}^n b_i'^{t_i} = g^{\sum_{i=1}^n S_i t_i} \prod_{i=1}^n y_i^{R_i t_i} \mod p. \tag{12}$$

This batch verification is a straightforward application of the principle in Theorem 4. Theorem 4 guarantees that if (11) and (12) are satisfied with a probability larger than 2^{-L} then (1) and (2) are satisfied. As explained in [3], the principle of batch verification is employment of small exponents. Bellare $et\ al$ notice that the exponentiation computations in cryptographic operations usually employ very large exponents (hundreds of bits long) and sometimes the exponents are not necessary to be so large. Actually in many practical applications the exponents can be much smaller (e.g. scores of bits long) but still large enough to guarantee very strong security (e.g. to control the probability of failure under one out of billions). So they employ many small exponents in verification of a batch of equations and estimate the computational cost in terms of the number of separate exponentiations with full-length exponents. More precisely, they set the computational cost of an exponentiation with a full-length exponent as the basic unit in efficiency analysis and estimate how many basic units cost the same as their operations[3]. In this way, they can clearly show advantages of batch verification in computational efficiency. More precisely, in our batch verification, 2^L can be much smaller than the integers in Z_q to improve efficiency, while correctness and soundness of the verification can still be guaranteed except for a probability of 2^{-L}.

Theorem 4. *Suppose $H, y_1, y_2, \ldots, y_n$ are in G_q, $t_1, t_2, \ldots, t_n$ are randomly chosen from $\{0, 1, \ldots, 2^L - 1\}$. If $\prod_{i=1}^{n} y_i^{t_i} = H^{\sum_{i=1}^{n} x_i t_i} \bmod p$ with a probability larger than 2^{-L}, then $y_i = H^{x_i} \bmod p$ for $i = 1, 2, \ldots, n$.*

Proof: $\prod_{i=1}^{n} y_i^{t_i} = H^{\sum_{i=1}^{n} x_i t_i} \bmod p$ with a probability larger than 2^{-L} implies that for any given integer v in $\{1, 2, \ldots, n\}$ there must exist integers $t_1, t_2, \ldots, t_n \in \{0, 1, \ldots, 2^L - 1\}$ and $t'_v \in \{0, 1, \ldots, 2^L - 1\}$ such that

$$\prod_{i=1}^{n} y_i^{t_i} = H^{\sum_{i=1}^{n} x_i t_i} \bmod p \tag{13}$$

$$\left(\prod_{i=1}^{v-1} y_i^{t_i}\right) y_v^{t'_v} \prod_{i=v+1}^{n} y_i^{t_i} = H^{\left(\sum_{i=1}^{v-1} x_i t_i\right) + x_v t'_v + \sum_{i=v+1}^{n} x_i t_i} \bmod p. \tag{14}$$

Otherwise, for any $(t_1, t_2, \ldots, t_{v-1}, t_{v+1}, \ldots, t_n)$ in $\{0, 1, \ldots, 2^L - 1\}^{n-1}$, there is at most one t_v in $\{0, 1, \ldots, 2^L - 1\}$ to satisfy $\prod_{i=1}^{n} y_i^{t_i} = H^{\sum_{i=1}^{n} x_i t_i} \bmod p$, which implies that among the 2^{nL} possible choices for $\{t_1, t_2, \ldots, t_n\}$ there are at most $2^{(n-1)L}$ choices to satisfy $\prod_{i=1}^{n} y_i^{t_i} = H^{\sum_{i=1}^{n} x_i t_i} \bmod p$ and leads to a contradiction to the assumption that $\prod_{i=1}^{n} y_i^{t_i} = H^{\sum_{i=1}^{n} x_i t_i} \bmod p$ with a probability larger than 2^{-L}.

(13)/(14) yields

$$y_v^{t_v - \hat{t}_v} = H^{(t_v - \hat{t}_v) x_v} \bmod p.$$

Note that t_v and $\hat{t}_v$ are L-bit integers and $2^L < q$. So $(t_v - \hat{t}_v)^{-1} \bmod q$ exists and thus

$$y_v = H^{x_v} \bmod p.$$

Therefore, $y_i = H^{x_i}$ for $i = 1, 2, \ldots, n$ as v can be any integer in $\{1, 2, \ldots, n\}$. $\square$

With this batch optimisation, computational efficiency of a verifier is greatly improved. For a verifier, the total computational cost includes two full-length exponentiation and four instances of computation of product of n exponentiations with L-bit exponents and one instance of computation of product of n exponentiations with $\log_2 q$-bit exponents. According to [2], computing each of the four instances of product of exponentiations with L-bit exponents costs about $2^{W-1}(n + 1) + L + nL/(W + 1)$ multiplications and computing the product of n exponentiations with $\log_2 q$-bit exponents costs about

---

[3] Namely, multiple exponentiations with small exponents are counted as one exponentiation with a full-length exponent, which has the same cost.

$2^{W-1}(n+1) + |q| + n|q|/(W+1)$ multiplications where $|q|$ is the bit-length of q and W is a parameter in the W-bit-sliding-window exponentiation method and is normally set as 3. When standard W-bit-sliding-window exponentiation method is employed, an exponentiation with a full-length exponent in Z_q costs $2^{W-1} + |q| + |q|/(W+1)$ multiplications. So the computational cost of a verifier is approximately equal to

$$(4(2^{W-1}(n+1) + L + nL/(W+1)) + 2^{W-1}(n+1) + |q| + n|q|/(W+1))$$
$$/(2^{W-1} + |q| + |q|/(W+1)) + 2$$

full-length exponentiations. When $L = 40$, 2^{-L} is smaller than one out of one trillion and thus negligible in any practical application. In this case, when $W = 3$ and $|q|$=1024, the computational cost of a verifier is very low.

5   Conclusion

As stated before, the new PVSS scheme needs no additional condition or assumption as it only needs the most basic assumptions absolutely needed in any PVSS scheme. The new PVSS scheme has advantages over the existing PVSS schemes in both security and efficiency. The extension of applicability in Section 4 shows that it can be widely applied to many distributed secure applications.

References

1. Adida, B., Wikström, D.: How to shuffle in public. In: Vadhan, S.P. (ed.) TCC 2007. LNCS, vol. 4392, pp. 555–574. Springer, Heidelberg (2007)
2. Avanzi, R., Cohen, H., Doche, C., Frey, G., Lange, T., Nguyen, K., Vercauteren, F.: Handbook of Elliptic and Hyperelliptic Curve Cryptography. In: HEHCC (2005)
3. Bellare, M., Garay, J.A., Rabin, T.: Fast batch verification for modular exponentiation and digital signatures. In: Nyberg, K. (ed.) EUROCRYPT 1998. LNCS, vol. 1403, pp. 236–250. Springer, Heidelberg (1998)
4. Boudot, F., Traoré, J.: Efficient publicly verifiable secret sharing schemes with fast or delayed recovery. In: Varadharajan, V., Mu, Y. (eds.) ICICS 1999. LNCS, vol. 1726, pp. 87–102. Springer, Heidelberg (1999)
5. Boudot, F.: Efficient proofs that a committed number lies in an interval. In: Preneel, B. (ed.) EUROCRYPT 2000. LNCS, vol. 1807, pp. 431–444. Springer, Heidelberg (2000)
6. Chandran, N., Ostrovsky, R., Skeith III, W.E.: Public-key encryption with efficient amortized updates. In: Garay, J.A., De Prisco, R. (eds.) SCN 2010. LNCS, vol. 6280, pp. 17–35. Springer, Heidelberg (2010)
7. Damgård, I., Thorbek, R.: Non-interactive proofs for integer multiplication. In: Naor, M. (ed.) EUROCRYPT 2007. LNCS, vol. 4515, pp. 412–429. Springer, Heidelberg (2007)
8. Feldman, P.: A practical scheme for non-interactive verifiable secret sharing. In: FOCS 1987, pp. 427–437 (1987)
9. Fouque, P., Poupard, G., Stern, J.: Sharing decryption in the context of voting or lotteries. In: Frankel, Y. (ed.) FC 2000. LNCS, vol. 1962, pp. 90–104. Springer, Heidelberg (2001)

10. Fujisaki, E., Okamoto, T.: A practical and provably secure scheme for publicly verifiable secret sharing and its applications. In: Nyberg, K. (ed.) EUROCRYPT 1998. LNCS, vol. 1403, pp. 32–46. Springer, Heidelberg (1998)

11. Ge, H., Tate, S.: A direct anonymous attestation scheme for embedded devices. In: Okamoto, T., Wang, X. (eds.) PKC 2007. LNCS, vol. 4450, pp. 16–30. Springer, Heidelberg (2007)

12. Groth, J.: Non-interactive zero-knowledge arguments for voting. In: Ioannidis, J., Keromytis, A.D., Yung, M. (eds.) ACNS 2005. LNCS, vol. 3531, pp. 467–482. Springer, Heidelberg (2005)

13. Juels, A., Catalano, D., Jakobsson, M.: Coercion-resistant electronic elections. In: Chaum, D., Jakobsson, M., Rivest, R.L., Ryan, P.Y.A., Benaloh, J., Kutylowski, M., Adida, B. (eds.) Towards Trustworthy Elections. LNCS, vol. 6000, pp. 37–63. Springer, Heidelberg (2010)

14. Kiayias, A., Yung, M.: Tree-homomorphic encryption and scalable hierarchical secret-ballot elections. In: Sion, R. (ed.) FC 2010. LNCS, vol. 6052, pp. 257–271. Springer, Heidelberg (2010)

15. Küpçü, A., Lysyanskaya, A.: Optimistic fair exchange with multiple arbiters. In: Gritzalis, D., Preneel, B., Theoharidou, M. (eds.) ESORICS 2010. LNCS, vol. 6345, pp. 488–507. Springer, Heidelberg (2010)

16. Paillier, P.: Public-key cryptosystems based on composite degree residuosity classes. In: Stern, J. (ed.) EUROCRYPT 1999. LNCS, vol. 1592, pp. 223–238. Springer, Heidelberg (1999)

17. Peng, K., Bao, F.: Efficient publicly verifiable secret sharing with correctness, soundness and ZK privacy. In: Youm, H.Y., Yung, M. (eds.) WISA 2009. LNCS, vol. 5932, pp. 118–132. Springer, Heidelberg (2009)

18. Peng, K.: Verifiable secret sharing with comprehensive and efficient public verification. In: Li, Y. (ed.) DBSec 2011. LNCS, vol. 6818, pp. 217–230. Springer, Heidelberg (2011)

19. Peng, K.: Impracticality of efficient PVSS in real life security standard (Poster). In: Parampalli, U., Hawkes, P. (eds.) ACISP 2011. LNCS, vol. 6812, pp. 451–455. Springer, Heidelberg (2011)

20. Saxenaa, N., Tsudikb, G., Yic, J.: Threshold cryptography in p2p and manets: The case of access control. In: Computer Networks, vol. 51(12), pp. 3632–3649 (2007)

21. Schoenmakers, B.: A simple publicly verifiable secret sharing scheme and its application to electronic voting. In: Wiener, M. (ed.) CRYPTO 1999. LNCS, vol. 1666, pp. 148–164. Springer, Heidelberg (1999)

22. Shamir, A.: How to share a secret. Communication of the ACM 22(11), 612–613 (1979)

23. Stadler, M.: Publicly verifiable secret sharing. In: Maurer, U.M. (ed.) EUROCRYPT 1996. LNCS, vol. 1070, pp. 190–199. Springer, Heidelberg (1996)

Studying a Range Proof Technique — Exception and Optimisation

Kun Peng[1] and Li Yi[2]

[1] Institute for Infocomm Research, Singapore
[2] Zhongnan University, China
dr.kun.peng@gmail.com

Abstract. A batch proof and verification technique is employed to design efficient range proof with practical small ranges in AFRICACRYPT 2010. It is shown in this paper that the batch proof and verification technique is not always sound in its application to range proof. We demonstrate that their batch proof and verification technique causes a concern such that in some cases a malicious prover without the claimed knowledge may pass the verification. As a result their range proof scheme to prove that a secret committed integer is in an interval range is not so reliable and cannot guarantee that the committed integer is in the range in some special cases. To ease the concern, we employ an efficient membership proof technique to replace the batch proof and verification technique in their range proof scheme and re-design it to achieve the claimed high efficiency with practical small ranges.

1 Introduction

Range proof is an applied cryptographic technique to enable a party to prove that a secret integer is in an interval range. The party chooses an integer from an interval range R, encrypts it or commits to it and publishes the ciphertext or commitment. Then he has to prove that the integer encrypted in the ciphertext or committed in the commitment is in R. The proof cannot reveal any information about the integer except that it is in the range. This proof operation is called range proof. The following security properties must be satisfied in a range proof protocol, while high efficiency is very important as well.

- Correctness: if the integer is in the range and the prover knows the integer and strictly follows the proof protocol, he can pass the verification in the protocol.
- Soundness: if the prover passes the verification in the protocol, the integer is guaranteed with an overwhelmingly large probability to be in the range.
- Privacy: no information about the integer is revealed in the proof except that it is in the range.

The most straightforward range proof technique is ZK (zero knowledge) proof of partial knowledge [6], which proves that the committed integer may be each

A. Youssef, A. Nitaj, A.E. Hassanien (Eds.): AFRICACRYPT 2013, LNCS 7918, pp. 328–341, 2013.

integer in the range one by one and then link the multiple proofs with OR logic. It has a drawback: the number of computations it needs is linear to the size of the range, which leads to very low efficiency. The range proof schemes in [2,8,7,11] improve efficiency of range proof by discarding the cyclic group with public order in [6]. They notice that non-negativity of an integer x is binded in g^x when the order of g is unknown. So in their range proofs they employ commitment of integers in Z instead of the traditional commitment of integers with a modulus. This special commitment function enables them to reduce a range proof in a range R to proof of non-negativity of integers. Although this method improves efficiency, it has two drawbacks. Firstly, its soundness depends on a computational assumption: when the multiplication modulus is a composite hard to factorize multiplication operation generates a large cyclic subgroup, whose order is secret and hard to calculate. So its soundness is only computational. Secondly, it cannot achieve perfect zero knowledge or simulatability. It reveals a statistically small (and intuitively-believed negligible) amount of information about the secret integer and only achieves the co-called statistical zero knowledge.

A special range proof scheme is proposed in [3]. On one hand it recognizes that sacrifice of unconditional soundness and perfect zero knowledge is necessary for high efficiency. On the other hand, it shows that asymptotical efficiency is not always the dominating factor in efficiency analysis. Actually, asymptotical efficiency in range proof is only important for large ranges. As the ranges in most practical applications of range proof are not large, an asymptotically higher cost may be actually lower in practice. So, the range proof scheme in [3] ignores asymptotical efficiency but focuses on the actual cost of range proof in practical small ranges. As a result, although its asymptotical efficiency is not the highest, it achieves the highest actual efficiency in practical small ranges and is more efficient in practical applications than the previous range proof solutions. However, [3] has its drawbacks as well. Besides conditional soundness like in [2,8,7], its has an additional limitation: its privacy depends on hardness of a special mathematical problem called $(\log_k)$-Strong Diffie Hellman assumption. Moreover, its efficiency advantage in practical small ranges is not great enough to dramatically improve efficiency of many applications.

The idea of range proof with actual high efficiency in practical small ranges is inherited by the most recent range proof scheme [10] in AFRICACRYPT 2010. It overcomes the drawbacks of the existing range proof schemes like computational soundness, statistical privacy and additional assumption and is much more efficient than them when applied to practical small ranges. It is based on a new batch proof and verification technique extended from a batch proof and verification protocol in [5]. Batch verification is first proposed by Bellare $et\ al$ [1] and then extended to batch proof and verification by Peng et al [9]. In [5], it is further extended to batch prove and verify multiple knowledge statements containing OR logic. The batch proof and verification protocol in [10] extends a batch proof and verification protocol in [5] and can prove and verify in a batch n instances of knowledge claims, each claiming knowledge of 1-out-of-k secret discrete logarithms. The batch proof and verification technique is claimed to be secure and much more efficient than separately proving and verifying the n instances of knowledge claims. In their new range proof protocol, the committed integer is represented in a k-base coding system so that range proof in a range with width $b - a$ is reduced to n instances of range proof in Z_k where $b - a = k^n$. Then their batch proof and verification technique is employed to batch the n instances of proof, so that efficiency of range proof in practical small ranges is greatly improved.

A concern is raised in this paper for soundness of the batch proof and verification technique in [10]. We show that although their batch proof and verification works in most cases, in some special cases a malicious prover without the claimed knowledge can pass the verification. Then it is illustrated that the concern affects the range proof in [10] and the range proof scheme cannot always guarantee that the committed integer is in the range. To ease the concern, we employ the membership proof technique in [12] to replace the batch proof and verification technique in the range proof scheme in question. The re-designed range proof technique can achieve the claimed high efficiency with practical small ranges in [10] without any compromise in security.

The rest of this paper is organized as follows. In Section 2, the batch proof and verification technique in [10] and its application to range proof are recalled. In Section 3, an exception is found for soundness of their batch proof and verification technique. In Section 4, the exception is shown to affect the range proof scheme in question such that it does not always guarantee a range for the secret integer. In Section 5, an efficient membership proof technique is employed to replace their batch proof and verification technique and re-design their range proof protocol to achieve their claimed efficiency without compromising security. The paper is concluded in Section 6.

2   The Batch Proof Technique and Its Application to Range Proof

In [10], a batch proof and verification protocol is proposed to batch prove and verify n knowledge statements, each of which claims knowledge of at least one of k discrete logarithms. Their proof protocol is recalled in Figure 1 where p, q are primes, $q|p - 1$, G is the cyclic subgroup of Z_p^* with order q and g is a generator of G.

The batch proof and verification protocol in Figure 1 actually proves and verifies knowledge of $\log_g \prod_{i=1}^{n} y_{i,1}^{c_{i,1}}$, $\log_g \prod_{i=1}^{n} y_{i,2}^{c_{i,2}}$, $\ldots$, $\log_g \prod_{i=1}^{n} y_{i,k}^{c_{i,k}}$. It is claimed in [10] that its proof guarantees that with an overwhelmingly large probability the prover knows at least one of $\log_g y_{i,1}$, $\log_g y_{i,2}$, $\ldots$, $\log_g y_{i,k}$ for $i = 1, 2, \ldots, n$.

In [10], the batch proof and verification protocol in Figure 1 is employed to design a range proof scheme. In their design, a secret integer x is represented in a base-k coding system and then is efficiently proved to be in a range $\{a, a + 1, \ldots, b\}$ where k is a parameter smaller than $b - a$. Their main idea is to reduce the range proof to $\log_k(b - a)$ instances of proof, which show that each digit of

Common input: (p, q, g), $\{(y_{i,1}, y_{i,2}, \ldots, y_{i,k})\}_{i=1,2,\ldots,n}$.
Knowledge to prove: $b_i \in \{1, 2, \ldots, k\}$, v_{i,b_i} s.t. $y_{i,b_i} = g^{v_{i,b_i}} \bmod p$ for $i = 1, 2, \ldots, n$.
Denotation: $v_i = \{1, 2, \ldots, b_i - 1, b_i + 1, \ldots, k\}$.

1. The prover randomly selects $r_1, r_2, \ldots, r_k$ from Z_q and $c_{i,j}$ for $i = 1, 2, \ldots, n$ and $j \in v_i$ from Z_q. Then he computes

$$R_1 = g^{r_1} \prod_{1 \leq i \leq n, \ b_i = 1} \prod_{j \in v_i} y_{i,j}^{c_{i,j}} \bmod p$$

$$R_2 = g^{r_2} \prod_{1 \leq i \leq n, \ b_i = 2} \prod_{j \in v_i} y_{i,j}^{c_{i,j}} \bmod p$$

$$\cdots\cdots$$
$$\cdots\cdots$$

$$R_k = g^{r_k} \prod_{1 \leq i \leq n, \ b_i = k} \prod_{j \in v_i} y_{i,j}^{c_{i,j}} \bmod p$$

$$c_i = H(CI\|c_{i-1}\|c_{i-1,1}\|c_{i-1,2}\|\ldots\|c_{i-1,k-1}) \text{ for } i = 1, 2, \ldots, n$$

$$c_{i,b_i} = c_i - \sum_{j \in v_i} c_{i,j} \bmod q \text{ for } i = 1, 2, \ldots, n$$

$$z_1 = r_1 - \sum_{\{i | b_i = 1\}} c_{i,1} v_{i,1} \bmod q$$

$$z_2 = r_2 - \sum_{\{i | b_i = 2\}} c_{i,2} v_{i,2} \bmod q$$

$$\cdots\cdots$$
$$\cdots\cdots$$

$$z_k = r_k - \sum_{\{i | b_i = k\}} c_{i,k} v_{i,k} \bmod q$$

where CI is a bit string comprising common inputs in a certain order and

$$c_0 = R_1$$
$$c_{0,1} = R_2$$
$$c_{0,2} = R_3$$
$$\cdots\cdots$$
$$\cdots\cdots$$
$$c_{0,k-1} = R_k.$$

It then sends
$(z_1, z_2, \ldots, z_k, c_1, c_{1,1}, c_{1,2} \ldots, c_{1,k-1}, c_{2,1}, c_{2,2} \ldots, c_{2,k-1}, \ldots\ldots$
$c_{n,1}, c_{n,2} \ldots, c_{n,k-1})$ to the verifier.

2. The verifier computes

$$c_{i,k} = c_i - \sum_{j=1}^{k-1} c_{i,j} \bmod q \text{ for } i = 1, 2, \ldots, n$$

$$c_i = H(CI\|c_{i-1}\|c_{i-1,1}\|c_{i-1,2}\|\ldots\|c_{i-1,k-1}) \text{ for } i = 1, 2, \ldots, n$$

and verifies

$$c_1 = H(CI\|g^{z_1} \prod_{i=1}^{n} y_{i,1}^{c_{i,1}} \bmod p \| g^{z_2} \prod_{i=1}^{n} y_{i,2}^{c_{i,2}} \bmod p$$

$$\|\ldots\|g^{z_k} \prod_{i=1}^{n} y_{i,k}^{c_{i,k}} \bmod p)$$

Fig. 1. Batch Proof and Verification of knowledge of 1-out-of-k Discrete Logarithms

the base-k representation of $x - a$ is in Z_k. Then the $\log_k(b-a)$ instances of proof can be batched using the batch proof and verification technique in Figure 1 to improve efficiency.

In their range proof scheme, a secret integer x chosen from an interval range $\{a, a+1, \ldots, b\}$ is committed to in $c = g^x h^r \bmod p$ where h is a generator of G, $\log_g h$ is unknown and r is a random integer in Z_q. A party with knowledge of x and r has to prove that the message committed in c is in $\{a, a+1, \ldots, b\}$. The proof protocol and the corresponding verification are as follows.

1. $c' = c/g^a \bmod p$ and the proof that the integer committed in c is in $\{a, a+1, \ldots, b\}$ is reduced to proof that the integer committed in c' is in $\{0, 1, \ldots, b-a\}$.
2. The prover calculates representation of $x - a$ in the base-k coding system $(x_1, x_2, \ldots, x_n)$ to satisfy $x - a = \sum_{i=1}^{n} x_i k^{i-1}$ where for simplicity of description it is assumed $(b - a) = k^n$.
3. The prover randomly chooses $r_1, r_2, \ldots, r_n$ in Z_q and calculates and publishes $e_i = g^{x_i} h^{r_i} \bmod p$ for $i = 1, 2, \ldots, n$.
4. The prover publicly proves that he knows a secret integer $r' = \sum_{i=1}^{n} r_i k^{i-1} - r \bmod q$ such that $h^{r'} c' = \prod_{i=1}^{n} e_i^{k^{i-1}} \bmod p$ using zero knowledge proof of knowledge of discrete logarithm [13].
5. The range proof is reduced to n smaller-scale range proofs: the integer committed in e_i is in Z_k for $i = 1, 2, \ldots, n$. Those n instances of proof can be implemented through n instances of proof of knowledge of 1-out-of-k discrete logarithms

$$KN(\log_h e_i) \vee KN(\log_h e_i/g) \vee KN(\log_h e_i/g^2) \vee \ldots$$
$$\vee KN(\log_h e_i/g^{k-1}) \text{ for } i = 1, 2, \ldots, n \quad (1)$$

where $KN(z)$ denotes knowledge of z.
6. Proof of (1) is implemented through batch proof and verification of knowledge of 1-out-of-k discrete logarithms in Figure 1.

It is claimed in [10] that this range proof technique is correct and sound. It achieves high efficiency when the range is not large and is suitable for applications needing to specify range proofs in practical small ranges.

3   Concern about the Batch Proof Protocol

In this section it is illustrated that with commitment $(y_{i,1}, y_{i,2}, \ldots, y_{i,k})$ for $i = 1, 2, \ldots, n$, $(R_1, R_2, \ldots, R_k)$ and hash-function-generated challenges $c_1, c_2, \ldots, c_n$ knowledge of $\log_g(R_j \prod_{i=1}^{n} y_{i,j}^{c_{i,j}})$ for $j = 1, 2, \ldots, k$ is not always enough to guarantee knowledge of at least one integer in each set $\{\log_g y_{i,1}, \log_g y_{i,2}, \ldots, \log_g y_{i,k}\}$ for $i = 1, 2, \ldots, n$. We show that soundness of this proof mechanism has an exception in some special cases. Firstly, a simple example of the exception is given as follows where $n = 3$ and $k = 2$, whose principle and effectiveness are proved in Theorem 1.

1. A malicious prover only knows $\log_g y_{2,2}$ and $\log_g y_{3,1}$.
2. The prover randomly chooses integers t_1, t_2 in Z_q and $y_{1,1}$, $y_{1,2}$ in G. Then he calculates $y_{2,1} = y_{1,1}^{t_1} \bmod p$ and $y_{3,2} = y_{1,2}^{t_2} \bmod p$.
3. The prover publishes $y_{1,1}, y_{1,2}, y_{2,1}, y_{2,2}, y_{3,1}, y_{3,2}$ and runs the proof protocol in Figure 1 in the case of $n = 3$ and $k = 2$.
4. In the proof protocol, the prover chooses v_1 and v_2 in Z_q and calculates $R_1 = g^{v_1} \bmod p$ and $R_2 = g^{v_2} \bmod p$.
5. The prover must provide c_1, c_2, c_3, $c_{1,1}$, $c_{1,2}$, $c_{2,1}$, $c_{2,2}$, $c_{3,1}$, $c_{3,2}$, $\log_g(R_1 y_{1,1}^{c_{1,1}} y_{2,1}^{c_{2,1}} y_{3,1}^{c_{3,1}})$ and $\log_g(R_2 y_{1,2}^{c_{1,2}} y_{2,2}^{c_{2,2}} y_{3,2}^{c_{3,2}})$ such that $c_1 = c_{1,1} + c_{1,2} \bmod q$, $c_2 = c_{2,1} + c_{2,2} \bmod q$, $c_3 = c_{3,1} + c_{3,2} \bmod q$, $c_1 = H(CI, R_1, R_2)$, $c_2 = H(CI, c_1, c_{1,1})$ and $c_3 = H(CI, c_2, c_{2,1})$ to pass the verification. This is feasible as illustrated in Theorem 1.

Theorem 1. *The malicious prover in the exception above does not need to know any $\log_g y_{i,j}$ other than $\log_g y_{2,2}$ and $\log_g y_{3,1}$ to pass the verification in Figure 1 when $n = 3$ and $k = 2$.*

Proof: As the prover knows $\log_g R_1$ and $\log_g y_{3,1}$ and his operations implies

$$\log_g(R_1 y_{1,1}^{c_{1,1}} y_{2,1}^{c_{2,1}} y_{3,1}^{c_{3,1}}) = \log_g R_1 + \log_g(y_{1,1}^{c_{1,1}} (y_{1,1}^{t_1})^{c_{2,1}}) + \log_g y_{3,1}^{c_{3,1}}$$
$$= \log_g R_1 + \log_g(y_{1,1}^{c_{1,1}} (y_{1,1}^{t_1})^{c_{2,1}}) + c_{3,1} \log_g y_{3,1}$$
$$= \log_g R_1 + (c_{1,1} + t_1 c_{2,1}) \log_g y_{1,1} + c_{3,1} \log_g y_{3,1} \bmod q,$$

the prover knows $\log_g(R_1 y_{1,1}^{c_{1,1}} y_{2,1}^{c_{2,1}} y_{3,1}^{c_{3,1}})$ if $c_{1,1} + t_1 c_{2,1} = 0 \bmod q$.
As the prover knows $\log_g R_1$, $\log_g y_1$ and $\log_g y_{2,1}$ and

$$\log_g(R_2 y_{1,2}^{c_{1,2}} y_{2,2}^{c_{2,2}} y_{3,2}^{c_{3,2}}) = \log_g R_2 + \log_g y_{2,2}^{c_{2,2}} + \log_g(y_{1,2}^{c_{1,2}} (y_{1,2}^{t_2})^{c_{3,2}})$$
$$= \log_g R_2 + c_{2,2} \log_g y_{2,2} + \log_g(y_{1,2}^{c_{1,2}} (y_{1,2}^{t_2})^{c_{3,2}})$$
$$= \log_g R_2 + c_{2,2} \log_g y_{2,2} + +(c_{1,2} + t_1 c_{3,2}) \log_g y_{1,2} \bmod q,$$

the prover knows $\log_g(R_2 y_{1,2}^{c_{1,2}} y_{2,2}^{c_{2,2}} y_{3,2}^{c_{3,2}})$ if $c_{1,2} + t_2 c_{3,2} = 0 \bmod q$.
So the prover can pass the verification if he can calculate $c_{1,1}$, $c_{1,2}$, $c_{2,1}$, $c_{2,2}$, $c_{3,1}$ and $c_{3,2}$ to satisfy

$$c_{1,1} + c_{1,2} = c_1 \bmod q$$
$$c_{2,1} + c_{2,2} = c_2 \bmod q$$
$$c_{3,1} + c_{3,2} = c_3 \bmod q$$
$$c_{1,1} + t_1 c_{2,1} = 0 \bmod q$$
$$c_{1,2} + t_2 c_{3,2} = 0 \bmod q$$
$$c_1 = H(CI\|R_1\|R_2)$$
$$c_2 = H(CI\|c_1\|c_{1,1})$$
$$c_3 = H(CI\|c_2\|c_{2,1})$$

where t_1, t_2 are chosen by him. A solution as follows can calculate such $c_{1,1}$, $c_{1,2}$, $c_{2,1}$, $c_{2,2}$, $c_{3,1}$ and $c_{3,2}$ such that the exception does exist.

1. The prover calculates $c_1 = H(CI||R_1||R_2)$.
2. The prover randomly chooses $c_{1,1}$ in Z_q and calculates $c_{1,2} = c_1 - c_{1,1} \bmod q$.
3. The prover calculates $c_2 = H(CI||c_1||c_{1,1})$.
4. The prover calculates $c_{2,1} = (-c_{1,1})/t_1 \bmod q$ and $c_{2,2} = c_2 - c_{2,1} \bmod q$.
5. The prover calculates $c_3 = H(CI||c_2||c_{2,1})$.
6. The prover calculates $c_{3,2} = (-c_{1,2})/t_2 \bmod q$ and $c_{3,1} = c_3 - c_{3,2} \bmod q$. □

Theorem 1 illustrates that a prover with knowledge of neither $\log_g y_{1,1}$ nor $\log_g y_{1,2}$ can always pass the verification in Figure 1 when $n = 3$ and $k = 2$. When n and k are larger, the exception is more variable and a malicious prover has many concrete implementations to use it to pass the verification in Figure 1 without the claimed knowledge. The following algorithm is an example of the exception as described in Figure 1 in general. It illustrates that a prover with knowledge of only $\log_g y_{n-1,1}$, $\log_g y_{n-1,2}$, $\ldots$, $\log_g y_{n-1,k-1}$, $\log_g y_{n,k}$ and no discrete logarithm of any other $y_{i,j}$ can pass the verification in Figure 1. The algorithm is proved to be effective in Theorem 2.

1. A malicious prover only knows $\log_g y_{n-1,1}$, $\log_g y_{n-1,2}$, $\ldots$, $\log_g y_{n-1,k-1}$, $\log_g y_{n,k}$ and no discrete logarithm of any other $y_{i,j}$.
2. The prover randomly chooses integers $t_{i,1}, t_{i,2}, \ldots, t_{i,k}$ for $i = 2, 3, \ldots, n-2$ and $t_{n-1,k}, t_{n,1}, t_{n,2}, \ldots, t_{n,k-1}$ in Z_q. He randomly chooses $y_{1,1}, y_{1,2}, \ldots, y_{1,k}$ in G. Then he calculates

$$y_{i,j} = y_{1,j}^{t_{i,j}} \bmod p \text{ for } i = 2, 3, \ldots, n-2 \text{ and } j = 1, 2, \ldots, k$$

$$y_{n-1,k} = y_{1,k}^{t_{n-1,k}} \bmod p$$

$$y_{n,j} = y_{1,j}^{t_{n,j}} \bmod p \text{ for } j = 1, 2, \ldots, k-1.$$

3. The prover publishes $y_{i,j}$ for $i = 1, 2, \ldots, n$ and $j = 1, 2, \ldots, k$ and runs the proof protocol in Figure 1.
4. In the proof protocol, the prover chooses $v_1, v_2, \ldots, v_k$ in Z_q and calculates $R_j = g^{v_j} \bmod p$ for $j = 1, 2, \ldots, k$.
5. The prover must provide c_1, c_2, $\ldots, c_n$ and $c_{i,j}$ for $i = 1, 2, \ldots, n$ and $j = 1, 2, \ldots, k$ and $\log_g(R_j \prod_{i=1}^{n} y_{i,j}^{c_{i,j}})$ for $j = 1, 2, \ldots, k$ such that $c_i = \sum_{j=1}^{k} c_{i,j} \bmod q$ for $i = 1, 2, \ldots, n$, $c_i = H(CI||c_{i-1}||c_{i-1,1}||c_{i-1,2}||\cdots||c_{i-1,k-1})$ for $i = 1, 2, \ldots, n$, $c_0 = R_1$ and $c_{0,j} = R_{j+1}$ for $j = 1, 2, \ldots, k-1$ to pass the verification. This is feasible as illustrated in Theorem 2.

Theorem 2. *The malicious prover in the algorithm above does not need to know any $\log_g y_{i,j}$ for $i = 1, 2, \ldots, n$ and $j = 1, 2, \ldots, k$ other than $\log_g y_{n-1,1}$, $\log_g y_{n-1,2}$, $\log_g y_{n-1,k-1}$, $\log_g y_{n,k}$ to pass the verification in Figure 1.*

Proof: As the prover knows $\log_g R_j$, $\log_g y_{n-1,1}$, $\log_g y_{n-1,2}$, $\ldots$, $\log_g y_{n-1,k-1}$, $\log_g y_{n,k}$ and his operations implies

$$\log_g(R_j \prod_{i=1}^{n} y_{i,j}^{c_{i,j}})$$

$$= \log_g R_j + \log_g(y_{1,j}^{c_{1,j}} \prod_{i=2}^{n-2}(y_{1,j}^{t_{i,j}})^{c_{i,j}}) + \log_g y_{n-1,j} + \log_g(y_{1,j}^{t_{n,j}})^{c_{n,j}}$$

$$= \log_g R_j + c_{1,j} \log_g y_{1,j} + \log_g y_{1,j}^{\sum_{i=2}^{n-2} t_{i,j}c_{i,j}} + \log_g y_{n-1,j} + c_{n,j}t_{n,j} \log_g y_{1,j}$$

$$= \log_g R_j + (c_{1,j} + \sum_{i=2}^{n-2} t_{i,j}c_{i,j} + c_{n,j}t_{n,j}) \log_g y_{1,j} + c_{n-1,j} \log_g y_{n-1,j}$$

$$\bmod q \text{ for } j = 1, 2, \ldots, k - 1$$

and

$$\log_g(R_k \prod_{i=1}^{n} y_{i,k}^{c_{i,k}})$$

$$= \log_g R_k + \log_g(y_{1,k}^{c_{1,k}} \prod_{i=2}^{n-1}(y_{1,k}^{t_{i,k}})^{c_{i,k}}) + \log_g y_{n,k}^{c_{n,k}}$$

$$= \log_g R_k + c_{1,k} \log_g y_{1,k} + \log_g y_{1,k}^{\sum_{i=2}^{n-1} t_{i,k}c_{i,k}} + \log_g y_{n,k}^{c_{n,k}}$$

$$= \log_g R_k + (c_{1,k} + \sum_{i=2}^{n-1} t_{i,k}c_{i,k}) \log_g y_{1,k} + c_{n,k} \log_g y_{n,k} \bmod q,$$

the prover knows $\log_g(R_j \prod_{i=1}^{n} y_{i,j}^{c_{i,j}})$ for $j = 1, 2, \ldots, k$ if

$$c_{1,j} + \sum_{i=2}^{n-2} t_{i,j}c_{i,j} + c_{n,j}t_{n,j} = 0 \bmod q \text{ for } j = 1, 2, \ldots, k - 1$$

$$c_{1,k} + \sum_{i=2}^{n-1} t_{i,k}c_{i,k} = 0 \bmod q.$$

So the prover can pass the verification if he can calculate $c_{i,j}$ for $i = 1, 2, \ldots, n$ and $j = 1, 2, \ldots, k$ to satisfy

$$\sum_{j=1}^{k} c_{1,j} = c_1 \bmod q$$

$$\sum_{j=1}^{k} c_{2,j} = c_2 \bmod q$$

$$\ldots \ldots$$

$$\ldots \ldots$$

$$\sum_{j=1}^{k} c_{n,j} = c_n \bmod q$$

$$c_{1,j} + \sum_{i=2}^{n-2} t_{i,j}c_{i,j} + c_{n,j}t_{n,j} = 0 \bmod q \text{ for } j = 1, 2, \ldots, k - 1$$

$$c_{1,k} + \sum_{i=2}^{n-1} t_{i,k}c_{i,k} = 0 \bmod q.$$

$$c_1 = H(CI||R_1||R_2||\ldots||R_k)$$

$$c_2 = H(CI||c_1||c_{1,1}||c_{1,2}, \ldots, c_{1,k-1})$$

$$\ldots \ldots$$

$$\ldots \ldots$$

$$c_n = H(CI||c_{n-1}||c_{n-1,1}||c_{n-1,2}||\ldots||c_{n-1,k-1})$$

where $t_{i,1}, t_{i,2}, \ldots, t_{i,k}$ for $i = 2, 3, \ldots, n - 2$ and $t_{n-1,k}, t_{n,1}, t_{n,2}, \ldots, t_{n,k-1}$ are chosen by him. A solution as follows can calculate such $c_{i,j}$s and thus the algorithm can succeed.

- The prover calculates $c_1 = H(CI||R_1||R_2||\ldots||R_k)$.
- The prover randomly chooses $c_{1,1}, c_{1,2}, \ldots, c_{1,k-1}$ in Z_q and calculates $c_{1,k} = c_1 - \sum_{j=1}^{k-1} c_{1,j} \bmod q$.

- The prover calculates $c_2 = H(CI||c_1||c_{1,1}||c_{1,2}, \ldots, c_{1,k-1})$.
- The prover randomly chooses $c_{2,1}, c_{2,2}, \ldots, c_{2,k-1}$ in Z_q and calculates $c_{2,k} = c_2 - \sum_{j=1}^{k-1} c_{2,j} \bmod q$.
- The prover calculates $c_3 = H(CI||c_2||c_{2,1}||c_{2,2}, \ldots, c_{2,k-1})$.
- The prover randomly chooses $c_{3,1}, c_{3,2}, \ldots, c_{3,k-1}$ in Z_q and calculates $c_{3,k} = c_1 - \sum_{j=1}^{k-1} c_{3,j} \bmod q$.
- $\ldots \ldots$
- $\ldots \ldots$
- The prover calculates $c_{n-2} = H(CI||c_{n-3}||c_{n-3,1}||c_{n-3,2}, \ldots, c_{n-3,k-1})$.
- The prover randomly chooses $c_{n-2,1}, c_{n-2,2}, \ldots, c_{n-2,k-1}$ in Z_q and calculates $c_{n-2,k} = c_{n-2} - \sum_{j=1}^{k-1} c_{n-2,j} \bmod q$.
- The prover calculates $c_{n-1} = H(CI||c_{n-2}||c_{n-2,1}||c_{n-2,2}, \ldots, c_{n-2,k-1})$.
- The prover calculates $c_{n-1,k} = (-c_{1,k} - \sum_{i=2}^{n-2} t_{i,k} c_{i,k})/t_{n-1,k} \bmod q$.
- The prover randomly chooses $c_{n-1,2}, c_{n-1,3}, \ldots, c_{n-1,k-1}$ in Z_q and calculates $c_{n-1,1} = c_{n-1} - \sum_{j=2}^{k} c_{n-1,j} \bmod q$.
- The prover calculates $c_n = H(CI||c_{n-1}||c_{n-1,1}||c_{n-1,2}|| \ldots ||c_{n-1,k-1})$.
- The prover calculates $c_{n,j} = (-c_{1,j} - \sum_{i=2}^{n-2} t_{i,j} c_{i,j})/t_{n,j} \bmod q$ for $j = 1, 2, \ldots, k-1$.
- The prover calculates $c_{n,k} = c_n - \sum_{j=1}^{k-1} c_{n,j} \bmod q$. □

According to Theorem 2, to pass the verification in Figure 1 the prover only needs to know k instances of $\log_g y_{i,j}$ instead of one discrete logarithm in each of the n sets $\{\log_g y_{i,1}, \log_g y_{i,2}, \ldots, \log_g y_{i,k}\}$ for $i = 1, 2, \ldots, n$ as claimed in [10]. Actually, a malicious prover can pass the verification in Figure 1 using knowledge of any k instances of $\log_g y_{i,j}$ on the condition that they are not in the same set. Moreover, the malicious prover can choose more $y_{i,j}$ randomly than in our example. If he likes, he can even choose $y_{i,j}$ for $i = 1, 2, \ldots, n-2$ and $j = 1, 2, \ldots, k$ randomly in G and calculate another k instances of $y_{i,j}$ from more than one set as their functions to pass the verification on the condition that he knows discrete logarithm of the left k instances of $y_{i,j}$.

4 Introducing the Concern to the Range Proof Scheme

In Section 3, it has been illustrated that the batch proof and verification protocol in [10] is not secure and cannot guarantee that the prover has the claimed knowledge. The exception found in Section 3 shows that the batch proof and verification technique is not always sound, as usually the $y_{i,j}$s are generated by the prover, otherwise he does not know the secret witness supposed to help him to pass the verification. However, when injecting the exception into the range proof scheme in [10], we need to notice a special requirement in (1): $y_{i,j} = e_i/g^{j-1} \bmod p$ and thus $y_{i,j} = y_{i,j+1} g \bmod p$. So, as $\log_g h$ is secret, any one including the prover only knows at most one $\log_g y_{i,j}$ for each i in $\{1, 2, \ldots, n\}$. So the prover cannot know $\log_g y_{n-1,1}$, $\log_g y_{n-1,2}$, $\log_g y_{n-1,k-1}$ no matter whether he is malicious or not as assumed in the algorithm in Section 3, which need to be adjusted to work in the range proof scheme as follow.

1. For simplicity of description, suppose a malicious prover only knows $\log_h e_{n-k+1}, \log_h e_{n-k+2}/g, \ldots, \log_h e_n/g^{k-1}$ instead of the knowledge statement in (1).

2. The malicious prover randomly chooses $e_1, e_2, \ldots, e_{n-k}$ from G.

3. He needs to choose $t_{i,j}$ for $i = 1, 2, \ldots, n-k$ and $j = n-k+1, n-k+2, \ldots, n$ to calculate $e_j = \prod_{i=1}^{n-k} e_i^{t_{i,j}} \bmod p$ for $j = n-k+1, n-k+2, \ldots, n$.

4. The prover publishes $y_{i,j}$ for $i = 1, 2, \ldots, n$ and $j = 1, 2, \ldots, k$ and runs the proof protocol in Figure 1.

5. In the proof protocol, the prover chooses $v_1, v_2, \ldots, v_k$ in Z_q and calculates $R_j = g^{v_j} \bmod p$ for $j = 1, 2, \ldots, k$.

6. The prover must provide $c_1, c_2, \ldots, c_n$ and $c_{i,j}$ for $i = 1, 2, \ldots, n$ and $j = 1, 2, \ldots, k$ and $\log_g(R_j \prod_{i=1}^{n} y_{i,j}^{c_{i,j}})$ for $j = 1, 2, \ldots, k$ such that $c_i = \sum_{j=1}^{k} c_{i,j} \bmod q$ for $i = 1, 2, \ldots, n$, $c_i = H(CI||c_{i-1}||c_{i-1,1}||c_{i-1,2}||\cdots||c_{i-1,k-1})$ for $i = 1, 2, \ldots, n$, $c_0 = R_1$ and $c_{0,j} = R_{j+1}$ for $j = 1, 2, \ldots, k-1$ to pass the verification where $y_{i,j} = e_i/g^{j-1} \bmod p$. This is feasible using the same method proposed in Theorem 1 and Theorem 2 due to a simple reason: although more equations are needed to satisfy in this algorithm to guarantee $y_{i,j} = e_i/g^{j-1} \bmod p$ we have $(n-k)k + nk$ integers (including $(n-k)k$ instances of $t_{i,j}$ and nk instances of $c_{i,j}$) to choose and only $k + k(k-1)$ linear equations to satisfy. According to the principle of linear algebra, we can always find $t_{i,j}$s and $c_{i,j}$s to pass the batch proof and verification and lead to an exception for the range proof scheme.

5   How to Ease the Concern in Efficient Range Proof in Practical Small Ranges

Although the range proof scheme in [10] has an exception in soundness, its idea of committing to the secret integer in a k-base coding system when proving it to be in a small range is useful. We can base the idea on an efficient membership proof technique instead of the original batch proof and verification protocol in [10]. We commit to the secret integer in n commitments, each of which contains an integer in Z_k. Then we can prove that each commitment really contains an integer in Z_k using membership proof.

5.1   The Membership Proof Technique in [12]

Membership proof is a cryptographic primitive to prove that a secret committed message m is in a finite set $S = \{s_1, s_2, \ldots, s_k\}$. We notice that the membership proof technique in [12] can prove that a committed secret integer is not a member of a set at a cost of $O(\sqrt{k})$ where k is the size of the set. It is recalled as follows where for simplicity of description it is supposed that S can be divided into μ subsets $S_1, S_2, \ldots, S_\mu$ and each S_t contains ν integers $s_{t,1}, s_{t,2}, \ldots, s_{t,\nu}$.

1. For each S_t the ν-rank polynomial $F_t(x) = \prod_{i=1}^{\nu}(x - s_{t,i}) \bmod q$ is expanded into

$$F_t(x) = \sum_{i=0}^{\nu} a_{t,i} x^i \bmod q$$

to obtain the $\nu+1$ coefficients of the polynomial $a_{t,0}, a_{t,1}, \ldots, a_{t,\nu}$. Therefore, functions $F_t(x) = \sum_{i=0}^{\nu} a_{t,i} x^i$ for $t = 1, 2, \ldots, \mu$ are obtained, each to satisfy

$$F_t(s_{t,i}) = 0 \text{ for } i = 1, 2, \ldots, \nu.$$

2. The prover calculates $c_i = c_{i-1}^m h^{r_i} \bmod p$ for $i = 2, 3, \ldots, \nu$ where $c_1 = c$ and r_i is randomly chosen from Z_q. The prover gives a zero knowledge proof that he knows m, r and r_i for $i = 2, 3, \ldots, \nu$ such that $c = g^m h^r \bmod p$ and $c_i = c_{i-1}^m h^{r_i} \bmod p$ for $i = 2, 3, \ldots, \nu$ using a simple combination of ZK proof of knowledge of discrete logarithm [13] and ZK proof of equality of discrete logarithms [4].

3. The prover proves that he knows $\log_h u_1$ or $\log_h u_2$ or $\ldots\ldots$ or $\log_h u_\mu$ using ZK proof of partial knowledge [6] where u_t can be publicly defined as

$$u_t = \prod_{i=0}^{\nu} c_i^{a_{t,i}} \bmod p$$

where $c_1 = c$ and $c_0 = g$. Actually the prover himself can calculate u_t more efficiently:

$$u_t = \begin{cases} h^{\sum_{i=1}^{\nu} a_{t,i} R_i} \bmod p & \text{if } m \in S_t \\ g^{\sum_{i=0}^{\nu} a_{t,i} m^i} h^{\sum_{i=1}^{\nu} a_{t,i} R_i} \bmod p & \text{if } m \notin S_t \end{cases} \tag{2}$$

where $R_i = m R_{i-1} + r_i \bmod q$ for $i = 2, 3, \ldots, \nu$ and $R_1 = r$.

4. Any verifier can publicly verify the prover's two zero knowledge proofs. He accepts the membership proof iff they are passed.

An interesting observation is that a verifier actually does not need to calculate

$$u_t = \prod_{i=0}^{\nu} c_i^{a_{t,i}} \bmod p \text{ for } t = 1, 2, \ldots, \mu$$

as it is costly. Instead, he only needs to verify validity of $u_1, u_2, \ldots, u_\mu$ calculated by the prover (through (2)) as follows.

1. He randomly chooses integers $\tau_1, \tau_2, \ldots, \tau_\mu$ from Z_q.
2. He verifies

$$\prod_{t=1}^{\mu} u_t^{\tau_t} = \prod_{i=0}^{\nu} c_i^{\sum_{t=1}^{\mu} \tau_t a_{t,i}} \bmod p,$$

which only costs $O(\mu + \nu)$ exponentiations.

This membership proof costs $O(\mu + \nu)$ in both computation (in terms of exponentiations) and communication (in terms of transfered integers). So it reduces the cost of general membership proof to $O(\sqrt{k})$ as $k = \mu\nu$. Its soundness is illustrated in Theorem 3.

Theorem 3. *The new membership proof is sound and the probability that the prover can pass its verification is negligible if $m \neq s_i$ mod q for $i = 1, 2, \ldots, k$.*

Proof: Suppose the prover commits to m in c where $m \neq s_i$ mod q for $i = 1, 2, \ldots, n$. If he passes the verification in the new membership proof with a non-negligible probability, it is guaranteed with a non-negligible probability that

$$c = g^m h^r \bmod p \tag{3}$$

$$c_i = c_{i-1}^m h^{r_i} \bmod p \text{ for } i = 2, 3, \ldots, \nu \tag{4}$$

where $c_1 = c$ and $c_0 = g$. As he passes the verification in the new membership proof with a non-negligible probability, there exists t in $\{1, 2, \ldots, \mu\}$ such that the prover knows R such that

$$h^R = \prod_{i=0}^{\nu} c_i^{a_{t,i}} \bmod p \tag{5}$$

with a non-negligible probability.

(3), (4) and (5) imply

$$h^R = g^{\sum_{i=0}^{\nu} a_{t,i} m^i} h^{\sum_{i=1}^{\nu} a_{t,i} R_i} \bmod p$$

where $R_i = m R_{i-1} + r_i$ mod q for $i = 2, 3, \ldots, \nu$ and $R_1 = r$. Namely

$$g^{\sum_{i=0}^{\nu} a_{t,i} m^i} h^{(\sum_{i=1}^{\nu} a_{t,i} R_i) - R} = 1 \bmod p.$$

So

$$\sum_{i=0}^{\nu} a_{t,i} m^i = 0 \bmod q \tag{6}$$

with a non-negligible probability as the employed commitment algorithm is binding and $g^0 h^0 = 1$. Note that $a_{t,0}, a_{t,1}, \ldots, a_{t,\nu}$ satisfy

$$\sum_{i=0}^{\nu} a_{t,i} s^{t,i} = 0 \bmod q \text{ for } i = 1, 2, \ldots, \nu. \tag{7}$$

So (6) and (7) imply

$$\begin{pmatrix} s_{t,1} & s_{t,1}^2 & \cdots & s_{t,1}^{\nu} \\ s_{t,2} & s_{t,2}^2 & \cdots & s_{t,2}^{\nu} \\ \cdots & \cdots & \cdots & \cdots \\ \cdots & \cdots & \cdots & \cdots \\ s_{t,\nu} & s_{t,\nu}^2 & \cdots & s_{t,\nu}^{\nu} \\ m & m^2 & \cdots, & m^{\nu} \end{pmatrix} \begin{pmatrix} a_{t,1} \\ a_{t,2} \\ \cdots \\ \cdots \\ \cdots \\ a_{t,\nu} \end{pmatrix} = \begin{pmatrix} -a_{t,0} \\ -a_{t,0} \\ \cdots \\ \cdots \\ -a_{t,0} \\ -a_{t,0} \end{pmatrix} \tag{8}$$

with a non-negligible probability. However, as $m \neq s_i$ mod q for $i = 1, 2, \ldots, k$ and all the calculations in the matrix is performed modulo q, $\begin{pmatrix} s_{t,1} & s_{t,1}^2 & \cdots & s_{t,1}^{\nu} \\ s_{t,2} & s_{t,2}^2 & \cdots & s_{t,2}^{\nu} \\ \cdots & \cdots & \cdots & \cdots \\ \cdots & \cdots & \cdots & \cdots \\ s_{t,\nu} & s_{t,\nu}^2 & \cdots & s_{t,\nu}^{\nu} \\ m & m^2 & \cdots, & m^{\nu} \end{pmatrix}$ is a non-singular matrix and thus (8) is not satisfied.

Therefore, a contradiction is found and the probability that a prover can pass the verification in the new membership proof must be negligible if the integer he commits to in c is not in S.   □

5.2   Range Proof Employing k-Base Coding and the Membership Proof

With the efficient membership proof present in Section 5.1, the k-base coding system in [10] can be inherited to design an efficient range proof protocol for practical small ranges as follows.

1. A party commits to a secret integer x in $c = g^x h^r \bmod p$ where h is a generator of G, $\log_g h$ is unknown and r is a random integer in Z_q. He then needs to prove that the message committed in c is in an interval range $\{a, a+1, \ldots, b\}$.
2. $c' = c/g^a \bmod p$ and the proof that the integer committed in c is in $\{a, a+1, \ldots, b\}$ is reduced to proof that the integer committed in c' is in $\{0, 1, \ldots, b-a\}$.
3. The prover calculates representation of $x - a$ in the base-k coding system $(x_1, x_2, \ldots, x_n)$ to satisfy $x - a = \sum_{i=1}^{n} x_i k^{i-1}$ where for simplicity of description it is assumed $(b - a) = k^n$.
4. The prover randomly chooses $r_1, r_2, \ldots, r_n$ in Z_q and calculates and publishes $e_i = g^{x_i} h^{r_i} \bmod p$ for $i = 1, 2, \ldots, n$.
5. The prover publicly proves that he knows a secret integer $r' = \sum_{i=1}^{n} r_i k^{i-1} - r \bmod q$ such that $h^{r'} c' = \prod_{i=1}^{n} e_i^{k^{i-1}} \bmod p$ using zero knowledge proof of knowledge of discrete logarithm [13].
6. The range proof is reduced to n smaller-scale membership proofs: the integer committed in e_i is in $\{0, 1, \ldots, k-1\}$ for $i = 1, 2, \ldots, n$. Those n instances of membership proof can be implemented through proof of

$$OPEN(e_i) = 0 \vee OPEN(e_i) = 1 \vee \ldots OPEN(e_i) = k - 1 \qquad (9)$$

   for $i = 1, 2, \ldots, n$ where $OPEN(z)$ denotes the opening to commitment z.
7. Proof of (9) is implemented through the efficient membership proof technique present in Section 5.1.

As explained in [10], such a design is very efficient for practical small ranges. For example, when $l = 10$, we can set $k = 4$ and $n = 2$ and it only costs $4\sqrt{k}n$ exponentiations and achieves high efficiency in practice. Correctness of the new range proof protocol is obvious and any reader can follow it step by step for verification. Its soundness depends on soundness of the employed membership proof protocol, which has been formally proved in Theorem 3. It is private as it employs standard zero knowledge proof primitives.

6   Conclusion

The batch proof and verification technique and its application to range proof in [10] have an exception in soundness and cause a security concern. Fortunately, after our modification, the range proof technique can still work securely and achieve the claimed high efficiency with practical small ranges.

References

1. Bellare, M., Garay, J.A., Rabin, T.: Fast batch verification for modular exponentiation and digital signatures. In: Nyberg, K. (ed.) EUROCRYPT 1998. LNCS, vol. 1403, pp. 236–250. Springer, Heidelberg (1998)
2. Boudot, F.: Efficient proofs that a committed number lies in an interval. In: Preneel, B. (ed.) EUROCRYPT 2000. LNCS, vol. 1807, pp. 431–444. Springer, Heidelberg (2000)
3. Camenisch, J.L., Chaabouni, R., Shelat, A.: Efficient protocols for set membership and range proofs. In: Pieprzyk, J. (ed.) ASIACRYPT 2008. LNCS, vol. 5350, pp. 234–252. Springer, Heidelberg (2008)
4. Chaum, D., Pedersen, T.: Wallet databases with observers. In: Brickell, E.F. (ed.) CRYPTO 1992. LNCS, vol. 740, pp. 89–105. Springer, Heidelberg (1993)
5. Chida, K., Yamamoto, G.: Batch processing for proofs of partial knowledge and its applications. Ieice Trans. Fundamentals E91CA(1), 150–159 (2008)
6. Cramer, R., Damgård, I.B., Schoenmakers, B.: Proof of partial knowledge and simplified design of witness hiding protocols. In: Desmedt, Y.G. (ed.) CRYPTO 1994. LNCS, vol. 839, pp. 174–187. Springer, Heidelberg (1994)
7. Groth, J.: Non-interactive zero-knowledge arguments for voting. In: Ioannidis, J., Keromytis, A.D., Yung, M. (eds.) ACNS 2005. LNCS, vol. 3531, pp. 467–482. Springer, Heidelberg (2005)
8. Lipmaa, H.: On diophantine complexity and statistical zero-knowledge arguments. In: Laih, C.-S. (ed.) ASIACRYPT 2003. LNCS, vol. 2894, pp. 398–415. Springer, Heidelberg (2003)
9. Peng, K., Boyd, C.: Batch zero knowledge proof and verification and its applications. ACM TISSEC 10(2), Article No. 6 (May 2007)
10. Peng, K., Bao, F.: Batch range proof for practical small ranges. In: Bernstein, D.J., Lange, T. (eds.) AFRICACRYPT 2010. LNCS, vol. 6055, pp. 114–130. Springer, Heidelberg (2010)
11. Peng, K., Bao, F.: An Efficient Range Proof Scheme. In: IEEE PASSAT 2010, pp. 826–833 (2010)
12. Peng, K.: A General, Flexible And Efficient Proof Of Inclusion And Exclusion. In: Kiayias, A. (ed.) CT-RSA 2011. LNCS, vol. 6558, pp. 33–48. Springer, Heidelberg (2011)
13. Schnorr, C.: Efficient signature generation by smart cards. Journal of Cryptology 4, 161–174 (1991)

Key-Leakage Resilient Revoke Scheme Resisting Pirates 2.0 in Bounded Leakage Model

Duong Hieu Phan and Viet Cuong Trinh

Université Paris 8, LAGA, CNRS, (UMR 7539), Université Paris 13,
Sorbonne Paris Cité, 2 rue de la Liberté, F-93526 Saint-Denis, Cedex

Abstract. Trace and revoke schemes have been widely studied in theory and implemented in practice. In the first part of the paper, we construct a fully secure key-leakage resilient identity-based revoke scheme. In order to achieve this goal, we first employ the dual system encryption technique to directly prove the security of a variant of the BBG − WIBE scheme under known assumptions (and thus avoid a loss of an exponential factor in hierarchical depth in the classical method of reducing the adaptive security of WIBE to the adaptive security of the underlying HIBE). We then modify this scheme to achieve a fully secure key-leakage resilient WIBE scheme. Finally, by using a transformation from a WIBE scheme to a revoke scheme, we propose the first fully secure key-leakage resilient identity-based revoke scheme.

In the classical model of traitor tracing, one assumes that a traitor contributes its entire secret key to build a pirate decoder. However, new practical scenarios of pirate has been considered, namely Pirate Evolution Attacks at Crypto 2007 and Pirates 2.0 at Eurocrypt 2009, in which pirate decoders could be built from sub-keys of users. The key notion in Pirates 2.0 is the anonymity level of traitors: they can rest assured to remain anonymous when each of them only contributes a very small fraction of its secret key by using a public extraction function. This scenario encourages dishonest users to participate in collusion and the size of collusion could become very large, possibly beyond the considered threshold in the classical model. In the second part of the paper, we show that our key-leakage resilient identity-based revoke scheme is immune to Pirates 2.0 in some special forms in bounded leakage model. It thus gives an interesting and rather surprised connection between the rich domain of key-leakage resilient cryptography and Pirates 2.0.

Keywords: Pirates 2.0, Leakage-resilience, wildcards, revocation.

1 Introduction

In a system of secure distribution of digital content, a center broadcasts encrypted content to legitimate recipients. *Broadcast encryption systems*, independently introduced by Berkovits [5] and Fiat-Naor [17], enable a center to encrypt a message for any subset of legitimate users while preventing any set of revoked users from recovering the broadcasted information. Moreover, even if all revoked users collude,

A. Youssef, A. Nitaj, A.E. Hassanien (Eds.): AFRICACRYPT 2013, LNCS 7918, pp. 342–358, 2013.
© Springer-Verlag Berlin Heidelberg 2013

they are unable to obtain any information about the content sent by the center. *Traitor tracing schemes*, introduced in [9], enable the center to trace users who collude to produce pirate decoders. *Trace and Revoke systems* [26,25] provide the functionalities of both broadcast encryption and traitor tracing.

In the classical model of tracing traitors, one assumes that a traitor contributes its entire secret key to build a pirate decoder. However, new practical scenarios of pirate has been considered, namely Pirate Evolution Attacks [21] and Pirates 2.0 [6], in which pirate decoders could be built from sub-keys of users. The notion of anonymity has been put forth in Pirates 2.0 and it is shown that if each user only contributes a very small fraction of its secret information by using a public extraction function, he can rest assured to remain anonymous. This scenario encourages dishonest users to participate in collusion and the size of collusion could becomes very large, beyond the considered threshold in the classical model.

Leakage resilient cryptography has been a very rich domain of research in the recent years, a non-exhaustive list of works can be found in [18,24,12,8,15,16] [30,27,19,4,7,22]. Under this framework, in the security game the adversary chooses an efficiently computable leakage function and learn the output of this function applied to the secret key and possibly other internal state information at specified moments.

1.1 Contribution

Construction. We first formalize the key-leakage resilient security for a revoke scheme, which enhances its classical security model, we then propose a concrete construction of key-leakage resilient revoke scheme. Our construction is based on the identity-based encryption with wildcards (WIBE) [2,1] in the similar way to [28], it turns out that we need to construct a key-leakage resilient WIBE which is inspired from the work of [22], and is achieved in successive steps:

- The security of a key-leakage resilient WIBE generalizes the full security of a WIBE by allowing the adversary to make additional leak queries. Our first step is then to construct an efficient fully secure WIBE. Fortunately, with the recent dual system encryption technique in [31] and changing the distribution of exponent of $\mathbb{G}_{p_2}$ part in the semi-functional key, we can construct a variant of the Boneh-Boyen-Goh's WIBE (BBG − WIBE) [2] scheme that is fully secure with a very efficient reduction that avoids a loss of an exponential factor in hierarchical depth as in the classical method of reducing the full security of WIBE to the full security of the underlying HIBE in [2].
- Inspired by the security proof technique of the key-leakage resilient HIBE in [22], our second step is to transform this variant of fully secure BBG − WIBE to a secure key-leakage resilient WIBE.

Fighting Pirates 2.0. We first define Pirates 2.0 attacks in bounded leakage model, and show that all existed methods for fighting Pirates 2.0 [10,11,28,32] only consider a particular form of Pirates 2.0 in bounded leakage model. We then present a theoretical result in which any key-leakage resilient revoke scheme satisfying the following conditions will resist Pirates 2.0 in bounded leakage model:

- any user's secret key is a high independent source, *i.e.*, it has a high entropy even under the condition that all the keys of the others users are known.
- resilience to a sufficient high level of leakage at secret keys of users.

Intuitively, the first condition assures that the secret keys of users are sufficiently independent each from the others and the second condition implies that the users should contribute a high information about its key to produce an useful decoder. Combining the two conditions, the users have to contribute high information of their own independent sources and thus lose their anonymity.

Finally, we prove that our key-leakage resilient identity-based revoke scheme resists Pirates 2.0 in bounded leakage model.

1.2 Related Works

Public key trace and revoke scheme [13] is the first paper which showed how IBE/HIBE can be used for broadcast encryption. *Identity-based traitor tracing scheme* was proposed by Abdalla et al [3] in which one can distribute content to various groups of users by taking as input the identity of the targeted group. *Identity-based trace and revoke schemes* (IDTR) in [28] extended this model to allow the center to be capable of revoking any subgroup of users.

Identity-based encryption with wildcards (or WIBE for short) was proposed by Abdalla *et al* [2] and can be seen as a generalization of HIBE. This primitive is related to broadcast encryption in the sense that the encryption is targeted to a group of users rather than to only one user. However, the targeted set of users in WIBE follows a pre-determined structure while a broadcast encryption should be able to target arbitrary group of users. Naturally, WIBE could then be used as a sub-structure to construct trace and revoke systems. This approach has been used in different ways, namely under the code-based framework [3,32], and under the tree-based framework [28]. Our construction is under the tree-based framework as in [28] but with a key-leakage resilient WIBE.

2 Key-Leakage Resilient Revoke Scheme

2.1 Definition

The definition of a key-leakage resilient revoke scheme is the same as a classical revoke scheme. Formally, it consists of four polynomial-time algorithms (**Setup, Keyder, Enc, Dec**):

Setup$(1^k, N_u)$: Takes as inputs the security parameter 1^k and the number of users N_u. This algorithm generates a master public key mpk and a master secret key msk.

Keyder(msk, i): Takes as inputs an indices i of user and the master secret key msk, the key extraction algorithm generates a user secret key SK_i.

Enc(mpk, $\mathcal{R}$, M): The encryption algorithm which on inputs of the master public key mpk, a revocation list $\mathcal{R}$ of revoked users in the scheme, and a message M outputs a ciphertext C.

Dec(SK_i, C): The decryption algorithm which on input of a user secret key SK_i and a ciphertext C outputs a plaintext message M, or $\perp$ to indicate a decryption error.

For correctness we require that **Dec**(SK_i, **Enc**(mpk, $\mathcal{R}$, M)) $= M$ with probability one for all $i \in \mathbb{N} \setminus \mathcal{R}$, $M \in \{0,1\}^*$, (mpk, msk) $\overset{\$}{\leftarrow}$ **Setup**(1^k, N_u) and $SK_i \overset{\$}{\leftarrow}$ **Keyder**(msk, i).

2.2 Security Model

We now present the security model for a (ℓ_{SK})-key-leakage resilient revoke scheme in bounded leakage model (each user leaks maximum ℓ_{SK} bits on his secret key SK).

Setup: The challenger takes as inputs a parameter k, a maximum number of users N_u and runs $setup(1^k, N_u)$ algorithm. The master public key mpk is passed to the adversary. Also, it sets the set of revoked users $\mathcal{R} = \emptyset$, $\mathcal{T} = \emptyset$, note that $\mathcal{R} \subseteq \mathcal{I}$, and $\mathcal{T} \subseteq \{\mathcal{I} \times \mathcal{SK} \times \mathcal{N}\}$ (users indices - secret key of users - leaked bits).

Phase 1: The adversary can adaptively request three types of query:

- **Create**(i): The challenger makes a call to **Keyder**(msk, i) $\rightarrow SK_i$ and adds the tuple $(i, SK_i, 0)$ to the set $\mathcal{T}$ if the indices i does not exists in $\mathcal{T}$.
- **Leak**(i, f) The challenger first finds the tuple (i, SK_i, L), then it checks if $L+ \mid f(SK_i)\mid \leq \ell_{SK}$. If true, it responds with $f(SK_i)$ and updates the $L = L+\mid f(SK_i)\mid$. If the checks fails, it returns $\perp$ to the adversary.
- **Reveal**(i): The challenger first finds the tuple (i, SK_i, L), then responds with SK_i and adds the indices i to the set $\mathcal{R}$.

Challenge: The adversary submits two equal length messages M_0, M_1. The challenger picks a random bit $b \in \{0,1\}$ and set $C = \text{Encrypt}(\text{msk}, \mathcal{R}, M_b)$. The ciphertext C is passed to the adversary.

Phase 2: This is identical to phase 1 except that the adversary is not allowed to ask **Reveal**(i) query in which $i \notin \mathcal{R}$.

Guess: The adversary outputs a guess b' and wins the game if $b' = b$.

Definition 1. *A revoke scheme is (ℓ_{SK})-key-leakage resilient secure if all probabilistic polynomial-time adversaries (called PPT adversaries for short) have at most a negligible advantage in winning the above security game.*

3   A Construction of Key-Leakage Resilient Revoke Scheme - KIDTR

3.1 Definition And Security Model

In [28], they proposed a generic construction of identity-based trace and revoke scheme - IDTR by integrating a WIBE scheme into a complete subtree scheme [25].

In our construction, KIDTR is the same as IDTR in [28] except we use a key-leakage resilient WIBE scheme instead of WIBE for encryption. Therefore, the definition and the security model of KIDTR follow closely to the ones in IDTR in [28], note that in the security model of KIDTR the adversary can ask leakage queries on all secret keys. We refer the definition and the security model of KIDTR to the full version of this paper [29].

The rest of section is now devoted to construct a key-leakage resilient revoke scheme. The construction is achieved via the following steps:

1. we first propose a variant of BBG − WIBE scheme which is proven fully secure by using the dual system encryption technique.
2. we then construct a key-leakage resilient BBG − WIBE scheme by employing the proof technique in [22] to the above BBG − WIBE.
3. we finally apply the generic transformation from a WIBE to an identity based trace and revoke scheme (denoted IDTR) in [28]. This results to a key-leakage resilient identity-based revoke scheme (denoted KIDTR).

3.2   BBG − WIBE in Composite Order Groups

In [23], Lewko and Waters apply the dual system encryption technique to prove the full security of the BBG − HIBE scheme. This technique first splits the security game into $q + 5$ games where q is the maximum number of queries that adversary makes. The first game is the real BBG − HIBE security game and the final game gives no advantage for the adversary. Second, based on the three complexity assumptions $1, 2, 3$ in [23], step by step they prove that these games are indistinguishable, this automatically avoids a loss of an exponential factor in hierarchical depth as in the classical method. This is achieved via the main concept of the nominal semi-functionality, in which a semi-functional key is *nominal* with a semi-functional ciphertext if the semi-functional key can decrypt the semi-functional ciphertext. If the challenger only can create a *nominal* semi-functional ciphertext with respect to the semi-functional challenge key, then he cannot test by himself whether the challenge key is semi-functional or not because the decryption always successes.

We follow their approach by applying the dual system encryption technique to construct a fully secure variant of the BBG − WIBE scheme. The problem here is that the transformation from the BBG − HIBE to the BBG − WIBE needs to introduce additional components $(C_{3,i})$ in the ciphertext, and these components demolish the *nominal* property. The reason is the challenger can create a *nominal* semi-functional ciphertext with respect to the semi-functional challenge key, then use $(C_{3,i})$ and the components (E_i) in the challenge key to test by himself whether the challenge key is semi-functional or not. In order to retain the *nominality*, we should manage to impose the distribution of exponents of $\mathbb{G}_2$ part in $C_{3,i}$ and in E_i in the semi-functional key and the corresponding semi-functional ciphertext in a compatible way such that they are always *nominal* with each other.

We provide the details about our construction of BBG − WIBE scheme in composite order groups and the proof of its full security in the full version of this paper [29].

3.3 KWIBE: Key-Leakage Resilient WIBE

In the construction of key-leakage resilient HIBE in [22], the user's secret key is constructed from elements in subgroups $\mathbb{G}_1$ and $\mathbb{G}_3$. This leads to secret keys that are relatively low independent sources because they are only in subgroups $\mathbb{G}_1$ and $\mathbb{G}_3$. In order to enhance the independent sources of each user's secret key, in our construction of KWIBE, the secret keys are in the semi-functional form and each user's secret key is now a high independent source since the main part of the secret key is in the whole group $\mathbb{G} = \mathbb{G}_1 \times \mathbb{G}_2 \times \mathbb{G}_3$. Fortunately, this slightly change doesn't affect the functionality and the security of the scheme.

Construction from BBG − WIBE. The main point in proving the key-leakage resilience of HIBE in [22] is to show that the adversary cannot distinguish between two games $\mathsf{KeyLeak}_0$ and $\mathsf{KeyLeak}_1$ which are briefly described as follow. In the game $\mathsf{KeyLeak}_b$ game (for both $b = 0$ and $b = 1$), the adversary can choose to receive a normal key or a semi-functional key from each leak and reveal query for all keys except one key- called the challenge key. Concerning the challenge key, it is set to be a normal key in the game $\mathsf{KeyLeak}_0$ and a semi-functional key in the game $\mathsf{KeyLeak}_1$. We can realize that, in this technique of proving the security, there is no significant difference between a HIBE attack and a WIBE attack. Indeed, the main difference between HIBE and WIBE is that an adversary against WIBE can ask more leak queries (for keys that *match* the challenge pattern) than an adversary against HIBE (who can only ask for keys which are *prefix* of the challenge identity). However, because the difference between two games $\mathsf{KeyLeak}_0$ and $\mathsf{KeyLeak}_1$ is only related to the challenge key which has the same form in both HIBE and WIBE, the proof in HIBE is well adapted to WIBE.

In order to make BBG − WIBE resilient to key-leakage, in the following construction, we first impose the distribution of exponents of $\mathbb{G}_2$ part in $C_{3,i}$ and in E_i in a compatible way such that they are nominal with each other, then we manage to choose compatibly some constants (as r_1, r_2, z_k, z_c) to keep the following properties:

− if $\overrightarrow{\Gamma}$ is orthogonal to $\overrightarrow{\delta}$ then the challenge key is nominally semi-functional and well-distributed.
− if $\overrightarrow{\Gamma}$ is not orthogonal to $\overrightarrow{\delta}$, then the challenge key is truly semi-functional and well-distributed.

The construction is detailed as follows.

Setup$(1^\lambda) \to (\mathsf{mpk}, \mathsf{msk})$ The setup algorithm chooses a bilinear group $\mathbb{G} = \mathbb{G}_1 \times \mathbb{G}_2 \times \mathbb{G}_3$ of order $N = p_1 p_2 p_3$ (each subgroup $\mathbb{G}_i$ is of order p_i). We will assume that users are associated with vectors of identities whose components are elements of $\mathbb{Z}_N$. If the maximum depth of the WIBE is D, the setup algorithm chooses a generator $g_1 \overset{\$}{\leftarrow} \mathbb{G}_1$, a generator $g_2 \overset{\$}{\leftarrow} \mathbb{G}_2$, and a generator $g_3 \overset{\$}{\leftarrow} \mathbb{G}_3$. It picks $b, a_1, \ldots, a_D \overset{\$}{\leftarrow} \mathbb{Z}_N^{D+1}$ and sets $h = g_1^b, u_1 = g_1^{a_1}, \ldots, u_D = g_1^{a_D}$. It also picks $n+1$ random exponents $\langle \alpha, x_1, x_2, \ldots, x_n \rangle \overset{\$}{\leftarrow} \mathbb{Z}_N^{n+1}$. The secret key is $\mathsf{msk} = (\alpha, a_1, \ldots, a_D)$, and the public parameters are:

$$\mathsf{mpk} = (N, g_1, g_3, h, u_1, \ldots, u_D, e(g_1, g_1)^\alpha, g_1^{x_1}, g_1^{x_2}, \ldots, g_1^{x_n})$$

KeyderSF$(\mathsf{msk}, (ID_1, ID_2, \ldots, ID_j), g_2, \mathsf{mpk})$ The key generation algorithm picks $n+1$ random exponents $\langle r, t_1, t_2, \ldots, t_n \rangle \overset{\$}{\leftarrow} \mathbb{Z}_N^{n+1}$, $\overrightarrow{\rho} \overset{\$}{\leftarrow} \mathbb{Z}_N^{n+2}$ and $z_k, \rho_{n+3}, \ldots, \rho_{n+2+D-j} \overset{\$}{\leftarrow} \mathbb{Z}_N$, and $\overrightarrow{\gamma} = (\gamma_1, \ldots, \gamma_{n+2})$ in which $(\gamma_1, \ldots, \gamma_n, \gamma_{n+2}) \overset{\$}{\leftarrow} \mathbb{Z}_N^{n+1}$, $\gamma_{n+1} = \gamma_{n+2}(z_k - \sum_{i=1}^j a_i ID_i)$. It outputs the secret key $SK = (\overrightarrow{K_1}, E_{j+1}, \ldots, E_D)$:

$$= \left(\left\langle g_1^{t_1}, g_1^{t_2}, \ldots, g_1^{t_n}, g_1^\alpha \left(h \cdot \prod_{i=1}^j u_i^{ID_i} \right)^{-r} \prod_{i=1}^n g_1^{-x_i t_i}, g_1^r \right\rangle * g_3^{\overrightarrow{\rho}} * g_2^{\overrightarrow{\gamma}}, \right.$$
$$\left. u_{j+1}^r g_3^{\rho_{n+3}} g_2^{\gamma_{n+2} a_{j+1}}, \ldots, u_D^r g_3^{\rho_{n+2+D-j}} g_2^{\gamma_{n+2} a_D} \right)$$

Note that, to run the **KeyderSF** algorithm one doesn't need to have g_2, he only need to have $X_2 \in \mathbb{G}_2$ or $X_2 X_3$ in which $X_2 \in \mathbb{G}_2, X_3 \in \mathbb{G}_3$.

Delegate $((ID_1, ID_2, \ldots, ID_j), \mathsf{SK'}, ID_{j+1})$ Given a secret key SK' $= (\overrightarrow{K'}, E'_{j+1}, \ldots, E'_D)$ for identity $(ID_1, ID_2, \ldots, ID_j)$, this algorithm outputs a key for $(ID_1, ID_2, \ldots, ID_{j+1})$. It works as follow:

It picks $n+1$ random exponents $\langle r', y_1, y_2, \ldots, y_n \rangle \overset{\$}{\leftarrow} \mathbb{Z}_N^{n+1}$, $\overrightarrow{\rho'} \overset{\$}{\leftarrow} \mathbb{Z}_N^{n+2}$, and $\rho'_{n+3}, \ldots, \rho'_{n+1+D-j} \overset{\$}{\leftarrow} \mathbb{Z}_N$. It outputs the secret key $SK = (\overrightarrow{K_1}, E_{j+2}, \ldots, E_D)$:

$$\left(\overrightarrow{K'_1} * \left\langle g_1^{y_1}, g_1^{y_2}, \ldots, g_1^{y_n}, h^{-r'} (E'_{j+1})^{-ID_{j+1}} \left(\prod_{i=1}^{j+1} u_i^{ID_i} \right)^{-r'} \prod_{i=1}^n g_1^{-x_i y_i}, g_1^{r'} \right\rangle * g_3^{\overrightarrow{\rho'}}, \right.$$
$$\left. E'_{j+2} u_{j+2}^{r'} g_3^{\rho'_{n+3}}, \ldots, E'_D u_D^{r'} g_3^{\rho'_{n+1+D-j}} \right)$$

Enc$(M, (P_1, P_2, \ldots, P_j))$ The encryption algorithm chooses $s \overset{\$}{\leftarrow} \mathbb{Z}_N$ and outputs the ciphertext:

$$CT = (C_0, \overrightarrow{C_1}, C_2) =$$

$$\left(M \cdot e(g_1, g_1)^{\alpha \cdot s}, \left\langle (g_1^{x_1})^s, \cdots, (g_1^{x_n})^s, g_1^s, \left(h \cdot \prod_{i \in \overline{W}(P)} u_i^{P_i} \right)^s \right\rangle, (C_{2,i} = u_i^s)_{i \in W(P)} \right)$$

Dec(CT, SK) Any other receiver with identity $ID = (ID_1, ID_2, \ldots, ID_j)$ matching the pattern P to which the ciphertext was created can decrypt the ciphertext $CT = (C_0, \overrightarrow{C_1}, C_2)$ as follows

First, he recovers the message by computing

$$\overrightarrow{C'_1} = \left\langle (g_1^{x_1})^s, \cdots, (g_1^{x_n})^s, g_1^s, (h \cdot \prod_{i \in \overline{W}(P)} u_i^{P_i})^s \cdot \prod_{i \in W(P)} (u_i^s)^{ID_i} \right\rangle$$

Finally, compute

$$e_{n+2}(\overrightarrow{K_1}, \overrightarrow{C_1'}) = e(g_1, g_1)^{\alpha s} \cdot e(g_1, u_1^{ID_1} \cdots u_j^{ID_j} h)^{-rs} \cdot e(g_1, u_1^{ID_1} \cdots u_j^{ID_j} h)^{rs}.$$

$$\cdot \prod_{i=1}^{n} e(g_1, g_1)^{-x_i t_i s} \cdot \prod_{i=1}^{n} e(g_1, g_1)^{x_i t_i s} = e(g_1, g_1)^{\alpha s}$$

Notice that the $\mathbb{G}_2$ and $\mathbb{G}_3$ parts do not contribute because they are orthogonal to the ciphertext under e.

Security of Key-Leakage Resilient BBG — WIBE Formally, the security model of a ℓ_{SK}-key-leakage resilient WIBE, we call Leak — WIBE security game, is defined as follows: We let I^* denote the set of all possible identity vectors, $\mathcal{R}$ denote the set of all revealed identities

Setup : The challenger makes a call to **Setup**(1^λ) and gets the master secret key msk and the public parameters mpk. It gives mpk to the attacker. Also, it sets $\mathcal{R} = \emptyset$ and $\mathcal{T} = \emptyset$, note that $\mathcal{R} \subseteq I^*, \mathcal{T} \subseteq \{I^*, \mathcal{SK}, \mathcal{N}\}$ (identity vectors - secret keys - leaked bits).

Phase 1 : The adversary can adaptively make three types of query:
- **Create**$(\overrightarrow{I})$: The challenger makes a call to **KeyderSF** to generate SK_I and adds the tuple $(\overrightarrow{I}, SK_I, 0)$ to the set $\mathcal{T}$ if this identity does not exist.
- **Leak**$(\overrightarrow{I}, f)$: The challenger first finds the tuple $(\overrightarrow{I}, SK_I, L)$, then it checks if $L+ \mid f(SK_I)\mid \leq \ell_{SK}$. If true, it responds with $f(SK_I)$ and updates the $L = L+\mid f(SK_I)\mid$. If the checks fails, it returns $\perp$ to the adversary.
- **Reveal**$(\overrightarrow{I})$: The challenger first finds the tuple $(\overrightarrow{I}, SK_I, L)$, then responds with SK_I and adds the identity vector $\overrightarrow{I}$ to the set $\mathcal{R}$.

Challenge : The adversary submits a challenge pattern $\overrightarrow{P^*}$ with the restriction that no identity vector in $\mathcal{R}$ *matches* $\overrightarrow{P^*}$. It also submits two messages M_0, M_1 of equal size. The challenger flips a uniform coin $c \xleftarrow{\$} \{0,1\}$ and encrypts M_c under $\overrightarrow{P^*}$ with a call to **Enc**$(M_c, \overrightarrow{P^*})$. It sends the resulting ciphertext CT^* to the adversary.

Phase 2 : This is the same as **Phase 1**, except the only allowed queries are **Create** queries for all identity vector, and **Reveal** queries for secret keys with identity vectors which do not *matches* $\overrightarrow{P^*}$.

Guess : The adversary outputs a bit $c' \xleftarrow{\$} \{0,1\}$. We say it succeeds if $c' = c$.

Definition 2. *A KWIBE scheme is* (ℓ_{SK})-*key-leakage secure if all PPT adversaries have at most a negligible advantage in the above security game.*

Theorem 1 (Security of Key-Leakage Resilient BBG — WIBE). *Under assumptions 1, 2, 3 in [23] and for* $\ell_{SK} = (n - 1 - 2c)\log(p_2)$, *where* $c > 0$ *is any fixed positive constant, our key-leakage resilient BBG — WIBE scheme is* (ℓ_{SK}) - *key-leakage secure.*

The condition for c is p_2^{-c} is negligible. The length of secret key sk at level i is $(n+2+D-i)(\log(p_1)+\log(p_2)+\log(p_3))$ where D is the depth of WIBE. As we can see, the *leakage fraction* of secret key at leaf node is the biggest. The proof of this theorem can be found in the full version of this paper [29].

3.4   Generic Construction of KIDTR

The construction of KIDTR closely follows to the construction of WIBE-IDTR in [28], using the new primitive KWIBE instead of WIBE for encryption. We integrate KWIBE into the complete subtree method: each group $ID \in \{0,1\}^*$ represents a binary tree and each user $id \in \{0,1\}^l$ ($id = id_1 id_2 \cdots id_l$, $id_i \in \{0,1\}$) in a group ID is assigned to be a leaf of the binary tree rooted at ID. For encryption, we will use a KWIBE of depth $l+1$, each user is associated with a vector $(ID, id_1, \cdots, id_l)$.

Setup$(1^k, N_u)$: Take a security parameter k and the maximum number in each group N_u (thus $l = \lceil \log_2 N_u \rceil$). Run the setup algorithm of KWIBE with the security parameter k and the hierarchical depth $L = l+1$ which returns $(\mathsf{mpk}, \mathsf{msk})$. The setup then outputs $(\mathsf{mpk}, \mathsf{msk})$. As in the complete subtree method, the setup also defines a data encapsulation method $E_K : \{0,1\}^* \to \{0,1\}^*$ and its corresponding decapsulation D_K. The session key K used will be chosen fresh for each message M as a random bit string. E_K should be a fast method and should not expand the plaintext.

Keyder(msk, ID, id): Run the key derivation of KWIBE for $l+1$ level identity $WID = (ID, id_1, \ldots, id_l)$ (the j-th component corresponds to the j-th bit of the identity id) and get the decryption key d_{WID}. Output $d_{ID,id} = d_{WID}$.

Enc$(\mathsf{mpk}, ID, \mathcal{R}_{ID}, M)$: A sender wants to send a message M to a group ID with the revocation list $\mathcal{R}_{ID}$. The revocation works as in the complete subtree scheme. Considering a group ID with its revocation list $\mathcal{R}_{ID}$, the users in $\mathcal{N}_{ID} \backslash \mathcal{R}_{ID}$ are partitioned into disjoint subsets $S_{i_1}, \ldots, S_{i_w}$ which are all the subtrees of the original tree (rooted at ID) that hang off the Steiner tree defined by the set $\mathcal{R}_{ID}$.

Each subset S_{i_j}, $1 \le j \le w$, is associated to an $l+1$ vector identity $ID_{S_{i_j}} = (ID, id_{i_j,1}, \ldots, id_{i_j,k}, *, .., *)$ where $id_{i_j,1}, \ldots, id_{i_j,k}$ is the path from the root ID to the node S_{i_j} and the number of wildcards $*$ is $l-k$. The encryption algorithm randomly chooses a session key K, encrypts M under the key K by using a symmetric encryption, and outputs as a header the encryption of KWIBE for each $ID_{S_{i_1}}, \ldots, ID_{S_{i_w}}$.

$$C = \langle [i_1, \ldots, i_w][\mathsf{KWIBE.Enc}(\mathsf{mpk}, ID_{S_{i_1}}, K), \ldots, \mathsf{KWIBE.Enc}(\mathsf{mpk}, ID_{S_{i_w}}, K)]$$

$$, E_K(M) \rangle$$

Dec$(d_{ID,id}, C)$: The user received the ciphertext C as above. First, find j such that $id \in S_{i_j}$ (in case $id \in \mathcal{R}_{ID}$ the result is null). Second, use private key $d_{ID,id}$ to decrypt $\mathsf{KWIBE.Enc}(\mathsf{mpk}, ID_{S_{i_j}}, K)$ to obtain K. Finally, compute $D_K(E_K(M))$ to recover the message M.

3.5 Security of KIDTR

Theorem 2 (Security of KIDTR). *If the KWIBE is (ℓ_{SK}) - key-leakage secure then our KIDTR is also (ℓ_{SK}) - key-leakage secure.*

The proof of this theorem can be found in the full version of this paper [29].

4 KIDTR is Immune to Pirates 2.0 in Bounded Leakage Model

4.1 Pirates 2.0 in Bounded Leakage Model

The basic idea behind Pirates 2.0 attacks is that traitors are free to contribute some piece of secret key as long as several users of the system could have contributed exactly the same information *following the same (public) strategy*: this way, they are able to remain somewhat anonymous. The leakage information is formalized via extraction function which is any efficiently computable function f on the space of the secret keys and a traitor u is said to be *masked* by a user u' for an extraction function f if $f(sk_u) = f(sk_{u'})$. The anonymity level is meant to measure exactly how anonymous they remain. This is defined in [6] as follows.

Definition 3 (Anonymity Level). *The level of anonymity of a traitor u after a contribution $\cup_{1 \le i \le t} f_i(sk_u)$ is defined as the number α of users masking u' for each of the t extraction functions f_i simultaneously:*

$$\alpha = \#\{u' \mid \forall i, \ f_i(sk_u) = f_i(sk_{u'})\} \ .$$

Definition 4 (Pirates 2.0 in Bounded Leakage Model). *We say that a Pirates 2.0 attack is in bounded leakage model if for every traitor u with his secret key sk_u, at each time i $(i = 1, \ldots, t)$, is free to choose any strategy f_i to contribute the bits information of sk_u to the public domain as long as*

$$\sum_{i=1}^{t} \mid f_i(sk_u) \mid \le \ell_{SK}$$

where ℓ_{SK} is the threshold.

4.2 Comparison to Other Methods

Until now, there have been many methods aiming to fight against Pirates 2.0 [10,11,28,32] but all of them only consider a particular form of leakage of secret key in bounded leakage model. In fact, it is assumed in these methods that by using a public extraction function the dishonest users leak the entire information of some sub-keys which could be used in the encryption procedure. Concretely, in [11] they describe the extraction function as a *projection function*, and the secret key of user SK is a vector $(SK_1, \ldots, SK_l)$ of elements, where each SK_i,

$i = 1, \ldots, l$, contains k bits information. The output of extraction function, at each time, is an i-th element of the vector. This require that a traitor, at each time, must contribute at least k bits information of his secret key where k must be bigger than the security parameter of the scheme. This kind of attack is thus a particular case of Pirates 2.0 in bounded leakage model in which the strategy of the traitor is limited: at each time each traitor has to choose a whole sub-key to contribute.

We consider the general form of Pirates 2.0 attack in bounded leakage model, by considering any strategy of the adversary. This generalizes thus all the previous consideration of Pirates 2.0. However, there is still a gap between the Pirates 2.0 attack in bounded leakage model and the general form of the Pirates 2.0 attack where the pirate can combine the information of the bits of the secret key and then contribute a particular form of information. This kind of attack could be captured by considering a general form of leakage for revoke schemes and this seems a very challenging problem.

4.3   Pirates 2.0 in Bounded Leakage Model Viewed from the Information Theory

We aim to re-explain the way Pirates 2.0 in bounded leakage model works under the information theory. This is also the basic starting point so that we can establish a sufficient condition for a scheme to resist Pirates 2.0 in bounded leakage model in the next sub-section. In a revoke scheme, when a user joins the system, its key is generated and has some entropy. However, as keys of users could be correlated, the user can contribute some correlated information without the risk being identified. The user really lose its anonymity when he contributes its independent secret information that the other users don't have. More formally, these are entropy conditioned on the information about the other users' keys. Let us first recall some classical definitions about entropy.

Definition 5. *Let X be a random variable. The min-entropy of X is*

$$H_\infty(X) = \min_x - \log(\Pr[X = x]) = -\log(\max_x \Pr[X = x])$$

We say that X is a k-source if $H_\infty(X) \geq k$.

The high min-entropy is used rather than the Shannon entropy in cryptography for describing good distributions for the keys. In fact, the conventional notion in cryptography is the intuitive notion of "guessability" and a distribution X has min-entropy of k bits if even an unbounded adversary cannot guess a sample from X with probability greater than 2^{-k}.

However, in context of Pirates 2.0 in bounded leakage model, a high min-entropy is not enough because the keys could be correlated. We should thus need to define how many information of the key a user has that is independent to the keys of the others users. This is quantified via the conditional min-entropy.

Definition 6. *Let X, E be a joint distribution. Then we define the min-entropy of X conditioned on E-denoted $H_\infty(X|E)$ as*

$$H_\infty(X|E) = -\log \max_e [\max_x \Pr[(X|E = e)]]$$

We say that X is a k-independent source of E if $H_\infty(X|E) \geq k$.

We note that Dodis et. al. [14] defines the conditional min-entropy as average entropy $\log \mathrm{E}[\max_x \Pr[(X|E = e)]]$. In our setting, we follow a conservative approach, taken in [20], and manage to deal with the above stronger notion. In fact, we will see later in our construction that the secret keys of users are sufficiently independent each from the others, the consideration of the conditional min-entropy can be justified. We first define the independence between the secret keys in a revoke system as follows.

Definition 7 (Independent Source). *In a revoke system of N_u users, let X_i be the distribution outputted by the key generation for a user i and let $E = (X_1, \ldots, X_{i-1}, X_{i+1}, \ldots, X_{N_u}, \mathsf{pub})$ where pub denotes the distribution of the public parameters in the system. Then we say that the key of user i is a k-independent source if $H_\infty(X_i|E) \geq k$.*

The key of user i is a k-independent source if it has k-bit entropy independently from the keys of the others users and from all the public information of the systems.

We now review the Pirates 2.0 in bounded leakage model in the context of Complete Subtree resumed in Figure 4.3. For a D-level tree, each user's key is a $(D \times \lambda)$-source but only a λ-independent source because each user only has an independent sub-key at the leaf. Therefore, even if a user contributes $((D-1) \times \lambda)$ entropy of its key, the remained information could still be a λ-independent source. Without leaking any independent entropy, the user could remain anonymous at a level $\alpha > 1$ (because at least two different users can have the same contributive information). In the example in Figure 4.3, the user U is assigned 5 sub-keys corresponding to the nodes from the root to the leaf. The user U can contribute a key S_4 and specifies the target set at S_4 that covers 4 users of the sub-tree rooted at S_4. A pirate decoder with only one key at S_4 can decrypt the ciphertext for the chosen target set S_4 with non-negligible probability while preserving an anonymity level $\alpha = 4$ for the contributor and therefore, the scheme is vulnerable against the Pirates 2.0 in bounded leakage model.

4.4   Key-Leakage Resilience vs. Pirates 2.0 in Bounded Leakage Model

We are now ready to prove a sufficient condition so that a key-leakage resilient revoke scheme is immune to Pirates 2.0 attacks in bounded leakage model. We first use the following lemma and give the proof in the full version [29].

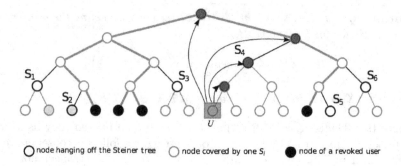

Fig. 1. An example of a complete subtree scheme where the center covers all non-revoked users with the nodes $S_1, \ldots, S_6$. A user is a leaf on the binary tree where each node is assigned to a long-lived randomly chosen key. Each user possesses all the long-lived keys of the nodes on the path from the user's leaf to the root.

Lemma 1. *For any function f, g, and any random variable X, Y, if $H_\infty(X|Y) \geq k$ and $H_\infty(X|f(X), Y) \leq k - \alpha$ then*

$$\Pr_{x \in X, y \in Y}[f(x) = g(y)] \leq \frac{1}{2^\alpha}$$

The following theorem gives a condition on the independence of the user's key under which we can relate the leakage resilience to the Pirates 2.0.

Theorem 3. *Let Π be a (ℓ_{SK})-key-leakage resilient revoke system of N_u users in which each user's key has length of m bit and is a m'-independent source. If $\alpha = \frac{N_u}{2^{\ell_{SK}+m'-m}} \leq 1$, then Π is immune to any Pirates 2.0 attack in bounded leakage model.*

Proof.

Proposition 1. *In a Pirates 2.0 attack in bounded leakage model, if a user leaks k bits of his secret key to the public domain then his anonymity level is at most $\frac{N_u}{2^{k+m'-m}}$.*

Proof. Intuitively, as the key of the user u is a high independent source even when the others users contribute their whole secret keys, if u leaks too much information on its key then it will also leak many independent information and loses its anonymity.

Formally, following the definition 3 of anonymity level in pirates 2.0, assume that a user u contributes k bits information L_u of his secret key sk_u to the public domain, we need to compute the probability for an user u' to contribute exactly the same information as the user u, at each period of time i.

– At time 0: u contributes nothing to the public domain. Let $E_i = (\cup_{w \neq u} sk_w,$ $\text{pub}_i)$ where pub_i denotes the public information at the time i which contains the publics parameters of the system plus contributed information of the users after the time $i-1$. Because each user's key is a m'-independent source: $H_\infty(sk_u|E_0) \geq m'$.

- At time i: u contributes his secret informations $L_u^i = f_i(sk_u, \mathsf{pub}_i)$ to the public domain by leaking k_i bits of his secret keys. If we denote k_i^{in} the number of independent bits that the user u losses in time i, i.e., $k_i^{in} = H_\infty(sk_u|E_i) - H_\infty(sk_u|E_{i-1})$. From the lemma 1, the probability that u' could contribute exactly the same information L_u^i is at most $\frac{1}{2^{k_i^{in}}}$. Note that E_0 and thus E_i already contain $\cup_{w \neq u} sk_w$, i.e., all the contributed information of the other users are already contained in E_i (for all i), the k_i^{in} independent bits are among k_i bit that the user u leaks at the time i.

At the end, after the time t, the user u contributes to the public domain by totally leaking $k = k_1 + \cdots + k_t$ bits of its secret information. By the above computation, the probability that an user u' can contribute exactly the same total information like u is at most $\prod_{j=1}^t \frac{1}{2^{k_j^{in}}}$, and

$$\sum_{j=1}^t k_j^{in} = H_\infty(sk_u|E_0) - H_\infty(sk_u|E_t)$$

Because the bit length of the secret key sk_u is m and the user u leaks k bits, we deduce that $H_\infty(sk_u|E_t) \leq m - k$ and therefore $\sum_{j=1}^t k_j^{in} \geq m' - (m - k) = k + m' - m$ which implies that the probability that an user u' can contribute exactly the same information like u as required in Pirates 2.0 is at most $\frac{1}{2^{k+m'-m}}$ and the anonymity level of u cannot be assured to be higher than $\frac{N_u}{2^{k+m'-m}}$. $\square$

Proposition 2. *Let Π be a (ℓ_{SK})-key-leakage resilient revoke scheme. If each user leaks no more than ℓ_{SK} bits of his secret key to the public domain, then one cannot produce a Pirates 2.0 decoder in bounded leakage model.*

Proof. We suppose by contradiction that there is a Pirates 2.0 $\mathcal{A}$ in bounded leakage model against Π in which each user leaks no more than ℓ_{SK} bits of his secret key to the public domain, then we build an algorithm $\mathcal{B}$ that breaks the security of Π in the context of key leakage resilience.

Algorithm $\mathcal{B}$ simulates $\mathcal{A}$ and makes use of the outputs of $\mathcal{A}$ to break the security of Π. It works as follows:

- At time 0: users contribute nothing to the public domain.
- At time 1: suppose that a user u decides to contribute $L_u^1 = f_1(sk_u)$ bits to the public domain by using a strategy f_1 where f_1 is a polynomial-time computable function, $\mathcal{B}$ requests the leak query $(u, g_1 := f_1)$ to his challenger and forwards the result to $\mathcal{A}$.
- At any time i: suppose that a user u decides to contribute $L_u^i = f_i(sk_u, I)$ bits to the public domain, where I is the public collected information after the time $i-1$. At this stage, $\mathcal{B}$ defines a polynomial-time computable function $g_{i,I}(sk_u) := f_i(sk_u, I)$, then requests the leak query $(u, g_{i,I})$ to his challenger and forwards the result to $\mathcal{A}$.

- When $\mathcal{A}$ outputs a pirate decoder and a target S so that the pirate decoder can decrypt ciphertexts for S with a non-negligible probability, $\mathcal{B}$ simply outputs $S^* = S$ and two different messages M_0, M_1 to his challenger. By using this pirate decoder, $\mathcal{B}$ can decrypt the challenge ciphertext with a non-negligible probability and thus break the security of the scheme.

We note that, since each user contributes maximum ℓ_{SK} bits to the public domain, $\mathcal{B}$ only need to ask in total at most ℓ_{SK} bits to his challenger. By definition, Π is then not ℓ_{SK}-key leakage resilient.

$\square$

The theorem immediately follows from the above two propositions. $\square$

Proposition 3. *In* KIDTR *scheme, if we call p_1, p_2, p_3 are primes of $\lambda_1, \lambda_2, \lambda_3$ bits, then each user's secret key with length $m = (n+2)(\lambda_1 + \lambda_2 + \lambda_3)$ is m'-independent source where $m' = ((n+1)(\lambda_1 + \lambda_2 + \lambda_3) + \lambda_2 + \lambda_3)$.*

Proof. In our KIDTR scheme, we make use of a KWIBE scheme in which each user's secret key is at leaf node 3.3, therefore an user's secret key is of the following form:

$$SK = \overrightarrow{K_1} = \left(\left\langle g_1^{t_1}, g_1^{t_2}, \ldots, g_1^{t_n}, g_1^\alpha \left(h \cdot \prod_{i=1}^{j} u_i^{ID_i} \right)^{-r} \prod_{i=1}^{n} g_1^{-x_i t_i}, g_1^r \right\rangle * g_3^{\overrightarrow{p}} * g_2^{\overrightarrow{\gamma}} \right)$$

where $r, t_1, t_2, \ldots, t_n, z_k \xleftarrow{\$} \mathbb{Z}_N$, $\overrightarrow{p} \xleftarrow{\$} \mathbb{Z}_N^{n+2}$, and $\overrightarrow{\gamma} = (\gamma_1, \ldots, \gamma_{n+2})$ in which $(\gamma_1, \ldots, \gamma_n, \gamma_{n+2}) \xleftarrow{\$} \mathbb{Z}_N^{n+1}$, $\gamma_{n+1} = \gamma_{n+2}(z_k - \sum_{i=1}^{j} a_i ID_i)$.
We realize that in each secret key, the elements corresponding to the indices $1, \ldots, n, n+2$ are randomly generated in the whole group $\mathbb{G} = \mathbb{G}_1 \times \mathbb{G}_2 \times \mathbb{G}_3$, the element corresponding to the indice $n+1$ is not independent in $\mathbb{G}_1$ but randomly generated in $\mathbb{G}_2 \times \mathbb{G}_3$. Therefore, it's easy to see that each user's secret key is of $(n+2)(\lambda_1 + \lambda_2 + \lambda_3)$ bit length and is a $((n+1)(\lambda_1 + \lambda_2 + \lambda_3) + \lambda_2 + \lambda_3)$-independent source.

Theorem 4. *The* KIDTR *scheme is immune to Pirates 2.0 attacks in bounded leakage model for any choice of parameters $n, c, \lambda_1, \lambda_2$ such that $2^{(n-1-2c)\lambda_2 - \lambda_1} > N_u$, where N_u is the number of subscribed users in the systems*

Proof. From the theorems 1 and theorem 2, we decude that the KIDTR scheme is ℓ_{SK}−leakage resilient with $\ell_{SK} = (n-1-2c)\lambda_2$ for any fixed positive constant $c > 0$ (such that p_2^{-c} is negligible). From the theorem 3, one cannot mount a Pirates 2.0 attack with an anonymity level larger than $\alpha = \frac{N_u}{2^{\ell_{SK}+m'-m}} = \frac{N_u}{2^{(n-1-2c)\lambda_2 - \lambda_1}} < 1$. $\square$

We note that there is no need to choose particular parameters for our system. For example, simply with $c = 1, n = 5$ and $\lambda_1 = \lambda_2 = 512$ ($p_2^{-c} = 2^{-512}$ is negligible) and suppose that there are $N_u = 2^{40}$ subscribed users, our system is immune to Pirates 2.0 in bounded leakage model because $2^{(n-1-2c)\lambda_2 - \lambda_1} = 2^{512} > N_u$ and the user's secret key contains only 7 elements in $\mathbb{G}$.

Acknowledgments. This work was partially supported by the French ANR-09-VERSO-016 BEST Project and partially conducted within the context of the International Associated Laboratory Formath Vietnam (LIAFV).

References

1. Abdalla, M., Birkett, J., Catalano, D., Dent, A.W., Malone-Lee, J., Neven, G., Schuldt, J.C.N., Smart, N.P.: Wildcarded identity-based encryption. Journal of Cryptology 24(1), 42–82 (2011)
2. Abdalla, M., Catalano, D., Dent, A.W., Malone-Lee, J., Neven, G., Smart, N.P.: Identity-based encryption gone wild. In: Bugliesi, M., Preneel, B., Sassone, V., Wegener, I. (eds.) ICALP 2006. LNCS, vol. 4052, pp. 300–311. Springer, Heidelberg (2006)
3. Abdalla, M., Dent, A.W., Malone-Lee, J., Neven, G., Phan, D.H., Smart, N.P.: Identity-based traitor tracing. In: Okamoto, T., Wang, X. (eds.) PKC 2007. LNCS, vol. 4450, pp. 361–376. Springer, Heidelberg (2007)
4. Alwen, J., Dodis, Y., Wichs, D.: Leakage-resilient public-key cryptography in the bounded-retrieval model. In: Halevi, S. (ed.) CRYPTO 2009. LNCS, vol. 5677, pp. 36–54. Springer, Heidelberg (2009)
5. Berkovits, S.: How to broadcast a secret. In: Davies, D.W. (ed.) EUROCRYPT 1991. LNCS, vol. 547, pp. 535–541. Springer, Heidelberg (1991)
6. Billet, O., Phan, D.H.: Traitors collaborating in public: Pirates 2.0. In: Joux, A. (ed.) EUROCRYPT 2009. LNCS, vol. 5479, pp. 189–205. Springer, Heidelberg (2009)
7. Brakerski, Z., Kalai, Y.T., Katz, J., Vaikuntanathan, V.: Overcoming the hole in the bucket: Public-key cryptography resilient to continual memory leakage. In: 51st FOCS Annual Symposium on Foundations of Computer Sciencei, pp. 501–510. IEEE Computer Society Press (2010)
8. Cash, D., Ding, Y.Z., Dodis, Y., Lee, W., Lipton, R.J., Walfish, S.: Intrusion-resilient key exchange in the bounded retrieval model. In: Vadhan, S.P. (ed.) TCC 2007. LNCS, vol. 4392, pp. 479–498. Springer, Heidelberg (2007)
9. Chor, B., Fiat, A., Naor, M.: Tracing traitors. In: Desmedt, Y.G. (ed.) CRYPTO 1994. LNCS, vol. 839, pp. 257–270. Springer, Heidelberg (1994)
10. D'Arco, P., Perez del Pozo, A.L.: Fighting Pirates 2.0. In: Lopez, J., Tsudik, G. (eds.) ACNS 2011. LNCS, vol. 6715, pp. 359–376. Springer, Heidelberg (2011)
11. D'Arco, P., del Pozo, A.L.P.: Toward tracing and revoking schemes secure against collusion and any form of secret information leakage. International Journal of Information Security 12 (2013)
12. Di Crescenzo, G., Lipton, R.J., Walfish, S.: Perfectly secure password protocols in the bounded retrieval model. In: Halevi, S., Rabin, T. (eds.) TCC 2006. LNCS, vol. 3876, pp. 225–244. Springer, Heidelberg (2006)
13. Dodis, Y., Fazio, N.: Public key trace and revoke scheme secure against adaptive chosen ciphertext attack. In: Desmedt, Y.G. (ed.) PKC 2003. LNCS, vol. 2567, pp. 100–115. Springer, Heidelberg (2002)
14. Dodis, Y., Reyzin, L., Smith, A.: Fuzzy extractors: How to generate strong keys from biometrics and other noisy data. In: Cachin, C., Camenisch, J.L. (eds.) EUROCRYPT 2004. LNCS, vol. 3027, pp. 523–540. Springer, Heidelberg (2004)
15. Dziembowski, S., Pietrzak, K.: Intrusion-resilient secret sharing. In: 48th FOCS Annual Symposium on Foundations of Computer Science, pp. 227–237. IEEE Computer Society Press (October 2007)

16. Dziembowski, S., Pietrzak, K.: Leakage-resilient cryptography. In: 49th Annual Symposium on Foundations of Computer Science, pp. 293–302. IEEE Computer Society Press (October 2008)
17. Fiat, A., Naor, M.: Broadcast encryption. In: Stinson, D.R. (ed.) CRYPTO 1993. LNCS, vol. 773, pp. 480–491. Springer, Heidelberg (1994)
18. Ishai, Y., Sahai, A., Wagner, D.: Private circuits: Securing hardware against probing attacks. In: Boneh, D. (ed.) CRYPTO 2003. LNCS, vol. 2729, pp. 463–481. Springer, Heidelberg (2003)
19. Katz, J., Vaikuntanathan, V.: Signature schemes with bounded leakage resilience. In: Matsui, M. (ed.) ASIACRYPT 2009. LNCS, vol. 5912, pp. 703–720. Springer, Heidelberg (2009)
20. Renner, R.S., Wolf, S.: Simple and Tight Bounds for Information Reconciliation and Privacy Amplification. In: Roy, B. (ed.) ASIACRYPT 2005. LNCS, vol. 3788, pp. 199–216. Springer, Heidelberg (2005)
21. Kiayias, A., Pehlivanoglu, S.: Pirate evolution: How to make the most of your traitor keys. In: Menezes, A. (ed.) CRYPTO 2007. LNCS, vol. 4622, pp. 448–465. Springer, Heidelberg (2007)
22. Lewko, A., Rouselakis, Y., Waters, B.: Achieving leakage resilience through dual system encryption. In: Ishai, Y. (ed.) TCC 2011. LNCS, vol. 6597, pp. 70–88. Springer, Heidelberg (2011)
23. Lewko, A., Waters, B.: New techniques for dual system encryption and fully secure HIBE with short ciphertexts. In: Micciancio, D. (ed.) TCC 2010. LNCS, vol. 5978, pp. 455–479. Springer, Heidelberg (2010)
24. Micali, S., Reyzin, L.: Physically observable cryptography (extended abstract). In: Naor, M. (ed.) TCC 2004. LNCS, vol. 2951, pp. 278–296. Springer, Heidelberg (2004)
25. Naor, D., Naor, M., Lotspiech, J.: Revocation and tracing schemes for stateless receivers. In: Kilian, J. (ed.) CRYPTO 2001. LNCS, vol. 2139, pp. 41–62. Springer, Heidelberg (2001)
26. Naor, M., Pinkas, B.: Efficient trace and revoke schemes. In: Frankel, Y. (ed.) FC 2000. LNCS, vol. 1962, pp. 1–20. Springer, Heidelberg (2001)
27. Naor, M., Segev, G.: Public-key cryptosystems resilient to key leakage. In: Halevi, S. (ed.) CRYPTO 2009. LNCS, vol. 5677, pp. 18–35. Springer, Heidelberg (2009)
28. Phan, D.H., Trinh, V.C.: Identity-based trace and revoke schemes. In: Boyen, X., Chen, X. (eds.) ProvSec 2011. LNCS, vol. 6980, pp. 204–221. Springer, Heidelberg (2011)
29. Phan, D.H., Trinh, V.C.: Key-Leakage Resilient Revoke Scheme Resisting Pirates 2. Full version available at, http://www.di.ens.fr/users/phan/2013-africa.pdf
30. Standaert, F.-X., Malkin, T.G., Yung, M.: A unified framework for the analysis of side-channel key recovery attacks. In: Joux, A. (ed.) EUROCRYPT 2009. LNCS, vol. 5479, pp. 443–461. Springer, Heidelberg (2009)
31. Waters, B.: Dual system encryption: Realizing fully secure IBE and HIBE under simple assumptions. In: Halevi, S. (ed.) CRYPTO 2009. LNCS, vol. 5677, pp. 619–636. Springer, Heidelberg (2009)
32. Zhao, X., Zhang, F.: Traitor tracing against public collaboration. In: Bao, F., Weng, J. (eds.) ISPEC 2011. LNCS, vol. 6672, pp. 302–316. Springer, Heidelberg (2011)

Fast Software Encryption Attacks on AES

David Gstir and Martin Schläffer

IAIK, Graz University of Technology, Austria

`martin.schlaeffer@iaik.tugraz.at`

Abstract. In this work, we compare different faster than brute-force single-key attacks on the full AES in software. Contrary to dedicated hardware implementations, software implementations are more transparent and do not over-optimize a specific type of attack. We have analyzed and implemented a black-box brute-force attack, an optimized brute-force attack and a biclique attack on AES-128. Note that all attacks perform an exhaustive key search but the latter two do not need to re-compute the whole cipher for all keys. To provide a fair comparison, we use CPUs with Intel AES-NI since these instructions tend to favor the generic black-box brute-force attack. Nevertheless, we are able to show that on Sandy Bridge the biclique attack on AES-128 is 17% faster, and the optimized brute-force attack is 3% faster than the black-box brute-force attack.

Keywords: fast software encryption, AES, brute-force attack, biclique attack, Intel AES-NI.

1 Introduction

In recent years, new attacks on the full Advanced Encryption Standard (AES) have been published [1,3]. Especially the single-key attacks are debatable due to their marginal complexity improvement compared to a generic exhaustive key search (brute-force). Therefore, Bogdanov et al. have implemented a variant of the biclique attack in hardware to show that their attack is indeed faster than brute-force [2].

However, in a dedicated hardware implementation it is less transparent how much effort has been put on optimizing each attack type. If the difference in complexity is very small, it may be possible to turn the result around by investing more optimization effort in the slower attack. Contrary to hardware implementations, the speed of well optimized software implementations tends to be more stable. This can also be observed when looking at different comparisons of the NIST SHA-3 candidates in hardware and in software [5,11]. Too many parameters can be optimized in hardware which can easily change a comparison in favor of one or the other primitive or attack.

In this work, we have implemented different single-key attacks on AES-128 using Intel AES-NI [7], which is the basis for the fastest software implementations of AES. We compare the generic black-box brute-force attack with an optimized

A. Youssef, A. Nitaj, A.E. Hassanien (Eds.): AFRICACRYPT 2013, LNCS 7918, pp. 359–374, 2013.

brute-force attack and with the (simplified) biclique attack used in the hardware implementation of Bogdanov et al. [2]. In the optimized brute-force attack we choose keys such that we do not need to recompute the full cipher for every new key guess.

Our results indicate that the biclique attack is indeed marginally faster than a black-box brute-force attack. However, also the optimized brute-force attack on AES is slightly faster than the black-box brute-force attack. We have concentrated our effort on AES-128 but the results are likely to be similar for the other variants as well. Nevertheless, these attacks do not threaten the security of AES since for any real world cipher, optimized brute-force attacks are most likely faster than black-box brute-force attacks.

Outline of the paper: In Section 2, we give a brief description of AES-128 and the implementation characteristics of Intel AES-NI. In Section 3, we describe and evaluate the theoretical complexity of the black-box brute-force attack and an optimized brute-force attack. Section 4 describes and analyzes the simplified biclique attack on AES-128 of Bogdanov et al. [2]. Then, our software implementations and results of these three attacks are given in Section 5. Finally, we conclude our work in Section 6.

2   Implementing AES in Software Using AES-NI

In this section, we briefly describe the Advanced Encryption Standard (AES) as well as the instructions and implementation characteristics of Intel AES-NI.

2.1   Description of AES-128

The block cipher Rijndael was designed by Daemen and Rijmen and standardized by NIST in 2000 as the Advanced Encryption Standard (AES) [9]. The AES consists of a key schedule and state update transformation. In the following, we give a brief description of the AES and for a more detailed description we refer to [9].

State Update. The block size of AES is 128 bits which are organized in a 4×4 state of 16 bytes. This AES state is updated using the following 4 round transformations with 10 rounds for AES-128:

- the non-linear layer SubBytes (SB) independently applies the 8-bit AES S-box to each byte of the state
- the cyclical permutation ShiftRows (SR) rotates the bytes of row r to the left by r positions with $r = \{0, ..., 3\}$
- the linear diffusion layer MixColumns (MC) multiplies each column of the state by a constant MDS matrix
- in round i, AddRoundKey (AK) adds the 128-bit round key rk_i to the AES state

A round key rk_0 is added prior to the first round and the MixColumns transformation is omitted in the last round of AES. All 4 round transformations plus the enumerations of individual state bytes are shown in Fig. 1

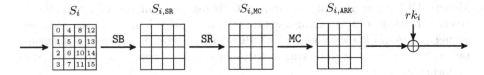

Fig. 1. Notation for state bytes and 4 round transformations of AES

Key Schedule. The key schedule of AES recursively generates a new 128-bit round key rk_i from the previous round key rk_{i-1}. In the case of AES-128, the first round key rk_0 is the 128-bit master key of AES-128. Each round of the key schedule consists of the following 4 transformations:

- an upward rotation of 4 column bytes by one position
- the nonlinear SubWord applies the AES S-box to 4 bytes of one column
- a linear part using XOR additions of columns
- a constant addition of the round constant RCON[i]

More specifically, each column $rk_{i,c}$ with $i = 1, \ldots, 10$ and $c = 0, \ldots, 3$ of round key rk_i is computed as follows:

$$rk_{i,0} = \text{SubWord}(rk_{i-1,3} \ggg 1) \oplus rk_{i-1,0} \oplus \text{RCON}[i] \qquad \text{for } c = 0$$

$$rk_{i,c} = rk_{i,c-1} \oplus rk_{i-1,c} \qquad \text{for } c = 1, \ldots, 3$$

2.2   Efficient Implementations of AES-128 Using AES-NI

For our software implementation we chose to use the Intel AES instruction set *AES-NI (AES New Instructions)* [7] because it provides the fastest way to implement AES on a standard CPU and makes the implementations easier to compare. Moreover, AES-NI gave us a fair basis for the brute-force and biclique implementations since all AES operations take exactly the same time throughout all implementations.

For the AES-NI instruction set, Intel integrated certain operations for AES directly into the hardware, thus, making them faster than any pure software implementation and providing more security against timing attacks due to constant time operations [7]. Overall, AES-NI adds the following operations for key schedule, encryption and decryption and a full description of all instructions can be found in [8]:

- aesenc, aesdec performs one full encryption or decryption round, respectively.
- aesenclast, aesdeclast performs the last encryption or decryption round.
- aeskeygenassist computes the SubWord, rotation and XOR with RCON operations required for the key schedule.
- aesimc performs InvMixColumns on the given 128-bit register and stores the result to another 128-bit register.

Modern CPUs use multiple techniques to boost performance of applications. Hence, creating software implementations with optimal performance requires some background knowledge on how CPUs operate. For our software implementations we took the following approaches into account to increase the performance:

High Pipeline Utilization: CPUs split a single instruction into multiple μops. This enables the CPU to start processing the next instruction before the previous has finished. Obviously, this only works if both instructions are independent of each other.

Minimal Memory Access: Fetching data from memory (registers) outside the CPU is slow and decreases performance. Since the attacks shown here, have minimal memory complexities, they can be implemented without almost any memory access.

Parallelized Encryption: For optimal pipeline utilization it is important to carefully utilize AES-NI and SSE instructions since they normally take more than one CPU cycle. Therefore, we compute each encryption round for multiple keys at the same time. This results in multiple independent instructions which are processed by the CPU. This in turn leads to higher pipeline utilization. For instance, on Intel Westmere the execution of `aesenc` takes 6 CPU cycles and the CPU can execute an instruction every second cycle. If we perform 4 independent instructions in parallel, it would require 6 cycles until the first operation is finished. After that, every second cycle another operation finishes. So, in total it requires only 12 cycles for those four operations to finish, instead of 24 cycles if we do not parallelize them. On Intel Sandy Bridge, one execution of `aesenc` takes 8 CPU cycles and the CPU can execute an instruction every cycle.

Reducing `aeskeygenassist` Instructions: The `aeskeygenassist` instruction performs suboptimal on current Sandy Bridge CPUs. Our tests have shown that this instruction has a latency of 8 cycles and a reciprocal throughput of 8 (independently shown in [4]). This is slower compared to the other AES-NI instructions. Since we have to compute the key schedule very often in all attacks, we need to compute round keys as fast as possible. Fortunately, `aeskeygenassist` is able compute to `SubWord` and the rotation for two words in parallel. Thus, we use a single `aeskeygenassist` instruction for two independent round keys. Another solution would be to avoid `aeskeygenassist` and compute `SubWord` using `aesenclast` and the rotation using an SSE byte shuffle instruction.

3   Brute-Force Key Recovery Attacks on AES-128

In this section we describe two brute-force attacks on AES-128 and evaluate their complexities. We show that an optimized brute-force attack which does not recompute all state bytes is in theory (marginally) faster than a black-box brute-force attack. A practical evaluation of these two attacks in software using AES-NI is given in Section 5.

3.1 Black-Box Brute-Force Attack

We call a generic key recovery attack which does not exploit any structural properties of a cipher a black-box brute-force attack. This is the only possible key recovery attack on an ideal cipher. For a cipher with key size n, 2^n keys have to be tested to find the unknown encryption key with probability 1. Thus, the generic complexity is determined by 2^n evaluations of the cipher. In such an attack, the data complexity is 1 and the key complexity is 2^n. Note that time-memory trade-offs apply if more keys are attacked at the same time [6], while the black-box brute-force attack and biclique attack need to be repeated for each key to attack. To compare a black-box brute-force attack on AES-128 with an optimized brute-force attack or any other brute-force attack, we need to determine its complexity in terms of AES-128 evaluations. Since most optimized attacks compute only parts of the cipher, we evaluate the complexity in terms of S-box computations, the most expensive part of most AES implementations. In total, one full AES-128 encryption including key schedule computation requires to compute 200 S-boxes.

3.2 Optimized Brute-Force Attack

Every practical cipher consists of non-ideal sub-functions or rounds. If the diffusion is not ideal, a flip in a single key bit does not immediately change all bits of the state. This effect can be exploited by an optimized brute-force attack. Instead of randomly testing keys in a key recovery attack, we can iterate the keys in a given order, such that only parts of the cipher need to be recomputed for each additional key. Hence, we save computations which may reduce the overall cost of performing a brute-force key recovery attack. Note that every optimized brute-force attack still needs to test all 2^n keys, which is not the case in a short-cut key recovery attack.

Such an optimized brute-force attack is possible for every real world cipher. However, the complexity reduction will only be marginal. In practical implementations of an attack it can even be worse, since computing a full round is usually more efficient than computing, extracting, restructuring and combining parts of a computation (see Section 5). After all, the final result depends heavily on the implementations and how well they are optimized themselves. Nevertheless, in the case of AES we can still count and compare S-box computations as an estimate for the optimized brute-force complexity.

In the following, we give a basic example of an optimized brute-force attack on AES-128. Instead of trying keys randomly, we first iterate over all values of a single key byte and fix the remaining 15 key bytes. Hence, we only compute the whole cipher once for a base key, and recompute only those parts of the state which change when iterating over the 2^8 values for a single byte. Fig. 2 shows those bytes in white, which do not need to be recomputed for every key candidate. The number of white bytes also roughly determines the complexity reduction compared to a black-box brute-force attack. To save additional recomputations on the last two rounds, we match the ciphertext only on four instead of all 16 bytes.

For each set of 2^8 keys, we save the computation of 15 S-boxes of state S_1, 9 S-boxes of state S_2, 12 S-boxes of state S_9 and 12 S-boxes of state S_{10}, in total 48 S-boxes. In the key schedule, we save the computation of 4 S-boxes in rk_0, 3 S-boxes in rk_1, rk_2 and rk_{10}, 2 S-boxes in rk_3 and 1 S-box in rk_4 in total 16 S-boxes. Hence, instead of 200 S-boxes we need to compute only 136 S-boxes or 0.68 full AES-128 evaluations. Therefore, the total complexity of this optimized brute-force attack is about

$$2^{120} \cdot (1 + 255 \cdot 0.68) = 2^{127.45}$$

full AES-128 evaluations.

4   Simplified Biclique Attack for Hardware Implementation

To evaluate the biclique attack [3] in hardware, Bogdanov et al. have proposed a simplified variant of the biclique attack [2] which is more suitable for practical implementations. To verify that the attack is practically faster than a black-box brute-force attack, they modified the biclique attack to reduce its data complexity and simplified the matching phase, where the full key is recovered. The main difference to the original biclique attack is that only a 2-dimensional biclique is applied on the first two rounds instead of the last three rounds. Furthermore, the key recovery phase matches on four bytes of the ciphertext instead of some intermediate state. This simplifies the matching phase since it is basically just a forward computation from state S_3 to these four bytes of the ciphertext. In this section, we briefly cover the theory of the modified attack from [2] and give a comparison of each step to a black-box brute-force attack on AES.

4.1   Biclique Construction

The modified biclique attack targets AES-128. The biclique originates from the idea of initial structures [10] and is placed on the initial two rounds and the meet-in-the-middle matching to recover the encryption key is done on the remaining eight rounds. The key space is divided into 2^{124} groups of 2^4 keys each. These key groups are constructed from the initial cipher key rk_0 (whitening key) and do not overlap. The partitioning of the key space defines the dimension d of the biclique which is $d = 2$.

Each key group is constructed from a base key. We retrieve the base keys by setting the two least significant bits of $rk_0[0]$ and $rk_0[6]$ to zero and iterating the remaining bits of rk_0 over all possible values. Hence, bytes 0 and 6 of the base key have the binary value $b = b_0b_1b_2b_3b_4b_500$. To get the 16 keys within a key group, we enumerate differences $i, j \in \{0, \ldots, 3\}$ and add them to bytes 0, 4, 6, and 10 of the base key (see rk_0 in Fig. 3). Note, that we add the same difference i to bytes 0 and 4 as well as difference j to bytes 6 and 10. This is done to cancel

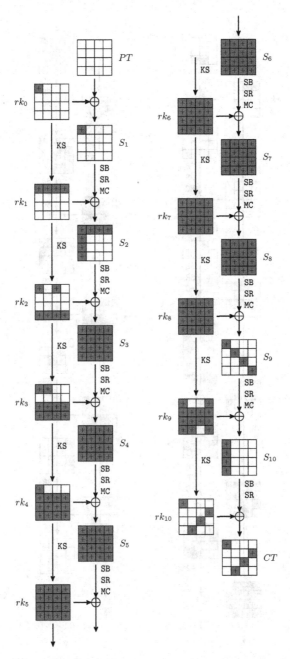

Fig. 2. Partial states which have to be recomputed for each set of 2^8 keys in the optimized brute-force attack. Only bytes ⊞ (input to S-boxes) need to be recomputed for every key guess. All empty bytes are recomputed only once for each set of 2^8 keys. In total, we save the computation of 48 S-boxes in the state update and 16 S-boxes in the key schedule.

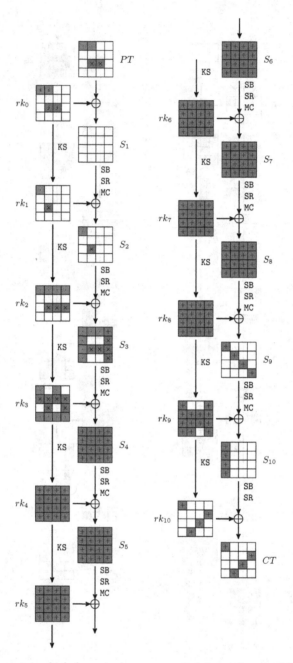

Fig. 3. Partial states which have to be recomputed for each set of 2^{16} keys in the hardware biclique attack. Blue bytes (⊞) need to be recomputed for every key guess. White bytes are recomputed only for each set of 2^{16} keys. Bytes indicated by ⬛ and ⊠ are used in the biclique structure and need to be recomputed 7 times for each set of 2^{16} keys.

some differences in the following round key such that only bytes 0 and 6 of rk_1 have non-zero differences.

Similar to the original biclique attack, we use these key differences to construct two differential trails Δ_i and $\triangledown_j$. The Δ_i-trail is constructed from the key difference Δ_i^K which covers all keys with $i \in \{1, \ldots, 3\}$ and $j = 0$. For the key difference $\triangledown_j^K$ ($\triangledown_j$-trail), we fix $i = 0$ and enumerate $j \in \{1, \ldots, 3\}$. Additionally, the key differences are also used as plaintext differences for the respective trail.

Finally, to construct the biclique, we combine both trails as shown in Fig. 3. This yields a mapping of 16 plaintexts to 16 values for the intermediate state S_3 under the 16 keys of a group[1]. Since both differential trails are not fully independent (they both have byte 12 active in S_3), we have to consider and recompute this byte separately as described below.

Until here, we have constructed the biclique from two (almost) independent differential trails. This yields 16 values for S_3 for the 16 plaintexts encrypted under the corresponding key from a key group. Thus, for each key group, we can retrieve a different set of 16 values for S_3. This enables an effective computation of 16 values for S_3 by performing the following steps for each key group:

1. Perform the base computation by taking the all-zero plaintext and encrypting it with the base key of the group. Store the resulting value for S_3 (S_3^0) and the value for rk_2 (rk_2^0).
2. Enumerate the difference $i \in \{1, 2, 3\}$, set $j = 0$ and recompute the active byte of the combined differential trail (indicated by ■ in Fig. 3). This yields three values for S_3, denoted by S_3^i.
3. Perform similar computations for $j \in \{1, 2, 3\}$ and $i = 0$ to get three more values for S_3, denoted by S_3^j (indicated by ■).
4. Combine the values for S_3^0 from the base computation, S_3^i and S_3^j to get the 16 values for S_3[2].
5. Since $S_3[12]$ is active in both differential trails, we consider this byte separately and retrieve its value by calculating $S_3^j[12] \oplus i$ or alternatively $S_3^i \oplus j$.

The advantage of the biclique construction over a black-box brute-force attack is that we save S-box computations by simply combining precomputed values for S_3. For a black-box brute-force attack, we perform 16 3-round AES computations to get 16 values for S_3. Here, we compute the full three rounds only for the base computation and then recompute only the required bytes for each differential trail for $i, j \in \{1, 2, 3\}$.

There is one possible improvement that enables us to save some S-box evaluations in the matching phase: Instead of computing values for S_3, we can include the following SubBytes and ShiftRows operations and compute $S_{3,\text{MC}}$ (the state after MixColumns in round 3) instead. In the base computation, we compute the

---

[1] Note that this does not exactly match the definition of a biclique as it maps 2^{2d} plaintexts to 2^{2d} states under 2^{2d} keys.

[2] To get the values for rk_2, we just add the i, j differences to the corresponding bytes of rk_2^0. This is possible because no S-box has to be computed for the active key bytes in the key schedule.

S-box for bytes 5,7,9,11, and for each of the differential trails we compute the S-box for their corresponding active bytes. This leaves only byte 12, which we have to compute for all 16 possible values of $S_{3,\text{MC}}$ (see also [2]).

4.2   Key Recovery

As already described, the matching phase is simplified to match on four bytes of the ciphertexts. Thus, we first take the 16 plaintexts and retrieve the corresponding ciphertexts from the encryption oracle (This has to be performed only once for the full attack). From the output of the biclique (16 values for $S_{3,\text{MC}}$), we then compute only the required bytes to get the four ciphertext bytes. As shown in Fig. 3, we compute the full states from S_4 up to $S_{7,\text{MC}}$. From this state on until the ciphertext, we only compute reduced rounds. For the resulting four ciphertext bytes, we check if they match the corresponding ciphertext bytes retrieved from the encryption oracle. If they do, we have a possible key candidate which we have to verify with one additional plaintext-ciphertext pair. However, we should get only about one false key candidate per 2^{32} keys.

The advantage of this phase over a black-box brute-force attack is that we avoid some S-box computations by matching on only four bytes of the ciphertext instead of all 16 bytes. Note that this part is identical to the optimized brute-force attack described in Section 3.2.

4.3   Complexity of the Attack

The data complexity is defined by the biclique and is thus 2^4. Concerning the time complexity, Bogdanov et al. estimated the computation of the 16 values for S_3 using the biclique (precomputation phase) to be at most 0.3 full AES-128 encryptions. For the matching phase they estimated an effort similar to 7.12 AES-128 encryptions. Thus, the time complexity is

$$2^{124} \cdot (0.3 + 7.12) = 2^{126.89}$$

AES-128 encryptions.

However, our calculation yields a slightly higher complexity: The biclique computation requires 8 S-boxes for key schedule up to rk_2, 32 S-boxes for the base computation and 6 S-boxes for the recomputations of both trails. Moreover, for computing $S_{3,\text{MC}}$, we have to compute 4 S-boxes for the inactive bytes, $4 \times 11 = 44$ for the active bytes of both differential trails in round 3, and 16 S-box evaluations for computing byte 12. This results in a total of 110 S-boxes. Computing three AES rounds (incl. the key schedule) for 16 keys, as it would be done in a brute force attack, takes $16 \times 3 \times 20 = 960$ S-Box computations. Thus, the precomputation phase is about the same as 0.11 3-round AES-128 encryptions or equivalently 0.55 full AES-128 executions.

The matching phase has to be performed for every one of the 16 keys in a key group. This phase requires 80 S-boxes for computing the full rounds 4-8, 8 S-boxes for the last two rounds and 29 S-boxes for the key schedule. This results in a total of $16 \times 117 = 1872$ S-box evaluations per key group. Since a brute-force attack for 16 keys on seven rounds requires $16 \times 7 \times 20 = 2240$ S-box computations, the matching phase is about the same as 0.84 7-round AES-128 encryptions or equivalently 9.36 full AES-128 executions. The resulting time complexity of the full modified biclique attack is thus

$$2^{124} \cdot (0.55 + 9.36) = 2^{127.31}.$$

5 Software Implementations and Benchmark Results

To compare the two brute-force attacks with the biclique attack, we have implemented all three variants in software using AES-NI. Of course, GPU or dedicated hardware implementations of these attacks will always perform better and are less costly than software implementations on standard CPUs. However, we use these software implementations to provide a more transparent and fair comparison of the attacks. The efficiency of hardware implementations depends a lot on the used device or underlying technology, which may be in favor of one or the other attack. Moreover, which attack performs better also depends a lot on the effort which has been spent in optimizing the attack.

In the case of software implementations using AES-NI all attacks have the same precondition. If AES-NI benefits one of the attacks, it is the black-box brute-force attack which needs to compute only complete AES rounds. In the following, we will show that nevertheless, both the optimized brute-force attack as well as the biclique attack are slightly faster than the generic attack.

For each attack, we have created and benchmarked a set of implementations to rule out less optimized versions:

- assembly implementations and C implementation using intrinsics
- parallel versions using 4x and 8x independent AES-128 executions
- benchmarked on Intel Westmere and Intel Sandy Bridge CPUs

Since Intel AES-NI has different implementation characteristics on Westmere and Sandy Bridge, we get slightly different results but the overall ranking of attacks does not change. We have also tried to used AVX instructions to save some mov operations. However, this has only a minor impact on the results. Note that in the C intrinsics implementations the compiler automatically uses AVX if it is available on the target architecture.

5.1 Black-Box Brute-Force Implementation

The implementation of the black-box brute-force attack is quite straightforward. Nevertheless, to find the fastest implementation we have evaluated several approaches. We have implemented two main variants, which test eight or four keys in parallel. In the 8x parallel variant, the 6-cycle (Westmere) or 8-cycles (Sandy Bridge) latency can be hidden. However, we need more memory accesses compared to the 4x variant, since we cannot store all keys, states and temporary values in the 16 128-bit registers. Therefore, the 4x variant may be faster in some cases.

The main bottleneck of AES implementations using AES-NI is the rather slow `aeskeygenassist` instruction. Therefore, the implementations do not reach the speed given by common AES-NI benchmarks without key schedule recomputations. Since the throughput of the instruction `aeskeygenassist` is much lower than the `aesenc` instruction, we compute the key schedule on-the-fly. This way, we also avoid additional memory operations.

The full performance measurements for all implementations are shown in Table 1. The 8x variants test eight keys in parallel but require memory access, the 4x variants test four keys in parallel without any memory access. The table shows nicely that the memory-less implementation can in fact be faster under certain circumstances. Overall, the performance of an implementation depends highly on the latency of the AES instructions. E.g. on the Sandy Bridge architecture, the instructions take longer and testing only four keys in parallel does not utilize the CPU pipeline optimally.

Table 1. Performance measurements for the various software implementations of the *black-box brute-force attack, optimized brute-force attack* and *biclique attack*. All values are given in cycles/byte. Best results per architecture and implementation are written in bold.

Approach	Black-Box Brute-F.		Optimized Brute-F.		Biclique (Modified)	
	Westmere	Sandy B.	Westmere	Sandy B.	Westmere	Sandy B.
C, 4x	3.09	**3.80**	3.20	3.70	2.71	**3.18**
C, 4x, rnd 1 full			**3.10**	3.75		
ASM, 4x	**3.00**	3.80			**2.61**	3.24
ASM-AVX, 4x		3.80				3.21
C, 8x	3.45	3.86	3.52	3.89	3.41	3.39
C, 8x, rnd 1 full			3.24	**3.67**		
ASM, 8x	3.40	3.95				
ASM-AVX, 8x		3.93				

5.2 Optimized Brute-Force Attack Implementation

The idea for the optimized brute-force attack is to avoid some computations by iterating all possible keys in a more structured way. As we have seen in Section 3.2, this slightly reduces the time complexity of the attack. To verify this, we implemented multiple variants of this attack (4 and 8 keys in parallel) using AES-NI.

However, since Intel AES-NI is optimized for computing full AES rounds, it is not possible to efficiently recompute only the required bytes and S-boxes.

Hence, we often compute the full state although we only need some parts of it. This is for instance the case for round 2 (see Fig. 2). The additional instructions required to perform only a partial encryption in this round take longer than just using `aesenc` to compute the full encryption round.

Table 1 lists the full measurements for this attack. We also included a comparison between computing the full round 1 and computing only the reduced round as it is the idea for this attack. The results clearly show that it is also faster (in most cases) to just compute the full round instead of a reduced round.

Nevertheless, the last 2 rounds can be computed as reduced rounds since one round does not contain `MixColumns` which makes the rearrangement of states less complex. For these last 2 reduced rounds we collect the required bytes of four states into one 128-bit register and use one `aesenc` (or `aesenclast`) instruction to compute the remaining bytes. In the matching phase, we save computations in the last two rounds. For testing 4 keys, we need to compute only 2 AES rounds and some recombinations instead of $4 \cdot 2$ AES rounds.

5.3   Biclique Attack Implementation

The modified biclique attack by Bogdanov et al. as covered in Section 4 has several advantages when implemented in software. Most notable are the low data complexity of only 16 plaintext-ciphertext pairs and the simple matching phase at the end of the cipher on four ciphertext bytes. These modifications allow us to implement the attack with almost no memory operations.

To better exploit the full-round `aesenc` instruction, we compute the biclique only until state S_3 and not $S_{3,\text{MC}}$. Since the biclique attack considers groups of 16 keys, our main loop consists of three steps:

1. Precompute values for S_3^0, rk_2 and active bytes of S_3^i, S_3^j.
2. Combine the precomputed bytes to 16 values for the full state S_3.
3. Encrypt the remaining rounds and match with the ciphertexts from the encryption oracle.

For the biclique computation we basically follow the steps given in Section 4.1. However, to compute the differential trails, we directly compute the combined trail for $i = j \in \{1, 2, 3\}$. Afterwards, we extract the values for S_3^i and S_3^j to construct 16 values for S_3 using the base computation. Consequently, we perform 4 full computations of the first 2 rounds per key group. This phase is equal for the 4x and 8x variants.

The 5 full-round computations of the third step are exactly the same as in the black-box brute-force attack. The 3 round computations at the end and the matching are the same as in the optimized brute-force attack. Similar to these attacks, we have implemented two variants, which compute and match either 4 or 8 keys in parallel. Again, we compute the key schedule on-the-fly since this results in faster code.

In general, our implementations of the biclique attack are quite similar to the brute-force attacks. Similar to the optimized brute-force attack, we also compute full round more often than necessary. For example, in round 3 and round 8 we compute full rounds although we would not need to. Especially computing only SubBytes by aesenclast followed by pshufb is more expensive than computing a whole round.

Overall, for the full matching phase (4x and 8x variants) we compute six full rounds with aesenc and then compute the remaining two rounds as reduced rounds. As can be seen in Fig. 2 and Fig. 3 the reduced rounds of the biclique attack and the optimized brute-force attack are in fact equal. Consequently, the implementation of the matching phase is equal to the last 8 rounds of the optimized brute-force attack.

5.4 Performance Results

We have tested our implementations of all three attacks on Intel Westmere (Mac-Book Pro with Intel Core i7 620M, running Ubuntu 11.10 and gcc 4.6.1) and Intel Sandy Bridge (Google Chromebox with Intel Core i5 2450M, running Ubuntu 12.04 and gcc 4.6.3). For the C implementations we used the following compiler flags: -march=native -O3 -finline-functions -fomit-frame-pointer -funroll-loops. All results are shown in Table 1.

Overall, the biclique attack is 13% faster on Westmere and 17% faster on Sandy Bridge, compared to the best black-box brute force on the same architecture. This clearly verifies that the biclique attack is faster than both brute-force attacks in all scenarios, although the advantage is smaller than in theory. Note that the Sandy Bridge implementations are slower in general but provide a larger advantage over the black-box brute-force attack.

The performance of the optimized brute-force attack varies depending on the CPU architecture and is actually slower than the black-box brute-force attack on Westmere CPUs. On Sandy Bridge, the optimized brute-force attack is 3% faster than the black-box brute-force attack. However, assembly implementations of the optimized brute-force attack may slightly improve the results.

If we compare Sandy Bridge implementations, the biclique attack results in a time complexity of about $2^{127.77}$ full AES-128 computations. For the optimized brute-force attack we get a complexity of $2^{127.95}$ in the best case. In the theoretical evaluation of the modified biclique attack, we have estimated an average complexity of 9.9 AES-128 encryptions to test 16 keys. However, using AES-NI we are able to get a complexity of only 14 AES-128 encryptions to test 16 keys.

6   Conclusions

In this work, we have analyzed three different types of single-key attacks on AES-128 in software using Intel AES-NI. The first attack is the black-box brute-force attack with a generic exhaustive key search complexity of 2^{128} AES computations. We have used this implementation as the base line for a comparisons of other attacks faster than brute-force. We get the best advantage of the faster than brute-force attacks on Sandy Bridge CPUs. In this case, the simplified bi-clique attack by Bogdanov et al. is 17% faster than the black-box brute-force attack, while our simple optimized brute-force attack is only 3% faster.

Note that we did not put much effort in the optimized brute-force attack. More clever tricks, better implementations or using a different platform may still improve the result. Nevertheless, neither the optimized brute-force attack nor the biclique attack threaten the security of AES in any way, since still all 2^n keys have to be tested. In this sense, both attacks just perform an exhaustive key search in a more or less optimized and clever way.

With this paper, we hope to provide a basis for a better comparison of close to brute force attacks. The open problem is to distinguish between clever brute-force attacks and worrisome structural attacks which may extend to attacks with less than a marginal improvement over the generic complexity. Based on our implementation and analysis, the biclique attack in its current form is probably not such an attack. Therefore, we believe that trying to improve the biclique attack itself is more important than merely applying it to every published cipher.

Acknowledgements. We thank Vincent Rijmen for valuable comments. This work has been supported in part by the Secure Information Technology Center-Austria (A-SIT) and by the Austrian Science Fund (FWF), project TRP251-N23.

References

1. Biryukov, A., Khovratovich, D.: Related-Key Cryptanalysis of the Full AES-192 and AES-256. In: Matsui, M. (ed.) ASIACRYPT 2009. LNCS, vol. 5912, pp. 1–18. Springer, Heidelberg (2009)
2. Bogdanov, A., Kavun, E.B., Paar, C., Rechberger, C., Yalcin, T.: Better than Brute-Force Optimized Hardware Architecture for Efficient Biclique Attacks on AES-128. In: Workshop records of Special-Purpose Hardware for Attacking Cryptographic Systems – SHARCS 2012, pp. 17–34 (2012), http://2012.sharcs.org/record.pdf
3. Bogdanov, A., Khovratovich, D., Rechberger, C.: Biclique Cryptanalysis of the Full AES. In: Lee, D.H., Wang, X. (eds.) ASIACRYPT 2011. LNCS, vol. 7073, pp. 344–371. Springer, Heidelberg (2011)
4. Fog, A.: Instruction tables – Lists of instruction latencies, throughputs and micro-operation breakdowns for Intel, AMD and VIA CPUs (2012), http://www.agner.org/optimize/instruction_tables.pdf (accessed September 2, 2012)
5. Gaj, K.: ATHENa: Automated Tool for Hardware EvaluatioN (2012), http://cryptography.gmu.edu/athenadb/fpga_hash/table_view (accessed February 1, 2013)
6. Hellman, M.E.: A cryptanalytic time-memory trade-off. IEEE Transactions on Information Theory 26(4), 401–406 (1980)
7. Intel Corporation:ntel® Advanced Encryption Standard (AES) Instruction Set, White Paper. Tech. rep., Intel Mobility Group, Israel Development Center, Israel (January 2010)

8. Intel Corporation: Intel® 64 and IA-32 Architectures Software Developer's Manual. Intel Corporation (March 2012)

9. NIST: Specification for the Advanced Encryption Standard (AES). National Institute of Standards and Technology (2001)

10. Sasaki, Y., Aoki, K.: Finding Preimages in Full MD5 Faster Than Exhaustive Search. In: Joux, A. (ed.) EUROCRYPT 2009. LNCS, vol. 5479, pp. 134–152. Springer, Heidelberg (2009)

11. SHA-3 Zoo Editors: SHA-3 Hardware Implementations (2012), http://ehash.iaik.tugraz.at/wiki/SHA-3_Hardware_Implementations (accessed February 1, 2013)

Sieving for Shortest Vectors in Ideal Lattices

Michael Schneider

Technische Universität Darmstadt, Germany
`mischnei@cdc.informatik.tu-darmstadt.de`

Abstract. Lattice based cryptography is gaining more and more importance in the cryptographic community. It is a common approach to use a special class of lattices, so-called ideal lattices, as the basis of lattice based crypto systems. This speeds up computations and saves storage space for cryptographic keys. The most important underlying hard problem is the shortest vector problem. So far there is no algorithm known that solves the shortest vector problem in ideal lattices faster than in regular lattices. Therefore, crypto systems using ideal lattices are considered to be as secure as their regular counterparts.

In this paper we present IdealListSieve, a variant of the ListSieve algorithm, that is a randomized, exponential time sieving algorithm solving the shortest vector problem in lattices. Our variant makes use of the special structure of ideal lattices. We show that it is indeed possible to find a shortest vector in ideal lattices faster than in regular lattices without special structure. The practical speedup of our algorithm is linear in the degree of the field polynomial. We also propose an ideal lattice variant of the heuristic GaussSieve algorithm that allows for the same speedup.

Keywords: Ideal Lattices, Shortest Vector Problem, Sieving Algorithms.

1 Introduction

Lattices are discrete additive subgroups of $\mathbb{R}^m$. Their elements can be considered to be vectors in the Euclidean vector space. One of the most important computational problems in lattices is the shortest vector problem (SVP). Roughly speaking, given a representation of a lattice, it asks to output a shortest non-zero element of the lattice.

In 2001, Ajtai, Kumar, and Sivakumar presented a randomized algorithm to solve the shortest vector problem in lattices. Unfortunately, the space requirement of the AKS *sieving algorithm* is exponential in the lattice dimension, and therefore the algorithm was not practical. In 2010, Micciancio and Voulgaris presented two variants of this sieving algorithm that still require exponential storage, but with much smaller constants. Their algorithm is the first sieving approach considered to be competitive to enumeration algorithms that solve SVP and only require polynomial space.

Lattices are widely used in cryptography. There, in order to save storage for cryptographic keys, it is common to use structured lattices. The NTRU crypto system [HPS98] for example uses so-called cyclic lattices, where for each lattice

A. Youssef, A. Nitaj, A.E. Hassanien (Eds.): AFRICACRYPT 2013, LNCS 7918, pp. 375–391, 2013.
© Springer-Verlag Berlin Heidelberg 2013

vector $\mathbf{v}$, its rotations $\mathsf{rot}(\mathbf{v})$ consisting of the rotated lattice vectors are also elements of the lattice. The work of [Mic07] initiated the usage of more general, structured lattices. These are used in the signature schemes [LM08, Lyu09], for the encryption systems [SSTX09, LPR10], the SWIFFTX hash function family [LMPR08, ADL+08], or the fully homomorphic encryption scheme of [Gen09], for example. They were called ideal lattices in [LM06]. The theory of ideal lattices is based on the work of [PR06, LM06]. Cyclic lattices are a special case of ideal lattices.

Micciancio proved a worst-case to average-case reduction for ideal lattices in [Mic07], where he showed that inverting his one-way function is as hard as solving ideal lattice problems in the worst case. Lattice problems are even considered unbroken in the presence of quantum computers, and so far they withstand sub-exponential attacks. Thus cryptographic schemes based on hard problems in ideal lattices are good candidates for security requirements in the near and far future.

So far there is no SVP algorithm making use of the special structure of ideal lattices. It is widely believed that solving SVP (and all other lattice problems) in ideal lattices is as hard as in regular lattices. Our intention is to show how sieving algorithms can be strengthened in ideal lattices using their circular structure. The idea was already presented in [MV10b]. There, the authors assume that the amount of storage required by their algorithm decreases with a factor of n, where n is the degree of the field polynomial. We show that even more not only the storage but as well the running time of sieving algorithms decreases by a factor of n. This is an important starting point for assessing the hardness of problems in ideal lattices.

1.1 Related Work

There are basically three approaches to solve the SVP in lattices: Voronoi-cell based algorithms, enumeration algorithms, and probabilistic sieving algorithms. The Voronoi cell based algorithms were presented in [MV10a]. It is the first deterministic single exponential algorithm for the shortest vector problem. So far, this type of algorithms is more-of theoretical interest. Enumeration algorithms solve the SVP deterministically in asymptotic time $2^{\mathcal{O}(m \log(m))}$, where m is the lattice dimension [HS07, PS08]. They perform exhaustive search by exploring all lattice vectors of a bounded search region. Enumeration algorithms can be rendered probabilistic using an extreme pruning strategy [GNR10], which allows for an exponential speedup and makes enumeration the fastest algorithm for solving SVP in practice. Enumeration algorithms only require polynomial space.

Sieving algorithms were first presented by Ajtai, Kumar, and Sivakumar [AKS01] in 2001. The runtime and space requirements were proven to be in $2^{\mathcal{O}(m)}$. Nguyen and Vidick [NV08] carefully analyzed this algorithm and presented the first competitive timings and results. They show that the runtime of AKS sieve is $2^{5.90m+o(m)}$ and the space required is $2^{2.95m+o(m)}$. The authors also presented a heuristic variant of AKS sieve without perturbations. Their runtime is $2^{0.41m+o(m)}$ and they require space $2^{0.21m+o(m)}$. In 2010, Micciancio and

Voulgaris [MV10b] presented a provable sieving variant called ListSieve and a more practical, heuristic variant called GaussSieve. ListSieve runs in time $2^{3.199m+o(m)}$ and requires space $2^{1.325m+o(m)}$. For GaussSieve, the maximum list size can be bounded by the kissing number τ_m, whereas, due to collisions, a runtime bound can not be proven. The practical runtime is $2^{0.48m}$ seconds, the space requirements is expected to be less than $2^{0.18m}$ and turns out to be even smaller in practice.

Pujol and Stehlé [PS09] improve the theoretical bounds of ListSieve [MV10b] using the birthday paradox to runtime $2^{2.465m+o(m)}$ and space $2^{1.233m+o(m)}$. Wang et at. [WLTB10] present a heuristic variant of the Nguyen-Vidick sieve running in $2^{0.3836m+o(m)}$ with space complexity of $2^{0.2557m}$. The work of [BN09] deals with all ℓ_p norms, generalizing the AKS sieve. There is only one public implementation of a sieving algorithm, namely `gsieve` [Vou10], which implements the GaussSieve algorithm of [MV10b].

Using heuristics like extreme pruning [GNR10], enumeration algorithms outperform sieving algorithms, as the SVP challenge [GS10] shows. We hope that it is possible to integrate heuristics such as extreme pruning to sieving algorithms, which would make them competitive to enumeration techniques again.

1.2 Our Contribution

Micciancio and Voulgaris already mention the possibility to speed up the sieving for ideal lattices [MV10b]. They propose to use the cyclic rotations of each sampled vector to reduce the size of the vectors. For ideal lattices, the "rotation" of each lattice vector is still an element of the lattice. Therefore, it can be used in the sieving process. The authors of [MV10b] expect a reduction of the list size linear in the degree of the field polynomial for ListSieve, and a substantial impact on the practical behaviour of the GaussSieve algorithm. In this paper, we present experimental results using this approach. We implement ListSieve and IdealListSieve without perturbations. Our experiments show that indeed the storage requirements decrease as expected. But even more, sieving in ideal lattices can find a shortest lattice vector much faster, with a speedup factor linear in the degree of the field polynomial in practice. To explain the results, we use the assumption that the number of reductions used in the sieving process stays the same in both the original and the ideal cases. We will show that this assumption conforms with our experiments.

To give an example from our experiments, the measured and fitted runtime of IdealListSieve in cyclic lattices is $2^{0.52m-21.9}$ seconds, compared to $2^{0.67m-26.3}$ seconds for ListSieve, where m is the lattice dimension. In dimension $m = 60$, the runtime difference is about 4 hours, which corresponds to a time advantage of 95%. The worst-case runtime of IdealListSieve remains the same as for ListSieve, since considering all rotations cancels out the linear factor in theory.

To our knowledge, this is the first exact SVP algorithm that can use the special structure of ideal lattices. (For cyclic NTRU lattices, there is a LLL-variant using the cyclic rotations [MS01], but this algorithm only finds vectors of size exponential in m, not shortest vectors.) It is often stated that solving problems in ideal lattices is as hard as in the general case, among others in [MR08, ADL+08, Lyu09]. Since the runtime of sieving algorithms is exponential, this linear speedup does not effect the asymptotic runtime of sieving algorithms. It only helps to speed up sieving in ideal lattices in practice noticeably. For the fully homomorphic encryption challenges for example, n is bigger than 2^{10}, which would result in a speedup of more than 1000 for sieving. [Lyu09] uses $n \geq 512$. These numbers show that, if one could run sieving in such high dimensions $m > 1000$, even linear speedup might lead to huge improvements in practice.

1.3 Organization of This Paper

In Section 2 we present some basic notation and facts on (ideal) lattices. In Section 3 we show the IdealListSieve algorithm. In Section 4 we give some theoretical analysis, and Section 5 shows experimental results of our implementation. Finally we give a conclusion, including some ideas for enumeration in ideal lattices.

2 Preliminaries

Define the index set $[n] = \{0, 1, \ldots, n-1\}$. $\mathbf{I}_m$ is the identity matrix in dimension m, $\mathbf{0}_m$ and $\mathbf{1}_m$ are m-dimensional column vectors consisting of 0 and 1 entries only. The scalar product of two vector elements $\mathbf{x}$ and $\mathbf{y}$ is written $\langle \mathbf{x} \mid \mathbf{y} \rangle$. Throughout this paper, n denotes the degree of polynomials and m is the dimension of lattices.

Let $m, d \in \mathbb{N}$, $m \leq d$, and let $\mathbf{b}_1, \ldots, \mathbf{b}_m \in \mathbb{R}^d$ be linearly independent. Then the set $\mathcal{L}(\mathbf{B}) = \{\sum_{i=1}^m x_i \mathbf{b}_i : x_i \in \mathbb{Z}\}$ is the lattice spanned by the basis column matrix $\mathbf{B} = [\mathbf{b}_1, \ldots, \mathbf{b}_m] \in \mathbb{Z}^{d \times m}$. $\mathcal{L}(\mathbf{B})$ is called m-dimensional. The basis $\mathbf{B}$ is not unique, unimodular transformations lead to a different basis of the same lattice. The first successive minimum $\lambda_1(\mathcal{L}(\mathbf{B}))$ is the length of a shortest vector of $\mathcal{L}(\mathbf{B})$. The (search) shortest vector problem (SVP) is formulated as follows. Given a basis $\mathbf{B}$, compute a vector $\mathbf{v} \in \mathcal{L}(\mathbf{B}) \setminus \{\mathbf{0}\}$ subject to $\|\mathbf{v}\| = \lambda_1(\mathcal{L})$. It can be formulated in every norm, we will consider the Euclidean norm ($\|\cdot\| = \|\cdot\|_2$) in the remainder of this paper.

The lattice determinant $det(\mathcal{L}(\mathbf{B}))$ is defined as $\sqrt{det(\mathbf{B}^t \mathbf{B})}$. It is invariant under basis changes. For full-dimensional lattices, where $m = d$, there is $det(\mathcal{L}(\mathbf{B})) = |det(\mathbf{B})|$ for every basis $\mathbf{B}$. In the remainder of this paper we will only be concerned with full-dimensional lattices ($m = d$).

2.1 Ideal Lattices

Ideal lattices are lattices with special structure. Let $\mathbf{f} = x^n + \mathbf{f}_n x^{n-1} + \ldots + \mathbf{f}_1 \in \mathbb{Z}[x]$ be a monic polynomial of degree n, and consider the ring $\mathbf{R} = \mathbb{Z}[x]/\langle \mathbf{f}(x) \rangle$. Elements in $\mathbf{R}$ are polynomials of maximum degree $n-1$. If $\mathbf{I} \subseteq \mathbf{R}$ is an ideal in $\mathbf{R}$, each element $\mathbf{v} = \sum_{i=1}^n \mathbf{v}_i x^{i-1} \in \mathbf{I}$ naturally corresponds to its coefficient vector $(\mathbf{v}_1, \ldots, \mathbf{v}_n) \in \mathbb{Z}^n$. Since ideals are additive subgroups, the set of all coefficient vectors corresponding to the ideal $\mathbf{I}$ forms a so-called *ideal lattice*.

For the sake of simplicity we can switch between the vector and the ideal notations and use the one that is more suitable in each case.

For each $\mathbf{v} \in \mathbf{R}$, the elements $x^i \cdot \mathbf{v}$ for $i \in [n]$ form a basis of an ideal lattice. We call this multiplication with x a *rotation*, since for special polynomials the vector $x \cdot \mathbf{v}$ consists of the rotated coefficients of $\mathbf{v}$. In vector notation, the multiplication of an element with x can be performed by multiplication with the matrix

$$\text{rot} = \left(\begin{array}{c|c} \mathbf{0}_{n-1}^t & -\bar{\mathbf{f}} \\ \mathbf{I}_{n-1} & \end{array} \right) , \tag{1}$$

where $\bar{\mathbf{f}}$ consists of the coefficients of the polynomial $\mathbf{f}$. If $\mathbf{f} \in \mathbf{R}$ is a monic, irreducible polynomial of degree n, then for any element $\mathbf{v} \in \mathbf{R}$, the elements $\mathbf{v}, \mathbf{v}x, \ldots, \mathbf{v}x^{n-1}$ are linearly independent (see for example the proof of Lemma 2.12 in [Lyu08]). For $\mathbf{f}(x) = x^n - 1$, which is not irreducible over $\mathbb{Z}$, it is easy to see that the vectors $\mathbf{v}x^i$ are also linearly independent, unless the vector has very special form.

The row matrices of the bases used in practice are of the form

$$\left(\begin{array}{cc} q\mathbf{I}_n & \mathbf{0} \\ (\text{rot}^i(\mathbf{v}))_{i \in [n]} & \mathbf{I}_n \end{array} \right) . \tag{2}$$

Here the lower part consists of the n rotations of $\mathbf{v}$, which correspond to the multiplications of the ring element $\mathbf{v}$ with x^i for $i \in [n]$. The upper part is necessary to make sure that every element in the lattice can be reduced modulo q. Bases for higher dimensional lattices can be generated using multiple points $\mathbf{v}_i$ and their rotations. The dimension m of the lattice is then a multiple of the field polynomial degree n.

The usage of ideal lattices reduces the storage amount for a basis matrix from nm elements to m elements, because every block of the basis matrix is determined by its first row. In addition, for an ideal basis $\mathbf{B}$, the computation $\mathbf{B} \cdot \mathbf{y}$ can be sped up using Fast Fourier transformation from $\mathcal{O}(mn)$ to $\tilde{\mathcal{O}}(m)$.

In this paper we are concerned with three types of ideal lattices, defined by the choice of $\mathbf{f}$:

- *Cyclic lattices*: Let $\mathbf{f}_1(x) = x^n - 1$, i.e., $\bar{\mathbf{f}} = (-1, 0, \ldots, 0)$. We call the ideal lattices of the ring $\mathbf{R}_1 = \mathbb{Z}[x]/\langle \mathbf{f}_1(x) \rangle$ cyclic lattices. $\mathbf{f}_1(x)$ is never irreducible over $\mathbb{Z}$ ($x - 1$ is always a divisor), therefore cyclic lattices do not guarantee worst-case collision resistance. The rotation of $\mathbf{v}$ is $\text{rot}(\mathbf{v}) = (\mathbf{v}_{n-1}, \mathbf{v}_0, \ldots, \mathbf{v}_{n-2})$.
- *Anti-cyclic lattices*: Let $\mathbf{f}_2(x) = x^n + 1$, i.e., $\bar{\mathbf{f}} = (1, 0, \ldots, 0)$. We call the ideal lattices of the ring $\mathbf{R}_2 = \mathbb{Z}[x]/\langle \mathbf{f}_2(x) \rangle$ anti-cyclic lattices. $\mathbf{f}_2(x)$ is irreducible over $\mathbb{Z}$ if n is a power of 2. The rotation of $\mathbf{v}$ is $\text{rot}(\mathbf{v}) = (-\mathbf{v}_{n-1}, \mathbf{v}_0, \ldots, \mathbf{v}_{n-2})$. Anti-cyclic lattices are the ones used most in cryptography.
- *Prime cyclotomic lattices*: Let $\mathbf{f}_3(x) = x^n + x^{n-1} + \ldots + 1$, i.e., $\bar{\mathbf{f}} = (1, \ldots, 1)$. We call the ideal lattices of the ring $\mathbf{R}_3 = \mathbb{Z}[x]/\langle \mathbf{f}_3(x) \rangle$ prime cyclotomic lattices. $\mathbf{f}_3(x)$ is irreducible over $\mathbb{Z}$ if $n + 1$ is prime. The rotation of $\mathbf{v}$ is

$\text{rot}(\mathbf{v}) = (-\mathbf{v}_{n-1}, \mathbf{v}_0 - \mathbf{v}_{n-1}, \dots, \mathbf{v}_{n-2} - \mathbf{v}_{n-1})$. We only consider cyclotomic polynomials of degree n where $n+1$ is prime.[1]

A nice and more detailed overview about ideal lattices is shown in [Lyu08].

3   IdealListSieve Algorithm

In this section we will present the ListSieve algorithm of [MV10b] and introduce the ideal lattice variant IdealListSieve. More details about the implementation will follow in Section 5.

3.1   ListSieve

Recall that the goal is to solve the shortest vector problem, i.e., given a basis $\mathbf{B}$ of a lattice find a non-zero vector $\mathbf{v} \in \mathcal{L}(\mathbf{B})$ with norm equal to $\lambda_1(\mathcal{L}(\mathbf{B}))$. The idea of ListSieve is the following. A list L of lattice vectors is stored. In each iteration of the algorithm, a new random vector $\mathbf{p}$ is sampled uniformly at random from a set of bounded vectors. This vector $\mathbf{p}$ is then reduced using the list L in the following manner. If a list vector $\mathbf{l} \in L$ can be subtracted from $\mathbf{p}$ lowering its norm more than a factor $\delta < 1$, $\mathbf{p}$ is replaced by $\mathbf{p} - \mathbf{l}$. With this, $\mathbf{p}$ gets smaller every time. When the vector has passed the list it is appended to L. It can be shown that in the end, L will contain a vector of minimal length with high probability. When the sampled vector $\mathbf{p}$ is a linear combination of smaller list vectors it will be reduced to $\mathbf{0}$ and not be appended. This rare case is called a *collision*. Collisions are important for runtime proofs (they avert a runtime proof for GaussSieve, for example). For practical issues, they are negligible, since they occur very seldom. Algorithm 1 shows a pseudo-code of ListSieve without perturbations.

Algorithm 1. ListSieve($\mathbf{B}$, targetNorm)

1 List $L \leftarrow LLL(\mathbf{B})$ ▷ *Pre-reduction with the LLL algorithm*
2 **while** *currentBestNorm > targetNorm* **do**
3     $\mathbf{p} \leftarrow$ sampleRandomLatticeVector($\mathbf{B}$)   ▷ *Sampling step*
4   ListReduce($\mathbf{p}, L, \delta = 1 - 1/m$)   ▷ *Reduction step*
5   **if** $\mathbf{p} \neq \mathbf{0}$ **then**
6   L.append($\mathbf{p}$)   ▷ *Append step*
7   **end**
8 **end**
9 **return** $\mathbf{l}_{best}$

Originally, ListSieve does not work with lattice points $\mathbf{p}$, but with a perturbed point $\mathbf{p}+\mathbf{e}$ with a small error $\mathbf{e}$. The use of perturbations is necessary in order to

---

[1] Other cyclotomic polynomials, where $n+1$ is not prime, have different structure, the rotations are hard to implement, and they are seldom used in practice.

upper bound the probability of collisions, which is essential for proving runtime bounds for the algorithm. Since in practice collisions play a minor role we will skip perturbations in our implementation. For the sampling of random vectors in Line 3, [MV10b] use Kleins randomized rounding algorithm [Kle00], which we will also apply for all our implementations.

3.2 IdealListSieve

One of the properties of ideal lattices is that for each lattice vector $\mathbf{v}$, rotations of this vector are also contained in the lattice. This is due to the property of the ideal $\mathbf{I}$ corresponding to the ideal lattice. Ideals in $\mathbf{R}$ are closed under multiplication with elements from $\mathbf{R}$, and since vectors in ideal lattices are the same as elements of the ideal, multiplications of these vectors are also elements of the lattice.

To compute the rotation of a vector $\mathbf{v}$ one has to rotate each block of length n of $\mathbf{v}$. If $m = 2n$, the first half of $\mathbf{v}$ (which belongs to the $q\mathbf{I}_n$ part in the first rows of the basis matrix (2)) is rotated and so is the second half. So when ListSieve tries to reduce the sample $\mathbf{p}$ with a vector $\mathbf{l} = (l_1, \ldots, l_n, l_{n+1}, \ldots, l_m)$, we can also use the vectors

$$\mathbf{l}^{(j)} = \left(\mathrm{rot}^j((l_1, \ldots, l_n)), \mathrm{rot}^j((l_{n+1}, \ldots, l_m))\right), \quad \text{for } j = 1 \ldots n-1,$$

where the first and the second half of the vector is rotated. Therefore, the sample $\mathbf{p}$ can be more reduced in each iteration. Instead of reducing with one single vector $\mathbf{l}$ per entry in the list L, n vectors can be used.

Function IdealListReduce shows a pseudo-code of the function that is responsible for the reduction part. Compared to the ListSieve algorithm of [MV10b], this function replaces the ListReduce function. Unfortunately, only the case where m is a multiple of n allows the usage of rotations of lattice vectors. For the case where $n \nmid m$, it is not possible to apply the rotation to the last block of a lattice vector $\mathbf{v}$.

Func. ListReduce$(\mathbf{p}, L, \delta)$	**Func.** IdealListReduce$(\mathbf{p}, L, \delta)$
1 **while** $(\exists \mathbf{l}' \in L : \|\mathbf{p} - \mathbf{l}'\| \leq \delta \|\mathbf{p}\|)$ **do**	1 **while** $(\exists j \in [n], \mathbf{l} \in L, \mathbf{l}' = \mathrm{rot}^j(\mathbf{l}) :$ $\|\mathbf{p} - \mathbf{l}'\| \leq \delta \|\mathbf{p}\|)$ **do**
2 $\quad \mathbf{p} \leftarrow \mathbf{p} - \mathrm{round}(\frac{\langle \mathbf{p} \mid \mathbf{l}' \rangle}{\langle \mathbf{l}' \mid \mathbf{l}' \rangle}) \cdot \mathbf{l}'$	2 $\quad \mathbf{p} \leftarrow \mathbf{p} - \mathrm{round}(\frac{\langle \mathbf{p} \mid \mathbf{l}' \rangle}{\langle \mathbf{l}' \mid \mathbf{l}' \rangle}) \cdot \mathbf{l}'$
3 **end**	3 **end**
4 **return p**	4 **return p**

The while loop condition in Line 1 introduces the rotation step (for all types of ideal lattices). The reduction step in Line 2 differs from the original ListSieve description in [MV10b]. It uses the reduction step known from the Gauss (respectively Lagrange) algorithm (an orthogonal projection), that is also used in the LLL algorithm [LLL82]. The step is not explained in [MV10b], whereas their implementation [Vou10] already uses this improvement. The slackness parameter $\delta = 1 - 1/m$ is used to ensure that the norm decrease is sufficient for each reduction in order to guarantee polynomial runtime in the list size.

3.3   IdealGaussSieve

For ListSieve, when a vector joined the list once it remains unchanged forever. GaussSieve introduces the possibility to remove vectors from the list if they can be more reduced by a newly sampled vector. The reduction process is twofold in GaussSieve. First, the new vector **p** is reduced as in ListSieve. Second, all list vectors are reduced using **p**. If a vector from the list is shortened, it will leave the list and pass it again in one of the next iterations. Therefore the list will consist of less and shorter vectors than in the ListSieve case.

It is easy to include the rotations into GaussSieve in the same manner as for ListSieve. We can replace the function `GaussReduce` of [MV10b] by a new function `IdealGaussReduce`, which uses the rotations twice. First it is used for the reduction of **p**, second for the reduction of list vectors. The rest of GaussSieve remains unchanged. IdealGaussSieve is also included in our implementation.

4   Predicted Advantage of IdealListSieve

In this section we theoretically analyze the IdealSieve algorithm and try to predict the results concerning number of iterations I, the total number of reductions R, and the maximum size L of the list L. For comparison of an algorithm and its ideal lattice variant we will always use the quotient of a characteristic of the non-ideal variant divided by the ideal variant. We will always denote it with *speedup*. For example, the speedup in terms of reductions is R_{orig}/R_{ideal}.

Recall that the only change we made in the algorithm is that in the reduction step, all rotations $rot^j(\mathbf{l})$ (for $j \in [n]$) of a vector **l** in the list L are considered, instead of only considering **l**. The runtime proof for ListSieve in [MV10b] uses the fact that the number of vectors of bounded norm can be bounded exponentially in the lattice dimension. Therefore, the list size L cannot grow unregulated. All list vectors have norm less than or equal $m \|\mathbf{B}\|$. For cyclic and anti-cyclic lattices, the norm of a vector remains unchanged when rotated. Therefore each list vector corresponds to n vectors of the same size, which results in a proven list size of factor n smaller. For prime cyclotomic lattices, the norm might increase when rotated (the expansion factor is > 1 in that case), therefore it is a bit harder to prove bounds on the size of the list.

Iterations. We assume that for finding a shortest vector in the lattice, the total number of *reductions* is the same. Our experiments show that this assumption is reasonable (cf. Section 5). In this case we predict the number of iterations of IdealListSieve compared to ListSieve. When ListSieve performs t iterations (sampling - reducing - appending), our assumption predicts that IdealListSieve takes t/n iterations, since in t/n steps it can use the same number of list vectors for reduction, namely $n \cdot t/n$. Therefore, we expect the number of iterations for IdealListSieve to be an n-th fraction of ListSieve.

Maximum List Size. Since in every iteration one single vector is sampled and appended to the list, the maximum list size will be in the order of magnitude as the number of iterations.

Runtime. The runtime of the whole algorithm grows linearly in the number of iterations. The runtime of ListReduce is quadratic in the size of the list $|L|$. As the list size is smaller by factor n for IdealListReduce, the IdealSieve algorithm saves a factor n^2 in each iteration here. In each call of IdealListReduce, n rotations instead of a single one are used for reduction, therefore the ideal lattice variant is factor n slower here. In total, each run of IdealListReduce is factor n faster than ListReduce. Overall we derive a speedup factor of n^2 for the ideal lattice variant concerning the runtime.

Recall that the speedups predicted in this section are asymptotical. They do not necessarily hold in practice, since we can only run experiments in small dimensions $m \leq 80$. In the next section, we present experimental results comparing the ListSieve and IdealListSieve, in order to show if our predictions hold in practice.

5   Experiments

The public implementation of [Vou10] (called `gsieve`) allows for running the GaussSieve algorithm. Based on this, we implemented ListSieve, IdealListSieve, and IdealGaussSieve. ListSieve is essentially the `gsieve` implementation without stack functionality. IdealListSieve uses the subroutine function `IdealListReduce` of Section 3 in addition. Both algorithms do not use perturbations. The Ideal-GaussSieve implements GaussSieve with the additional function. All three implemented algorithms are published online.[2]

Since we are using the NTL-library [Sho11], it would be possible to implement a generic function `IdealReduce` for all polynomials f. However, specializing on a special class of polynomials allows some code improvements and leads to a huge speed-up in practice. Therefore, we have implemented three different functions for reduction, namely `AntiCyclicReduce`, `CyclicReduce`, and `CyclotomicReduce`. These functions can be used for sieving in anti-cyclic, cyclic, or prime cyclotomic lattices, respectively. Here we present experimental results for cyclic and prime cyclotomic lattices.

All experiments were performed on an AMD Opteron (2.3 GHz) quad core processor, using one single core, with 64 GB of RAM available. We only apply LLL as pre-reduction, not BKZ. This is due to the fact that BKZ-reduction is too strong in small dimensions, and the sieving algorithms are not doing any work

---

[2] https://www.cdc.informatik.tu-darmstadt.de/de/cdc/personen/
michael-schneider

if BKZ already finds the shortest vector. Interestingly, we encountered in our experiments that the effect of pre-reduction for sieving is much less noticeable as in the enumeration case.

The results shown in this section are average values of 10 randomly generated lattices in each dimension. Since we do not know the length of a shortest vector in these lattices, we ran an SVP algorithm first to assess the norm. So we can stop our sieving algorithms as soon as we have found a vector below that bound. For cyclic and prime cyclotomic lattice we chose $n \in \{10, 12, 16, 18, 20, 22, 28, 30, 32\}$ and $m = 2n$. These are the values where $n + 1$ is prime, which is important for prime cyclotomic lattices. We chose these values for cyclic lattices as well in order to have results for both lattice types in the same dimensions. The generator of the ideal lattices is included in Sage [S$^+$10] since version 4.5.2. The modulus q was fixed as 257. Naturally, the determinant of the lattices is q^n, i.e., 257^n. For a second series of experiments, we generate cyclic and prime cyclotomic lattices with $m = 4n$. We choose $n \in \{6 \ldots 15\}$ (cyclic) and $n \in \{6, 10, 12, 16\}$ (prime cyclotomic), q is again 257.

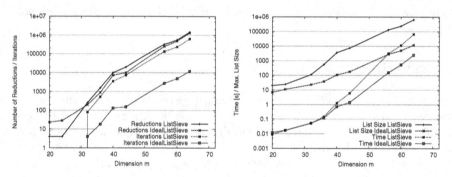

Fig. 1. Results for cyclic lattices. Left: The number of reductions is comparable for ListSieve and IdealListSieve, whereas the number of iterations differs. Right: The maximum list size as well as the runtime decrease for IdealListSieve.

Figure 1 shows the results concerning R, I, L, and the runtime for cyclic lattices. The speedups for cyclic lattices are shown in Figure 2 and for prime cyclotomic lattices in Figure 3. Figure 2(a) shows the speedups of IdealListSieve compared to ListSieve. More exactly it shows the values for the number of iterations I, the maximum list size L, the runtime, and the total number of reductions R of ListSieve divided by the same values for IdealListSieve in the same lattices. Figure 2(b) shows the same data for $m = 4n$. Figures 3(a) and (b) show the same data using cyclotomic lattices. All graphs contain a line for the identity function $f(m) = m$, and a line for $f(m) = m/2$ or $f(m) = m/4$, in order to ease comparison with the prediction of Section 4.

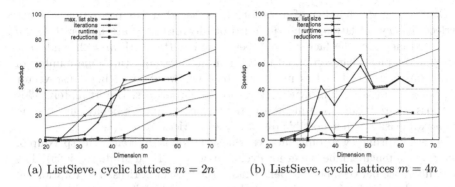

(a) ListSieve, cyclic lattices $m = 2n$ (b) ListSieve, cyclic lattices $m = 4n$

Fig. 2. Speedup of IdealListSieve compared to ListSieve, for cyclic lattices

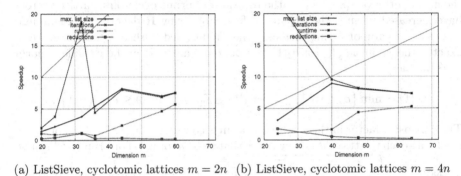

(a) ListSieve, cyclotomic lattices $m = 2n$ (b) ListSieve, cyclotomic lattices $m = 4n$

Fig. 3. Speedup of IdealListSieve compared to ListSieve, for cyclotomic lattices

5.1   Interpretation

In small dimensions, the results are kind of abnormal. Sometimes one of the ideal lattice variants finds shortest vectors very fast, which results in speedups of more than 100, e.g. in dimension $m = 36$ in Figure 2(b). Therefore, small dimensions of (say) less than 40 should be taken into account only carefully. Testing higher dimensions failed due to time reasons. Neither better pre-reduction nor searching for longer vectors helped decreasing the runtime noticeably.

A first thing that is apparent is that the number of reductions R stays nearly the same in all cases. With increasing dimension the speedup tends to 1 in all cases. Therefore our assumption was reasonable, namely that the number of reductions required to find a shortest vector is the same for the ideal and the non-ideal variant of ListSieve.

The higher the dimension gets, the closer the list size L and the iteration counter I get. Again this is how we expected the algorithms to behave. The runtime grows slower than the number of iterations. In dimension $m = 64$ for example, IdealListSieve finds a shortest vector 53 times faster than ListSieve.

Considering the number of iterations I, we see that our prediction was not perfect. For cyclic lattices, the speedups of IdealListSieve are higher than the predicted factor n; the factor is between n and m (for both $m = 2n$ and $m = 4n$). This implies that compared to the non-ideal variant, the same number of reductions is reached in less iterations. In other words, rotating a list vector I is better than sampling new vectors, for cyclic lattices. Unfortunately, it is not possible from our experiments to predict the asymptotic behaviour. Testing higher dimension is not possible due to time restrictions.

In case of prime cyclotomic lattices, the situation is different. The speedup of iterations is much smaller than for cyclic lattices (≤ 10 in all dimensions). The only difference between both experiments is the type of lattices. The rotations of prime cyclotomic lattices are less useful than those of cyclic lattices. A possible explanation for this is that rotating a vector of a cyclic lattice does not change the norm of the vector, whereas the rotations of prime cyclotomic lattice vectors have increased norms. The expansion factor of a ring $\mathbf{R}$ denotes the maximum "blow up" factor of a ring element when multiplied with a second one. More exactly, the expansion factor $\theta_2(\mathbf{f})$ of a polynomial $\mathbf{f} \in \mathbf{R}$ in the Euclidean norm is defined as

$$\theta_2(\mathbf{f}) = \min \left\{ c : \left\| \mathbf{a} x^i \right\|_2 \leq c \left\| \mathbf{a} \right\|_2 \forall \mathbf{a} \in \mathbb{Z}[x]/\langle \mathbf{f} \rangle \text{ for } 0 \leq i \leq n-1 \right\} .$$

The expansion factor in the Euclidean norm is considered in [SSTX09]. For cyclic (and anti-cyclic) lattices it is easy to see that this factor is 1. For prime cyclotomic lattices, it is $\sqrt{\frac{n+1}{2} + \sqrt{\left(\frac{n+1}{2}\right)^2 - 1}} \approx \sqrt{n}$ (for a proof see Appendix A). So when the norm of the rotated list vectors I increases, this lowers the probability of a vector to be useful for reduction of the new sample. Therefore, compared to cyclic lattices, the speedup for iterations decreases. But still, sieving in prime cyclotomic lattices using the IdealListSieve is up to 10 times faster than in the original case. In order to check if the expansion factor really is that crucial, it is necessary to start experiments with different ideal lattices equipped with higher expansion factor.

5.2   IdealGaussSieve

We also performed experiments using the GaussSieve implementation of Voulgaris and our IdealGaussSieve version. The results are shown in Figure 4. The speedup factors are comparable to those of IdealListSieve.

5.3   Anti-cyclic Lattices

Lattices corresponding to ideals in the ring factored with $\mathbf{f}(x) = x^n + 1$ behave exactly as cyclic lattices. The algebra of both rings differs, but the algorithmic behaviour is exactly the same. In order to have the polynomial f irreducible, we choose $n \in \{2, 4, 8, 16, 32\}$ and $m = 2n$.

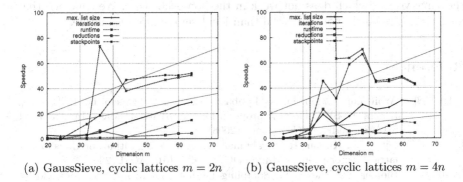

(a) GaussSieve, cyclic lattices $m = 2n$ (b) GaussSieve, cyclic lattices $m = 4n$

Fig. 4. Speedup of IdealGaussSieve compared to GaussSieve, for cyclic lattices

6   Conclusion and Further Work

We have shown that it is indeed possible to make use of the special structure of ideal lattices when searching for shortest vectors. The gained speedup does not affect the asymptotic runtime of the SVP algorithms, but it allows for some improvements in practice. Our experiments show that runtime speedups of up to 60 are possible in suitable lattice dimensions. With this we also propose the fastest sieving implementation for ideal lattices.

The projection of an ideal lattice is not an ideal lattice any more. This prevents the usage of IdealSieve in block-wise reduction algorithms like the BKZ algorithm.

6.1   Ideal Enumeration

The enumeration algorithm for exhaustive search for shortest lattice vectors can also exploit the special structure of cyclic lattices. In the enumeration tree, linear combinations $\sum_{i=1}^{n} x_i \mathbf{b}_i$ in a specified search region are considered. For cyclic (and also anti-cyclic) lattices, a coefficient vector $\mathbf{x} = (\mathbf{x}_1, \ldots, \mathbf{x}_n)$ and its rotations $\mathrm{rot}^i \cdot \mathbf{x}$ for $i \in [n]$ specify the same vector. Therefore it is sufficient to enumerate the subtree predefined by one of the rotations. It is for example possible to choose only the coefficient vectors where the top coordinate $\mathbf{x}_n$ is the biggest entry, i.e., $\mathbf{x}_n = \max_i(\mathbf{x}_i)$. This would decrease the number of subtrees in the enumeration tree with a factor of up to n.

Unfortunately, this approach is only applicable if the input matrix has circular structure. When LLL-reducing the basis, usually the special structure of the matrix is destroyed. Therefore, when applying enumeration for ideal lattices one would lose the possibility of pre-reducing the lattice. Even the symplectic LLL [GHGN06] does not maintain the circulant structure of the basis.

A second flaw of the ideal enumeration is that it is not applicable to ideal, non-cyclic lattices. For cyclic lattices it is easy to specify which rotations predefine the same vector, which does not work in the non-cyclic case. We conclude that ideal sieving is much more practical than ideal enumeration would be.

References

[ADL+08]   Arbitman, Y., Dogon, G., Lyubashevsky, V., Micciancio, D., Peikert, C., Rosen, A.: SWIFFTX: A proposal for the SHA-3 standard. In: The First SHA-3 Candidate Conference (2008)

[AKS01]   Ajtai, M., Kumar, R., Sivakumar, D.: A sieve algorithm for the shortest lattice vector problem. In: STOC, pp. 601–610. ACM (2001)

[BN09]   Blömer, J., Naewe, S.: Sampling methods for shortest vectors, closest vectors and successive minima. Theor. Comput. Sci. 410(18), 1648–1665 (2009)

[Gen09]   Gentry, C.: Fully homomorphic encryption using ideal lattices. In: STOC, pp. 169–178. ACM (2009)

[GHGN06]   Gama, N., Howgrave-Graham, N., Nguyên, P.Q.: Symplectic lattice reduction and NTRU. In: Vaudenay, S. (ed.) EUROCRYPT 2006. LNCS, vol. 4004, pp. 233–253. Springer, Heidelberg (2006)

[GNR10]   Gama, N., Nguyen, P.Q., Regev, O.: Lattice Enumeration Using Extreme Pruning. In: Gilbert, H. (ed.) EUROCRYPT 2010. LNCS, vol. 6110, pp. 257–278. Springer, Heidelberg (2010)

[GS10]   Gama, N., Schneider, M.: SVP Challenge (2010), http://www.latticechallenge.org/svp-challenge

[HPS98]   Hoffstein, J., Pipher, J., Silverman, J.H.: NTRU: A ring-based public key cryptosystem. In: Buhler, J.P. (ed.) ANTS 1998. LNCS, vol. 1423, pp. 267–288. Springer, Heidelberg (1998)

[HS07]   Hanrot, G., Stehlé, D.: Improved analysis of kannan's shortest lattice vector algorithm. In: Menezes, A. (ed.) CRYPTO 2007. LNCS, vol. 4622, pp. 170–186. Springer, Heidelberg (2007)

[Kle00]   Klein, P.N.: Finding the closest lattice vector when it's unusually close. In: SODA 2000, pp. 937–941. ACM (2000)

[LLL82]   Lenstra, A., Lenstra, H., Lovász, L.: Factoring polynomials with rational coefficients. Mathematische Annalen 4, 515–534 (1982)

[LM06]   Lyubashevsky, V., Micciancio, D.: Generalized compact knapsacks are collision resistant. In: Bugliesi, M., Preneel, B., Sassone, V., Wegener, I. (eds.) ICALP 2006. LNCS, vol. 4052, pp. 144–155. Springer, Heidelberg (2006)

[LM08]   Lyubashevsky, V., Micciancio, D.: Asymptotically efficient lattice-based digital signatures. In: Canetti, R. (ed.) TCC 2008. LNCS, vol. 4948, pp. 37–54. Springer, Heidelberg (2008)

[LMPR08]   Lyubashevsky, V., Micciancio, D., Peikert, C., Rosen, A.: Swifft: A modest proposal for FFT hashing. In: Nyberg, K. (ed.) FSE 2008. LNCS, vol. 5086, pp. 54–72. Springer, Heidelberg (2008)

[LPR10]   Lyubashevsky, V., Peikert, C., Regev, O.: On ideal lattices and learning with errors over rings. In: Gilbert, H. (ed.) EUROCRYPT 2010. LNCS, vol. 6110, pp. 1–23. Springer, Heidelberg (2010)

[Lyu08]   Lyubashevsky, V.: Towards practical lattice-based cryptography. Phd thesis, University of California, San Diego (2008)

[Lyu09]   Lyubashevsky, V.: Fiat-Shamir with aborts: Applications to lattice and factoring-based signatures. In: Matsui, M. (ed.) ASIACRYPT 2009. LNCS, vol. 5912, pp. 598–616. Springer, Heidelberg (2009)

[Mic07]   Micciancio, D.: Generalized compact knapsacks, cyclic lattices, and efficient one-way functions. Computational Complexity 16(4), 365–411 (2007); Preliminary version in FOCS (2002)

[MR08]   Micciancio, D., Regev, O.: Lattice-based cryptography. In: Bernstein, D.J., Buchmann, J.A., Dahmen, E. (eds.) Post-Quantum Cryptography, pp. 147–191. Springer (2008)

[MS01]   May, A., Silverman, J.H.: Dimension reduction methods for convolution modular lattices. In: Silverman, J.H. (ed.) CaLC 2001. LNCS, vol. 2146, pp. 110–125. Springer, Heidelberg (2001)

[MV10a]   Micciancio, D., Voulgaris, P.: A deterministic single exponential time algorithm for most lattice problems based on Voronoi cell computations. In: STOC, pp. 351–358. ACM (2010)

[MV10b]   Micciancio, D., Voulgaris, P.: Faster exponential time algorithms for the shortest vector problem. In: SODA, pp. 1468–1480. ACM/SIAM (2010)

[NV08]   Nguyen, P.Q., Vidick, T.: Sieve algorithms for the shortest vector problem are practical. J. of Mathematical Cryptology 2(2) (2008)

[PR06]   Peikert, C., Rosen, A.: Efficient collision-resistant hashing from worst-case assumptions on cyclic lattices. In: Halevi, S., Rabin, T. (eds.) TCC 2006. LNCS, vol. 3876, pp. 145–166. Springer, Heidelberg (2006)

[PS08]   Pujol, X., Stehlé, D.: Rigorous and efficient short lattice vectors enumeration. In: Pieprzyk, J. (ed.) ASIACRYPT 2008. LNCS, vol. 5350, pp. 390–405. Springer, Heidelberg (2008)

[PS09]   Pujol, X., Stehlé, D.: Solving the shortest lattice vector problem in time $2^{2.465n}$. Cryptology ePrint Archive, Report 2009/605 (2009)

[S+10]   Stein, W.A., et al.: Sage Mathematics Software (Version 4.5.2). The Sage Development Team (2010), http://www.sagemath.org

[Sho11]   Victor Shoup. Number theory library (NTL) for C++ (2011), http://www.shoup.net/ntl/

[SSTX09]   Stehlé, D., Steinfeld, R., Tanaka, K., Xagawa, K.: Efficient Public Key Encryption Based on Ideal Lattices. In: Matsui, M. (ed.) ASIACRYPT 2009. LNCS, vol. 5912, pp. 617–635. Springer, Heidelberg (2009)

[Vou10]   Voulgaris, P.: Gauss Sieve alpha V. 0.1 (2010), http://cseweb.ucsd.edu/~pvoulgar/impl.html

[WLTB10]   Wang, X., Liu, M., Tian, C., Bi, J.: Improved Nguyen-Vidick heuristic sieve algorithm for shortest vector problem. Cryptology ePrint Archive, Report 2010/647 (2010), http://eprint.iacr.org/

A   Expansion Factor in the Euclidean Norm

The expansion factor $\theta_2(\mathbf{f})$ of a polynomial $\mathbf{f} \in \mathbf{R} = \mathbb{Z}[x]/\langle \mathbf{f} \rangle$ in the Euclidean norm is defined as

$$\theta_2(\mathbf{f}) = \min \left\{ c : \left\| \mathbf{a}x^i \right\|_2 \leq c \left\| \mathbf{a} \right\|_2 \forall \mathbf{a} \in \mathbb{Z}[x]/\langle \mathbf{f} \rangle \text{ for } i \in [n] \right\} .$$

It is easy to see that $\theta_2(x^n - 1) = \theta_2(x^n + 1) = 1$. It is an easy adaption of the proof of [Lyu08, Lemma 2.6], already mentioned in [SSTX09]. Here, we will present the expansion factor of $\mathbf{f} = x^{n-1} + x^{n-2} + \ldots + x + 1$ in the Euclidean norm.

Let λ_i be the eigenvalues of the matrix $\mathbf{M} \in \mathbb{R}^{n \times n}$, for $i \in [n]$. The spectral radius of $\mathbf{M}$ is the maximum of the absolute values of the eigenvalues of the matrix $\mathbf{M}$, i.e. $\rho(\mathbf{M}) = \max_{i \in [n]} \{|\lambda_i|\}$.

The rotation matrix is $\mathbf{A} = \begin{pmatrix} \mathbf{0}_{n-1}^T & -\mathbf{1}_n \\ \mathbf{I}_{n-1} & \end{pmatrix}$. Lemma 1 will help us later in the proof of the expansion factor of $\mathbf{f}$.

Lemma 1. *For every $i \in [m]$, the spectral radius of the matrix $(\mathbf{A}^i)^T(\mathbf{A}^i)$ is* $\rho = \frac{m+1}{2} + \sqrt{\left(\frac{m+1}{2}\right)^2 - 1}$.

Proof. We are looking for the eigenvalues of the matrix $(\mathbf{A}^i)^T(\mathbf{A}^i)$, i.e. the values λ where $(\mathbf{A}^i)^T(\mathbf{A}^i) \cdot \mathbf{x} = \lambda \mathbf{x}$ for an $\mathbf{x} \in \mathbb{R}^n$. It is

$$\mathbf{A}^i = \begin{pmatrix} \mathbf{0} & -\mathbf{1}_n & \mathbf{I}_{i-1} \\ \mathbf{I}_{n-i} & & \mathbf{0} \end{pmatrix} \Rightarrow (\mathbf{A}^i)^T \cdot (\mathbf{A}^i) = \begin{pmatrix} \mathbf{I}_{n-i} & -\mathbf{1}_{n-i} & \mathbf{0} \\ -\mathbf{1}_{n-i}^T & n & -\mathbf{1}_{i-1}^T \\ \mathbf{0} & -\mathbf{1}_{i-1} & \mathbf{I}_{i-1} \end{pmatrix}.$$

The equation system $\left((\mathbf{A}^i)^T(\mathbf{A}^i) - \lambda \mathbf{I}_n\right) \cdot \mathbf{x} = 0$ is equal to

$$\begin{pmatrix} (1-\lambda)\mathbf{I}_{n-i} & -\mathbf{1}_{n-i} & \mathbf{0} \\ -\mathbf{1}_{n-i}^T & n - \lambda & -\mathbf{1}_{i-1}^T \\ \mathbf{0} & -\mathbf{1}_{i-1} & (1-\lambda)\mathbf{I}_{i-1} \end{pmatrix} \cdot \mathbf{x} = \mathbf{0}.$$

We consider two different cases for the value of λ.
Case 1 ($\lambda \neq 1$): Dividing the first $n - 1$ and the last $i - 1$ rows by $1 - \lambda$ leads to

$$\begin{pmatrix} \mathbf{I}_{n-i} & -\mathbf{1}_{n-i}/(1-\lambda) & \mathbf{0} \\ -\mathbf{1}_{n-i}^T & n - \lambda & -\mathbf{1}_{i-1}^T \\ \mathbf{0} & -\mathbf{1}_{i-1}/(1-\lambda) & \mathbf{I}_{i-1} \end{pmatrix} \cdot \mathbf{x} = \mathbf{0}.$$

The sum of all rows leads to 0 in the first $n - i$ and in the last $i - 1$ columns. In the $(n - i + 1)$th column, we compute

$$\left(\frac{(n-1)\cdot(-1)}{1-\lambda} + n - \lambda\right) \cdot \mathbf{x}_{n-i+1} = 0 \quad \Leftrightarrow \quad \frac{\lambda^2 - (n+1)\lambda + 1}{1-\lambda} = 0.$$

Therefore we derive the first two eigenvalues $\lambda_{1,2} = \frac{n+1}{2} \pm \sqrt{\left(\frac{n+1}{2}\right)^2 - 1}$.
Case 2 ($\lambda = 1$): The equation system is in this case

$$\begin{pmatrix} & -\mathbf{1}_{n-i} & \\ -\mathbf{1}_{n-i}^T & n - 1 & -\mathbf{1}_{i-1}^T \\ & -\mathbf{1}_{i-1} & \end{pmatrix} \cdot \mathbf{x} = \mathbf{0}$$

which is equivalent to $x_{n-i+1} = 0 \quad \wedge \quad \sum_{j=1, j \neq n-i+1}^n x_j = 0$. This defines an eigenspace of dimension $n - 2$. Therefore, the eigenvalue 1 has geometric multiplicity $n - 2$. Since the geometric multiplicity is smaller or equal to the algebraic multiplicity, and we have already found 2 eigenvalues, the algebraic multiplicity of the eigenvalue 1 is also $n - 2$. Since we have found all eigenvalues we can compute the maximum $\max(1, \lambda_1, \lambda_2)$, which is always $\lambda_1 = \frac{n+1}{2} + \sqrt{\left(\frac{n+1}{2}\right)^2 - 1}$. □

Now we can proof the expansion factor of prime cyclotomic polynomials in the Euclidean norm.

Lemma 2. *If* $\mathbf{f} = x^{n-1} + x^{n-2} + \ldots + x + 1$, *then the expansion factor of* $\mathbf{f}$ *is*
$$\theta_2(\mathbf{f}) = \sqrt{\frac{n+1}{2} + \sqrt{\left(\frac{n+1}{2}\right)^2 - 1}}.$$

Proof. The operator norm of an operator between normed vector spaces X and Y over the same base field, $A : X \to Y$, is defined as

$$\|A\| = \sup\left\{\frac{\|Ax\|}{\|x\|} : x \in X, \|x\| \neq 0\right\} = \inf\left\{c : \|Ax\| \leq c\|x\| \, \forall x \in X\right\}.$$

This can be defined for every norm of the base field of the vector spaces. The expansion factor defined for the ring $\mathbb{Z}/\langle\mathbf{f}\rangle$ is the maximum of the operator norms of the operators that perform multiplication with x^i, for $i \in [n]$. Using the usual embedding from the ring to a vector space, the operators can be described by the matrices $\mathbf{A}^i$.

For the Euclidean norm the operator norm of a matrix $\mathbf{M}$ is equal to the square root of the spectral radius of the matrix $\mathbf{M}^T\mathbf{M}$. In our case, we have to compute the spectral radius of the matrix $(\mathbf{A}^i)^T(\mathbf{A}^i)$. Therefore, the expansion factor equals the square root of the value computed in Lemma 1. □

An Identity-Based Key-Encapsulation Mechanism Built on Identity-Based Factors Selection

Sebastian Staamann

Intercrypt GmbH
Westkorso 44, 15747 Wildau, Germany
sebastian.staamann@intercrypt.com

Abstract. A new approach to identity-based encryption (IBE), called identity-based factors selection (IBFS), allows to build efficient and fully collusion-resistant IBE schemes without the need for pairings or the use of lattices. The security of these constructions (in the random oracle model) rests on the hardness of a new problem which combines the computational Diffie-Hellman problem with the fact that linear equation systems with more variables than given equations do not have unambiguous solutions. The computational efficiency of the resulting IBE schemes is (for values of the security parameter not smaller than 80) better than in previous IBE schemes. The construction of these schemes may be seen as an extension of the ElGamal public-key encryption scheme. The sender of a message computes the ElGamal-like public key of the message receiver by first selecting, uniquely determined by the identity of the receiver, from a set of group elements $\{g^{e_1}, ..., g^{e_z}\}$ made available as public parameters a subset, and then multiplying the selected elements.

Keywords: Identity-based encryption, IBE, Identity-based key-encapsulation mechanism, ID-KEM, Identity-based factors selection, IBFS.

1   Introduction

The identity-based encryption (IBE) schemes [25] [8] [7] [19] [12] and identity-based key-encapsulation mechanisms (ID-KEM, [5] [13]) usually considered for practical application today are based on pairings [17]. Their security generally relies on the bilinear Diffie-Hellman assumption [21] [8] [26].

In order to not put all eggs in one basket concerning the dependence on one (not yet very long-seasoned) hardness assumption, it seems prudent to have some alternatives to pairing-based constructions for IBE and ID-KEMs ready. Such alternatives already exist, but concerning the resources needed they may not always fit for resource-constrained environments. Cocks' IBE scheme [14], based on the quadratic residuosity problem (QR), causes very high message expansion. Boneh, Gentry, and Hamburg's space-efficient pairing-free IBE scheme [9] (also based on QR) requires an extraordinarily high computation effort for the encryption. Identity-based crypto schemes based on the use of discrete log

A. Youssef, A. Nitaj, A.E. Hassanien (Eds.): AFRICACRYPT 2013, LNCS 7918, pp. 392–405, 2013.

trapdoors [24] [22] require a very high computation effort for the generation of a user's private key. Lattice-based IBE schemes [18] [3] have unusually large private keys. This paper introduces a novel pairing-free construction of IBE schemes that avoids such unfavorable characteristics. The resulting IBE schemes are very efficient concerning all relevant aspects, except possibly one: the size of the public parameters is, like in lattice-based IBE schemes, significantly bigger than in other schemes.

The construction may be seen as an extension of the ElGamal public-key encryption scheme [16]. What makes the construction identity-based is that the sender of a message does not have to retrieve the public key of the message receiver from somewhere else. The sender himself computes the receiver's ElGamal-like public key. The sender first selects, uniquely determined by the identity of the receiver, from a set of group elements $\{g^{e_1}, ..., g^{e_z}\}$ (made available as public parameters, whereas the e_i's are part of the master key) a subset, and then multiplies[1] the elements of this subset. The result is the receiver's ElGamal-like public key. The computation of the receiver's ElGamal-like public key by means of *identity-based factors selection* is the core idea presented in this paper how to construct an identity-based key-encapsulation mechanism (ID-KEM). On top of the ID-KEM, one can build complete and fully collusion-resistant IBE schemes in well-known ways, for instance, analogously to how the public-key encryption scheme DHIES [1] builds on the Diffie-Hellman protocol [15]. The computational efficiency of the resulting IBE schemes is for practical values of the security parameter (i.e., not smaller than 80) better than in previous IBE schemes (native or built on top of previous ID-KEMs). The message expansion caused by the use of our ID-KEM is (in suitable implementation groups) at most double of what it is in pairing based ID-KEMs and IBE schemes.

The presentation is organized as follows: Section 2 introduces the idea of identity-based factors selection in more detail. Section 3 introduces and discusses the foundational problem, called PX. PX combines the computational Diffie-Hellman problem [15] with the fact that linear equation systems with more variables than given equations do not have unambiguous solutions. Section 4 presents our ID-KEM in detail. The section shows in the random oracle model [4] that the scheme is ID-IND-CCA2 secure under the assumption that PX is hard. The section also discusses the efficiency of the scheme and compares it with other ID-KEMs and IBE schemes. Finally, Section 5 summarizes the work presented.

2   Identity-Based Factors Selection – The Idea in Detail

In discrete-logarithm based schemes for public key encryption, such as ElGamal encryption [16] or DHIES/ECIES [1] [11], the receiver of encrypted messages makes available for potential senders: (i) the description of a finite cyclic group G of order q (q typically being a prime number) chosen such that the most efficient algorithm to solve the discrete-logarithm problem (DL, see, e.g., [23])

---

[1] In this paper, generic groups are generally written multiplicatively.

in G has an average complexity $\geq 2^l$ (where l is the security parameter), (ii) a generator g of G, and (iii) another generator y of G such that the receiver (and *only* the receiver) knows the $a \in \mathbb{Z}_q^*$ for which $g^a = y$. In most systems used in practice, G and g are the same for all potential receivers of encrypted messages in the system, so g and the parameters that define G, including the order q, are public system parameters. The public key of a receiver is then y ($= g^a$).

In IBE schemes and ID-KEMs, the public key of the receiver must be easily derivable from the receiver's identity (most generally, a bit string) and the public system parameters. The question how a discrete-log based conventional public-key encryption scheme, as characterized above, can be extended to become an IBE scheme can be asked as: How can it be arranged that the sender of an encrypted message can (easily and without any interaction with other parties) compute the receiver's public key $y = g^a$ from the receiver's identity and the public system parameters, while the receiver (and only the receiver) knows the a (the receiver's personal, private key) for which $g^a = y$.

The core of our idea is the following: As is typical for identity-based cryptography (IBC), a *Trusted Authority* (TA) computes the private keys (and additional public system parameters) from the already existing public system parameters and a *master key* (which only the TA knows). First, the TA defines G of order q and a generator g of G. Second, as its master key, the TA creates an indexed array (for short: *vector*) $\mathbf{e}$ of z different random numbers, each one independently chosen from $\mathbb{Z}_q^*$. The number of elements z is at least $2l$, which means z is big enough such that the average complexity to find a collision of a typical cryptographic hash function of output length z is not smaller than 2^l. From the secret vector $\mathbf{e}$, the TA computes a corresponding vector $\mathbf{f}$ of z elements (as for $\mathbf{e}$, the index running from 1 to z). The elements of $\mathbf{f}$ are in G. The TA computes them as follows: $\forall i \in \{1...z\} : \mathbf{f}[i] \in G \leftarrow g^{\mathbf{e}[i]}$. The TA publishes z, q, G, g, and $\mathbf{f}$ as public system parameters. As further public system information, the TA defines a (collision-resistant) cryptographic hash function *hid* which maps an identity to a vector consisting of z bits.

From the public system parameters $\langle z, q, G, g, \mathbf{f}, \text{hid} \rangle$ and a user Alice's identity ID_A (which, as usual for IBC, must be unique in the system), another user Bob (who wants to encrypt a message to Alice) can now compute Alice's public key y_A by himself. First, Bob computes a cryptographic hash of the identity ID_A: $\mathbf{s}_A \in \{0,1\}^z \leftarrow \text{hid}(\text{ID}_A)$. Using the resulting *selection vector* $\mathbf{s}_A$, Bob then computes Alice's public key as $y_A \in G \leftarrow \prod_{i=1}^z (\mathbf{f}[i])^{\mathbf{s}_A[i]}$. Note that the exponentiations with the $\mathbf{s}_A[i]$'s (which can only have values 0 or 1) are practically just a selection of *actual* factors for the multiplication from the set (precisely, the *vector*) of *potential* factors, namely from $\mathbf{f}$.

Alice's personal key, i.e., the private key a_A corresponding to the identity ID_A, is generated as follows: The TA first computes the selection vector $\mathbf{s}_A \in \{0,1\}^z \leftarrow \text{hid}(\text{ID}_A)$. Then, the TA computes, using its master key $\mathbf{e}$, Alice's private key $a_A \in \mathbb{Z}_q \leftarrow \sum_{i=1}^z (\mathbf{s}_A[i] \cdot \mathbf{e}[i]) \mod q$. (Note that the multiplications with the $\mathbf{s}_A[i]$'s are practically just a selection of summands from the vector $\mathbf{e}$). The TA gives a_A to Alice as her private key. For other receivers of encrypted messages, the TA generates their private keys in the same way.

When Bob wants to encrypt a message to Alice, he first computes Alice's public key y_A from the public system parameters and Alice's identity ID_A, as described above. Second, Bob randomly picks an r from $\mathbb{Z}_q^*$. Third, Bob computes $\gamma \in G \leftarrow (y_A)^r$ and $c \in G \leftarrow g^r$. γ is the "raw message key", and c is the encapsulation of the message key sent to Alice (typically along with the encapsulated data).

Alice computes the "raw message key" as $\gamma' \in G \leftarrow c^{a_A}$. Looking at how a_A and c were computed, we see that $\gamma' = c^{a_A} = g^{r \cdot a_A} = \left(g^{\sum_{i=1}^{z}(\mathbf{s}_A[i] \cdot \mathbf{e}[i])}\right)^r = \left(\prod_{i=1}^{z}(g^{\mathbf{e}[i]})^{\mathbf{s}_A[i]}\right)^r = \left(\prod_{i=1}^{z}(\mathbf{f}[i])^{\mathbf{s}_A[i]}\right)^r = (y_A)^r = \gamma$. Obviously, both sides, Alice and Bob, have finally obtained the same "raw message key".

The scheme sketched above should introduce the core idea of building an ID-KEM around identity-based factors selection. However, this scheme is still rudimentary and not yet secure against collusion attacks: A group of $t \geq z$ colluding users who have received their own private keys a_j $(1 \leq j \leq t)$ could compute the private keys of other users, i.e., break the security of the scheme: If the selection vector $\mathbf{s}_T \in \{0,1\}^z \leftarrow \text{hid}(ID_T)$ resulting from a target identity ID_T is a linear combination in $\mathbb{Z}_q$ of the selection vectors resulting from the identities ID_j of the t colluders $\mathbf{s}_1 \in \{0,1\}^z = \text{hid}(ID_1), ..., \mathbf{s}_t \in \{0,1\}^z = \text{hid}(ID_t)$ (as is likely for $t \geq z$), then the colluders could first efficiently compute (for instance, using Gaussian elimination in $\mathbb{Z}_q$), the coefficients $c_j \in \mathbb{Z}_q$ that satisfy the vector equation $\sum_{j=1}^{t} c_j \mathbf{s}_j \mod q = \mathbf{s}_T$, and then compute the target identity's private key as $a_T \in \mathbb{Z}_q \leftarrow \sum_{j=1}^{t}(c_j \cdot a_j) \mod q$. Alternatively, a group of $t \geq z$ colluding users (having their t private keys a_j) could probably solve the system of t linear equations (in $\mathbb{Z}_q$) with z variables (the elements of $\mathbf{e}$) given by the private keys for the elements of $\mathbf{e}$ as the unknown variables, thus completely breaking the security of the system.

In order to achieve collusion resistance, we extend the rudimentary scheme above such that the number of unknowns in the linear equation system in $\mathbb{Z}_q$ resulting from the private keys of the colluders is always higher than the number of equations given: When computing Alice's private key a_A, the TA adds a further summand. This summand δ_A is randomly chosen from $\mathbb{Z}_q^*$ and is specific for this private key. The TA now computes $a_A \in \mathbb{Z}_q$ as $\delta_A + \sum_{i=1}^{z}(\mathbf{s}_A[i] \cdot \mathbf{e}[i]) \mod q$. In order to enable Alice to "neutralize" the "blinding addend" δ when using a_A, the TA gives Alice a second component of her private key b_A, and the TA has initially set up a second component w of the master key and an additional public parameter x. The TA has w randomly chosen from $\mathbb{Z}_q^*$, has $x \in G$ computed as $g^{\left(\frac{1}{w} \mod q\right)}$, and computes $b_A \in \mathbb{Z}_q^*$ as $-w \cdot \delta \mod q$.

When Bob wants to encrypt a message to Alice, he first computes Alice's public key y_A, exactly as before. Second, Bob randomly picks an r from $\mathbb{Z}_q^*$, as before. Third, Bob computes the "raw message key" $\gamma \in G \leftarrow (y_A)^r$ and $c \in G \leftarrow g^r$ (both as before), and additionally $d \in G \leftarrow x^r$. $\langle c, d \rangle$ is the encapsulated message key.

Alice computes the "raw message key" as $\gamma' \in G \leftarrow c^{a_A} \cdot d^{b_A}$. Looking at how a_A, b_A, c and d were computed, we see that $\gamma' = c^{a_A} \cdot d^{b_A} = g^{r \cdot a_A} \cdot x^{r \cdot b_A}$

$$= g^{r \cdot a_A} \cdot g^{\left(\frac{r \cdot b_A}{w} \mod q\right)} = \left(g^{\left(a_A + \frac{b_A}{w} \mod q\right)}\right)^r = \left(g^{\left(\delta_A + \sum_{i=1}^{z}(s_A[i] \cdot e[i]) - \delta_A\right)}\right)^r$$

$$= \left(g^{\sum_{i=1}^{z}(s_A[i] \cdot e[i])}\right)^r = \left(\prod_{i=1}^{z}(f[i])^{s_A[i]}\right)^r = (y_A)^r = \gamma.$$ Obviously, Alice and Bob have again finally obtained the same "raw message key".

3   The Foundational Problem

The security of ID-KEMs and IBE schemes constructed based on the idea sketched above is determined by the hardness of a new problem, which we call *PX*. PX can be seen as two interwoven problems, *P0* and the problem of unambiguously solving a linear equation system where the number of equations is less than the number of variables. P0 is related to the computational Diffie-Hellman problem (CDH). First, we define P0 and show that this problem is as hard as CDH in the same group, and can thus reasonably be assumed to be as hard as the discrete logarithm problem (DL, see, e.g., [23]) in this group. Then, we define problem PX and argue that also PX is as hard as CDH in the same group.

We think that problems like PX may provide the foundation for cryptographic protocols beyond the identity-based key-encapsulation mechanism presented further below. With the stepwise introduction of PX, we would like to encourage the study of this kind of problems also in its own right. We believe that, just like the bilinear Diffie-Hellman problem [21] [8] and its relatives (see, e.g., [7]) have, after extensive study, allowed to substantially extend the tool set for the construction of cryptographic protocols by the use of pairings (although none of these problems has ever been *proved* to be as hard as one of the "classical" problems, such as the actual Diffie-Hellman problem), problems like PX may after thorough study allow to extend the cryptographic tool set by the method of identity-based factors selection.

Problem P0

Problem P0 is: Let G be a finite cyclic group of prime order q, and let g be a generator of G. Given q, G, g, a set F of z different group elements $\{f_1 = g^{e_1}, ..., f_z = g^{e_z}\}$, a non-empty subset $S = \{s_1, ..., s_x\}$ of $\{1, ..., z\}$, and a group element $c = g^r$, find the group element $\gamma = g^{(r \cdot \sum_{\forall i \in S} e_i)}$.

P0 is as hard as CDH in the same group G. We recall CDH (in prime-order groups): Let G be a finite cyclic group of prime order q, and let φ be a generator of G. Given q, G, φ, and the two group elements $\mu = \varphi^a$ and $\nu = \varphi^b$, find the group element $\lambda = \varphi^{ab}$.

CDH reduces to P0. From an algorithm $\mathcal{A}_{P0}$ that solves P0 in G for some $\langle z, S \rangle$ one can, adding only polynomial complexity, construct an algorithm $\mathcal{A}_{CDH}$ that solves CDH in G, as shown below.

$\mathcal{A}_{\mathsf{CDH}}(q, G, \varphi, \mu, \nu)$

1: Creates an ordered set $F = \{f_1, ..., f_z\}$ of z group elements, each element having an arbitrarily chosen value.
2: Splits μ arbitrarily into x factors in G and assigns for each i in S to the element f_i a new value, namely one of these factors (each time a different one).
3: Runs $\mathcal{A}_{\mathsf{P0}}$ with input $\langle q, G, \varphi, F, S, \nu \rangle$ and returns with $\mathcal{A}_{\mathsf{P0}}$'s result.

Problem PX

Problem PX is: Let G be a finite cyclic group of prime order q, and let g be a generator of G. Let $\{e_1, ..., e_z\}$ be a set of z different elements of $\mathbb{Z}_q^*$, and let w be another element of $\mathbb{Z}_q^*$. Given are z, q, G, a set F of group elements $\{f_1 = g^{e_1}, ..., f_z = g^{e_z}\}$, another group element $x = g^{\left(\frac{1}{w} \mod q\right)}$, a non-empty subset S of $\{1, ..., z\}$, and two further group elements $c = g^r$ and $d = x^r$. Given are further n tuples $\{K_1, ..., K_n\}$ where $K_j = \langle S_j, \langle a_j, b_j \rangle \rangle$ where S_j is a non-empty subset of $\{1, ..., z\}$, $a_j \in \mathbb{Z}_q^* = \delta_j + \sum_{\forall i \in S_j} e_i \mod q$, and $b_j \in \mathbb{Z}_q^* = -w \cdot \delta_j$ mod q where $\delta_j \in \mathbb{Z}_q^*$ is non-repeating. No two sets in $\{S_1, ..., S_n, S\}$ are equal. Find $\gamma = g^{\left(r \cdot \sum_{\forall i \in S} e_i\right)}$.

We conjecture that PX is not easier than CDH, based on the following observations: One may look at PX as P0 with the same $\langle q, G, g, F, S, r, c \rangle$, with the difference that the n tuples $\{K_1, ..., K_n\}$ are given as additional information. Note that $\gamma = g^{\left(r \cdot \sum_{\forall i \in S} e_i\right)}$ is $c^{\left(\sum_{\forall i \in S} e_i \mod q\right)}$. Apparently, the additionally given pairs $\{K_1, ..., K_n\}$ might help to solve the instance of P0 "embedded" in PX, but only if one could deduce $\sum_{\forall i \in S} e_i \mod q$ from them. Indeed, the components S_j and a_j of the K_j's give a linear equation system in $\mathbb{Z}_q$ which might raise expectations that one could learn the e_i's by solving the linear equation system and finally compute $\sum_{\forall i \in S} e_i \mod q$ from them. The n linear equations given by the S_j's and a_j's are:

$$\delta_1 + s_{1,1} e_1 + ... + s_{1,z} e_z \mod q = a_1$$

$$\vdots$$

$$\delta_n + s_{n,1} e_1 + ... + s_{n,z} e_z \mod q = a_n$$

where the $s_{j,i} \in \{0, 1\}, j \in \{1, ..., n\}, i \in \{1, ..., z\}$ and the $a_j \in \mathbb{Z}_q, j \in \{1, ..., n\}$ are known and the $e_i \in \mathbb{Z}_q^*, i \in \{1, ..., z\}$ and $\delta_j^* \in \mathbb{Z}_q, j \in \{1, ..., n\}$ are not. Obviously, the number of unknowns is for all $n \in \mathbb{N}^*$ greater than the number of equations. Each e_i can have any value in $\mathbb{Z}_q^*$, making each value in $\mathbb{Z}_q$ look equally likely to be $\sum_{\forall i \in S} e_i \mod q$. The b_j $(= -w \cdot \delta_j \mod q)$ given as the third component of the K_j's would only be helpful if they revealed δ_j. However, each $\delta_j \in \mathbb{Z}_q^*$ is multiplicatively blinded with $w \in \mathbb{Z}_q^*$ so that if each value in $\mathbb{Z}_q^*$ looks equally likely to be w then also each value in $\mathbb{Z}_q^*$ looks equally likely to be δ_j. For w, each value in $\mathbb{Z}_q^*$ does indeed look equally likely since the only further information about w available is x $\left(= g^{\left(\frac{1}{w} \mod q\right)}\right)$ and learning w from x (in order to learn δ_j from b_j and w) means solving DL in G (which is at least as hard as CDH in G).

4   ID-KEM Built on Identity-based Factors Selection

Our identity-based key-encapsulation mechanism presented now completely builds on identity-based factors selection as introduced in Section 2. Its security (in the random oracle model) rests on the hardness of problem PX and the collision resistance of a cryptographic hash function.

4.1   The Algorithms

According to the common definition of ID-KEM schemes (see, e.g., [5]), the scheme consists of four algorithms. In order to set the system up, i.e., to generate the master key and to define the system parameters, the TA first executes algorithm Setup. When requested by a user for the (long-term) private key corresponding to this user's identity, the TA executes algorithm Extract to generate this key. The algorithm Encapsulate is executed by a user who wants to generate a message key and the encapsulation of this message key. The algorithm Decapsulate is executed by a user who wants to retrieve from the encapsulation of a message key the respective message key.

Setup Algorithm. Given the security parameter $l \in \mathbb{N}^*$ (e.g., 80, 128, or 256), the algorithm performs the following steps:

1. Determines the *potential-factors number* $z \in \mathbb{N}^* \leftarrow 2l$.

2. Defines a cryptographic hash function (suitable to be modeled as a random oracle [4]) hid: $\{0,1\}^* \longrightarrow \{0,1\}^z$, for instance, SHA-3/Keccak [6] with output length z.

3. Chooses a a finite cyclic group G of prime order q such that the most efficient algorithm for solving the discrete logarithm problem in G has an average computational complexity not smaller than 2^l. For instance, G could be the group of points on an elliptic curve over a finite field $E(\mathbb{F}_p)$ (where p is another large prime number of the same bitlength as q) that are generated by some point $P \in E(\mathbb{F}_p)$ which is chosen such that the smallest $i \in \mathbb{N}^*$ for which $iP = O$ is q (see, e.g., [20]). Another instance of G is the (multiplicative) subgroup of order q of $\mathbb{Z}_p^*$ where p is a sufficiently large prime number (e.g., of bitlength 1024, for $l = 80$) with $p - 1 = cq$ where c is an even positive integer.

4. Chooses a generator g of group G.

5. Generates the *exponents vector* $\mathbf{e}[1...z] \in \left(\mathbb{Z}_q^*\right)^z$ of z different elements $\in \mathbb{Z}_q^*$, each element being an independently chosen random number.

6. Generates the *potential-factors vector* $\mathbf{f}[1...z] \in G^z$, its z elements being computed:
$$\forall i \in \{1, ..., z\}: \ \mathbf{f}[i] \in G \leftarrow g^{\mathbf{e}[i]}$$

7. Randomly picks w from $\mathbb{Z}_q^*$.

8. Computes $x \in G \leftarrow g^{\left(\frac{1}{w} \bmod q\right)}$

9. Defines another cryptographic hash function (suitable to be modeled as a random oracle) $hel: G \longrightarrow \{0,1\}^l$. For instance, if G is an elliptic-curve group as mentioned as the first example in step 3, then hel may return the output of a standard cryptographic hash function with output length l (such as SHA-3/Keccak [6] with output length l) executed on a standard binary representation of the x-coordinate of the input point. If G is, as in the other example given in step 3, a subgroup of the multiplicative group of a number field, then hel may return the output of a standard cryptographic hash function with output length l executed on a standard binary representation of the input.

The identity space is $\{0,1\}^*$. The private-key space is $(\mathbb{Z}_q)^2$. The message-key space is $\{0,1\}^l$. The encapsulated-key space is G^2. The public system parameters are: $\mathsf{params} = \langle z, q, G, g, \mathbf{f}, x, \mathsf{hid}, \mathsf{hel} \rangle$. The master key is: $\mathsf{masterkey} = \langle \mathbf{e}, w \rangle$.

Extract Algorithm. Given $\mathsf{params}, \mathsf{masterkey}$ and an identity $\mathsf{ID} \in \{0,1\}^*$, the algorithm performs the following steps:[2]

1. Randomly picks δ from $\mathbb{Z}_q^*$. (δ is kept secret. It can be deleted after step 5).

2. Computes the ID-specific *selection vector* $\mathbf{s}$, a bit vector of z elements

$$\mathbf{s}[1...z] \in \{0,1\}^z \leftarrow \mathsf{hid}(\mathsf{ID})$$

   where hid is the cryptographic hash function defined as a public parameter.

3. Computes $a \in \mathbb{Z}_q \leftarrow \delta + \sum_{i=1}^{z}(\mathbf{s}[i] \cdot \mathbf{e}[i]) \bmod q$ where $\mathbf{e}$ is from the master key, and z and q are from the public parameters.

4. Verifies that $d \neq \delta$. If the test fails, denies the generation of a private key for identity ID.[3]

5. Computes $b \in \mathbb{Z}_q^* \leftarrow -w \cdot \delta \bmod q$ where w is from the master key.

Returns ID's private key $\langle a, b \rangle$.

---

[2] As usual in identity-based cryptography, it is assumed that the executor of algorithm Extract has authenticated the identity of the user that requests a private key, and that the extracted private key is transferred to the user in a way preserving the confidentiality and the integrity of the key.

[3] Note that the probability that $\sum_{i=1}^{z}(\mathbf{s}[i] \cdot \mathbf{e}[i]) \bmod q = 0$ is negligibly small. If $\mathbf{s} = 0^z$, then one has accidentally found hid's preimage for 0^z, an accident of probability $2^{-z} = 2^{-2l}$. Otherwise, one has accidentally solved the instance of the subset sum problem in $\mathbb{Z}_q$ where the elements of $\mathbf{e}$ are the elements of the problem, an accident also of probability $2^{-z} = 2^{-2l}$.

Encapsulate Algorithm. Given params and an identity ID $\in \{0,1\}^*$, the algorithm performs the following steps:

1. Computes the ID-specific selection vector **s**, a bit vector of z elements:

$$\mathbf{s}[1...z] \in \{0,1\}^z \leftarrow \mathrm{hid}(\mathrm{ID})$$

2. Computes ID's public key $y \in G \leftarrow \prod_{i=1}^{z} (\mathbf{f}[i])^{\mathbf{s}[i]}$ where **f** is from the public parameters.

3. Verifies that $y \neq 1$. If the test fails, denies the key encapsulation for ID.[4]

4. Randomly picks r from $\mathbb{Z}_q^*$. (r is kept secret and shall not be reused, it can be deleted after step 7).

5. Computes the "raw message key" $\gamma \in G \leftarrow y^r$
   (γ is kept secret; it can be deleted after step 9).

6. Computes the first part of the encapsulated key $c \in G \leftarrow g^r$ where g is from the public parameters.

7. Computes the second part of the encapsulated key $d \in G \leftarrow x^r$ where x is from the public parameters.

8. Verifies that $c \neq 1$ and $d \neq 1$. If any one of the two tests fails, repeats from step 4.

9. Computes the message key $k \in \{0,1\}^l \leftarrow \mathrm{hel}(\gamma)$ where hel is the cryptographic hash function defined as part of the public parameters.

Returns the message key k and its encapsulation $\langle c, d \rangle$.

Decapsulate Algorithm. Given params, a private key $\langle a \in \mathbb{Z}_q, b \in \mathbb{Z}_q \rangle$, and an encapsulation $\langle c \in G, d \in G \rangle$ of a message key, the algorithm performs the following steps:

1. Computes the "raw message key" $\gamma' \in G \leftarrow c^a \cdot d^b$

2. Computes the message key $k' \in \{0,1\}^l \leftarrow \mathrm{hel}(\gamma')$

Returns k'.

4.2   Consistency

The scheme is consistent if it satisfies the following constraint: $\forall \mathrm{ID} \in \{0,1\}^*$: If, first, $\langle a, b \rangle$ is the private key generated by algorithm Extract when given ID and, second, $\langle k, c, d \rangle$ is a tuple resulting from Encapsulate(params, ID), then Decapsulate(params, a, b, c, d) = k. In order to see that this is indeed the case, we look at algorithm Decapsulate's result k' and what it actually is, step by step.

---

[4] Note that a private key for ID would not be issued, see footnote 3.

As k' is computed in step 2 of algorithm Decapsulate, as γ' was computed in step 1, as c was computed in step 6 of algorithm Encapsulate, and as d was computed in step 7:

$$k' = \text{hel}\,(\gamma') = \text{hel}\,(c^a \cdot d^b) = \text{hel}\,(g^{ra} \cdot x^{rb}) = \text{hel}\,\Big(\,(g^a \cdot x^b)^r\,\Big)$$

As a is computed in step 3 of algorithm Extract, and as b is computed in step 5:

$$= \text{hel}\,\Big(\,\Big(g^{(\delta\, +\, \sum_{i=1}^{z}(\mathbf{s}[i]\,\cdot\,\mathbf{e}[i]))\ \text{mod}\ q}\,\cdot\,x^{(-w\,\cdot\,\delta\ \text{mod}\ q)}\Big)^r\,\Big)$$

As x is computed in step 8 of algorithm Setup:

$$= \text{hel}\,\Big(\,\Big(g^{(\delta\, +\, \sum_{i=1}^{z}(\mathbf{s}[i]\,\cdot\,\mathbf{e}[i]))\ \text{mod}\ q}\,\cdot\,\big(g^{(\frac{1}{w}\ \text{mod}\ q)}\big)^{(-w\,\cdot\,\delta\ \text{mod}\ q)}\Big)^r\,\Big)$$

$$= \text{hel}\,\Big(\,\Big(g^{(\delta\, +\, \sum_{i=1}^{z}(\mathbf{s}[i]\,\cdot\,\mathbf{e}[i]))\ \text{mod}\ q}\,\cdot\,g^{-\delta}\Big)^r\,\Big)$$

$$= \text{hel}\,\Big(\,\big(g^{(\sum_{i=1}^{z}(\mathbf{s}[i]\,\cdot\,\mathbf{e}[i]))}\big)^r\,\Big) = \text{hel}\,\Big(\,\Big(\textstyle\prod_{i=1}^{z}\big(g^{\mathbf{e}[i]}\big)^{\mathbf{s}[i]}\Big)^r\,\Big)$$

As $\mathbf{f}$ is computed in step 6 of algorithm Setup, as y is computed in step 2 of algorithm Encapsulate, as γ is computed in step 5, and k in step 9:

$$= \text{hel}\,\Big(\,\Big(\textstyle\prod_{i=1}^{z}(\mathbf{f}[i])^{\mathbf{s}[i]}\Big)^r\,\Big) = \text{hel}\,(\,y^r\,) = \text{hel}\,(\,\gamma\,) = k$$

Obviously, $k' = k$, which means the algorithms of the scheme are consistent.

4.3   Security

The strictest widely accepted definition of the security of ID-KEMs (ID-IND-CCA2, see, e.g., [5]) is stated in terms of a game between an adversary $\mathcal{A}$ and a challenger $\mathcal{C}$. At the start, $\mathcal{C}$ takes as its input the security parameter l, runs algorithm Setup and gives $\mathcal{A}$ the public parameters $\mathsf{params} = \langle z, q, G, g, \mathbf{f}, x, \text{hid}, \text{hel}\rangle$. $\mathcal{C}$ keeps the master key $\mathsf{masterkey} = \langle \mathbf{e}, w\rangle$ to itself. $\mathcal{A}$ then makes queries of the following three types:

Extract($\mathsf{ID}_j \in \{0,1\}^*$): $\mathcal{C}$ responds by running algorithm Extract with input (params, masterkey, ID_j) to generate the private key $\langle a_{\mathsf{ID}_j}, b_{\mathsf{ID}_j}\rangle$. $\mathcal{C}$ gives $\langle a_{\mathsf{ID}_j}, b_{\mathsf{ID}_j}\rangle$ to $\mathcal{A}$.

Decapsulate($\mathsf{ID}_j \in \{0,1\}^*, \langle c_j \in \mathbb{Z}_q, d_j \in \mathbb{Z}_q^*\rangle$): $\mathcal{C}$ responds by running algorithm Extract(params, masterkey, ID_j) to obtain the private key $\langle a_{\mathsf{ID}_j}, b_{\mathsf{ID}_j}\rangle$ and then running algorithm Decapsulate(params, $\langle a_{\mathsf{ID}_j}, b_{\mathsf{ID}_j}\rangle, \langle c_j, d_j\rangle$) to generate the message key k_j. $\mathcal{C}$ gives k_j to $\mathcal{A}$.

Test($\mathsf{ID} \in \{0,1\}^*$): $\mathcal{C}$ responds by first randomly selecting ω from $\{0,1\}$. Then $\mathcal{C}$ computes $\langle k, \langle c, d\rangle\rangle \leftarrow$ Encapsulate(ID). If $\omega = 1$, then $\mathcal{C}$ gives $\langle k, \langle c, d\rangle\rangle$ to $\mathcal{A}$, else $\mathcal{C}$ randomly picks u from $\{0,1\}^l$ and gives $\langle u, \langle c, d\rangle\rangle$ to $\mathcal{A}$.

$\mathcal{A}$'s queries may be made adaptively and are arbitrary in number, except that $\mathcal{A}$ is allowed only one Test query. Two further constraints are that (i) Decapsulate queries with the tuple $\langle \mathsf{ID}, \langle c, d\rangle\rangle$ from the Test query as the input tuple are not allowed, and that (ii) the input argument of the Test query is not identical with the input argument of any Extract query.

   Finally, $\mathcal{A}$ outputs a bit ω', and wins the game if $\omega' = \omega$.

$\mathcal{A}$'s advantage in winning the game (and thus in attacking the security of the scheme) is defined as: $\mathbf{Adv}_{\mathsf{ID-KEM}}^{\mathsf{ID-IND-CCA2}}(\mathcal{A}) = |2 \cdot \Pr[\omega' = \omega] - 1|$. (The probability Pr is over the random bits used by $\mathcal{C}$ and $\mathcal{A}$). We say that the ID-KEM scheme is secure if $\mathbf{Adv}_{\mathsf{ID-KEM}}^{\mathsf{ID-IND-CCA2}}(\mathcal{A}) \leq 2^{\epsilon} 2^{-l}$ with ϵ being negligibly small ($\epsilon << l$), say 1.

Proof that the scheme is secure as defined above under the assumption that the hash functions hid and hel are random oracles [4] and the assumption that that PX is not easier than CDH (see Section 3):

Assume $\mathcal{A}$ has, making m queries, including n Extract queries, won the game. Two cases can be distinguished: (i) The set of selection vectors $\{\mathbf{s}_1, ..., \mathbf{s}_t\}$ computed by $\mathcal{A}$ or $\mathcal{C}$ (using the hash function hid) from the t different identities $\{\mathsf{ID}_{t1}, ..., \mathsf{ID}_{tt}\} \subseteq \{\mathsf{ID}_1, ..., \mathsf{ID}_m, \mathsf{ID}\}$ involved in the course of the game contains at least one pair of identical elements. In this case, $\mathcal{A}$ has found a collision of hid. (ii) All elements in the set $\{\mathbf{s}_1, ..., \mathbf{s}_t\}$ are different. In this case, $\mathcal{A}$ has solved for the tuple $\langle n, z, q, G, g, \mathbf{e} \rangle$ a randomly chosen instance of problem PX, as follows: If hel is a random oracle, then $\mathcal{A}$'s advantage $\mathbf{Adv}_{\mathsf{ID-KEM}}^{\mathsf{ID-IND-CCA2}}(\mathcal{A})$ in winning the game above is identical with $\mathcal{A}$'s advantage in finding γ. For finding γ, the responses of the Decapsulate queries do not help since these responses are the output of a random oracle (namely hel) and due to its random nature this output cannot provide any helpful information, such as about γ, about the elements of $\mathbf{e}$, or about relations between these elements. Finding γ by playing the game above without using the responses of Decapsulate queries (as it effectively would have to be done then) means solving an instance of PX, namely the instance where the $\langle n, z, q, G, g \rangle$ are the same, the e_i's are the $\mathbf{e}[i]$'s (with $i \in \{1, ..., z\}$), PX's n is the number of Extract queries, PX's S_j's contain exactly the i's for which $\mathbf{s}_{\mathsf{ID}_j}[i] = 1$ (where $\mathbf{s}_{\mathsf{ID}_j} = hid(\mathsf{ID}_j)$ with ID_j being the input argument of the j-th Extract query), and S contains exactly the i's for which $\mathbf{s}_{\mathsf{ID}}[i] = 1$ (where $\mathbf{s}_{\mathsf{ID}} = hid(\mathsf{ID})$. If hid is a random oracle, then the $\langle n, z, q, G, g, \mathbf{e} \rangle$-instance of problem PX is in effect randomly chosen, having the average complexity of $\langle n, z, q, G, g, \mathbf{e} \rangle$-instances of PX.

Obviously, $\mathcal{A}$'s advantage in winning the game $\mathbf{Adv}_{\mathsf{ID-KEM}}^{\mathsf{ID-IND-CCA2}}(\mathcal{A})$ has an upper bound in the sum of (i) the advantage of the most efficient algorithm to break the collision-resistance of a random oracle with z bits output and (ii) the advantage of the most efficient algorithm to solve random $\langle n, z, q, G, g, \mathbf{e} \rangle$-instances of problem PX. The first advantage is $\sqrt{2^{-z}} = \sqrt{2^{-2l}} = 2^{-l}$. The second advantage is, following the two assumptions mentioned above (according to our conjecture about the hardness of PX in Section 3) and following also the usual assumptions about the relation between CDH and DL (see, e.g., [23]), $\leq 2^{-l}$. Thus, $\mathbf{Adv}_{\mathsf{ID-KEM}}^{\mathsf{ID-IND-CCA2}}(\mathcal{A}) \leq 2^{-l} + 2^{-l} = 2 \cdot 2^{-l}$, as required ($\epsilon << l$).

4.4 Efficiency

The computational complexity of a key encapsulation is essentially the complexity of three exponentiations in G (two of them with a fixed base) plus, on average, l additional group operations ($2l$ in the worst case). The computational complexity of a key decapsulation is essentially the complexity of two exponentiations in G. So, when applying the usual methods for performance optimization of fixed exponentiations, a key encapsulation costs, on average, $6l$ group operations ($10l$ in the worst case). A key decapsulation costs, on average, $6l$ group operations ($8l$ in the worst case).

Compared to the pairing-based ID-KEMs (and IBE schemes) with the most efficient key-encapsulation (most notably BB_1 [7], for a comparative analysis of their performances, see, e.g., [10]), a key encapsulation costs, on average, $2k$ group operations more. However, since our ID-KEM can be fully implemented in efficient standard elliptic-curve groups (see, e.g., [11]) so that, in contrast to the pairing-based schemes, the performance of *all* exponentiations and all group operations benefits from the efficiency of these elliptic-curve groups, the overall computational cost for these group operations (and thus for the whole key-encapsulation) is somewhat less than in pairing-based schemes. On the key-decapsulation side, the computational complexity is significantly lower, because no pairing is necessary. Both differences grow over-polynomially with the security parameter.

Compared to Cocks' IBE scheme [14], the complexity of the key-decapsulation is lower as soon as $6l$ group operations in our scheme's group G cost less than l computations of the Jacobi symbol, l modular inversions, and l modular multiplications together (all these operations with a composite modulus that resists factorization according to l). The complexity of the key-encapsulation is lower as soon as $6k$ group operations in G cost less than l computations of the Jacobi symbol with the same modulus. Compared to Boneh, Gentry and Hamburg's space-efficient pairing-free IBE scheme [9], our scheme is computationally much less costly.

Our scheme requires the transfer of two group elements (as the encapsulated key), i.e., typically about $4l$ bits (if G is a suitable elliptic-curve group and point compression is used). Compared to expansion-efficient pairing-based IBE schemes (see, e.g., [10]), this is at most double of what it is there. It is much less than what is required in Cocks' IBE scheme [14].

The complexity of the generation of a private key is essentially one multiplication plus, on average, l additions (at most about $2l$), all modulo q, which means it is very low compared to other ID-KEMs and IBE schemes.

The size of the private keys is small, about $4l$ bits.

The size of the public system parameters is dominated by **f**. It is about $2l$ times q. If G is a suitable elliptic-curve group, this is about $4l^2$. This is in the same range as in lattice-based ID-KEMs or IBE schemes, but significantly higher than in schemes based on pairings (e.g., [7]) or on the quadratic-residuosity problem [14] [9]. While this size of the system parameters should in most scenarios still not be a problem, it must be considered in environments with tight memory constraints.

5   Summary

Identity-based factors selection (IBFS) enables identity-based key-encapsulation mechanisms (ID-KEMs) and identity-based encryption schemes (IBE schemes) where the computational complexities of all three relevant operations, i.e., key-encapsulation, key-decapsulation, and private-key generation, are lower than in previous ID-KEMs and IBE schemes. The only possible drawback concerning efficiency is that the public parameters have typically a size of about $4l^2$ bits (where l is the security parameter). IBFS allows the sender of an encrypted message to *compute* the same type of public key as used in conventional discrete-logarithm based public-key encryption schemes, such as ECIES [11]. The security (ID-IND-CCA2) of these constructions (in the random oracle model) rests on the hardness of a new problem which combines the computational Diffie-Hellman problem with the fact that linear equation systems with more variables than given equations do not have unambiguous solutions.

References

1. Abdalla, M., Bellare, M., Rogaway, P.: DHIES: An encryption scheme based on the Diffie-Hellman Problem. In: Extended Version of [2], September 18 (2001), http://www.cs.ucdavis.edu/~rogaway/papers/dhies.pdf

2. Abdalla, M., Bellare, M., Rogaway, P.: The oracle Diffie-Hellman assumptions and an analysis of DHIES. In: Naccache, D. (ed.) CT-RSA 2001. LNCS, vol. 2020, pp. 143–158. Springer, Heidelberg (2001)

3. Agrawal, S., Boneh, D., Boyen, X.: Efficient lattice (H)IBE in the standard model. In: Gilbert, H. (ed.) EUROCRYPT 2010. LNCS, vol. 6110, pp. 553–572. Springer, Heidelberg (2010)

4. Bellare, M., Rogaway, P.: Random oracles are practical: A paradigm for designing efficient protocols. In: Proc. of 1st ACM Conference on Computer and Communications Security, pp. 62–73. ACM (1993)

5. Bentahar, K., Farshim, P., Malone-Lee, J., Smart, N.: Generic constructions of identity-based and certificateless KEMs. Journal of Cryptology 21(2), 178–199 (2008)

6. Bertoni, G., Daemen, J., Peeters, M., van Assche, G.: The Keccak reference, version 3.0, January 14 (2011), http://keccak.noekeon.org/Keccak-reference-3.0.pdf

7. Boneh, D., Boyen, X.: Efficient selective identity-based encryption without random oracles. Journal of Cryptology 24(4), 659–693 (2011)

8. Boneh, D., Franklin, M.: Identity-based encryption from the Weil pairing. SIAM Journal on Computing 32(3), 586–615 (2003)

9. Boneh, D., Gentry, C., Hamburg, M.: Space-efficient identity based encryption without pairings. In: Proc. of 48th Annual IEEE Symposium on Foundations of Computer Science (FOCS 2007), pp. 647–657 (2007) ISBN 0-7695-3010-9

10. Boyen, X.: A tapestry of identity-based encryption: practical frameworks compared. Int. Journal of Applied Cryptography (1), 3–21 (2008)

11. Certicom Research: Standards for efficient cryptography, SEC 1: Elliptic Curve Cryptography, Version 2.0, May 21 (2009), http://www.secg.org/download/aid-780/sec1-v2.pdf

12. Chatterjee, S., Sarkar, P.: Identity-Based Encryption. Springer (2011) ISBN 978-1-44199-382-3
13. Chen, L., Cheng, Z., Malone-Lee, J., Smart, N.P.: An efficient ID-KEM based on the Sakai–Kasahara key construction. IEE Proc. Information Security 153, 19–26 (2006)
14. Cocks, C.: An identity based encryption scheme based on quadratic residues. In: Honary, B. (ed.) Cryptography and Coding 2001. LNCS, vol. 2260, pp. 360–363. Springer, Heidelberg (2001)
15. Diffie, W., Hellman, M.: New directions in cryptography. IEEE Trans. on Information Theory 22, 644–654 (1976)
16. ElGamal, T.: A public-key cryptosystem and a signature scheme based on discrete logarithms. IEEE Trans. on Inform. Theory 31(4), 469–472 (1985)
17. Galbraith, S.D., Paterson, K.G., Smart, N.P.: Pairings for cryptographers. Discrete Applied Mathematics 156(16), 3113–3121 (2008)
18. Gentry, C., Peikert, C., Vaikuntanathan, V.: Trapdoors for hard lattices and new cryptographic constructions. In: Proc. of the 40th ACM Symposium on Theory of Computing (STOC 2008), pp. 197–206. ACM (2008)
19. Gentry, C., Silverberg, A.: Hierarchical ID-based cryptography. In: Zheng, Y. (ed.) ASIACRYPT 2002. LNCS, vol. 2501, pp. 548–566. Springer, Heidelberg (2002)
20. Hankerson, D., Menezes, A., Vanstone, S.: Guide to elliptic curve cryptography. Springer (2004) ISBN 0-387-95273-X
21. Joux, A.: A one round protocol for tripartite Diffie-Hellman. Journal of Cryptology 17(4), 263–276 (2004)
22. Maurer, U., Yacobi, Y.: A non-interactive public-key distribution system. Designs, Codes and Cryptography 9(3), 305–316 (1996)
23. Menezes, A., van Oorschot, P., Vanstone, S.: Handbook of applied cryptography, 5th edn. CRC Press (2001) ISBN 0-8493-8523-7
24. Paterson, K.G., Srinivasan, S.: On the relations between non-interactive key distribution, identity-based encryption and trapdoor discrete log groups. Designs, Codes and Cryptography 52(2), 219–241 (2009)
25. Shamir, A.: Identity-based cryptosystems and signature schemes. In: Blakely, G.R., Chaum, D. (eds.) CRYPTO 1984. LNCS, vol. 196, pp. 47–53. Springer, Heidelberg (1985)
26. Yacobi, Y.: A note on the bilinear Diffie-Hellman assumption, IACR Cryptology ePrint Archive, Report 2002/113 (2002), http://eprint.iacr.org/2002/113

A Comparison of Time-Memory Trade-Off Attacks on Stream Ciphers

Fabian van den Broek and Erik Poll

Institute for Computing and Information Sciences,
Radboud University Nijmegen, The Netherlands

Abstract. Introduced by Hellman, Time-Memory Trade-Off (TMTO) attacks offer a generic technique to reverse one-way functions, where one can trade off time and memory costs and which are especially effective against stream ciphers. Hellman's original idea has seen many different improvements, notably the Distinguished Points attack and the Rainbow Table attack. The trade-off curves of these approaches have been compared in literature, but never leading to a satisfying conclusion. A new TMTO attack was devised for the A5/1 cipher used in GSM, which combines both distinguished points and rainbow tables, which we refer to as the Kraken attack. This paper compares these four approaches by looking at concrete costs of these attacks instead of comparing their trade-off curves. We found that when multiple samples are available the Distinguished Points attack has the lowest costs. The Kraken attack is an alternative to save more disk space at the expense of attack time.

1 Introduction

An attacker trying to break a cryptographic function can always try to either brute force the function, or precompute all possible values beforehand and store them in a large table, so every subsequent attack is a simple look-up. Most cryptographic functions are protected from these attacks by having a large enough key size or state size, which makes the time complexity or the storage requirements of such attacks too large in practice.

In 1980 Hellman caused a breakthrough by suggesting a Time-Memory Trade-Off attack which is probabilistic and falls somewhere in between a brute force attack and a precomputation attack. Hellman showed that using his attack he could reverse an n-bit key cipher, in $2^{2n/3}$ time complexity, by precomputing 2^n values and storing these in $2^{2n/3}$ values [1]. This made ciphers using keys that until then were thought large enough to prevent a brute-force attack suddenly susceptible to this new Time-Memory Trade-Off attack.

Later research into TMTO attacks led to many improvements on Hellman's attack. First came the Distinguished Points method, which reduced the number of disk seeks and is referenced to Rivest [2]. Later Oechslin [3] devised a competing method with a slight speed-up, called Rainbow Table. The Rainbow Table attack seems to be better known, presumably due to its colorful name. Biryukov and Shamir [4] found that TMTO attacks were especially useful against stream

A. Youssef, A. Nitaj, A.E. Hassanien (Eds.): AFRICACRYPT 2013, LNCS 7918, pp. 406–423, 2013.

ciphers, since an attacker can then make generic TMTO tables for a cipher which can be matched against any large enough sample of keystream, increasing the success chance for every sample. This new understanding directly led to new proposed attacks against one of the most widely deployed stream ciphers in the world: GSM's A5/1 cipher [5,6].

In 2010 researchers demonstrated a TMTO attack to break the A5/1 cipher of GSM [7]. This attack uses a new, unresearched, TMTO method which combines two important, but very different TMTO improvements; namely Distinguished Points and Rainbow Tables [8]. This new attack is called Kraken in this paper, after the name of the tool used to perform the actual attack.

It seems rather strange for these researchers to have chosen a new approach for their attack, so the question arises whether this new Kraken attack improves on the already existing attacks. This paper aims to research how much, if any, of an improvement this Kraken attack brings to the area of TMTO attacks.

Section 2 introduces the general idea of TMTO attacks. Section 3 introduces and analyzes the four TMTO attacks: Hellman's original attack [1] with Biryukov and Shamir's improvement for stream ciphers [4], Rivest's Distinguished Points approach [2], Oechslin's Rainbow Tables [3], and the first theoretical analysis of the Kraken attack (Section 3.4). We compare the TMTO attacks in Section 4, including an informal analysis on the chances of chain merges. Finally some ideas for future research are given and conclusions are drawn.

Related Work. Some of the discussed TMTO methods have previously been compared with each other. Most of these publications compare the trade-off curves for these attacks [4,9], which give the rate at which extra memory can be traded in for a reduced attack time. Such as $M^2T = N^2$ for both the Hellman and Distinguished Point attack, with M the memory cost, T the time cost of the online phase, and N the size of the state space. Our comparisons are not based on trade-off curves, because we feel that these curves hide too much of the real costs such attacks have, such as the seek times in the online attack, or the precomputation effort. Biryukov and Shamir compare Hellman's attack with Distinguished Points [4] and Erguler et al. compare Hellman's attack with Rainbow Tables in [9]. Barkan et al. [10] make the most complete comparison; within a new theoretic framework they find the Distinguished Points attack better than the Rainbow Table attack, mainly based on the possibility to shorten the stored values of a Distinguished Points attack. However, this comparison is still very broad and the question on which TMTO attack has the lowest costs, in terms of time and memory, seems to still be open to debate.

In 2008 Hong et al. [11] already combined Distinguished Points with Rainbow Tables, but in a different way than in the Kraken approach. Their combination does not improve on just Distinguished Points or Rainbow Table attacks.

To the best of our knowledge there has been no analysis of the Kraken approach. The attack is considered in work by Krhovjak et al. [12], where they use it as a practical example for an attack against A5/1, but they make no comparison with the earlier attacks.

2   Typical TMTO

Assume a scenario in which an attacker tries to break a known cryptographic function f for which he has obtained at least one sample of ciphertext y. His goal is to reverse the function f, i.e. to find an input x for which $y = f(x)$. This model covers different scenarios:

- Finding the preimage x of a hash function f for the hash value y.
- Finding the key x used to encrypt a known plaintext p to produce y, i.e. a key x such that $y = f(x) = encrypt_x(p)$, with $encrypt$ e.g. a DES encryption.
- Finding the internal state used to encrypt a known plaintext p with a stream cipher. Here x is the internal state of cipher f and y is the corresponding keystream. So $f(x) = y$ and y is obtained by XORing cipherstream and known plaintext.

This paper is concerned with the third scenario, finding the "internal state" x of a stream cipher and not the key. Note that in many stream ciphers it is possible to retrieve the key that was used from a given internal state. The essential difference is that when reversing a stream cipher an attacker can construct tables which are more generic, so they can accept multiple samples from different plaintexts as explained in Section 3.1.

A typical TMTO attack consists of two phases: the first is the precomputation phase, often called the *offline phase*, while the second is referred to as the real-time, or *online phase*. In the offline phase, the attacker precomputes a large table (or sets of tables) using the function f he is trying to break, while in the online phase the attacker captures a sample of keystream and checks if this happens to be in his tables. If this attack is successful the attacker can learn the internal state x for which $y = f(x)$. We can evaluate these kinds of attacks by looking at different parameters and costs:

- N: the size of the state space.
- T (Attack time): This can be subdivided between the time for the offline phase, T_{pre}, in orders of magnitude, and the time for the online phase, which in turn can be subdivided into computation time T_c, measured in computation steps of f and seek time T_s, measured in number of disk seeks.
- M (Memory): memory cost of the attack.
- C (Coverage): the number of points from N covered by the tables.
- D (Data): number of usable data samples (y's) during the online phase.
- $\mathbb{P}$ (Chance of success): the chances of a collision between the observed ciphertext and the precomputed tables.
- ρ (Precomputation ratio): the ratio between the number of precomputed points from N and the total number of points N.

Intuitively, the chance of success $\mathbb{P}$ seems equal to the precomputation ratio, $\rho = C/N$, i.e. the number of points covered by the tables divided by the number of points in the search space. However, this is not exactly true for a number of reasons. Firstly, the tables can contain duplicate values. A certain number of duplicate values is to be expected when the coverage increases, however duplicates within the same table can lead to so called *chain merges*, which cause large parts of table rows to overlap. These chain merges will be discussed in more detail in the next section, but will for the most part be ignored in the analysis until Section 4, which details why it is hard to give an estimate on the occurrences of these chain merges. To stress this difference we introduce $\bar{C}$ and $\bar{\rho}$ as variants of the respective variables that do take chain merges into account.

Secondly, a definition of $\mathbb{P} = \rho$ assumes that all outputs of the cryptographic function f are equally likely, so all points in N have the same chance of occurring. This difference between $\mathbb{P}$ and ρ does not matter for our comparisons, but we will see in the practical example of Section 3.4 that this assumption does not always hold in practice.

Lastly, if an attacker has multiple samples, as he might have for a stream cipher, then the chance of success increases by a factor D, the number of samples.

3   The TMTO Attacks

This section compares the costs of the four attacks: Hellman's original attack, Distinguished Points, Rainbow Tables, and Kraken. For each we give a theorem that states the cost for the general case with an arbitrary number of tables, followed by a corollary where we align some of the parameters to allow for an easy comparison. For these corollaries we assume an attacker abides by the $mt^2 = N$ rule, which will be introduced in Section 3.1, and precomputes enough points so that $D\rho = 1$. So, the corollaries normalize the attack costs, for easier comparison.

3.1   Hellman's Original TMTO Attack on Stream Ciphers

TMTO attacks were introduced by Hellman for attacking block ciphers [1]. In Hellman's attack the precomputation tables were created using a single piece of known plaintext. During the online phase an attacker needs to retrieve an encryption of that exact same piece of known plaintext in order to match it against his precomputed tables and have a chance on a successful attack. These precomputed tables are useless for other known plaintext/ciphertext pairs.

In 2000 Biryukov and Shamir [4] found that TMTO attacks against stream ciphers have an extra benefit: an attacker can create tables which are more generic, so any piece of known key stream can be matched to them. These samples can even be overlapping. If an attacker has created TMTO tables to look for keystream occurrences of n bits and he obtains e.g. $n + 6$ consecutive bits, this gives him 7 different keystream samples of length n to match with the results in his tables. Since every sample of known keystream has its independent chance of matching with the precomputed values, every sample increases the success chance of the attack. Alternatively, an attacker can make an estimate, D, on the expected number of samples he will be able to obtain in the real-time phase, this enables him to save a factor D on precomputation (both time and storage)

$$x_0 \rightarrow f_i(x_0) \rightarrow f_i(f_i(x_0)) \rightarrow \ldots \rightarrow f_i^t(x_0)$$
$$x_1 \rightarrow f_i(x_1) \rightarrow f_i(f_i(x_1)) \rightarrow \ldots \rightarrow f_i^t(x_1)$$
$$x_2 \rightarrow f_i(x_2) \rightarrow f_i(f_i(x_2)) \rightarrow \ldots \rightarrow f_i^t(x_2)$$
$$\vdots \qquad \vdots \qquad \vdots \qquad \ddots \qquad \vdots$$
$$x_m \rightarrow f_i(x_m) \rightarrow f_i(f_i(x_m)) \rightarrow \ldots \rightarrow f_i^t(x_m)$$

Fig. 1. A single $m \times t$ matrix of function f_i. Only the first and last points of each chain are stored

to achieve the same success probability as that obtained with an attack on a cipher with $D = 1$. This effectively transforms the time-memory trade-off into a time-memory-'number of data samples' trade-off.

Hellman's attack on stream ciphers then goes as follows. In order to reverse the function f, a table is precomputed in the offline phase, for a single known plaintext. In order to cover as much of the N points of the search space as possible, an $m \times t$ matrix is computed, where the m rows consist of chains of length t and where each point in the chain is a new iteration of f on the result of the previous point (see Figure 1). Finally, only the begin point and end point of each chain are stored (ordered by the endpoints) as the precomputation table. In the rest of this article we will talk about precomputation *matrices* and *tables*, where matrices denote the temporary $m \times t$ precomputation chains and tables refer to the end product, essentially the compressed storage of the matrices. During the online phase, the attacker obtains keystream samples (e.g. by sniffing a known plaintext encryption, or because he can perform a chosen-plaintext attack). He then makes another chain of at most t iterations of applying the function f and for each iteration checks if the result matches one of the endpoints stored in his table. If this happens, he recomputes the chain starting from the corresponding begin point until the preimage of the ciphertext, thereby reversing function f in an attack time of order t in the online phase.

Adding more rows to the matrix computed in the offline phase will eventually cause duplicates, two duplicate points in different chains will cause the rest of these chains to cover the exact same points: the chains merge. Merging chains waste storage and precomputation effort on duplicate points. Hellman shows [1] that the probability of success is bounded by:

$$(1/N) \sum_{i=1}^{m} \sum_{j=0}^{t-1} [(N - it)/N)]^{j+1} \leq \mathbb{P} \leq (mt/N). \qquad (1)$$

Hellman proves that this lower bound can be approximated to $3/4$ for tables for which $mt^2 = N$. He argues that increasing m and t beyond $mt^2 = N$ is ineffective, since the chance of overlap only increases as m and t increase. Therefore Hellman continues his analysis of using $m \times t$ matrices satisfying $mt^2 = N$. Most of the subsequent work on time-memory trade-offs copies this choice, although there is no real reason for this.

A single $m \times t$ matrix satisfying $mt^2 = N$ covers only $1/t$-th of the search space N. So, in order to cover a larger part of the search space, Hellman proposed to construct l different $m \times t$ matrices each using a variant of the f function, f_i. The function f_i is defined as $f_i(x) = h_i(f(x))$ where h_i is a simple output modification that is different for each i. In this way, all l tables only have a small chance of duplicate chains (only within a single table). Naturally there are still chances of duplicate points between different tables, but these will not cause chain merges and are thus not so costly.

Theorem 1. *The general costs for Hellman's attack adapted for stream ciphers are:*
$M = 2ml$ *entries,*
$T_c = tlD$ f_i-*computations,*
$T_s = tlD$ *seeks in tables of m entries.*

Proof. *The memory costs equals the costs of one table, $2m$ since it only stores the starting and endpoints, times the number of tables, l. Having l different tables also carries additional costs in terms of attack time during the online phase, since the attacker will now have to create l different chains of length t for every sample, so both T_c and T_s are in the order of tlD f_i-computations or seeks, respectively.*

In this general case, it might seem that the factor D only has a negative impact on the costs, however, the value for l, the number of tables, can be reduced with a factor D when attacking stream ciphers while the success chance remains the same.

Corollary 1. *When reversing a stream cipher, using D samples and the $m \times t$ matrices satisfy $mt^2 = N$ and precomputing enough points to satisfy $D\rho = 1$, the costs are:*
$T_{pre} = \mathcal{O}(N/D)$, $T_c = t^2$ f_i-*computations,*
$M = 2mt/D$ *entries,* $T_s = t^2$ *seeks in tables of m entries.*

Proof. *The attacker makes l tables, each with a different f_i. Since each table covers $1/t$-th of the search space ($mt^2 = N$) and the attacker expects D samples, he needs t/D different tables to cover an area of equal size to the search space. So, there are $l = t/D$ tables each covering mt points, which means the precomputation time T_{pre} is in the order of N/D, since $mt^2 = N$ (assuming that $D \leq t$). The costs for M, T_c and T_s follow by simply substituting l with t/D in Theorem 1*

The memory costs M are measured in entries. We are assuming two entries are needed per chain, which is an overestimate, since some bits can be spared by clever storage methods. The seek time T_s is measured in the number of disk seeks necessary for the attack. In his original analysis Hellman ignores the effect that the size of the tables might have on the time of an individual disk seek. In order to achieve a more accurate measure we take the size of the tables into account, but we ignore the way the tables are organized on disk in our analysis.

Hellman's attack provides a time-memory trade-off controlled by choice of the chain length t. The table only stores two points for each chain, the begin and end point. As Theorem 1 shows, increasing t reduces the memory cost, but increases the time needed in the online phase, as more time is needed for computing the chain. Conversely, reducing t reduces the time in the online phase at the expense of higher memory cost. Note that if we choose $t = 1$ we have a dictionary attack, while if we choose $t = N$ we have a part of a brute-force attack.

3.2   Distinguished Points

The use of distinguished points was the first improvement on Hellman's approach. Hellman's analysis has a practical problem: there is a huge time difference between computing f_i and a disk seek to see if any $f_i(x)$ is stored in the precomputation table. In fact, Hellman's t^2 seeks in the precomputation tables are extremely more expensive than the t^2 f_i-computations [2]. Since Hellman's analysis counted only the computation steps ($T = t^2$) the difference between theory and practice is very big.

In 1982, a solution was proposed referenced to Ron Rivest [2, page 100], namely to identify a subset of special points, called distinguished points. These points should be easily recognized, usually by a fixed prefix, such as the first k bits being '0'. In the offline phase, chains are computed until such a distinguished point is reached, and that point is then stored as the endpoint. If no distinguished point is reached for a certain number of maximum computation steps, the entire chain is dropped and a new one is computed. In the online phase, the attacker starts developing a chain from captured ciphertext until he reaches a distinguished point, and only then does he need to perform an expensive disk seek. If no distinguished point is encountered in the development of this chain after a predetermined number of steps, than this captured piece of ciphertext is not covered by the tables.

Rivest's approach reduces the number of disk seeks, since now only a single disk seek is needed for every chain that is computed during the online attack, instead of one disk seek for every link. This leads to matrices with chains of varying length. However on average the chain length will be $t = 2^k$.

Using distinguished points has one other benefit. When the precomputation tables are finished it is possible to remove all chain merges from the tables, simply by looking for identical end points. After all, if two chains within a table merge, they will end in the same distinguished point. There is not really an easy way to decide which chain to drop from the table, although an attacker could record the number of points in each chain, while precomputing, in order to keep the longest one. Alternatively, keeping both chains will increase the coverage of the search space (assuming different start points where chosen, then a least a single unique point is added to the coverage by keeping merging chains), at the cost of using storage for duplicate points.

Theorem 2. *The general costs for a Distinguished Points attack are:*
$M = 2ml$ entries,
$T_c = tlD$ f_i-computations,
$T_s = lD$ seeks in tables of m entries.

Proof. The memory costs remain exactly the same as in the previous theorem. The computation costs will also remain the same since a distinguished point will

*on average be encountered after t steps. The disk-seek cost is now lowered to one
disk seek per chain. Since the attacker needs to make l chains —one for each
table— for every data sample, the seek time is $T_s = lD$.*

Corollary 2. *For the Distinguished Points attack, where the $m \times t$ matrices
satisfy $mt^2 = N$ and precomputing enough points to satisfy $D\rho = 1$, the costs
are:*
$T_{pre} = \mathcal{O}(N/D),$ $T_c = t^2$ *f_i-computations,*
$M = 2mt/D$ *entries,* $T_s = t$ *seeks in tables of m entries.*

*Proof. The attacker again needs to create $l = t/D$ tables, so both the precompu-
tational work and memory storage remain the same. The costs for T_c and T_s are
determined by substituting t/D for l in the preceding theorem.*

This approach can actually save some memory in practice, since k bits of every
endpoint are constant and need not be stored. This makes the entries smaller,
but the number of entries remains $2mt$. The time cost in the online phase also
remains t^2 evaluations of an f_i, but now only t disk seeks are expected, instead of
t^2 for Hellman's original attack: a disk seek is only needed when a distinguished
point is encountered, which happens once for each chain (on average after per
$t = 2^k$ computations), whereas in Hellman's original attack it has to be done for
all points in the chain.

3.3   Rainbow Table

A different improvement on Hellman's approach, called Rainbow Table, was
proposed by Oechslin in 2003 [3], with a factor-2 speed-up in the online phase,
for an attack with single samples. Additionally, it has none of the overhead that
Distinguished Points causes with its variable length, sometimes even unending,
chains. However, the Rainbow Table attack is mostly known for its smaller chance
of chain merge when less than N points are precomputed.

Oechslin suggested to precompute one large matrix (instead of t different ones)
with a different f_i for every link in the chain. The name Rainbow Table stems
from the idea of calling each simple output modification h_i a different color; each
column has its own color, so the entire table looks like a rainbow. This prevents
some chain merges, since now two chains can only merge if they reach the same
value in the same column (i.e. while applying the same f_i). Duplicate points can,
of course, still occur, but the penalty for these is not as severe since the chains
will not merge if a duplicate happens in a different column.

Theorem 3. *The general costs for a Rainbow Table attack are:*
$M = 2ml$ *entries,*
$T_c = \frac{t(t+1)}{2}D$ *f_i-computations,*
$T_s = tD$ *seeks in a table of lm entries.*

*Proof. In a Rainbow Table attack there is a single rainbow table which has ml
chains, of which only the first and last point are stored, so 2ml entries. These ml
chains are for comparisons sake, so ml chains of length t have the same coverage*

$$x_0 \rightarrow_1 f_0(x_0) \rightarrow_2 f_1(f_0(x_0)) \rightarrow_3 \ldots \rightarrow_t f_t(f_{t-1}(\ldots f_0(x_0)\ldots))$$
$$x_1 \rightarrow_1 f_0(x_1) \rightarrow_2 f_1(f_0(x_1)) \rightarrow_3 \ldots \rightarrow_t f_t(f_{t-1}(\ldots f_0(x_1)\ldots))$$
$$x_2 \rightarrow_1 f_0(x_2) \rightarrow_2 f_1(f_0(x_2)) \rightarrow_3 \ldots \rightarrow_t f_t(f_{t-1}(\ldots f_0(x_2)\ldots))$$
$$\vdots \qquad \vdots \qquad \vdots \qquad \ddots \qquad \vdots$$
$$x_{ml} \rightarrow_1 f_0(x_{ml}) \rightarrow_2 f_1(f_0(x_{ml})) \rightarrow_3 \ldots \rightarrow_t f_t(f_{t-1}(\ldots f_0(x_{ml})\ldots))$$

Fig. 2. A $ml \times t$ rainbow matrix using t different f_i functions. Only the first and last points of each chain are stored

as the l $m \times t$ matrices of other attacks. The online attack time becomes $\frac{t(t+1)}{2}D$ instead of tlD, because a different f_i is used for every link in the chain. So instead of computing a single chain $(y, f(y), f^2(y), ..)$ for every data sample an attacker now needs to evaluate t chains of a length ascending from 1 to t f_i calculations, with a different f_i for every link:

$$y \qquad \rightarrow_{f_t} f_t(y) \qquad \uparrow$$
$$y \rightarrow_{f_{t-1}} f_{t-1}(y) \rightarrow_{f_t} f_t(f_{t-1}(y)) \qquad t$$
$$\vdots \qquad \vdots \qquad \vdots$$
$$y \rightarrow_{f_0} \cdots \rightarrow_{f_{t-1}} f_{t-1}(\ldots) \rightarrow_{f_t} f_t(f_{t-1}(\ldots(f_0(y)\ldots))) \downarrow$$

For each of these t chains, the end point needs to be looked up in the table, for each of the D data samples, which results in tD disk seeks.

In order to compare this attack to the other approaches we need the matrix to cover an equal number of points. The other approaches use t $m \times t$ matrices. With a rainbow table there is only a single table, so this needs to cover mt^2 points. Keeping the chain length t, means the attacker will need mt entries in his table to cover mt^2 points. So we assume an $mt \times t$ matrix, with t different f_i's, as Figure 2 shows.

Corollary 3. *For the Rainbow Table attack, where the $ml \times t$ matrices satisfy $mt^2 = N$ and precomputing enough points to satisfy $D\rho = 1$, the costs are:*
$T_{pre} = \mathcal{O}(N/D)$, $\qquad T_c = \frac{t(t+1)}{2}D$ f_i-computations,
$M = 2mt/D$ entries, $\qquad T_s = tD$ seeks in a table with mt/D entries.

Proof. In the Rainbow Table case there is little difference in costs between the general case and the case where an attacker chooses m and t to satisfy $mt^2 = N$, since there is only a single table and only the chain size determines the attack time. For comparison's sake, we use a rainbow table of dimensions $(mt/D) \times t$, which covers an equal number of points as the previous stream cipher TMTO attacks, and keeps the values for T_{pre} and M equal. By substituting l with t/D, the memory costs are also fixed.

Since D will generally be smaller than t, the number of disk seeks is an improvement when compared to a Hellman style attack, though not as much as the use of distinguished points. Also keep in mind that every table seek could be more costly when using a rainbow table, because of its larger size than the l tables used in the other approaches.

The Rainbow Table attack is most known for a smaller chance of chain merges, but the table defined in Corollary 3 will have a similar chance of chain merges than the previous attacks for $D = 1$. Because the same amount of points are precomputed in every f_i (all the points in a single rainbow table column, or all the points covered in 1 Hellman or distinguished point table) and a duplicate between those points causes a chain merge. When assuming more samples, or when precomputing fewer points, i.e. $D\rho < 1$, then the Rainbow Table attack will probably have fewer chain merges than the other TMTO attacks.

The online attack time of the Rainbow Table attack is only dependent on the chain length, ignoring the number of entries in the table, which causes the slight speed up for attacks where $D = 1$.

3.4   Generalized Kraken Approach

In 2009, researchers started a project to break GSM's standard encryption cipher A5/1 in practice, using a combination of time-memory trade-off techniques. They proposed the joint creation of a set of TMTO tables to which everyone could contribute [13]. The idea was to share the intense computing burden of a TMTO's precomputation step by having everyone willing to participate perform a part of the computation on modern GPUs, and share their results over the Internet. In the end however, the project ended up using a set of tables being computed on a single computer. This set was dubbed "The Berlin Set" and its parameters are discussed in detail later in this section. First we focus on the general approach that was used in this attack.

In order to find the internal state of a generic stream cipher, the Kraken approach combines both distinguished points and rainbow tables in the table layout. This is done by first choosing distinguished points as bit strings starting with k zeros. Then, normal TMTO chains are computed by repeatedly applying f_i to random start points until the output is a distinguished point. The chain is then continued but now with a different f_i; in essence changing the rainbow color. This is repeated for a predetermined number, s, of rainbow colors ($f_0 \ldots f_s$ functions), until a distinguished point is found while using the final f_s of this chain. This point is the endpoint of a chain and is stored together with the corresponding start point in the TMTO table. Figure 3 shows such a precomputation matrix. In order to match a sample y against a table during the online phase, s different chains need to be developed ranging in size from t to st f_i-computations, analogously to the Rainbow Table online attack. On average this will lead to one distinguished point per f_i subchain, assuming that each possible output of f_i is equally likely. While applying f_s, the attacker can see if the resulting distinguished point matches the stored endpoint. The distinguished point found in the chain while applying f_{s-1}, needs to be developed further by applying f_s until the last distinguished point is found. This continues to the distinguished point found using the first f_0, which should require a chain of around st computation steps to match the final distinguished point with the

$$x_0 \to f_0(x_0) \to f_0(f_0(x_0)) \to^* k||y_{00} \to f_1(k||y_{00}) \to^* \ldots \to f_s(\ldots) \to k||y_{0s}$$
$$x_1 \to f_0(x_1) \to f_0(f_0(x_1)) \to^* k||y_{10} \to f_1(k||y_{10}) \to^* \ldots \to f_s(\ldots) \to k||y_{1s}$$
$$x_2 \to f_0(x_2) \to f_0(f_0(x_2)) \to^* k||y_{20} \to f_1(k||y_{20}) \to^* \ldots \to f_s(\ldots) \to k||y_{2s}$$

$$\vdots$$

$$x_m \to f_0(x_m) \to f_0(f_0(x_m)) \to^* k||y_{m0} \to f_1(k||y_{m0}) \to^* \ldots \to f_s(\ldots) \to k||y_{ms}$$

Fig. 3. A Kraken matrix, where $k||y$ denotes a distinguished point with the first k bits '0'. Only the first and last points of each chain are stored. Note that each Kraken chain consists of s Distinguished Points chains.

stored endpoint, as Figure 4 shows. This approach can boil down to compressing s different Distinguished Points tables into one: Each chain basically consists of s subchains, depending on the choice of s and t. Intuitively, this means that the memory costs will be lowered by a factor s, but the attack time will increase by a factor s. This attack should keep all other advantages of a Distinguished Points attack, such as the easily identifiable chain merges. When compared with a Rainbow Table attack the number of chain merges should rise with a factor $t = t'/s$, where t' is the new total chain length, because the same f_i is used for each subchain.

The average length of each subchain, t, can be adjusted by choosing a different length of k for the k-bit distinguished point. The length of one full chain is equal to $t' = st$.

Theorem 4. *The general costs for the Generalized Kraken attack are:*
$M = 2ml$ *entries,*
$T_c = \frac{s(s+1)}{2}tlD$ f_i*-computations,*
$T_s = sl\bar{D}$ *seeks in tables of m entries.*

Proof. The costs for memory use and disk seeks remain the same as in the Distinguished Points case. The computation costs are still based on the costs for matching a single sample against a single table multiplied with the number of tables and the number of samples. The attacker needs to make s chains of sizes increasing from t to st, so in total $\frac{s(s+1)}{2}t$ f_i-computations, to match a single sample against a single table.

$$
\begin{array}{ccccc}
 & & y & \to^*_{f_s} k||x_{0s} \uparrow \\
 & y & \to^*_{f_{s-1}} k||x_{1s-1} & \to^*_{f_s} k||x_{1s} \\
y & \to^*_{f_{s-2}} k||x_{2s-2} & \to^*_{f_{s-1}} k||x_{2s-1} & \to^*_{f_s} k||x_{2s} \; s \\
\vdots & & \vdots & \vdots \\
y \to^*_{f_0} \ldots & \to^*_{f_{s-2}} k||x_{ss-2} & \to^*_{f_{s-1}} k||x_{ss-1} & \to^*_{f_s} k||x_{ss} \downarrow
\end{array}
$$

Fig. 4. The online phase of a Kraken attack. Here $k||X$ denotes a distinguished point with the first k bits '0'. Only the last point of every chain is matched against the precomputation table.

In the Kraken attack we are faced with an additional variable s, which introduces a new Time-Memory Trade-Off within a TMTO attack. It also complicates matters when creating the accompanying corollary by increasing the possible choices. Here we choose two of the most obvious scenario's:

- the full chain length of the Kraken tables is as large as in the previous attacks, which leads to s more tables (Corollary 4, more tables),
- the sub chain length of the Kraken tables is equal to the chain length of the previous attacks, the full chains are s times larger than the previous attacks (Corollary 5, bigger tables).

Of course these only show two possible choices for s, t and m, of which the first scenario coincides with $m(st)^2 = N$ and the second with the familiar $mt^2 = N$.

Corollary 4 (more tables). *When reversing a stream cipher with the Kraken approach, using D samples and the $m \times st$ matrices satisfy $m(st)^2 = N$ and precomputing enough points to satisfy $D\rho = 1$, the costs are:*
$$T_{pre} = \mathcal{O}(N/D), \qquad T_c = \tfrac{(s+1)}{2}(st)^2 \; f_i\text{-computations},$$
$$M = \tfrac{2mst}{D} \text{ entries}, \; T_s = s^2 t \text{ seeks in } a \text{ tables of } m \text{ entries}.$$

Proof. A single table will cover $m \times st$ points, or $1/st$ of the key space N (since $m(st)^2 = N$), so st tables are needed to achieve enough coverage for $D\rho = 1$. Given that f is a stream cipher, an attacker can reduce the number of required tables to $l = \tfrac{st}{D}$. This means the memory costs will be $M = 2m \times \tfrac{st}{D} = \tfrac{2mst}{D}$ entries.

The attacker needs a total of $\tfrac{s(s+1)}{2}t$ f_i-computations, to match a single sample against a single table. Since there are D samples and $\tfrac{st}{D}$ tables, the total attack time T_c equals: $\tfrac{(s+1)}{2}(st)^2$. The attacker must create s separate chains for each table. During the online attack this comes down to s disk seeks per table, on $\tfrac{st}{D}$ tables and D samples gives $T_s = s^2 t$ disk seeks within tables of m value pairs.

Corollary 5 (bigger tables). *When reversing a stream cipher with the Kraken approach, using D samples and the $m \times st$ matrices satisfy $mt^2 = N$ and precomputing enough points to satisfy $D\rho = 1$, the costs are:*
$$T_{pre} = \mathcal{O}(N/D), \qquad T_c = \tfrac{(s+1)}{2}t^2 \; f_i\text{-computations},$$
$$M = \tfrac{2mt}{sD} \text{ entries}, \; T_s = t \text{ seeks in } a \text{ tables of } m \text{ entries}.$$

Proof. A single table will cover $m \times st$ points, or s/t of the key space N (assuming $s < t$), so t/s tables are needed to achieve enough coverage for $D\rho = 1$. Given that f is a stream cipher, an attacker can reduce the number of required tables to $l = \tfrac{t}{sD}$. This means the memory costs will be $M = 2m \times \tfrac{t}{sD} = \tfrac{2mt}{sD}$ entries.

The attacker needs a total of $\tfrac{s(s+1)}{2}t$ f_i-computations, to match a single sample against a single table. Since there are D samples and $\tfrac{t}{sD}$ tables, the total attack time T_c equals: $\tfrac{(s+1)}{2}t^2$. The attacker must create s separate chains for each table. During the online attack this comes down to s disk seeks per table, on $\tfrac{t}{sD}$ tables and D samples gives $T_s = t$ disk seeks, but again disk seeks within tables of M value pairs.

The scenario of Corollary 4 for Kraken uses the same sized matrices as that of Corollary 2 for the Distinguished Points. It needs s^2 more tables than the scenario of Corollary 5, which is reflected in all the costs. However, keep in mind that the value of t in Corollary 4 is s times higher than the value of t in Corollary 5.

The costs in Corollary 5 confirm our intuition of Kraken using s compressed distinguished points tables. So it can reduce memory costs a factor s at the price of increasing the computation cost in the online phase by a factor $\frac{s+1}{2}$ in comparison with Distinguished Points approach, which is better than our initial intuition.

Kraken in Practice. The Kraken approach was devised by researchers from the hacker community to demonstrate the weakness of the encryption used in GSM; the stream cipher A5/1. This stream cipher has an internal state of 64 bits, which is initiated with a 64 bit session key and a 22-bit, publicly known, frame number. The cipher then produces 328 bits of keystream of which the first 100 are discarded. Of the remaining 228 bits, the first half are used for the encryption of a packet on the uplink (mobile phone to cell tower) and the second half is used for encryption on the downlink (cell tower to mobile phone).

The natural assumption here is that the state space has size 2^{64}, but careful examination of the clocking function shows that a large part of the possible internal states are unreachable from any valid state. Several studies have measured the decline of possible states in the A5/1 cipher [14,8], and all of these find that only around 15% of all possible states are still viable after the initial 100 clockings. This means in practice that an attacker only needs to cover around 15% of the state space: $N \approx 2^{61.26}$.

In the attack against A5/1 the $f_i(x)$ is setting x in the internal state of A5/1, clocking it a 100 steps forward and then producing 64 bits of keystream, combined with some trivial output modification (the rainbow colors). These 64 output bits are then used to set the new internal state for the next round.

In the precomputation phase 40 independent tables ($l \approx 2^{5.3}$) were created, dubbed the Berlin set. As distinguished points were chosen those points starting with 12 zeros ($k = 12$). Eight rainbow colors were used per table ($s = 8$) and they differ for each table, so in total there are 320 different colors.

Every chain consists of eight subchains of average length t. Assuming each possible outcome of an f_i is equally likely: $t = 2^{12}$. So $t' = 8 \times 2^{12} = 2^{15}$.

Initially, every table was computed with 8,662,000,000, approximately 2^{33}, rows ($m = 2^{33}$). After which one of every two chains with duplicate end points was removed and the current set contains around 6,000,000,000 entries per table.

This means that for the Berlin set around $2^{15} \times 2^{33} \times 2^{5.3} = 2^{53.3}$ points were precomputed. The set ended up covering around $2^{15} \times 2^{32.5} \times 2^{5.3} = 2^{52.8}$ distinct points, so over 29% of the chains ended up merging with an earlier chain. This surprisingly high percentage can, in part, be explained by the state-space collapse of A5/1. However, it still shows that chain merges can indeed have a significant impact on a TMTO attack performance in practice.

Table 1. Comparison of the different attacks, for $D\rho = 1$ and $mt^2 = n$

TMTO technique	M	T_c	T_s
Hellman's attack	$2mt/D$	t^2	t^2 in m entries
Distinguished Points	$2mt/D$	t^2	t in m entries
Rainbow Table	$2mt/D$	$\frac{t(t+1)}{2}D$	tD in mt/D entries
Kraken (more tables, $t' = st$)	$2mt'/D$	$\frac{(s+1)}{2}t'^2$	st' in m entries
Kraken (bigger tables)	$2mt/sD$	$\frac{(s+1)}{2}t^2$	t in m entries

With $N \approx 2^{61.26}$, the Berlin set has its parameters between the two discussed options in Corollaries 4 and 5: $mt^2 < N < m(st)^2$. The tables in the Berlin set take up around 1.6TB on disk and one attack with 51 samples (one packet in GSM is 114 bits, so 51 samples of 64 bits) can be performed within several seconds on high-end, but off-the-shelf hardware. Experiments with self-generated bursts put the success chance of the attack with 51 samples to around 20%.

4   Comparison

The main idea behind Kraken is to combine the benefits of both Distinguished Points (i.e. low number of disk seeks) and Rainbow Tables (i.e. fewer duplicates). The question is whether this really turns out beneficial. The cost of the different attacks are compared in Table 1, which lists the costs given in Corollaries 1 to 5. The two possible Kraken approaches are both shown, but Corollary 4 has st substituted for t', so its t' is comparable to the value of t for the other attacks.

The Three Classic Attacks. Hellman's attack (adapted for stream ciphers) is added in this table as a baseline, since the other attacks all improve on almost all costs. Between Distinguished Points and Rainbow Table it seems that a Rainbow Table is the best choice for $D = 1$, with only the disk seeks being more expensive due to the larger table. This makes Rainbow Tables the best choice for attacking block ciphers.

However, when attacking stream ciphers with multiple samples, $D > 1$, the comparison is not so simple. The online attack time and the number of disk seeks for Rainbow Table, T_c and T_s, both increase beyond those of the Distinguished Points attack. Of course increasing D also decreases the size of the rainbow table used, making each table search cheaper, but generally seek time will be in the order of the logarithm of the table size, so this benefit is smaller than the increase in the number of disk seeks. Based on these calculations, the more samples are expected during the online attack, the more attractive the Distinguished Points approach becomes, compared to Rainbow Tables.

The Kraken Attacks. Our initial intuition that the Kraken approach is comparable to s Distinguished Points tables stored as one, seems validated when looking at the respective costs of both Kraken attacks and the Distinguished Points attack. If we take $s = 1$, then the Kraken approaches are the same as Distinguished Points, and their costs are identical. Another way to look at the Kraken approach is as a "bloated" rainbow table, with every rainbow color expanded from one column to t columns (on average). Looking at the costs for the online attack time for the Kraken approach, if we choose $t = 1$ and $s = t'$ (essentially a rainbow table), it almost compares to the online attack costs of a Rainbow Table attack, where the difference can be explained by a Rainbow Table attack having a single table instead of the $\frac{st}{D}$ or $\frac{t}{sD}$ tables of the respective Kraken attacks.

It is clear from this comparison that having Kraken tables of the more-tables variant is not the best choice. It has more disk seeks than the bigger-tables approach and the Distinguished Points attack, without the benefit of smaller memory costs.

Kraken vs. the Classics. The Kraken attack can be tuned further by changing the s parameter. Basically, the Kraken attack moves in between the Distinguished Points and Rainbow Table attacks, guided by the value of s, where a higher choice of s will save memory costs, but increase the online attack costs.

From the two realistic Kraken approaches shown in the comparison table, only the bigger-tables approach is competitive in this analysis. If the memory costs are the single limiting factor for using the Distinguished Points or Rainbow Table attack, then this Kraken attack seems a good choice.

Although, with the continuous drop in the prices of memory such a scenario seems unlikely, so depending on the number of expected samples a Rainbow Table or a Distinguished Points attack is probably the better choice.

Comparing Chain Merges. The comparison above is based on all attacks satisfying $D\rho = 1$, in other words the costs are compared when all attacks precompute the same amount of points. However, due to duplicates and chain merges not all precomputed points will be unique. It would be more fair if we compared the costs of the attacks when they all satisfy $D\bar{\rho} = 1$, so the number of precomputed unique values would be in the order of N/D.

However, it is hard to estimate the chances on chain merges in a general case for the different approaches. In essence this problem boils down to the expected overlap between two paths (of length t for most approaches) within a digraph consisting of the N points of the search space as nodes and the current f_i as the edges between these nodes. This digraph is a directed pseudo forest, so every node has out-degree 1, meaning there can exist source nodes, but no sinks and from every node there exists a path leading to a cycle. An analysis of the number of expected duplicates, or analogously the number of unique values for a certain TMTO attack seems hard [15,16,17,10] and to our knowledge this is in fact an open problem for most TMTO attacks.

We can however make some assumptions over the chain merges for the different approaches, when they all precompute the same amount of points. We ignore single duplicate points and only look at chain merges, so only duplicates under the same f_i. Then by looking at the number of precomputed points per different f_i function, although ignoring many of the subtle differences between the TMTO attacks, can give an indication on the chances of chain merges.

When we assume that the distinguished points from a Distinguished Points attack are uniformly spread over the iterative function graph, then in general the number of duplicates in the Distinguished Points tables will be about the same as those in the Hellman attack. The number of duplicates in the Rainbow Table attack will in general be smaller than for the Hellman and the Distinguished Points attack, as long as the number of records in a rainbow table is smaller than the $m \times t$ points in a Hellman or distinguished point table.

When we look at the two possible approaches for the Kraken attack, then they are most easily compared to a Distinguished Points attack. The first Kraken approach ($m(st)^2 = N$, Corollary 4) uses s different colors inside an almost standard Distinguished Points matrix. So, only $1/s$ th of the points in a single table have the chance of leading to a chain merge, and we would therefore expect s less chain merges in this Kraken approach than in a Distinguished Points approach. The second Kraken approach ($mt^2 = N$, Corollary 5) compresses s different Distinguished Points tables into one, but because the end point of one of those Distinguished Points tables is the start point for the next, chain merges in one of these subchains will carry through the rest of the chain. Therefore, we can roughly estimate that this Kraken attack has around $s/2$ more duplicates due to chain merges than a Distinguished Points attack.

When we keep chain merges in mind, the comparison from Table 1 becomes more subtle, since it seems that the factor s extra costs in memory and disk seeks of the first Kraken approach compared to the second, is somewhat mitigated by having less duplicates, and thus more unique points in its tables and a higher success chance. However, since we have no hard way of quantifying the number of chain merges in these attacks, this analysis remains very tentative.

Both Distinguished Points and Kraken have one extra benefit when comparing chain merges. Both approaches have identifiable chain merges, which means that every chain merge will automatically end in the same end point. So by simply comparing end points all chain merges can be identified. With extra precomputation effort both approaches are able to replace one chain of every merging chain pair, with a new one, thereby increasing their coverage $\bar{C}$. Naturally, both the increase in coverage and the amount of extra precomputation work are dependent on the number of chain merges.

5 Conclusions and Directions for Future Research

We have presented the first analysis of the cost of the generalized form of the TMTO attack used to break the A5/1 cipher, which we have called Kraken. We have also given a first comparison of the costs of Kraken and three older TMTO attacks: Hellman's original attack, Distinguished Points, and Rainbow Tables.

Our comparison is more detailed than earlier work comparing these three older forms of attack. Most [4,9] earlier work compared the trade-off curves of these well known attacks. This tells us the rate at which extra memory can be traded in for a reduced time, but completely ignores some important costs, namely the precomputation, seek times, and the number of unique points covered by an attack. We do consider these costs in our comparison: for each attack we give the memory and time costs, split into precomputation time, online computation time, and number of disk seeks.

In our comparison in Section 4 the new Kraken attack performed fine, with the lowest memory cost of all attacks and the ability to identify chain merges as its major benefits. Only Distinguished Points seems a better choice in comparison, having a higher memory cost, but the lowest online attack costs. The more well-known Rainbow Tables are only interesting for attacks with only a single sample of plaintext-ciphertext known, as it is outperformed by Distinguished Points for multiple samples.

Another limitation of comparisons of trade-off curves for the different approaches is that these curves are invariably made under the assumption that the table sizes are always chosen so that $mt^2 = N$. We see no convincing reason to constraint the choice in parameters in this way. Hellman used the constraint $mt^2 = N$ to compute a nice bound for the chance of success of his attack, but other choices for m and t that do not satisfy this constraint might perform better in concrete instances.

One factor that we still have not been able to quantify precisely in our comparison is the chance of duplicates during the precomputation of the tables.

We conjecture that the effectiveness of the Kraken attack is in fact lower than our current results suggest when this number of duplicates values is taken into account. The informal analysis of the expected number of chain merges in Section 4 shows that Kraken has a higher chance of chain merges than the other attacks when the number of rainbow colors in the Kraken approach is chosen to achieve lower memory cost.

Estimating the chance of duplicates during precomputation is the most difficult aspect in achieving a fair comparison. Over 29% of all chains created in the Kraken tables ended up merging with existing chains, showing that chain merges can indeed be a significant factor when comparing TMTO attacks. We know no way to compute the expected number of chain merges for the general case, or indeed for any non-trivial practical cipher. Since theoretical analysis of the chance of duplicates seems very difficult, we think that further research which collects empirical data of practical experiments in constructing TMTO tables may be the best way to shed light on this.

References

1. Hellman, M.: A cryptanalytic time-memory trade-off. IEEE Transactions on Information Theory 26(4), 401–406 (1980)
2. Denning, D.: Cryptography and Data Security. Addison-Wesley (1992)
3. Oechslin, P.: Making a faster cryptanalytic time-memory trade-off. In: Boneh, D. (ed.) CRYPTO 2003. LNCS, vol. 2729, pp. 617–630. Springer, Heidelberg (2003)

4. Biryukov, A., Shamir, A.: Cryptanalytic time/memory/data tradeoffs for stream ciphers. In: Okamoto, T. (ed.) ASIACRYPT 2000. LNCS, vol. 1976, pp. 1–13. Springer, Heidelberg (2000)
5. Biryukov, A., Shamir, A., Wagner, D.: Real time cryptanalysis of A5/1 on a PC. In: Schneier, B. (ed.) FSE 2000. LNCS, vol. 1978, pp. 1–18. Springer, Heidelberg (2001)
6. Barkan, E., Biham, E., Keller, N.: Instant ciphertext-only cryptanalysis of GSM encrypted communication. In: Boneh, D. (ed.) CRYPTO 2003. LNCS, vol. 2729, pp. 600–616. Springer, Heidelberg (2003)
7. Nohl, K., Munaut, S.: Wideband GSM sniffing. Presentation at 26C3 (2010), http://events.ccc.de/congress/2010/Fahrplan/events/4208.en.html
8. Nohl, K.: A5/1 decrypt website (November 2012), http://opensource.srlabs.de/projects/a51-decrypt/
9. Erguler, I., Anarim, E.: A new cryptanalytic time-memory trade-off for stream ciphers. In: Yolum, p., Güngör, T., Gürgen, F., Özturan, C. (eds.) ISCIS 2005. LNCS, vol. 3733, pp. 215–223. Springer, Heidelberg (2005)
10. Barkan, E., Biham, E., Shamir, A.: Rigorous bounds on cryptanalytic time/Memory tradeoffs. In: Dwork, C. (ed.) CRYPTO 2006. LNCS, vol. 4117, pp. 1–21. Springer, Heidelberg (2006)
11. Hong, J., Jeong, K.C., Kwon, E.Y., Lee, I.-S., Ma, D.: Variants of the distinguished point method for cryptanalytic time memory trade-offs. In: Chen, L., Mu, Y., Susilo, W. (eds.) ISPEC 2008. LNCS, vol. 4991, pp. 131–145. Springer, Heidelberg (2008)
12. Krhovjak, J., Siler, O., Leyland, P., Kur, J.: TMTO attacks on stream ciphers theory and practice. Security and Protection of Information 2011 (2011)
13. Nohl, K.: Cracking A5 GSM encryption. Presentation at HAR 2009 (2009), https://har2009.org/program/events/187.en.html
14. Golic, J.: Cryptanalysis of Alleged A5 Stream Cipher (1997), http://jya.com/a5-hack.htm
15. Hong, J.: The cost of false alarms in Hellman and rainbow tradeoffs. Designs, Codes and Cryptography 57, 293–327 (2010)
16. Flajolet, P., Odlyzko, A.M.: Random mapping statistics. In: Quisquater, J.-J., Vandewalle, J. (eds.) EUROCRYPT 1989. LNCS, vol. 434, pp. 329–354. Springer, Heidelberg (1990)
17. Fiat, A., Naor, M.: Rigorous time/space tradeoffs for inverting functions. In: STOC 1991, pp. 534–541. ACM (1991)

On the Expansion Length
of Triple-Base Number Systems*

Wei Yu[1,2], Kunpeng Wang[2], Bao Li[2], and Song Tian[2]

[1] Department of Electronic Engineering and Information Science,
University of Science and Technology of China, Hefei, 230027, China
yuwei_1_yw@163.com
[2] Institute of Information Engineering, Chinese Academy of Sciences, Beijing, 100093

Abstract. Triple-base number systems are mainly used in elliptic curve cryptography to speed up scalar multiplication. We give an upper bound on the length of the canonical triple-base representation with base $\{2, 3, 5\}$ of an integer x, which is $\mathcal{O}(\frac{\log x}{\log \log x})$ by the greedy algorithm, and show that there are infinitely many integers x whose shortest triple-base representations with base $\{2, 3, 5\}$ have length greater than $\frac{c \log x}{\log \log x \log \log \log x}$, where c is a positive constant, using the universal exponent method. This analysis gives a limit how much scalar multiplication on elliptic curves may be made faster.

Keywords: Elliptic Curve Cryptography, Triple-Base Number System, Upper Bound, Greedy Algorithm, Universal Exponent.

1 Introduction

Double-base number systems (DBNSs) were first introduced in [1,2] because of their sparseness, and their main use is in cryptography and digital filter implementation. Mishra and Dimitrov [3] first introduced triple-base number systems (TBNSs), an extension of DBNSs, for computing scalar multiplication on elliptic curves more efficiently. Since then, TBNSs have been used by Longa [4] and Purohit and Rawat [5] to speed up scalar multiplication on elliptic curves.

The theoretical analysis of a number system gives a bound on the average Hamming weight, which provides a limit on how much the speed can be increased and determines whether the system is sparse. In [6,7,8], DBNSs were analyzed, leading to the wide use of DBNSs in elliptic curve cryptography (ECC) [9,10,11,12,13,14].

Although TBNSs have been used to speed up scalar multiplication on elliptic curves, a theoretical analysis of the expansion length has not appeared in any literature. A theoretical analysis of the expansion length of TBNSs will give a

* Supported in part by National Basic Research Program of China(973) under Grant No.2013CB338002, in part by National Research Foundation of China under Grant No. 61272040 and 61070171, and in part by the Strategic Priority Research Program of Chinese Academy of Sciences under Grant XDA06010702.

A. Youssef, A. Nitaj, A.E. Hassanien (Eds.): AFRICACRYPT 2013, LNCS 7918, pp. 424–432, 2013.
© Springer-Verlag Berlin Heidelberg 2013

limit on how much the speed of scalar multiplication on elliptic curves may be increased, which may lead to a wider use of TBNSs in scalar multiplication for ECC. In fact, the complexity of TBNSs is a number-theoretic question.

The span $s(x)$ of an integer x is the smallest r such that x has a triple-base representation of length r. In this paper, we mainly focus on the upper bound of $s(x)$. The big oh ($\mathcal{O}$) needed in the complexity analysis is defined as follows. Let f and g be functions from the set of natural numbers $\mathbb{N}$ to $\mathbb{N}$. $f(n) = \mathcal{O}(g(n))$ means that there are positive integers c and n_0 such that, for all $n \geqslant n_0$, $f(n) \leqslant c \cdot g(n)$. Informally, f grows as fast as g or slower.

Our contributions in this paper are:

1. The upper bound of $s(x)$ is $\mathcal{O}(\frac{\log x}{\log \log x})$, which we show using the greedy algorithm.
2. There are infinitely many integers x whose shortest triple-base representations have length greater than $\frac{c \log x}{\log \log x \log \log \log x}$, where c is a positive constant using the universal exponent method (that is, the upper bound of $s(x)$ is greater than or equal to $\frac{\log x}{\log \log x \log \log \log x}$).

This paper is organized as follows. Section 2 describes some basic facts about TBNSs. Sections 3 and 4 prove the above two results. Section 5 concludes the paper.

2   Triple-Base Number Systems

Definition 1. *An S-integer is a positive integer whose prime factors all belong to a given set of primes S.*

Following de Weger's definition of s-integers [15], for which the largest prime factor does not exceed the s-th prime number, {2, 3}-integers are called 2-integers, and {2, 3, 5}-integers are called 3-integer.

A DBNS [2] is a representation scheme in which every integer x is represented as the sum of 2-integers:

$$x = \sum_{i=1}^{l} 2^{e_{1i}} 3^{e_{2i}}, \text{ where } e_{1i}, e_{2i} \text{ are nonnegative integers.}$$

A TBNS, which is an extension of a DBNS, is defined as follows.

Definition 2. *(TBNS). Given three relatively prime positive integers b_1, b_2, b_3, the TBNS is a representation scheme in which every positive integer x is represented as the sum of $\{b_1, b_2, b_3\}$-integers, that is, numbers of the form $b_1^{e_1} b_2^{e_2} b_3^{e_3}$: $x = \sum_{i=1}^{l} b_1^{e_{1i}} b_2^{e_{2i}} b_3^{e_{3i}}$, where e_{1i}, e_{2i}, e_{3i} are nonnegative integers.*

TBNSs mainly use bases 2, 3 and 5 in scalar multiplication on elliptic curves. In this paper, we focus primarily on the same system. The triple-base representation of x in this system is:

$$x = \sum_{i=1}^{l} 2^{e_{1i}} 3^{e_{2i}} 5^{e_{3i}}, \text{ where } e_{1i}, e_{2i}, e_{3i} \text{ are nonnegative integers.}$$

In this representation, l is called the expansion length. A representation of a given integer as the sum of the minimal number 3-integers will be called a canonical triple-base number representation (CTBNR). Such an integer can be represented as the sum of l 3-integers, but cannot be represented with $l-1$ or fewer 3-integers. These so-called canonical representations are extremely sparse. The span $s(n)$ is the expansion length of a CTBNR.

We take 127 for example.

127 has six canonical DBNS representations with span 3:

$$127 = 2^2 3^3 + 2^1 3^2 + 2^0 3^0$$
$$= 2^2 3^3 + 2^4 3^0 + 2^0 3^1$$
$$= 2^5 3^1 + 2^0 3^3 + 2^2 3^0$$
$$= 2^3 3^2 + 2^1 3^3 + 2^0 3^0$$
$$= 2^6 3^0 + 2^1 3^3 + 2^0 3^2$$
$$= 2^6 3^0 + 2^2 3^2 + 2^0 3^3.$$

127 has two canonical TBNS representations with span 2:

$$127 = 2^2 5^2 + 3^3$$
$$= 5^3 + 2^1.$$

Because double-base representation is a special case of triple-base representation, a TBNS is usually more sparse than a DBNS.

For each $i > 0$, let $S(i)$ denote the smallest positive integer x with canonical span $s(x) = i$. Calculating the value of $S(i)$ is useful for analyzing the properties of TBNSs.

Dimitrov, Imbert, and Mishra [8] gave the following $S(i)$ for the DBNS: $S(1) = 1, S(2) = 5, S(3) = 23, S(4) = 431, S(5) = 18431, S(6) = 3448733, S(7) = 1441896119$.

We use an increasing search to calculate $S(i)$ in the TBNS: $S(1) = 1, S(2) = 7, S(3) = 71, S(4) = 30359$.

Our increasing search runs as follows. Choose a positive integer M as a boundary. Let S_i denote all numbers in the interval $[1, M]$ with the length i of a CTBNR. Considering $S_i = \{a+b | a \in S_1, b \in S_{i-1}, a+b \leq M\} - S_1 \cup S_2 \cup \ldots \cup S_{i-1}$, the increasing search first calculates S_1, which only contains all 3-integers. Based on S_1, $S_2 = \{a + b | a, b \in S_1, a + b \leq M, a + b \notin S_1\}$. Next we calculate $S_3 = \{a + b | a \in S_1, b \in S_2, a + b \leq M, a + b \notin S_1 \cup S_2\}$. ... Based on $S_1, S_2, \ldots, S_{i-1}$, $S_i = \{a + b | a \in S_1, b \in S_{i-1}, a + b \leq M, a + b \notin S_1 \cup S_2 \cup \ldots \cup S_{i-1}\}$. $S(i)$ is equal to the smallest number in S_i. If S_i is empty for $[1, M]$, we can increase M until we find the value of $S(i)$. This method is very efficient when we want to know the span and the CTBNR of all numbers in the given interval $[1, M]$.

3   The Upper Bound

We first give the distance between two adjacent 3-integers.

3.1   The Distance between 3-Integers

Lemma 1. *There exist two absolute constants $D, N > 0$ such that there is a 3-integer between $x - \frac{x}{(\log x)^D}$ and x when $x > N$.*

Proof. Let $n_1 = 1 < n_2 < \dots$ be the sequence of all 3-integers. In [16], Tijdeman proved that there exist efficiently computable constants $0 < D_1 \leqslant 2$ and N such that

$$n_{i+1} - n_i < \frac{n_i}{(\log n_i)^{D_1}} \text{ for } n_i \geqslant N.$$

The function $\frac{x}{(\log x)^{D_1}}$ is an increasing function in $[e, \infty)$ when the differential $\frac{\log x - D_1}{(\log x)^{D_1+1}} > 0$, where e is the natural constant.

We deduce the following:

$$n_{i+1} < n_i + \frac{n_i}{(\log n_i)^{D_1}} < n_i + \frac{n_{i+1}}{(\log n_{i+1})^{D_1}}$$

$$\Rightarrow n_{i+1} - \frac{n_{i+1}}{(\log n_{i+1})^{D_1}} < n_i < n_{i+1}.$$

Because the function $x - \frac{x}{(\log x)^{D_1}}$ is increasing when $x > e$, if we let $n_i < x \leqslant n_{i+1}$, then $n_i > n_{i+1} - \frac{n_{i+1}}{(\log n_{i+1})^{D_1}} \geqslant x - \frac{x}{(\log x)^{D_1}}$. Thus, there exists an absolute constant $0 < D = D_1 \leqslant 2$ such that there is a 3-integer between $x - \frac{x}{(\log x)^D}$ and x for any x when $x > N = \max\{e, e^D, N_1\}$. ∎

3.2   Greedy Algorithm

We now use the greedy algorithm to prove that every integer x has a triple-base representation whose length is at most $C \frac{\log x}{\log \log x}$, where C is a positive constant.

The greedy algorithm is shown as Algorithm 1, with the input being a positive integer x, and the output being a set of 3-integers, a_i, such that $\sum_i a_i = x$. The algorithm finds the largest 3-integer, z, smaller than or equal to x, and recursively applies the same for $x - z$ until reaching zero.

Theorem 1. *The above greedy algorithm terminates after $k = \mathcal{O}(\frac{\log x}{\log \log x})$ steps.*

Algorithm 1. Greedy Algorithm

Input: an n-bit positive integer x
Output: $x = \sum_i a_i$, a_i is a 3-integer
1. $i \leftarrow 0$
2. while $x > 0$ do
3.　　z, the largest 3-integer smaller than or equal to x
4.　　$a_i \leftarrow z$
5.　　$i++$
6.　　$x \leftarrow x - z$
7. return $x = \sum_i a_i$

Proof. Lemma 1 shows that there exists an absolute constant $0 < D \leqslant 2$ such that there is always a number of the form $2^{e_1} 3^{e_2} 5^{e_3}$ between $x - \frac{x}{(\log x)^D}$ and x. Let $n_0 = x$. Then there exists a sequence

$$n_0 > n_1 > n_2 > \ldots > n_l > n_{l+1},$$

such that $n_i = 2^{e_1} 3^{e_2} 5^{e_3} + n_{i+1}$ and $n_{i+1} < \frac{n_i}{(\log n_i)^D}$ for $i = 0, 1, 2, \ldots, l$. Obviously the sequence of integers n_i obtained via the greedy algorithm satisfies these conditions. Let $l = l(x)$, so that

$$n_{l+1} \leqslant f(x) < n_l$$

for some function f to be chosen later.

We mention that $k = \mathcal{O}(\log x)$, simply by taking the 2-adic, 3-adic and 5-adic expansion of x. n_i can be represented as the sum of k 3-integers, where $k = l(x) + \mathcal{O}(\log f(x))$.

We have to determine $l(x), f(x)$. If $i < l$, then $f(x) < n_l < n_i$, thus $n_{i+1} < \frac{n_i}{(\log n_i)^D} < \frac{n_i}{(\log f(x))^D}$. Then the following inequality holds:

$$f(x) < n_l < \frac{x}{(\log f(x))^{l(x)D}}$$

$$\Rightarrow (\log f(x))^{l(x)D} < \frac{x}{f(x)}.$$

Taking the logarithm of both sides,

$$l(x)D \log\log f(x) < \log x - \log f(x).$$

Thus $l(x) < \frac{\log x - \log f(x)}{D \log\log f(x)}$.

The function $f(x) = \exp\frac{\log x}{\log\log x}$ is the largest possible (apart from constant) such that any number in the interval $[1, f(x)]$ can be written as the sum of $\mathcal{O}\left(\frac{\log x}{\log\log x}\right)$ terms that are 3-integers. We now show that $l(x) = \mathcal{O}\left(\frac{\log x}{\log\log x}\right)$, that is, there is a constant $C > 0$ such that

$$l(x) < C \frac{\log x}{\log\log x}.$$

It suffices to prove that

$$\frac{\log x - \frac{\log x}{\log\log x}}{D \log\frac{\log x}{\log\log x}} < C \frac{\log x}{\log\log x}.$$

Then

$$\log\log x - 1 < CD(\log\log x - \log\log\log x),$$

and thus

$$\log\log x + CD\log\log\log x < CD\log\log x + 1,$$

which is true if $C > \frac{1}{D}$ and x is large enough. Thus, $k = \mathcal{O}(\frac{\log x}{\log\log x})$. ∎

The length of the CTBNR $s(x)$ is less than or equal to the expansion of the greedy algorithm, giving an upper bound on $s(x)$ of $\mathcal{O}\left(\frac{\log x}{\log\log x}\right)$.

4   The Compactness of the Upper Bound

We next recall the notion of a universal exponent, which will be used in the following. Let $\lambda(n)$ be the universal exponent of the multiplicative group $(Z/nZ)^*$. A more explicit definition of λ is

$$\lambda(p^e) = \phi(p^e) = p^{e-1} \quad \text{if } p \text{ is an odd prime,}$$
$$\lambda(2^e) = \phi(2^e) \quad\qquad \text{if } e = 0, 1, \text{ or } 2,$$
$$\lambda(2^e) = \frac{1}{2}\phi(2^e) \qquad \text{if } e \geqslant 3,$$

and $\lambda(n) = \text{l.c.m.}(\lambda(p_1^{e_1}), \ldots, \lambda(p_v^{e_v}))$ if $n = p_1^{e_1} \ldots p_v^{e_v}$ (p_i are distinct primes). This is Carmichael's function [17].

Dimitrov and Howe [7] show that there are infinitely many integers x whose shortest signed double-base representations have length greater than

$$\frac{c_0 \log x}{\log \log x \log \log \log x}, \text{ where } c_0 \text{ is a positive constant.}$$

Although a TBNS has one more base than a DBNS, we can still show that there are infinitely many integers x whose shortest unsigned triple-base representations have length greater than

$$\frac{c \log x}{\log \log x \log \log \log x}, \text{ where } c \text{ is a positive constant.}$$

For each positive integer m, let $T(m)$ be the image in Z/mZ of the set of 3-integers, and $t(m)$ denote the cardinality of $T(m)$. For every $r \geqslant 2$, we define the expected degree-r density $D_r(m)$ to be the smaller of 1 and

$$\frac{1}{m}\left[\binom{t(m)}{1} + \binom{t(m)}{2} + \ldots + \binom{t(m)}{r}\right].$$

Lemma 2. *Suppose that $r \geqslant 2$ and that m is an integer whose expected degree-r density is less than 1. Then not every element of Z/mZ can be expressed as a sum of r elements of $T(m)$ or less.*

Proof. There are $\binom{t(m)}{i}$ ways of choosing i elements from $T(m)$ without repetition, so the number of sums of r or fewer elements of $T(m)$ is at most $\binom{t(m)}{1} + \binom{t(m)}{2} + \ldots + \binom{t(m)}{r}$. If $D_r(m)$ is less than 1, then the number of such sums is less than m, so some element of Z/mZ is not such a sum. That is, not every element of Z/mZ can be represented as a sum of r or fewer elements of $T(m)$. ∎

Lemma 3. *Suppose that $1 \leqslant r \leqslant n$. Then*

$$\binom{n}{1} + \binom{n}{2} + \ldots + \binom{n}{r} \leqslant \binom{n+r-1}{r}.$$

Proof. 1. When $r = 1, 2$, it is easy to check that

$$\binom{n}{1} + \ldots + \binom{n}{r} = \binom{n+r-1}{r}.$$

2. When $3 \leqslant r \leqslant n$,

$$\binom{n}{1} + \binom{n}{2} + \ldots + \binom{n}{r}$$

$$= \binom{n+1}{2} + \binom{n}{3} + \ldots + \binom{n}{r}$$

$$< \binom{n+1}{2} + \binom{n+1}{3} + \ldots + \binom{n}{r}$$

$$= \binom{n+2}{3} + \binom{n}{4} + \ldots + \binom{n}{r}$$

$$\vdots$$

$$< \binom{n+r-2}{r-1} + \binom{n+r-2}{r}$$

$$= \binom{n+r-1}{r}.$$

∎

We prove the following theorem mainly using universal exponent, then the process is denoted by universal exponent method.

Theorem 2. *There is a constant $c > 0$ such that, for infinitely many values of n, we have*

$$s(n) > \frac{c \log n}{\log \log n \log \log \log n}$$

Proof. Erdös, Pomerance, and Schmutz [18] show that there is a constant $d > 0$ such that there are infinitely many squarefree numbers m such that

$$\lambda(m) < (\log m)^{d \log \log \log m}. \tag{1}$$

We will prove that $c = \frac{1}{4d}$. Note that the function $\frac{\log x}{\log \log x \log \log \log x}$ is increasing for $x \geqslant 12006$ where the differential of this function is $\frac{\log \log x \log \log \log x - \log \log \log x - 1}{x (\log \log x \log \log \log x)^2}$, and that every integer $x < 30359$ has a triple-base representation of length 4 or less. Let m be one of the infinite number of squarefree integers that satisfy equation (1), with $\lambda(m) \geqslant 4$ ($x \geqslant 30359$) and $\frac{c \log m}{\log \log m \log \log \log m} \geqslant 4$.

Since m is squarefree, there are at most $\lambda(m) + 1$ distinct powers of 2 in Z/mZ, at most $\lambda(m) + 1$ distinct powers of 3 in Z/mZ, and at most $\lambda(m) + 1$ distinct powers of 5 in Z/mZ. It follows that

$$t(m) \leqslant (\lambda(m) + 1)^3 < \lambda(m)^4.$$

Let $r = \lfloor \frac{c \log m}{\log \log m \log \log \log m} \rfloor$, so that $r \geqslant 4$. Then

$$D_r(m) = \frac{1}{m} \left[\binom{t(m)}{1} + \binom{t(m)}{2} + \ldots + \binom{t(m)}{r} \right]$$

$$\leqslant \frac{1}{m} \binom{t(m) + r - 1}{r} \qquad \text{by Lemma 3}$$

$$< \frac{t(m)^r}{m}$$

$$< \lambda(m)^{4r}/m$$

$$< e^{4rd \log \log m \log \log \log m}/m$$

$$\leqslant e^{\log m}/m$$

$$= 1,$$

so by Lemma 2, there is a nonnegative integer $n < m$ such that the image of n in Z/mZ cannot be written as the sum of r elements of $T(m)$. It follows that

$$s(n) \geqslant r + 1 > \frac{c \log m}{\log \log m \log \log \log m} > \frac{c \log n}{\log \log n \log \log \log n}.$$

The final inequality depends on the fact that $n \geqslant 12006$, but we know that $n \geqslant 30359$ because the span of n is at least $r + 1 \geqslant 5$.   ∎

There are infinitely many integers x whose shortest unsigned triple-base representations have length greater than

$$\frac{c \log x}{\log \log x \log \log \log x}, \text{ where } c \text{ is a positive constant.}$$

Therefore, the upper bound is greater than or equal to $\frac{c \log x}{\log \log x \log \log \log x}$ and less than or equal to $\frac{C \log x}{\log \log x}$. Because the difference between these two numbers is small, the presented upper bound on $s(x)$ is meaningful.

5   Conclusion

In this paper, we have shown that an integer x can be represented using at most $k = \mathcal{O}(\frac{\log x}{\log \log x})$ 3-integers. We have also shown that there are infinitely many integers x whose shortest triple-base representations have length greater than $\frac{c \log x}{\log \log x \log \log \log x}$, where c is a positive constant.

Acknowledgments. We are very grateful to anonymous reviewers for their helpful comments.

References

1. Dimitrov, V.S., Jullien, G.A., Miller, W.C.: Theory and applications for a double-base number system. In: IEEE Symposium on Computer Arithmetic, pp. 44–53 (1997)
2. Dimitrov, V.S., Jullien, G.A.: Loading the bases: A new number representation with applications. IEEE Circuits and Systems Magazine 3(2), 6–23 (2003)
3. Mishra, P.K., Dimitrov, V.S.: Efficient Quintuple Formulas for Elliptic Curves and Efficient Scalar Multiplication Using Multibase Number Representation. In: Garay, J.A., Lenstra, A.K., Mambo, M., Peralta, R. (eds.) ISC 2007. LNCS, vol. 4779, pp. 390–406. Springer, Heidelberg (2007)
4. Longa, P.: Accelerating the Scalar Multiplication on Elliptic Curve Cryptosystems over Prime Fields, Master Thesis, University of Ottawa (2007)
5. Purohit, G.N., Rawat, A.S.: Fast Scalar Multiplication in ECC Using The Multi base Number System, http://eprint.iacr.org/2011/044.pdf
6. Dimitrov, V.S., Jullien, G.A., Miller, W.C.: An algorithm for modular exponentiation, Inform. Process. Lett. 66(3), 155–159 (1998)
7. Dimitrov, V.S., Howe, E.W.: Lower bounds on the lengths of double-base representations. Proceedings of the American Mathematical Society 139(10), 3423–3430 (2011)
8. Dimitrov, V., Imbert, L., Mishra, P.K.: The double-base number system and its application to elliptic curve cryptography. Math. Comp. 77(262), 1075–1104 (2008)
9. Doche, C., Kohel, D.R., Sica, F.: Double-Base Number System for multi-scalar multiplications. In: Joux, A. (ed.) EUROCRYPT 2009. LNCS, vol. 5479, pp. 502–517. Springer, Heidelberg (2009)
10. Dimitrov, V.S., Imbert, L., Mishra, P.K.: Efficient and Secure Elliptic Curve Point Multiplication Using Double-Base Chains. In: Roy, B. (ed.) ASIACRYPT 2005. LNCS, vol. 3788, pp. 59–78. Springer, Heidelberg (2005)
11. Doche, C., Imbert, L.: Extended Double-Base Number System with Applications to Elliptic Curve Cryptography. In: Barua, R., Lange, T. (eds.) INDOCRYPT 2006. LNCS, vol. 4329, pp. 335–348. Springer, Heidelberg (2006)
12. Doche, C., Habsieger, L.: A Tree-Based Approach for Computing Double-Base Chains. In: Mu, Y., Susilo, W., Seberry, J. (eds.) ACISP 2008. LNCS, vol. 5107, pp. 433–446. Springer, Heidelberg (2008)
13. Méloni, N., Hasan, M.A.: Elliptic Curve Scalar Multiplication Combining Yao's Algorithm and Double Bases. In: Clavier, C., Gaj, K. (eds.) CHES 2009. LNCS, vol. 5747, pp. 304–316. Springer, Heidelberg (2009)
14. Suppakitpaisarn, V., Edahiro, M., Imai, H.: Fast Elliptic Curve Cryptography Using Optimal Double-Base Chains. eprint.iacr.org/2011/030.ps
15. de Weger, B.M.M.: Algorithms for Diophantine equations. CWI Tracts, vol. 65. Centrum voor Wiskunde en Informatica, Amsterdam (1989)
16. Tijdeman, R.: On the maximal distance between integers composed of small primes. Compositio Mathematica 28, 159–162 (1974)
17. Carmichael, R.D.: On composite numbers p which satisfy the Fermat congruence $a^{p-1} \equiv 1 \mod p$. Amer. Math. Monthly 19, 22–27 (1912)
18. Erdös, P., Pomerance, C., Schmutz, E.: Carmichael's lambda function. Acta Arith. 58(4), 363–385 (1991)

Triple-Base Number System
for Scalar Multiplication[*]

Wei Yu[1,2], Kunpeng Wang[2], Bao Li[2], and Song Tian[2]

[1] Department of Electronic Engineering and Information Science,
University of Science and Technology of China, Hefei, 230027, China
yuwei_1_yw@163.com
[2] Institute of Information Engineering, Chinese Academy of Sciences, Beijing, 100093

Abstract. The triple-base number system is used to speed up scalar multiplication. At present, the main methods to calculate a triple-base chain are greedy algorithms. We propose a new method, called the add/sub algorithm, to calculate scalar multiplication. The density of such chains gained by this algorithm with base $\{2, 3, 5\}$ is $\frac{1}{5.61426}$. It saves 22% additions compared with the binary/ternary method; 22.1% additions compared with the multibase non-adjacent form with base $\{2, 3, 5\}$; 13.7% additions compared with the greedy algorithm with base $\{2, 3, 5\}$; 20.9% compared with the tree approach with base $\{2, 3\}$; and saves 4.1% additions compared with the add/sub algorithm with base $\{2, 3, 7\}$, which is the same algorithm with different parameters. To our knowledge, the add/sub algorithm with base $\{2, 3, 5\}$ is the fastest among the existing algorithms. Also, recoding is very easy and efficient and together with the add/sub algorithm are very suitable for software implementation. In addition, we improve the greedy algorithm by plane search which searches for the best approximation with a time complexity of $\mathcal{O}(\log^3 k)$ compared with that of the original of $\mathcal{O}(\log^4 k)$.

Keywords: Elliptic Curve Cryptography, Scalar Multiplication, Hamming Weight, Triple Base Chain, Density.

1 Introduction

Because there are no general-purpose sub-exponential algorithms known for the elliptic curve discrete logarithm problem, greater attention have focused on elliptic curves in the public key cryptography, for which speeding up scalar multiplications is significant.

The double-base number system (DBNS) was first introduced in [1,2] for its inherent sparseness in representing binary integers. M. Ciet, M. Joye, K. Lauter

---

[*] Supported in part by National Basic Research Program of China(973) under Grant No.2013CB338002, in part by National Research Foundation of China under Grant No. 61272040 and 61070171, and in part by the Strategic Priority Research Program of Chinese Academy of Sciences under Grant XDA06010702.

A. Youssef, A. Nitaj, A.E. Hassanien (Eds.): AFRICACRYPT 2013, LNCS 7918, pp. 433–451, 2013.

and P. L. Montgomery [3] first used this system in elliptic curve cryptography (ECC). They gave a binary/ternary method to calculate scalar multiplications with a chain of density $\frac{1}{4.3774}$. Following B.M.M.de Weger's definition of s-integers [4], for which the largest prime factor does not exceed the s-th prime number, the integer of the form $2^b 3^t$ is called a 2-integer. In 2005, V. Dimitrov, L. Imbert and P. K. Mishra [5] introduced the concept of a double-base chain for computing scalar multiplication. They used a greedy algorithm to calculate double-base chains; the objective is to find the best approximation with a 2-integer at each step. Several methods exist [6,7,8]. Later, C. Doche and L. Imbert [9] used extended DBNS to perform scalar multiplications that also uses the greedy algorithm. C. Doche and L. Habsieger [10] proposed a tree-based approach for computing double-base chains. The density of the result is $\frac{1}{4.6419}$. N. Méloni and M. A. Hasan [11] introduced an algorithm combining Yao's algorithm and double bases to calculate scalar multiplication. V. Suppakitpaisarn, M. Edahiro, and H. Imai [12] proposed a method to calculate optimal double-base chains although recoding is time consuming. For more about DBNS, we refer readers to [13].

To compute scalar multiplications efficiently, the triple-base number system (TBNS) was first introduced by P. K. Mishra and V. S. Dimitrov [14]. This system is even more sparse than DBNS. P. Longa and A. Miri [15] proposed a multibase non-adjacent form (mbNAF) to compute scalar multiplication. Recently, G. N. Purohit and A. S. Rawat [16] calculated scalar multiplication with base $\{2, 3, 7\}$.

In this paper, we improve the implementation of the greedy algorithm using plane search to speed up the recoding process. We present an add/sub algorithm, a new method to compute triple-base chains (TBCs) that is suitable for software implementation. Theoretical analysis of the add/sub algorithm reveals that the recoding time is very quick and the TBC density returned by the add/sub algorithm with base $\{2, 3, 5\}$ is $\frac{1}{5.61426}$. The average values of the biggest powers of 2, 3, and 5 in the corresponding chain are approximately equal to $0.454 \log_2 k$, $0.216 \log_2 k$ and $0.0876 \log_2 k$ respectively for any given scalar k. The density of the add/sub algorithm using base $\{2, 3, 7\}$ is $\frac{1}{5.38543}$. The average values of the biggest powers of 2, 3, and 7 in the corresponding chain are approximately equal to $0.4702 \log_2 k$, $0.2264 \log_2 k$ and $0.0609 \log_2 k$ respectively for given scalar k.

In calculating scalar multiplication, the TBC obtained by the add/sub algorithm with base $\{2, 3, 5\}$ has the smallest Hamming weight among existing algorithms. Add/sub algorithm with base $\{2, 3, 5\}$ saves 22% additions compared with binary/ternary method; 22.1% additions compared to mbNAF with base $\{2, 3, 5\}$; 13.7% additions compared to greedy algorithms with restriction $0.5 \log k, 0.3 \log k, 0.2 \log k$ with base $\{2, 3, 5\}$; 20.9% compared to tree approach using base $\{2, 3\}$; and saves 4.1% additions compared to add/sub algorithm with base $\{2, 3, 7\}$ which is the same algorithm with different parameters. To our knowledge, the algorithm add/sub algorithm with base $\{2, 3, 5\}$ is the fastest among the existing algorithms.

The paper is organized as follows: In the next section, we recall some basic facts about double-base number systems, multi-base number systems, and some algorithms to compute TBCs. In section 3, we introduce the greedy algorithm and give plane search to implement the greedy algorithm efficiently. In section 4, we introduce the add/sub algorithm to compute TBCs. In section 5, we give the algorithm using TBCs to compute scalar multiplication and give the add/sub recursive algorithm to recode and calculate scalar multiplication simultaneously. In section 6, we provide a theoretical analysis of the add/sub algorithm with base $\{2, 3, 5\}$ and with base $\{2, 3, 7\}$. In section 7, we compare the add/sub algorithm with other methods from different aspects such as Hamming weight, bit cost, and recoding time. Finally, we conclude the paper.

2   Preliminary

2.1   Double-Base Chain

An important result of DBNS is that every integer k can be represented as at most $\mathcal{O}\left(\frac{\log k}{\log \log k}\right)$ 2-integers [13]:

$$k = \sum_{i}^{m} s_i 2^{b_i} 3^{t_i}, \text{ where } s_i \in \{-1, 1\}. \tag{1}$$

Scalar multiplication in ECC using double-base integers requires constructing the double-base chain [5]. The concept of double-base chain is a special type of DBNS where b_i, t_i in equation (1) satisfy $b_1 \geqslant b_2 \geqslant \ldots \geqslant b_m \geqslant 0$, $t_1 \geqslant t_2 \geqslant \ldots \geqslant t_m \geqslant 0$. This representation is not unique and highly redundant. The average length of double-base chains generated by existing algorithms are all $\mathcal{O}(\log k)$ [10], not $\mathcal{O}\left(\frac{\log k}{\log \log k}\right)$.

2.2   Multibase Non-adjacent Form

P. Longa and A. Miri [15] proposed the mbNAF, the window multibase non-adjacent form (wmbNAF) and the extended wmbNAF to compute scalar multiplications. The respective average densities of mbNAF, wmbNAF, and extended wmbNAF are $\frac{1}{2 \log_2 a_1 + \sum_{i=1}^{J} \frac{1}{a_i-1} \log_2 a_i}$, $\frac{1}{w \log_2 a_1 + \sum_{i=1}^{J} \frac{1}{a_i-1} \log_2 a_i}$, and $\frac{1}{\sum_{i=1}^{J} \left(w_i + \frac{1}{a_i-1}\right) \log_2 a_i}$, where $a_i, 1 \leqslant i \leqslant J$ are the bases, and $\{w_i, 1 \leqslant i \leqslant J\}$ the corresponding window widths [17].

The precomputed table for the mbNAF consists of $\frac{a_1^2 - a_1 - 2}{2}$ points. If $a_1 = 2$, the requirement of mbNAF's precomputations is 0. If $a_2 = 3$, $a_3 = 5$, the average density is $\frac{1}{3 + \frac{1}{2} \log_2 3 + \frac{1}{4} \log_2 5} = \frac{1}{4.3730} = 0.2287$, whereas if $a_2 = 3$, $a_3 = 7$, the average density is $\frac{1}{3 + \frac{1}{2} \log_2 3 + \frac{1}{6} \log_2 7} = \frac{1}{4.2604} = 0.2347$.

2.3   Triple-Base Chains

Any integer k can be represented as a TBC with base $\{a_1, a_2, a_3\}$:

$$k = \sum_{i=1}^{m} s_i a_1^{e_{1i}} a_2^{e_{2i}} a_3^{e_{3i}},$$

where $s_i \in \{-1,1\}$, $e_{11} \geqslant e_{12} \geqslant e_{13} \geqslant \ldots \geqslant e_{1m} \geqslant 0$, $e_{21} \geqslant e_{22} \geqslant e_{23} \geqslant \ldots \geqslant$ $e_{2m} \geqslant 0$, $e_{31} \geqslant e_{32} \geqslant e_{33} \geqslant \ldots \geqslant e_{3m} \geqslant 0$. In the following, we assume that a_1, a_2, a_3 are three prime numbers and $a_1 < a_2 < a_3$. For example, with base $\{2, 3, 5\}$, k is represented as

$$ k = \sum_{i=1}^{m} s_i 2^{b_i} 3^{t_i} 5^{q_i}, $$

where $s_i \in \{-1, 1\}$, $b_1 \geqslant b_2 \geqslant b_3 \geqslant \ldots \geqslant b_m \geqslant 0$, $t_1 \geqslant t_2 \geqslant t_3 \geqslant \ldots \geqslant t_m \geqslant 0$, $q_1 \geqslant q_2 \geqslant q_3 \geqslant \ldots \geqslant q_m \geqslant 0$.

P. K. Mishra, V. S. Dimitrov [14] used the greedy algorithm to calculate a TBC by finding the best approximation of a 3-integer at each step. They mentioned that one can generate a random integer directly in a TBC given an arbitrary scalar. In this instance, the TBC may not be the shortest. In other situations, if the scalar is known beforehand, the conversion can be done offline. G. N. Purohit and A. S. Rawat [16] performed efficient scalar multiplications using base $\{2, 3, 7\}$. They also used the greedy method finding the best approximation for a $\{2, 3, 7\}$-base integer at each step.

3   Plane Search

W. Yu, K. Wang, and B. Li [8] proposed a line algorithm to compute the best approximation of a $\{a, b\}$-integers. They calculated the $\{a, b\}$-integers near the line $x \log_2 a + y \log_2 b = \log_2 k$ where k is the scalar at each step. We generalize the line algorithm to find the TBC representation of a $\{a_1, a_2, a_3\}$-integer. In this way, the line becomes a plane described by $x \log_2 a_1 + y \log_2 a_2 + z \log_2 a_3 = \log_2 k$. The greedy algorithm [14,16], which is outlined in Table 1, works by finding the best approximation with a $\{a_1, a_2, a_3\}$-integer at each step. We give the plane search method for finding the best approximation faster.

The plane search method scans all the points with integer coordinates near the plane $x \log_2 a_1 + y \log_2 a_2 + z \log_2 a_3 = \log_2 k$ and keeps only the best approximation. It returns the same $\{a_1, a_2, a_3\}$-integers as the method in [14] searching the best approximation in cube $[0, e_{1max}] \times [0, e_{2max}] \times [0, e_{3max}]$. The complexity [14] of searching the best approximation in this cube is $\mathcal{O}(\log^3 k)$. The complexity of a plane search is $\mathcal{O}(\log^2 k)$ where k is the scalar.

4   Add/Sub Algorithm

In this section, we propose a new method to compute scalar multiplications that is more efficient than the greedy algorithm. Let $v_p(n)$ denote the p-adic valuation of the integer n.

Definition 1. *The gather* $\mathcal{C}(a_1, a_2, a_3)$ *denotes the set of all positive integers* x *such that* $v_{a_1}(x) = v_{a_2}(x) = v_{a_3}(x) = 0$. *We simplify the symbol as* $\mathcal{C}$ *if* a_1, a_2, a_3 *are known.*

Table 1. Greedy Algorithm

Input: a n-bit positive integer k; three bases a_1, a_2, a_3;
the largest allowed a_1, a_2 and a_3 exponents e_{1max}, e_{2max}, e_{3max}

Output: The sequence $(s_i, e_{1i}, e_{2i}, e_{3i})_{i>0}$,
such that $k = \sum_i s_i a_1^{e_{1i}} a_2^{e_{2i}} a_3^{e_{3i}}$, with $e_{11} \geqslant e_{12} \geqslant \cdots \geqslant e_{1m} \geqslant 0$,
$e_{21} \geqslant e_{22} \geqslant \cdots \geqslant e_{2m} \geqslant 0$, $e_{31} \geqslant e_{32} \geqslant \cdots \geqslant e_{3m} \geqslant 0$

---

1. $s_1 \leftarrow 1$, i$\leftarrow 1$
2. while $k > 0$ do
3. $z \leftarrow a_1^{e_{1i}} a_2^{e_{2i}} a_3^{e_{3i}}$, the best approximation of k with
        $0 \leqslant e_{1i} \leqslant e_{1\,max}$, $0 \leqslant e_{2i} \leqslant e_{2\,max}$, $0 \leqslant e_{3i} \leqslant e_{3\,max}$
4. $e_{1\,max} \leftarrow e_{1i}$
5. $e_{2\,max} \leftarrow e_{2i}$
6. $e_{3\,max} \leftarrow e_{3i}$
7. $i++$
8.   if $k < z$ then
9. $s_i \leftarrow -s_{i-1}$
10.   else
11. $s_i \leftarrow s_{i-1}$
12. $k \leftarrow |k - z|$
13. return $k = \sum_i s_i a_1^{e_{1i}} a_2^{e_{2i}} a_3^{e_{3i}}$

Table 2. Plane Search

Input: a positive integer k; three bases a_1, a_2, a_3;
the largest allowed a_1, a_2 and a_3 exponents e_{1max}, e_{2max}, $e_{3max} \geqslant 0$

Output: (e_1, e_2, e_3), $e_1 \leqslant e_{1\,max}$, $e_2 \leqslant e_{2\,max}$, $e_3 \leqslant e_{3\,max}$
such that $z = a_1^{e_1} a_2^{e_2} a_3^{e_3}$ the best approximation of k

---

1. $spare \leftarrow k$, $j \leftarrow 0$
2. for r from 0 to $e_{3\,max}$
3. $z \leftarrow a_1^{e_{1\,max}} \cdot a_3^r$, $i \leftarrow e_{1\,max}$
4.   while $z < k$
5. $z \leftarrow z * a_2$, $j \leftarrow j + 1$.
6.   while$(i \geqslant 0)$
7.   if$(|k - z| < spare)$
8. $spare \leftarrow |k - z|$
9. $e_1 \leftarrow i$, $e_2 \leftarrow j$, $e_3 \leftarrow r$
10.   if$(z < k)$
11. $z \leftarrow z * a_2$, $j \leftarrow j + 1$
12.   if$(j > e_{2max})$ *break*
13.   else $z \leftarrow \frac{z}{a_1}$, $i \leftarrow i - 1$
14. return (e_1, e_2, e_3)

The main method, the add/sub algorithm, is a generalization of the binary/ternary approach [3]. The main operations in this algorithm are add and sub. At each iteration of the add/sub algorithm, for a given positive integer $k \in \mathcal{C}$, remove the powers of a_1, a_2, a_3 from $k-1$, $k+1$; select the smaller one as the initial k of the next iteration; repeat the iteration until $k = 1$. The algorithm is

Table 3. Add/Sub Triple Base Chain Algorithm

Input: scalar $k > 0$; three bases a_1, a_2, a_3
Output: A triple base chain
1. $i \leftarrow 1$
2. $e_{1i} \leftarrow v_{a_1}(k)$, $e_{2i} \leftarrow v_{a_2}(k)$, $e_{3i} \leftarrow v_{a_3}(k)$
3. $k \leftarrow \frac{k}{a_1^{e_{1i}} a_2^{e_{2i}} a_3^{e_{3i}}}$
4. while $k > 1$ do
5.　　$g_1 \leftarrow v_{a_1}(k-1)$, $g_2 \leftarrow v_{a_2}(k-1)$, $g_3 \leftarrow v_{a_3}(k-1)$
6.　　$h_1 \leftarrow v_{a_1}(k+1)$, $h_2 \leftarrow v_{a_2}(k+1)$, $h_3 \leftarrow v_{a_3}(k+1)$
7.　　if$(a_1^{g_1} a_2^{g_2} a_3^{g_3} > a_1^{h_1} a_2^{h_2} a_3^{h_3})$
8.　　　　$s_i \leftarrow 1, i \leftarrow i+1, e_{1i} \leftarrow e_{1(i-1)} + g_1, e_{2i} \leftarrow e_{2(i-1)} + g_2,$
$e_{3i} \leftarrow e_{3(i-1)} + g_3$
9.　　　　$k = \frac{k-1}{a_1^{g_1} a_2^{g_2} a_3^{g_3}}$
10.　　else
11.　　　　$s_i \leftarrow -1, i \leftarrow i+1, e_{1i} \leftarrow e_{1(i-1)} + h_1, e_{2i} \leftarrow e_{2(i-1)} + h_2,$
$e_{3i} \leftarrow e_{3(i-1)} + h_3$
12.　　　　$k \leftarrow \frac{k+1}{a_1^{h_1} a_2^{h_2} a_3^{h_3}}$
13. $s_i \leftarrow k$, $e_{1i} \leftarrow 0, e_{2i} \leftarrow 0, e_{3i} \leftarrow 0$
14. return $\sum_i (s_i a_1^{e_{1i}} a_2^{e_{2i}} a_3^{e_{3i}})$

shown in Table 3. In the application of the add/sub algorithm for this paper, we mainly focus on bases $\{2, 3, 5\}$ and base $\{2, 3, 7\}$. An example is presented in section 4.1.

4.1   An Example of Add/Sub Algorithm

Let $k = 895712$ the same number as in [16]. We show the output of Add/Sub algorithm when the base is $\{2, 3, 5\}$ and $\{2, 3, 7\}$.

Base $\{2, 3, 5\}$
The processes of add/sub algorithm are run as:
$e_{11} = 5$, $e_{21} = 0$, $e_{31} = 0$, $s_1 = 1$,
$e_{12} = 5 + 1 = 6$, $e_{22} = 2$, $e_{32} = 1$, $s_2 = -1$,
$e_{13} = 6 + 3 = 9$, $e_{23} = 2 + 1 = 3$, $e_{33} = 1$, $s_3 = 1$,
$e_{14} = 9 + 2 = 11$, $e_{24} = 3 + 1 = 4$, $e_{34} = 1$, $s_4 = 1$.
895712 is then represented as:

$$895712 = 2^5 - 2^6 3^2 5^1 + 2^9 3^3 5^1 + 2^{11} 3^4 5^1$$
$$= 2^{11} 3^4 5^1 + 2^9 3^3 5^1 - 2^6 3^2 5^1 + 2^5.$$

In the add/sub algorithm, we produce sequences $k = \sum_{i=1}^m s_i a_1^{e_{1i}} a_2^{e_{2i}} a_3^{e_{3i}}$, where $s_i \in \pm 1$, $e_{11} \leqslant e_{12} \leqslant e_{13} \leqslant \ldots \leqslant e_{1m}$, $e_{21} \leqslant e_{22} \leqslant e_{23} \leqslant \ldots \leqslant e_{2m}$, $e_{31} \leqslant e_{32} \leqslant e_{33} \leqslant \ldots \leqslant e_{3m}$. The chain gained by Add/Sub Triple Base Chain is a Nondecreasing sequence. Let $s_i' = s_{m+1-i}$, $e_{1i}' = e_{1(m+1-i)}$, $e_{2i}' = e_{2(m+1-i)}$, $e_{3i}' =$

$e_{3(m+1-i)}$, $1 \leqslant i \leqslant m$, i.e. Reversing the sequence, k can be rewritten as $k = \sum_{i=1}^{m} s_i' a_1^{e_{1i}'} a_2^{e_{2i}'} a_3^{e_{3i}'}$, which is a TBC.

Remark: Using the greedy algorithm, 895712 has representation

$$895712 = 2^5 3^2 5^5 - 2^2 3^2 5^3 - 2^2 3^2 5^3 + 3^2 5^2 - 3^1 5^1 + 3 - 1,$$

which is 3 terms longer than that given by the add/sub algorithm, whereas using mbNAF, 895712 is represented as

$$895712 = 2^{12} 5^3 - 2^9 5^3 - 2^8 - 2^5.$$

Base $\{2, 3, 7\}$

Similarly, the processes of the add/sub algorithm are run with:

$e_{11} = 5$, $e_{21} = 0$, $e_{31} = 0$, $s_1 = 1$,
$e_{12} = 5 + 1 = 6$, $e_{22} = 2$, $e_{32} = 0$, $s_2 = 1$,
$e_{13} = 6 + 1 = 7$, $e_{23} = 2 + 1 = 3$, $e_{33} = 1$, $s_3 = 1$,
$e_{14} = 7 + 2 = 9$, $e_{24} = 3 + 2 = 5$, $e_{34} = 1$, $s_4 = 1$.

895712 is with $\{2,3,7\}$-integer representations:

$$\begin{aligned}
895712 &= 2^5 + 2^6 3^2 + 2^7 3^3 7^1 + 2^9 3^5 7^1 \\
&= 2^9 3^5 7^1 + 2^7 3^3 7^1 + 2^6 3^2 + 2^5.
\end{aligned}$$

4.2   The Correctness of Add/Sub Algorithm

For the two bases focused on, a_1 is equal to 2 in both. In every iteration of add/sub algorithm, $v_2(k) = 0$, i.e. k_i is an odd number. Then $k_i - 1$, $k_i + 1$ are even numbers. In every iteration of the add/sub algorithm, k_{i+1} is at most $\frac{k_i+1}{2}$. Then, the add/sub algorithm requires at most $\log_2(k)$ iterations. Thus, the correction of add/sub algorithm is proved when $a_1 = 2$.

If $a_1 = 3$, the add/sub algorithm can also process successfully. Notice that one of $k + 1$, $k - 1$ can be divided by 3. In the same way as analyzing $a_1 = 2$, the add/sub algorithm can process at most $\log_3(k)$ iterations.

Thus, the correction of the add/sub algorithm is demonstrated if $a_1 = 2$ or $a_1 = 3$.

Not all bases can use the add/sub algorithm to compute the TBC. For example, take $a_1 = 7$, $a_2 = 11$, $a_3 = 17$, if $k = 5$, the add/sub algorithm to find the TBC does not work.

5   Scalar Multiplication

5.1   Triple Base Chain Method to Calculate Scalar Multiplication

In Table 4, we show how to calculate scalar multiplication using a TBC. The implementation is easy to understand and perform.

Table 4. TBC to Calculate Scalar Multiplication

Input: scalar $k = \sum_i (s_i a_1^{e_{1i}} a_2^{e_{2i}} a_3^{e_{3i}})$, such that a_1, a_2, a_3 are three bases, $s_i \in \{\pm1\}$, $e_{11} \geqslant e_{12} \geqslant e_{13} \geqslant \ldots e_{1m} \geqslant 0$, $e_{21} \geqslant e_{22} \geqslant e_{23} \geqslant \ldots e_{2m} \geqslant 0$, $e_{31} \geqslant e_{32} \geqslant e_{33} \geqslant \ldots e_{3m} \geqslant 0$; a point P
Output: kP
1. $i \leftarrow 1$, $Q \leftarrow 0$
2. while(i<m)
3. $Q \leftarrow (Q + s_i P) a_1^{e_{1i}-e_{1(i+1)}} a_2^{e_{2i}-e_{2(i+1)}} a_3^{e_{3i}-e_{3(i+1)}}$
4. $i{+}{+}$
5. $Q \leftarrow (Q + s_m P) a_1^{e_{1m}} a_2^{e_{2m}} a_3^{e_{3m}}$
6. return Q

Table 5. Recursive TBC to Calculate Scalar Multiplication

Input: scalar k; a point P; three bases a_1, a_2, a_3
Output: kP
1. $e_{11} \leftarrow v_{a_1}(k)$, $e_{21} \leftarrow v_{a_2}(k)$, $e_{31} \leftarrow v_{a_3}(k)$
2. $k \leftarrow \frac{k}{a_1^{e_{11}} a_2^{e_{21}} a_3^{e_{31}}}$
3. return $a_1^{e_{11}} a_2^{e_{21}} a_3^{e_{31}} (Recursive(k, P, a_1, a_2, a_3))$

Table 6. Recursive algorithm

Input: scalar k; a point P; three bases a_1, a_2, a_3
Output: kP
1. if $k = 1$
2.   return P
3. else
4. $g_1 \leftarrow v_{a_1}(k-1)$, $g_2 \leftarrow v_{a_2}(k-1)$, $g_3 \leftarrow v_{a_3}(k-1)$
5. $h_1 \leftarrow v_{a_1}(k+1)$, $h_2 \leftarrow v_{a_2}(k+1)$, $h_3 \leftarrow v_{a_3}(k+1)$
6.   if($a_1^{g_1} a_2^{g_2} a_3^{g_3} > a_1^{h_1} a_2^{h_2} a_3^{h_3}$)
7. $k \leftarrow \frac{k-1}{a_1^{g_1} a_2^{g_2} a_3^{g_3}}$
8.   return $a_1^{g_1} a_2^{g_2} a_3^{g_3} (Recursive(k, P, a_1, a_2, a_3) + P)$
9.   else
10. $k \leftarrow \frac{k+1}{a_1^{h_1} a_2^{h_2} a_3^{h_3}}$
11.   return $a_1^{h_1} a_2^{h_2} a_3^{h_3} (Recursive(k, P, a_1, a_2, a_3) - P)$

5.2   Recursive Algorithm

In Table 5, we give the algorithm using recursive algorithm, given in Table 6, to calculate scalar multiplication. This recursive algorithm is a method that performs recoding and calculating scalar multiplication simultaneously.

6 Complexity Analysis

Lemma 1. *Given three integers α_1, α_2, α_3, the cardinality of $\mathcal{C} \cap [1, \ a_1^{\alpha_1} a_2^{\alpha_2} a_3^{\alpha_3}]$ is equal to $(a_1 - 1)(a_2 - 1)(a_3 - 1)a_1^{\alpha_1 - 1} a_2^{\alpha_2 - 1} a_3^{\alpha_3 - 1}$.*

Proof. The cardinality of $[1, \ a_1^{\alpha_1} a_2^{\alpha_2} a_3^{\alpha_3}]$ is $a_1^{\alpha_1} a_2^{\alpha_2} a_3^{\alpha_3}$. The cardinality of $\mathcal{C} \cap [1, \ a_1^{\alpha_1} a_2^{\alpha_2} a_3^{\alpha_3}]$ is equal to $a_1^{\alpha_1} a_2^{\alpha_2} a_3^{\alpha_3} \times (1 - \frac{1}{a_1}) \times (1 - \frac{1}{a_2}) \times (1 - \frac{1}{a_3}) = (a_1 - 1)(a_2 - 1)(a_3 - 1)a_1^{\alpha_1 - 1} a_2^{\alpha_2 - 1} a_3^{\alpha_3 - 1}$. ∎

The cardinalities of the two bases of focus, $\{2, 3, 5\}$ and $\{2, 3, 7\}$, are $2^{\alpha_1 + 2} 3^{\alpha_2 - 1} 5^{\alpha_3 - 1}$ and $2^{\alpha_1 + 1} 3^{\alpha_2} 7^{\alpha_3 - 1}$ respectively.

Lemma 2. *Suppose that k is an integer satisfying $v_{a_1}(k) = 0$, $v_{a_2}(k) = 0$ and $v_{a_3}(k) = 0$. Let $g_1 = v_{a_1}(k-1)$, $g_2 = v_{a_2}(k-1)$, $g_3 = v_{a_3}(k-1)$, $h_1 = v_{a_1}(k+1)$, $h_2 = v_{a_2}(k+1)$, $h_3 = v_{a_3}(k+1)$. Let $x_1 = g_1 + h_1$, $x_2 = g_2 + h_2$ and $x_3 = g_3 + h_3$. Then*

$$x_1 = v_{a_1}(k - 1 + ia_1^{x_1+1} a_2^{x_2+1} a_3^{x_3+1}) + v_{a_1}(k + 1 + ia_1^{x_1+1} a_2^{x_2+1} a_3^{x_3+1}),$$
$$x_2 = v_{a_2}(k - 1 + ia_1^{x_1+1} a_2^{x_2+1} a_3^{x_3+1}) + v_{a_2}(k + 1 + ia_1^{x_1+1} a_2^{x_2+1} a_3^{x_3+1}), \quad \forall i \in \mathbb{Z}$$
$$x_3 = v_{a_3}(k - 1 + ia_1^{x_1+1} a_2^{x_2+1} a_3^{x_3+1}) + v_{a_3}(k + 1 + ia_1^{x_1+1} a_2^{x_2+1} a_3^{x_3+1}),$$

Proof. Let $k - 1$ be equal to $a_1^{g_1} a_2^{g_2} a_3^{g_3} \times y$, where $y \in \mathcal{C}$. Then

$$k - 1 + i \cdot a_1^{x_1+1} a_2^{x_2+1} a_3^{x_3+1}$$
$$= a_1^{g_1} a_2^{g_2} a_3^{g_3}(y + a_1^{x_1-g_1+1} a_2^{x_2-g_2+1} a_3^{x_3-g_3+1}),$$

$$v_{a_1}(k - 1 + i \cdot a_1^{x_1+1} a_2^{x_2+1} a_3^{x_3+1}) = g_1,$$
$$v_{a_2}(k - 1 + i \cdot a_1^{x_1+1} a_2^{x_2+1} a_3^{x_3+1}) = g_2,$$
$$v_{a_3}(k - 1 + i \cdot a_1^{x_1+1} a_2^{x_2+1} a_3^{x_3+1}) = g_3.$$

In the same way, let $k + 1$ be equal to $a_1^{h_1} a_2^{h_2} a_3^{h_3} \times z$, where $z \in \mathcal{C}$. Then

$$k + 1 + i \cdot a_1^{x_1+1} a_2^{x_2+1} a_3^{x_3+1}$$
$$= a_1^{g_1} a_2^{g_2} a_3^{g_3}(z + a_1^{x_1-h_1+1} a_2^{x_2-h_2+1} a_3^{x_3-h_3+1}),$$

$$v_{a_1}(k + 1 + i \cdot a_1^{x_1+1} a_2^{x_2+1} a_3^{x_3+1}) = h_1,$$
$$v_{a_2}(k + 1 + i \cdot a_1^{x_1+1} a_2^{x_2+1} a_3^{x_3+1}) = h_2,$$
$$v_{a_3}(k + 1 + i \cdot a_1^{x_1+1} a_2^{x_2+1} a_3^{x_3+1}) = h_3.$$

Thus,

$$x_1 = v_{a_1}(k - 1 + ia_1^{x_1+1} a_2^{x_2+1} a_3^{x_3+1}) + v_{a_1}(k + 1 + ia_1^{x_1+1} a_2^{x_2+1} a_3^{x_3+1}),$$
$$x_2 = v_{a_2}(k - 1 + ia_1^{x_1+1} a_2^{x_2+1} a_3^{x_3+1}) + v_{a_2}(k + 1 + ia_1^{x_1+1} a_2^{x_2+1} a_3^{x_3+1}), \quad \forall i \in \mathbb{Z}$$
$$x_3 = v_{a_3}(k - 1 + ia_1^{x_1+1} a_2^{x_2+1} a_3^{x_3+1}) + v_{a_3}(k + 1 + ia_1^{x_1+1} a_2^{x_2+1} a_3^{x_3+1}),$$

∎

If $a_1 = 2$, $a_2 = 3$, $a_3 = 5$, $v_2(k) = 0$, thus $v_2(k+1) \geq 1$, $v_2(k-1) \geq 1$. As one of the numbers for two adjacent even numbers must be divisible by 4, then $x_1 = v_2(k+1) + v_2(k-1) \geq 3$. With $v_3(k) = 0$, one of the three $k-1$, k, $k+1$ is divisible by 3, then $v_3(k-1) + v_3(k) + v_3(k+1) \geq 1$, i.e. $x_2 = v_3(k-1) + v_3(k+1) \geq 1$. Then $x_3 = v_5(k-1) + v_5(k+1) \geq 0 + 0 = 0$.

In the same way, $x_1 \geq 3$, $x_2 \geq 1$, $x_3 \geq 0$ if $a_1 = 2$, $a_2 = 3$, $a_3 = 7$.

6.1   With Base $\{2, 3, 5\}$

Definition 2. *The probability associated with* x_1, x_2, x_3 *is* $\lim\limits_{n \to \infty} \frac{|T(n)|}{n}$ *where* $T(n) = \{k | k \in \mathbb{Z}^+, \ k < n \ \text{and} \ k \ \text{satisfies equation (2)}\ \}$

$$
\begin{aligned}
x_1 &= v_{a_1}(k-1) + v_{a_1}(k+1), \\
x_2 &= v_{a_2}(k-1) + v_{a_2}(k+1), \quad \forall \, i \in \mathbb{Z} \\
x_3 &= v_{a_3}(k-1) + v_{a_3}(k+1),
\end{aligned}
\tag{2}
$$

Lemma 3. *For every* $x_1 \geq 3$, $x_2 \geq 1$, $x_3 > 0$, *there are eight cases, listed in Table 7, for the values of* g_1, g_2, g_3, h_1, h_2, h_3 *(defined in Table 6); each case gives the same probability* $2^{-x_1+1}3^{-x_2}5^{-x_3}$.

Table 7. Relationship between x_i and g_i, h_i, $1 \leqslant i \leqslant 3$, $x_3 \neq 0$

case	g_1	g_2	g_3	h_1	h_2	h_3	probability
1	$x_1 - 1$	x_2	x_3	1	0	0	$2^{-x_1+1}3^{-x_2}5^{-x_3}$
2	1	0	0	$x_1 - 1$	x_2	x_3	$2^{-x_1+1}3^{-x_2}5^{-x_3}$
3	$x_1 - 1$	x_2	0	1	0	x_3	$2^{-x_1+1}3^{-x_2}5^{-x_3}$
4	1	0	x_3	$x_1 - 1$	x_2	0	$2^{-x_1+1}3^{-x_2}5^{-x_3}$
5	$x_1 - 1$	0	x_3	1	x_2	0	$2^{-x_1+1}3^{-x_2}5^{-x_3}$
6	1	x_2	0	$x_1 - 1$	0	x_3	$2^{-x_1+1}3^{-x_2}5^{-x_3}$
7	$x_1 - 1$	0	0	1	x_2	x_3	$2^{-x_1+1}3^{-x_2}5^{-x_3}$
8	1	x_2	x_3	$x_1 - 1$	0	0	$2^{-x_1+1}3^{-x_2}5^{-x_3}$

Lemma 4. *For every* $x_1 \geq 3$, $x_2 \geq 1$, $x_3 = 0$, *there are four cases (listed in Table 8) for the values of* g_1, g_2, g_3, h_1, h_2, h_3. *The probability associated with each case is* $2^{-x_1}3^{-x_2}$.

Table 8. Relationship between x_i and g_i, h_i, $1 \leqslant i \leqslant 3$, $x_3 = 0$

case	g_1	g_2	g_3	h_1	h_2	h_3	probability
1	$x_1 - 1$	x_2	0	1	0	0	$2^{-x_1}3^{-x_2}$
2	1	0	0	$x_1 - 1$	x_2	0	$2^{-x_1}3^{-x_2}$
3	$x_1 - 1$	0	0	1	x_2	0	$2^{-x_1}3^{-x_2}$
4	1	x_2	0	$x_1 - 1$	0	0	$2^{-x_1}3^{-x_2}$

The proofs of lemma 3, 4 are given in appendix A and B respectively.

Theorem 1. *Given base* $\{2,3,5\}$*, the TBC returned by add/sub algorithm exhibit the following features:*

1. *The average density is* 5.61426.
2. *The average decreasing number of powers of* 2, 3, 5 *are* 2.54888, 1.21364, 0.491749 *respectively.*
3. *The average values of the biggest powers of* 2, 3, 5 *in the corresponding chain are approximately equal to* $0.454 \log_2 k$, $0.216 \log_2 k$ *and* $0.0876 \log_2 k$ *respectively.*

Proof. Let K denote the average number of bits eliminated at each step in the add/sub algorithm.

$$
\begin{aligned}
K =&2 \sum_{x_1=3}^{\infty} \sum_{x_2=1}^{\infty} 2^{-x_1} 3^{-x_2}[(x_1-1+x_2\log_2 3+\max\{(x_1-1),1+x_2\log_2 3\}] \\
&+2\sum_{x_1=3}^{\infty}\sum_{x_2=1}^{\infty}\sum_{x_3=1}^{\infty} 2^{-x_1+1}3^{-x_2}5^{-x_3}\,[(x_1-1+x_2\log_2 3+x_3\log_2 5) \\
&\quad+\max\{(x_1-1)+x_2\log_2 3,1+x_3\log_2 5\} \\
&\quad+\max\{(x_1-1)+x_3\log_2 5,1+x_2\log_2 3\} \\
&\quad+\max\{(x_1-1),1+x_2\log_2 3+x_3\log_2 5\}] \\
\approx&2\sum_{x_1=3}^{200}\sum_{x_2=1}^{100} 2^{-x_1}3^{-x_2}[(x_1-1+x_2\log_2 3+\max\{(x_1-1),1+x_2\log_2 3\}] \\
&+2\sum_{x_1=3}^{200}\sum_{x_2=1}^{100}\sum_{x_3=1}^{100} 2^{-x_1+1}3^{-x_2}5^{-x_3}\,[(x_1-1+x_2\log_2 3+x_3\log_2 5) \\
&\quad+\max\{(x_1-1)+x_2\log_2 3,1+x_3\log_2 5\} \\
&\quad+\max\{(x_1-1)+x_3\log_2 5,1+x_2\log_2 3\} \\
&\quad+\max\{(x_1-1),1+x_2\log_2 3+x_3\log_2 5\}] \\
=&2.54888+1.21364\log_2 3+0.491749\log_2 5 \\
=&5.61426.
\end{aligned}
$$

Then the average values of the biggest powers of 2, 3, 5 in the corresponding chain are approximately equal to $\frac{2.54888}{5.61426}\log_2 k = 0.454\log_2 k$, $\frac{1.21364}{5.61426}\log_2 k = 0.216\log_2 k$ and $\frac{0.491749}{5.61426}\log_2 k = 0.0876\log_2 k$ respectively. ∎

6.2 With Base $\{2, 3, 7\}$

Theorem 2. *With base* $\{2, 3, 7\}$*, the probability of* x_1, x_2, x_3 *is* $2^{-x_1+1}3^{-x_2}7^{-x_3}$ *for* $x_3 \neq 0$ *and* $2^{2-x_1}3^{-1-x_2}$ *if* $x_3 = 0$*. The TBC returned by the add/subtraction algorithm with base* $\{2, 3, 7\}$ *has the following features:*

1. *The average density is 5.38543.*
2. *The average decreasing number of powers of 2, 3, 7 are 2.53212, 1.21944, 0.327904 respectively.*
3. *The average values of the biggest powers of 2, 3, 7 in the corresponding chain are approximately equal to $0.4702 \log_2 k$, $0.2264 \log_2 k$ and $0.0609 \log_2 k$ respectively.*

Proof. Let K' denote the average number of bits eliminated at each step of add/sub algorithm using base $\{2, 3, 7\}$.

$$K' \approx 2 \sum_{x_1=3}^{200} \sum_{x_2=1}^{100} 2^{2-x_1} 3^{-1-x_2} \left[(x_1 - 1 + x_2 \log_2 3 + \max\{(x_1 - 1), 1 + x_2 \log_2 3\} \right]$$

$$+ 2 \sum_{x_1=3}^{200} \sum_{x_2=1}^{100} \sum_{x_3=1}^{100} 2^{-x_1+1} 3^{-x_2} 7^{-x_3} \left[(x_1 - 1 + x_2 \log_2 3 + x_3 \log_2 7) \right.$$

$$+ \max\{(x_1 - 1) + x_2 \log_2 3, 1 + x_3 \log_2 7\}$$

$$+ \max\{(x_1 - 1) + x_3 \log_2 7, 1 + x_2 \log_2 3\}$$

$$\left. + \max\{(x_1 - 1), 1 + x_2 \log_2 3 + x_3 \log_2 7\} \right]$$

$$= 2.53212 + 1.21944 \log_2 3 + 0.327904 \log_2 7$$

$$= 5.38543. \qquad \blacksquare$$

7   Comparison

In appendix C, the computation cost associated elliptic curve point operations is given for Weierstrass forms using Jacobian coordinates (Jacobian), Jacobi quartic forms using extended coordinates (JQuartic), and binary fields (Bfield).

Definition 3. *The bit cost is the time cost per bit to perform a scalar multiplication, i.e. the total scalar multiplication cost divided by the length of the scalar.*

If this cost estimate is not a constant number, then the bit cost does not exist. In the algorithms we are analyzing, the bit cost always exists.

In Table 9, method binary/ternary, tree(2,3), mbNAF(2,3,5), Add/Sub(2,3,5), Add/Sub(2,3,7) refer respectively to the binary/ternary method [3], tree-based approach [10], mbNAF with base $\{2, 3, 5\}$, add/sub algorithm with base $\{2, 3, 5\}$, add/sub algorithm with base $\{2, 3, 7\}$. The average density of these five algorithms are $\frac{1}{4.3774}$, $\frac{1}{4.6419}$, $\frac{1}{4.373}$, $\frac{1}{5.61426}$, $\frac{1}{5.38543}$ which are the values in column mA in Table 9. For the greedy algorithms, we mainly focus on the algorithm with base $\{2, 3, 5\}$. In fact, the greedy algorithm with base $\{2, 3, 7\}$ will be slightly slower than that with base $\{2, 3, 5\}$. Greedy(c_1, c_2, c_3) refers to the greedy algorithm for which the initial refinements of $e_{1\max}$, $e_{2\max}$, $e_{3\max}$ in Table 1 are $c_1 \log_2 k$, $c_2 \log_2 k$, $c_3 \log_2 k$.

Table 9. Theoretical time cost of different methods with base {2,3,5} or base {2,3,7} for per bit

Method	D	T	Q	ST	mA	Jacobian	JQuartic	Bfield
binary/ternary	0.4569	0.3427	-	-	0.2284	9.846	8.498	11.99
tree(2,3)	0.5569	0.2795	-	-	0.2154	9.617	8.281	11.92
mbNAF(2,3,5)	0.686	0.1143	0.0572	-	0.2287	9.696	8.301	11.89
Add/Sub(2,3,5)	0.454	0.216	0.0876	-	0.1781	9.433	8.146	11.14
Add/Sub(2,3,7)	0.4702	0.2264	-	0.0609	0.1857	9.622	8.281	-

Table 10. Experimental time cost of different Greedy methods using base {2,3,5} for per bit

Method	D	T	Q	mA	Jacobian	JQuartic	Bfield
Greedy(1,1,1)	0.32826	0.20188	0.15031	0.26228	10.4627	9.019	11.639
Greedy(0.4,0.3,0.25)	0.24947	0.202666	0.183702	0.240612	10.3547	8.94751	11.2475
Greedy(0.3,0.3,0.25)	0.22166	0.25302	0.161307	0.22803	10.2272	8.85374	11.1958
Greedy(0.5,1/3,0.2)	0.3348	0.227156	0.130238	0.209412	9.89441	8.55208	11.1937
Greedy(0.5,0.3,0.2)	0.3518	0.206854	0.136776	0.206456	9.85558	8.51431	11.1473

Because the greedy algorithm is hard to analyze, we ran every greedy algorithm 100,000 times for 160 bits, obtaining the averaged data shown in Table 10. Greedy(0.4,0.3,0.25) and Greedy(0.3,0.3,0.25) were selected by [14]. They did not find better refinements of the greedy algorithm. Greedy(0.5,0.3,0.2) is the fastest among these algorithms with the chosen refinements we have made. In the following, we use specifically Greedy(0.5,0.3,0.2) as representative of greedy algorithms.

7.1 Hamming Weight Comparison

The Hamming weight, listed in Table 11 for different algorithms, is one of the most important factors influencing the speed of scalar multiplications.

From Table 11, the add/sub algorithm with base {2, 3, 5} is seen to have the smallest Hamming weight. The Hamming weight is the addition number

Table 11. Hamming Weight of different methods

Method	Hamming Weight
binary/ternary	$\frac{1}{4.3774} \approx 0.2284$
tree(2,3)	$\frac{1}{4.6419} \approx 0.2154$
mbNAF(2,3,5)	$\frac{1}{4.373} \approx 0.2287$
Add/Sub(2,3,5)	$\frac{1}{5.61426} \approx 0.1781$
Add/Sub(2,3,7)	$\frac{1}{5.38543} \approx 0.1857$
Greedy(0.5,0.3,0.2)	0.206456

in the scalar multiplication. Add/Sub(2,3,5) saves 22% in additions compared with binary/ternary; 22.1% in additions compared to mbNAF(2,3,5); 13.7% in additions compared to Greedy(0.5,0.3,0.2) with base $\{2,3,5\}$; 20.9% compared to tree(2,3); and saves 4.1% in additions compared to Add/Sub(2,3,7) which is also proposed in this paper.

7.2   Bit Cost Comparison

The total cost for scalar multiplications kP is equal to the bit cost multiplied by the length of scalar k. The bit cost comparison is rational, i.e., bit cost reflects the total cost in performing scalar multiplication.

The National Institute of Standards and Technology suggests that elliptic curves over prime fields are suitable for implementation using Jacobian coordinates. Jquartic is the fastest among the different forms for the elliptic curves, with Bfield being the more suitable for hardware implementation. We then analyzed different algorithms for scalar multiplication using these typical forms of the elliptic curves. Add/Sub(2, 3, 5) is the fastest among these algorithms for Jacobian, JQuartic, and Bfield.

In Tables 9 and 10, $D + T \log_2 3 + Q \log_2 5 + ST \log_2 7 \approx 1$. The Hamming weight is very significant in the performance of multi-scalar multiplications. When we used the Jacobian coordinates, the cost of a doubling is only $7M$ which is effective, but a quintupling needs $19.6M > \log_2 5 \times$ the cost of double $= 7 \log_2 5M$. The time cost reduction for scalar multiplication with the multi-base method is slower than that for the Hamming weight. Thus, the Hamming weight is not the only factor that influences scalar multiplication. In single-base methods, such as the binary method and the NAF, the Hamming weight is the only factor influencing the speed of scalar multiplication.

From Tables 9 and 10, the Hamming weight still plays a significant role in the scalar multiplication.

7.3   Recoding Time Comparison

Although our plain search, used in the greedy algorithm at each step, has tremendously sped up the recoding time, the recoding time needed is $\mathbb{O}(\log q)^3$. The recoding time of the add/sub algorithm is $\mathbb{O}(\log q)$ which is very efficient on software implementation.

8   Conclusions

With a focus on TBC, we proposed a new method, called add/sub algorithm, to compute scalar multiplication. According with a comparison of the Hamming weight, bit cost, and recoding time, the add/sub algorithm is very efficient. Also this algorithm is very suitable for software implementation. To our knowledge, the algorithm add/sub algorithm with base $\{2, 3, 5\}$ is fastest among the existing algorithms.

$$k_i + 1 = 2 \cdot 3^{x_2}[n + i \cdot 2^{x_1 - 2}],$$

After a long and interesting analysis, there are total eight items satisfying the conditions $v_5(k_i) = 0$, $v_2(m + i \cdot 3^{x_2}) = 0$, $v_5(m + i \cdot 3^{x_2}) = 0$, $v_3(n + i \cdot 2^{x_1 - 2}) = 0$ and $v_5(n + i \cdot 2^{x_1 - 2}) = 0$. According to Lemma 1, the cardinality of $\mathcal{C} \cap [1, \ 2^{x_1+1}3^{x_2+1}5^1]$ is $2^{x_1+3}3^{x_2}$. Thus the probability for case 3 of Table 8 is $\frac{8}{2^{x_1+3}3^{x_2}} = 2^{-x_1}3^{-x_2}$.

Similar to case 3, the probability for case 4 of Table 8 is $\frac{8}{2^{x_1+3}3^{x_2}} = 2^{-x_1}3^{-x_2}$. ∎

Appendix C: Cost of Elliptic Curve Point Operations

C.1: Elliptic Curve of Weierstrass Form over Prime Fields

We first consider the standard elliptic curves E over a prime field $\mathbb{F}_q$, (denoted $E(\mathbb{F}_q)$), where $q = p^n$, p is a prime, and $p > 3$. These are defined by the Weierstrass equation [18]:

$$y^2 = x^3 + ax + b,$$

where a, $b \in \mathbb{F}_q$ and $\Delta = 4a^3 + 27b^2 \neq 0$.

We set $a = -3$ and choose Jacobian coordinates. The point representation using (x, y) determines the affine coordinates introduced by Jacobian. Another point representation with the form $(X : Y : Z)$ defines the projective coordinates which has very efficient point operations that is inversion-free. Jacobian coordinates are a special case of projective coordinates. where the equivalence class of a Jacobian projective point $(X : Y : Z)$ is

$$(X : Y : Z) = \{(\lambda^2 X, \lambda^3 Y, \lambda Z) : \lambda \in \mathbb{F}_p^*\}$$

Table 12 shows the cost, as summarized by P. Longa and C. Gebotys [19], associated with elliptic curve point operations using Jacobian coordinates with no stored values. For the remaindering, doubling (2P), tripling (3P), quintupling (5P), septupling(7P), addition (P+Q) and mixed addition (P+Q) are denoted D, T, Q, ST, A and mA respectively, where mixed addition means that one of the addends is given in affine coordinates [20]. Explicit formulae for these point operations can be found in [19]. Costs are expressed in terms of field multiplication (M) and field squaring (S). In the prime fields, we make the usual assumption that $1S = 0.8M$ and for purposes of simplification disregard field additions/subtractions and discard multiplications/divisions by small constants.

C.2: Elliptic Curve of Jacobi Quartic Form over Prime Fields

The elliptic curve in the Jacobi quartic form is another form of elliptic curve which is defined by the projective curve

$$Y^2 = X^4 + 2aX^2Z^2 + Z^4,$$

where $a \in \mathbb{F}_q$ and $a^2 \neq 1$. The projective point $(X : Y : Z)$ corresponds to the affine point $(X/Z, Y/Z^2)$.

Acknowledgments. We are very grateful to anonymous reviewers for their helpful comments.

References

1. Dimitrov, V.S., Jullien, G.A., Miller, W.C.: Theory and applications for a double-base number system. In: IEEE Symposium on Computer Arithmetic, pp. 44–53 (1997)
2. Dimitrov, V.S., Jullien, G.A.: Loading the bases: A new number representation with applications. IEEE Circuits and Systems Magazine 3(2), 6–23 (2003)
3. Ciet, M., Joye, M., Lauter, K., Montgomery, P.L.: Trading inversions for multiplications in elliptic curve cryptography. Designs, Codes and Cryptography 39(6), 189–206 (2006)
4. de Weger, B.M.M.: Algorithms for Diophantine equations. CWI Tracts, vol. 65. Centrum voor Wiskunde en Informatica, Amsterdam (1989)
5. Dimitrov, V.S., Imbert, L., Mishra, P.K.: Efficient and Secure Elliptic Curve Point Multiplication Using Double-Base Chains. In: Roy, B. (ed.) ASIACRYPT 2005. LNCS, vol. 3788, pp. 59–78. Springer, Heidelberg (2005)
6. Dimitrov, V.S., Jullien, G.A., Miller, W.C.: An algorithm for modular exponentiation. Information Processing Letters 66(3), 155–159 (1998)
7. Berthé, V., Imbert, L.: On converting numbers to the double-base number system. In: Luk, F.T. (ed.) Proceedings of SPIE Advanced Signal Processing Algorithms, Architecture and Implementations XIV, vol. 5559, pp. 70–78 (2004)
8. Yu, W., Wang, K., Li, B.: Fast Algorithm Converting integer to Double Base Chain. In: Information Security and Cryptology, Inscrypt 2010, pp. 44–54 (2011)
9. Doche, C., Imbert, L.: Extended Double-Base Number System with Applications to Elliptic Curve Cryptography. In: Barua, R., Lange, T. (eds.) INDOCRYPT 2006. LNCS, vol. 4329, pp. 335–348. Springer, Heidelberg (2006)
10. Doche, C., Habsieger, L.: A Tree-Based Approach for Computing Double-Base Chains. In: Mu, Y., Susilo, W., Seberry, J. (eds.) ACISP 2008. LNCS, vol. 5107, pp. 433–446. Springer, Heidelberg (2008)
11. Méloni, N., Hasan, M.A.: Elliptic Curve Scalar Multiplication Combining Yao's Algorithm and Double Bases. In: Clavier, C., Gaj, K. (eds.) CHES 2009. LNCS, vol. 5747, pp. 304–316. Springer, Heidelberg (2009)
12. Suppakitpaisarn, V., Edahiro, M., Imai, H.: Fast Elliptic Curve Cryptography Using Optimal Double-Base Chains. eprint.iacr.org/2011/030.ps
13. Dimitrov, V., Imbert, L., Mishra, P.K.: The double-base number system and its application to elliptic curve cryptography. Mathematics of Computation 77, 1075–1104 (2008)
14. Mishra, P.K., Dimitrov, V.S.: Efficient Quintuple Formulas for Elliptic Curves and Efficient Scalar Multiplication Using Multibase Number Representation. In: Garay, J.A., Lenstra, A.K., Mambo, M., Peralta, R. (eds.) ISC 2007. LNCS, vol. 4779, pp. 390–406. Springer, Heidelberg (2007)
15. Longa, P., Miri, A.: New Multibase Non-Adjacent Form Scalar Multiplication and its Application to Elliptic Curve Cryptosystems. Cryptology ePrint Archive, Report 2008/052 (2008)
16. Purohit, G.N., Rawat, A.S.: Fast Scalar Multiplication in ECC Using The Multi base Number System, http://eprint.iacr.org/2011/044.pdf

17. Li, M., Miri, A., Zhu, D.: Analysis of the Hamming Weight of the Extended wmb-NAF, http://eprint.iacr.org/2011/569.pdf

18. Hankerson, D., Menezes, A., Vanstone, S.: Guide to Elliptic Curve Cryptography. Springer-Verlag (2004)

19. Longa, P., Gebotys, C.: Fast multibase methods and other several optimizations for elliptic curve scalar multiplication. In: PKC: Proceedings of Public Key Cryptography, pp. 443–462 (2009)

20. Cohen, H., Miyaji, A., Ono, T.: Efficient elliptic curve exponentiation using mixed coordinates. In: Ohta, K., Pei, D. (eds.) ASIACRYPT 1998. LNCS, vol. 1514, pp. 51–65. Springer, Heidelberg (1998)

21. Hisil, H., Wong, K., Carter, G., Dawson, E.: Faster Group Operations on Elliptic Curves. In: Proceedings of the 7th Australasian Information Security Conference on Information Security 2009, pp. 11–19. Springer, Heidelberg (2009)

22. Hisil, H., Wong, K., Carter, G., Dawson, E.: An Intersection Form for Jacobi-Quartic Curves. Personal communication (2008)

Appendix A: Proof of Lemma 3

Proof. In the add/sub algorithm, one of the pair g_1, h_1 is 1 whereas the other is greater than 1; one of the pair g_2, h_2 is 0 and one of the pair g_3, h_3 is 0. The lowest common multiple (lcm) of $2^{g_1}3^{g_2}5^{g_3}$ and $2^{h_1}3^{h_2}5^{h_3}$ is $2^{x_1-1}3^{x_2}5^{x_3}$.

In the interval $[1, 2^{x_1+1}3^{x_2+1}5^{x_3+1}]$, if k, k' in the case g_1, g_2, g_3, h_1, h_2, h_3, then k, k' satisfy $k-1 = 2^{g_1}3^{g_2}5^{g_3} \times y$, $k+1 = 2^{h_1}3^{h_2}5^{h_3} \times z$, $k'-1 = 2^{g_1}3^{g_2}5^{g_3} \times y'$, $k'+1 = 2^{h_1}3^{h_2}5^{h_3} \times z'$, where y, z, y', $z' \in \mathbb{Z}$. Then $k'-k_0 = 2^{g_1}3^{g_2}5^{g_3} \times (y'-y) = 2^{h_1}3^{h_2}5^{h_3} \times (z-z')$ where k_0 satisfies $k_0-1 = 2^{g_1}3^{g_2}5^{g_3} \times y_0$, $k_0+1 = 2^{h_1}3^{h_2}5^{h_3} \times z_0$, y_0, $z_0 \in \mathbb{Z}$, $0 \leqslant k_0 < 2^{x_1-1}3^{x_2}5^{x_3}$.

Thus $k'-k_0$ is exactly divisible by $\mathrm{lcm}(2^{g_1}3^{g_2}5^{g_3}, 2^{h_1}3^{h_2}5^{h_3})$, $k_i = k_0 + i \cdot \mathrm{lcm}(2^{g_1}3^{g_2}5^{g_3}, 2^{h_1}3^{h_2}5^{h_3}) = k_0 + i \cdot 2^{x_1-1}3^{x_2}5^{x_3}$ satisfy $k_i-1 = 2^{g_1}3^{g_2}5^{g_3} \times (y + i \cdot 2^{x_1-g_1-1}3^{x_2-g_2}5^{x_3-g_3})$, $k_i+1 = 2^{h_1}3^{h_2}5^{h_3} \times (z + i \cdot 2^{x_1-h_1-1}3^{x_2-h_2}5^{x_3-h_3})$, where $0 \leqslant i < 60$.

1. $g_1 = x_1 - 1$, $g_2 = x_2$, $g_3 = x_3$; $h_1 = 1$, $h_2 = 0$, $h_3 = 0$. $k_0 = 1$, $k_i = i \cdot 2^{x_1-1}3^{x_2}5^{x_3} + 1$, $0 \leqslant i < 60$, there exist $60 - \frac{60}{2} - \frac{60}{3} - \frac{60}{5} + \frac{60}{2\times3} + \frac{60}{2\times5} + \frac{60}{3\times5} - \frac{60}{2\times3\times5} = 16$ numbers i coprime to 30.
   According to Lemma 1, the cardinality of $\mathcal{C} \cap [1, 2^{x_1+1}3^{x_2+1}5^{x_3+1}]$ is $2^{x_1+3}3^{x_2}5^{x_3}$. Thus the probability of case 1 in Table 7 is $\frac{16}{2^{x_1+3}3^{x_2}5^{x_3}} = 2^{-x_1+1}3^{-x_2}5^{-x_3}$.

2. $g_1 = 1$, $g_2 = 0$, $g_3 = 0$; $h_1 = x_1-1$, $h_2 = x_2$, $h_3 = x_3$. $k_0 = 2^{x_1-1}3^{x_2}5^{x_3}-1$, $k_i = (i+1) \cdot 2^{x_1-1}3^{x_2}5^{x_3} - 1$, $0 \leqslant i < 60$, there are then $60 - \frac{60}{2} - \frac{60}{3} - \frac{60}{5} + \frac{60}{2\times3} + \frac{60}{2\times5} + \frac{60}{3\times5} - \frac{60}{2\times3\times5} = 16$ numbers $i+1$ coprime to 30.
   According to Lemma 1, the cardinality of $\mathcal{C} \cap [1, 2^{x_1+1}3^{x_2+1}5^{x_3+1}]$ is $2^{x_1+3}3^{x_2}5^{x_3}$. Thus the probability of case 2 in Table 7 is $\frac{16}{2^{x_1+3}3^{x_2}5^{x_3}} = 2^{-x_1+1}3^{-x_2}5^{-x_3}$.

3. In the same way as items 1 and 2, the probability for cases 3 through 6 in Table 7 is $\frac{16}{2^{x_1+3}3^{x_2}5^{x_3}} = 2^{-x_1+1}3^{-x_2}5^{-x_3}$.

4. $g_1 = x_1 - 1$, $g_2 = 0$, $g_3 = 0$, $h_1 = 1$, $h_2 = x_2$, $h_3 = x_3$. $k_0 = m \cdot 2^{x_1} \cdot n \cdot 2 \cdot 3^{x_2} \cdot 5^{x_3} - 1$. $k_i = k_0 + i \cdot 2^{x_1-1}3^{x_2}5^{x_3}$ where $k_0 < 2^{x_1-1}3^{x_2}5^{x_3}$
   $k_i - 1 = 2^{x_1-1}(m + i \cdot 3^{x_2}5^{x_3})$ requires that i is even.
   $k_i + 1 = 2 \cdot 3^{x_2} \cdot 5^{x_3}(n + i \cdot 2^{x_1-2})$ requires that $v_3(n + i \cdot 2^{x_1-2})$
   $v_5(n + i \cdot 2^{x_1-2}) = 0$.
   Then there exist $60 \times \frac{1}{2} \times \frac{2}{3} \times \frac{4}{5} = 16$ items.
   According to Lemma 1, the cardinality of $\mathcal{C} \cap [1, 2^{x_1+1}3^{x_2+}$ $2^{x_1+3}3^{x_2}5^{x_3}$. Thus the probability of case 7 in Table 7 is $\frac{}{2^{x_1+}}$ $2^{-x_1+1}3^{-x_2}5^{-x_3}$.

5. Similar to the previous item, there are $60 \times \frac{1}{2} \times \frac{2}{3} \times \frac{4}{5} = 16$ items
   According to Lemma 1, the cardinality of $\mathcal{C} \cap [1, 2^{x_1+1}3^{x_2+}$ $2^{x_1+3}3^{x_2}5^{x_3}$. Thus the probability of case 8 in Table 7 is $\frac{}{2^{x_1+}}$ $2^{-x_1+1}3^{-x_2}5^{-x_3}$.

Appendix B: Proof of Lemma 4

Proof. In the interval $[1, 2^{x_1+1}3^{x_2+1}5^{x_3+1}]$, if k, k' in the case g_1, g_2 h_2, h_3, then k, k' satisfy $k-1 = 2^{g_1}3^{g_2}5^{g_3} \times y$, $k+1 = 2^{h_1}3^{h_2}5^{h_3} \times z$ $2^{g_1}3^{g_2}5^{g_3} \times y'$, $k'+1 = 2^{h_1}3^{h_2}5^{h_3} \times z'$, where y, z, y', $z' \in \mathbb{Z}$. Then $2^{g_1}3^{g_2}5^{g_3} \times (y'-y) = 2^{h_1}3^{h_2}5^{h_3} \times (z-z')$ where k_0 satisfies $k_0-1 = 2^{g_1}3$ $k_0+1 = 2^{h_1}3^{h_2}5^{h_3} \times z_0$, y_0, $z_0 \in \mathbb{Z}$, $0 \leqslant k_0 < 2^{x_1-1}3^{x_2}$.

Then $k'-k_0$ is exactly divisible by $\mathrm{lcm}(2^{g_1}3^{g_2}5^{g_3}, 2^{h_1}3^{h_2}5^{h_3})$, k_i $\mathrm{lcm}(2^{g_1}3^{g_2}5^{g_3}, 2^{h_1}3^{h_2}5^{h_3}) = k_0 + i \cdot 2^{x_1-1}3^{x_2}5^{x_3}$ satisfy $k_i-1 = 2^{g_1}3^{g_2}$ $i \cdot 2^{x_1-g_1-1}3^{x_2-g_2}5^{x_3-g_3})$, $k_i+1 = 2^{h_1}3^{h_2}5^{h_3} \times (z + i \cdot 2^{x_1-h_1-1}3^{x_2-}$ where $0 \leqslant i < 60$.

1. $k_0 = 1$, $k_i = i \cdot 2^{x_1-1}3^{x_2}+1$, $\gcd(i, 30) = 1$, $0 \leqslant i < 60$, there are 16 satisfy $\gcd(i, 30) = 1$ such as 1, 7, 11, 13, 17, 19, 23, 29, 31, 37, 43, 47, 49, 53, 59. Among these k_i, there are 4 items which are not prime to 5. $k_i + 1 = i \cdot 2^{x_1-1}3^{x_2} + 2 = 2(i \cdot 2^{x_1-2}3^{x_2} + 1)$ have an items which is not relatively prime to 5. There are $16 - 4 - 4 =$ satisfying $g_1 = x_1 - 1$, $g_2 = x_2$, $g_3 = 0$, $h_1 = 1$, $h_2 = 0$, $h_3 = 0$. to Lemma 1, the cardinality of $\mathcal{C} \cap [1, 2^{x_1+1}3^{x_2+1}5^1]$ is $2^{x_1+3}3^{x_2}$. probability for case 1 of Table 8 is $\frac{8}{2^{x_1+3}3^{x_2}} = 2^{-x_1}3^{-x_2}$.

2. $k_0 = 2^{x_1-1}3^{x_2} - 1$, $k_i = (i+1) \cdot 2^{x_1-1}3^{x_2} - 1$, $\gcd(i+1, 30) = 1$, 0 there are 16 numbers $i+1$ such as 1, 7, 11, 13, 17, 19, 23, 29, 31 43, 47, 49, 53, 59. k_i has four items which are not relatively pr $k_i - 1 = (i+1) \cdot 2^{x_1-1}3^{x_2} - 2 = 2((i+1) \cdot 2^{x_1-2}3^{x_2} - 1)$ have and items which are not relatively prime to 5. There are eight items $g_1 = x_1 - 1$, $g_2 = x_2$, $g_3 = 0$, $h_1 = 1$, $h_2 = 0$, $h_3 = 0$, According t 1, the cardinality of $\mathcal{C} \cap [1, 2^{x_1+1}3^{x_2+1}5^1]$ is $2^{x_1+3}3^{x_2}$. Thus the pr for case 2 of Table 8 is $\frac{8}{2^{x_1+3}3^{x_2}} = 2^{-x_1}3^{-x_2}$.

3. $g_1 = x_1 - 1$, $g_2 = 0$, $g_3 = 0$, $h_1 = 1$, $h_2 = 0$, $h_3 = 0$; $k_0 = 2^{x_1} \cdot 2 \cdot 3^{x_2} \cdot n - 1$. $0 \leqslant m < 3^{x_2}$, $0 < j \leqslant 2^{x_1-2}$, $k_0 < 2^{x_1-1}3^{x_2}$.
   $k_i = k_0 + i \cdot 2^{x_1-1}3^{x_2}$,
   $k_i - 1 = 2^{x_1-1}[m + i \cdot 3^{x_2}]$,

Table 12. Cost of Jacobian operations over prime fields

computation	cost	M
A	11M+5S	15
mA	7M+4S	10.2
D	3M+5S	7
T	7M+7S	12.6
Q	10M+12S	19.6
ST	14M+15S	26

Table 13 shows a similar cost assessment for point operations on the Jacobi quartic curve with no stored values using an extended coordinate system of the form $(X : Y : Z : X^2 : Z^2)$ called a JQuartic. We refer to [21,22] for the explicit formulae for JQuartics.

Table 13. Cost of JQuartic over prime fields

computation	cost	M
A	7M+4S	10.2
mA	6M+3S	8.4
D	2M+5S	6
T	8M+4S	11.2
Q	14M+4S	17.2
ST	16M+8S	22.4

C.3: Cost of Elliptic Curve Point Operations over Binary Fields

G. N. Purohit and A. S. Rawat [16] have given formulae and cost associated with elliptic curve point operations over the binary field, denoted Bfield (see Table 14). Nevertheless, the ST formulae do not speed up, because we can calculate $7P = 5P + 2P$ that needs $1A+1D+1Q=3[i]+7[s]+17[m]$ which is faster than the ST formulae in [16]. As the new formulae are not faster, we do not analysis the add/sub algorithm with bases containing 7 in binary fields. Generally, $I = 3 \sim 10M, S = 0M$ in Bfields. For this study, we assume $I = 8M$ which is the same as [14].

Table 14. Cost of elliptic curve point operations over binary fields

computation	cost	M
A	1[i]+1[s]+2[m]	10
D	1[i]+1[s]+2[m]	10
T	1[i]+4[s]+7[m]	15
Q	1[i]+5[s]+13[m]	18
ST	3[i]+7[s]+18[m]	42

Author Index

Acknowledgments. We are very grateful to anonymous reviewers for their helpful comments.

References

1. Dimitrov, V.S., Jullien, G.A., Miller, W.C.: Theory and applications for a double-base number system. In: IEEE Symposium on Computer Arithmetic, pp. 44–53 (1997)
2. Dimitrov, V.S., Jullien, G.A.: Loading the bases: A new number representation with applications. IEEE Circuits and Systems Magazine 3(2), 6–23 (2003)
3. Ciet, M., Joye, M., Lauter, K., Montgomery, P.L.: Trading inversions for multiplications in elliptic curve cryptography. Designs, Codes and Cryptography 39(6), 189–206 (2006)
4. de Weger, B.M.M.: Algorithms for Diophantine equations. CWI Tracts, vol. 65. Centrum voor Wiskunde en Informatica, Amsterdam (1989)
5. Dimitrov, V.S., Imbert, L., Mishra, P.K.: Efficient and Secure Elliptic Curve Point Multiplication Using Double-Base Chains. In: Roy, B. (ed.) ASIACRYPT 2005. LNCS, vol. 3788, pp. 59–78. Springer, Heidelberg (2005)
6. Dimitrov, V.S., Jullien, G.A., Miller, W.C.: An algorithm for modular exponentiation. Information Processing Letters 66(3), 155–159 (1998)
7. Berthé, V., Imbert, L.: On converting numbers to the double-base number system. In: Luk, F.T. (ed.) Proceedings of SPIE Advanced Signal Processing Algorithms, Architecture and Implementations XIV, vol. 5559, pp. 70–78 (2004)
8. Yu, W., Wang, K., Li, B.: Fast Algorithm Converting integer to Double Base Chain. In: Information Security and Cryptology, Inscrypt 2010, pp. 44–54 (2011)
9. Doche, C., Imbert, L.: Extended Double-Base Number System with Applications to Elliptic Curve Cryptography. In: Barua, R., Lange, T. (eds.) INDOCRYPT 2006. LNCS, vol. 4329, pp. 335–348. Springer, Heidelberg (2006)
10. Doche, C., Habsieger, L.: A Tree-Based Approach for Computing Double-Base Chains. In: Mu, Y., Susilo, W., Seberry, J. (eds.) ACISP 2008. LNCS, vol. 5107, pp. 433–446. Springer, Heidelberg (2008)
11. Méloni, N., Hasan, M.A.: Elliptic Curve Scalar Multiplication Combining Yao's Algorithm and Double Bases. In: Clavier, C., Gaj, K. (eds.) CHES 2009. LNCS, vol. 5747, pp. 304–316. Springer, Heidelberg (2009)
12. Suppakitpaisarn, V., Edahiro, M., Imai, H.: Fast Elliptic Curve Cryptography Using Optimal Double-Base Chains. eprint.iacr.org/2011/030.ps
13. Dimitrov, V., Imbert, L., Mishra, P.K.: The double-base number system and its application to elliptic curve cryptography. Mathematics of Computation 77, 1075–1104 (2008)
14. Mishra, P.K., Dimitrov, V.S.: Efficient Quintuple Formulas for Elliptic Curves and Efficient Scalar Multiplication Using Multibase Number Representation. In: Garay, J.A., Lenstra, A.K., Mambo, M., Peralta, R. (eds.) ISC 2007. LNCS, vol. 4779, pp. 390–406. Springer, Heidelberg (2007)
15. Longa, P., Miri, A.: New Multibase Non-Adjacent Form Scalar Multiplication and its Application to Elliptic Curve Cryptosystems. Cryptology ePrint Archive, Report 2008/052 (2008)
16. Purohit, G.N., Rawat, A.S.: Fast Scalar Multiplication in ECC Using The Multi base Number System, http://eprint.iacr.org/2011/044.pdf

17. Li, M., Miri, A., Zhu, D.: Analysis of the Hamming Weight of the Extended wmb-NAF, http://eprint.iacr.org/2011/569.pdf
18. Hankerson, D., Menezes, A., Vanstone, S.: Guide to Elliptic Curve Cryptography. Springer-Verlag (2004)
19. Longa, P., Gebotys, C.: Fast multibase methods and other several optimizations for elliptic curve scalar multiplication. In: PKC: Proceedings of Public Key Cryptography, pp. 443–462 (2009)
20. Cohen, H., Miyaji, A., Ono, T.: Efficient elliptic curve exponentiation using mixed coordinates. In: Ohta, K., Pei, D. (eds.) ASIACRYPT 1998. LNCS, vol. 1514, pp. 51–65. Springer, Heidelberg (1998)
21. Hisil, H., Wong, K., Carter, G., Dawson, E.: Faster Group Operations on Elliptic Curves. In: Proceedings of the 7th Australasian Information Security Conference on Information Security 2009, pp. 11–19. Springer, Heidelberg (2009)
22. Hisil, H., Wong, K., Carter, G., Dawson, E.: An Intersection Form for Jacobi-Quartic Curves. Personal communication (2008)

Appendix A: Proof of Lemma 3

Proof. In the add/sub algorithm, one of the pair g_1, h_1 is 1 whereas the other is greater than 1; one of the pair g_2, h_2 is 0 and one of the pair g_3, h_3 is 0. The lowest common multiple (lcm) of $2^{g_1}3^{g_2}5^{g_3}$ and $2^{h_1}3^{h_2}5^{h_3}$ is $2^{x_1-1}3^{x_2}5^{x_3}$.

In the interval $[1, 2^{x_1+1}3^{x_2+1}5^{x_3+1}]$, if k, k' in the case g_1, g_2, g_3, h_1, h_2, h_3, then k, k' satisfy $k - 1 = 2^{g_1}3^{g_2}5^{g_3} \times y$, $k + 1 = 2^{h_1}3^{h_2}5^{h_3} \times z$, $k' - 1 = 2^{g_1}3^{g_2}5^{g_3} \times y'$, $k' + 1 = 2^{h_1}3^{h_2}5^{h_3} \times z'$, where y, z, y', $z' \in \mathbb{Z}$. Then $k' - k_0 = 2^{g_1}3^{g_2}5^{g_3} \times (y'-y) = 2^{h_1}3^{h_2}5^{h_3} \times (z-z')$ where k_0 satisfies $k_0-1 = 2^{g_1}3^{g_2}5^{g_3} \times y_0$, $k_0 + 1 = 2^{h_1}3^{h_2}5^{h_3} \times z_0$, y_0, $z_0 \in \mathbb{Z}$, $0 \leqslant k_0 < 2^{x_1-1}3^{x_2}5^{x_3}$.

Thus $k' - k_0$ is exactly divisible by $\mathrm{lcm}(2^{g_1}3^{g_2}5^{g_3}, 2^{h_1}3^{h_2}5^{h_3})$, $k_i = k_0 + i \cdot \mathrm{lcm}(2^{g_1}3^{g_2}5^{g_3}, 2^{h_1}3^{h_2}5^{h_3}) = k_0 + i \cdot 2^{x_1-1}3^{x_2}5^{x_3}$ satisfy $k_i - 1 = 2^{g_1}3^{g_2}5^{g_3} \times (y + i \cdot 2^{x_1-g_1-1}3^{x_2-g_2}5^{x_3-g_3})$, $k_i + 1 = 2^{h_1}3^{h_2}5^{h_3} \times (z + i \cdot 2^{x_1-h_1-1}3^{x_2-h_2}5^{x_3-h_3})$, where $0 \leqslant i < 60$.

1. $g_1 = x_1 - 1$, $g_2 = x_2$, $g_3 = x_3$; $h_1 = 1$, $h_2 = 0$, $h_3 = 0$. $k_0 = 1$, $k_i = i \cdot 2^{x_1-1}3^{x_2}5^{x_3} + 1$, $0 \leqslant i < 60$, there exist $60 - \frac{60}{2} - \frac{60}{3} - \frac{60}{5} + \frac{60}{2\times3} + \frac{60}{2\times5} + \frac{60}{3\times5} - \frac{60}{2\times3\times5} = 16$ numbers i coprime to 30.
   According to Lemma 1, the cardinality of $\mathcal{C} \cap [1, \ 2^{x_1+1}3^{x_2+1}5^{x_3+1}]$ is $2^{x_1+3}3^{x_2}5^{x_3}$. Thus the probability of case 1 in Table 7 is $\frac{16}{2^{x_1+3}3^{x_2}5^{x_3}} = 2^{-x_1+1}3^{-x_2}5^{-x_3}$.

2. $g_1 = 1$, $g_2 = 0$, $g_3 = 0$; $h_1 = x_1-1$, $h_2 = x_2$, $h_3 = x_3$. $k_0 = 2^{x_1-1}3^{x_2}5^{x_3}-1$, $k_i = (i+1) \cdot 2^{x_1-1}3^{x_2}5^{x_3} - 1$, $0 \leqslant i < 60$, there are then $60 - \frac{60}{2} - \frac{60}{3} - \frac{60}{5} + \frac{60}{2\times3} + \frac{60}{2\times5} + \frac{60}{3\times5} - \frac{60}{2\times3\times5} = 16$ numbers $i + 1$ coprime to 30.
   According to Lemma 1, the cardinality of $\mathcal{C} \cap [1, \ 2^{x_1+1}3^{x_2+1}5^{x_3+1}]$ is $2^{x_1+3}3^{x_2}5^{x_3}$. Thus the probability of case 2 in Table 7 is $\frac{16}{2^{x_1+3}3^{x_2}5^{x_3}} = 2^{-x_1+1}3^{-x_2}5^{-x_3}$.

3. In the same way as items 1 and 2, the probability for cases 3 through 6 in Table 7 is $\frac{16}{2^{x_1+3}3^{x_2}5^{x_3}} = 2^{-x_1+1}3^{-x_2}5^{-x_3}$.

4. $g_1 = x_1 - 1$, $g_2 = 0$, $g_3 = 0$, $h_1 = 1$, $h_2 = x_2$, $h_3 = x_3$. $k_0 = m \cdot 2^{x_1-1} + 1 = n \cdot 2 \cdot 3^{x_2} \cdot 5^{x_3} - 1$. $k_i = k_0 + i \cdot 2^{x_1-1}3^{x_2}5^{x_3}$ where $k_0 < 2^{x_1-1}3^{x_2}5^{x_3}$.
   $k_i - 1 = 2^{x_1-1}(m + i \cdot 3^{x_2}5^{x_3})$ requires that i is even.
   $k_i + 1 = 2 \cdot 3^{x_2} \cdot 5^{x_3}(n + i \cdot 2^{x_1-2})$ requires that $v_3(n + i \cdot 2^{x_1-2}) = 0$ and $v_5(n + i \cdot 2^{x_1-2}) = 0$.
   Then there exist $60 \times \frac{1}{2} \times \frac{2}{3} \times \frac{4}{5} = 16$ items.
   According to Lemma 1, the cardinality of $\mathcal{C} \cap [1, \; 2^{x_1+1}3^{x_2+1}5^{x_3+1}]$ is $2^{x_1+3}3^{x_2}5^{x_3}$. Thus the probability of case 7 in Table 7 is $\frac{16}{2^{x_1+3}3^{x_2}5^{x_3}} = 2^{-x_1+1}3^{-x_2}5^{-x_3}$.

5. Similar to the previous item, there are $60 \times \frac{1}{2} \times \frac{2}{3} \times \frac{4}{5} = 16$ items.
   According to Lemma 1, the cardinality of $\mathcal{C} \cap [1, \; 2^{x_1+1}3^{x_2+1}5^{x_3+1}]$ is $2^{x_1+3}3^{x_2}5^{x_3}$. Thus the probability of case 8 in Table 7 is $\frac{16}{2^{x_1+3}3^{x_2}5^{x_3}} = 2^{-x_1+1}3^{-x_2}5^{-x_3}$. ∎

Appendix B: Proof of Lemma 4

Proof. In the interval $[1, \; 2^{x_1+1}3^{x_2+1}5^{x_3+1}]$, if k, k' in the case g_1, g_2, g_3, h_1, h_2, h_3, then k, k' satisfy $k - 1 = 2^{g_1}3^{g_2}5^{g_3} \times y$, $k + 1 = 2^{h_1}3^{h_2}5^{h_3} \times z$, $k' - 1 = 2^{g_1}3^{g_2}5^{g_3} \times y'$, $k' + 1 = 2^{h_1}3^{h_2}5^{h_3} \times z'$, where y, z, y', $z' \in \mathbb{Z}$. Then $k' - k_0 = 2^{g_1}3^{g_2}5^{g_3} \times (y' - y) = 2^{h_1}3^{h_2}5^{h_3} \times (z - z')$ where k_0 satisfies $k_0 - 1 = 2^{g_1}3^{g_2}5^{g_3} \times y_0$, $k_0 + 1 = 2^{h_1}3^{h_2}5^{h_3} \times z_0$, y_0, $z_0 \in \mathbb{Z}$, $0 \leqslant k_0 < 2^{x_1-1}3^{x_2}$.

Then $k' - k_0$ is exactly divisible by $\operatorname{lcm}(2^{g_1}3^{g_2}5^{g_3}, 2^{h_1}3^{h_2}5^{h_3})$, $k_i = k_0 + i \cdot \operatorname{lcm}(2^{g_1}3^{g_2}5^{g_3}, 2^{h_1}3^{h_2}5^{h_3}) = k_0 + i \cdot 2^{x_1-1}3^{x_2}5^{x_3}$ satisfy $k_i - 1 = 2^{g_1}3^{g_2}5^{g_3} \times (y + i \cdot 2^{x_1-g_1-1}3^{x_2-g_2}5^{x_3-g_3})$, $k_i + 1 = 2^{h_1}3^{h_2}5^{h_3} \times (z + i \cdot 2^{x_1-h_1-1}3^{x_2-h_2}5^{x_3-h_3})$, where $0 \leqslant i < 60$.

1. $k_0 = 1$, $k_i = i \cdot 2^{x_1-1}3^{x_2} + 1$, $\gcd(i, 30) = 1$, $0 \leqslant i < 60$, there are 16 numbers satisfy $\gcd(i, 30) = 1$ such as 1, 7, 11, 13, 17, 19, 23, 29, 31, 37, 41, 43, 47, 49, 53, 59. Among these k_i, there are 4 items which are not relatively prime to 5. $k_i + 1 = i \cdot 2^{x_1-1}3^{x_2} + 2 = 2(i \cdot 2^{x_1-2}3^{x_2} + 1)$ have another four items which is not relatively prime to 5. There are $16 - 4 - 4 = 8$ items satisfying $g_1 = x_1 - 1$, $g_2 = x_2$, $g_3 = 0$, $h_1 = 1$, $h_2 = 0$, $h_3 = 0$. According to Lemma 1, the cardinality of $\mathcal{C} \cap [1, \; 2^{x_1+1}3^{x_2+1}5^1]$ is $2^{x_1+3}3^{x_2}$. Thus the probability for case 1 of Table 8 is $\frac{8}{2^{x_1+3}3^{x_2}} = 2^{-x_1}3^{-x_2}$.

2. $k_0 = 2^{x_1-1}3^{x_2} - 1$, $k_i = (i+1) \cdot 2^{x_1-1}3^{x_2} - 1$, $\gcd(i+1, 30) = 1$, $0 \leqslant i < 60$, there are 16 numbers $i + 1$ such as 1, 7, 11, 13, 17, 19, 23, 29, 31, 37, 41, 43, 47, 49, 53, 59. k_i has four items which are not relatively prime to 5. $k_i - 1 = (i+1) \cdot 2^{x_1-1}3^{x_2} - 2 = 2((i+1) \cdot 2^{x_1-2}3^{x_2} - 1)$ have another four items which are not relatively prime to 5. There are eight items satisfying $g_1 = x_1 - 1$, $g_2 = x_2$, $g_3 = 0$, $h_1 = 1$, $h_2 = 0$, $h_3 = 0$, According to Lemma 1, the cardinality of $\mathcal{C} \cap [1, \; 2^{x_1+1}3^{x_2+1}5^1]$ is $2^{x_1+3}3^{x_2}$. Thus the probability for case 2 of Table 8 is $\frac{8}{2^{x_1+3}3^{x_2}} = 2^{-x_1}3^{-x_2}$.

3. $g_1 = x_1 - 1$, $g_2 = 0$, $g_3 = 0$, $h_1 = 1$, $h_2 = 0$, $h_3 = 0$; $k_0 = 2^{x_1} \cdot m + 1 = 2 \cdot 3^{x_2} \cdot n - 1$. $0 \leqslant m < 3^{x_2}$, $0 < j \leqslant 2^{x_1-2}$, $k_0 < 2^{x_1-1}3^{x_2}$.
   $k_i = k_0 + i \cdot 2^{x_1-1}3^{x_2}$,
   $k_i - 1 = 2^{x_1-1}[m + i \cdot 3^{x_2}]$,

$k_i + 1 = 2 \cdot 3^{x_2}[n + i \cdot 2^{x_1-2}]$,

After a long and interesting analysis, there are total eight items satisfying the conditions $v_5(k_i) = 0$, $v_2(m + i \cdot 3^{x_2}) = 0$, $v_5(m + i \cdot 3^{x_2}) = 0$, $v_3(n + i \cdot 2^{x_1-2}) = 0$ and $v_5(n + i \cdot 2^{x_1-2}) = 0$. According to Lemma 1, the cardinality of $\mathcal{C} \cap [1, 2^{x_1+1}3^{x_2+1}5^1]$ is $2^{x_1+3}3^{x_2}$. Thus the probability for case 3 of Table 8 is $\frac{8}{2^{x_1+3}3^{x_2}} = 2^{-x_1}3^{-x_2}$.

4. Similar to case 3, the probability for case 4 of Table 8 is $\frac{8}{2^{x_1+3}3^{x_2}} = 2^{-x_1}3^{-x_2}$.   ∎

Appendix C: Cost of Elliptic Curve Point Operations

C.1: Elliptic Curve of Weierstrass Form over Prime Fields

We first consider the standard elliptic curves E over a prime field $\mathbb{F}_q$, (denoted by $E(\mathbb{F}_q)$), where $q = p^n$, p is a prime, and $p > 3$. These are defined by the Weierstrass equation [18]:

$$y^2 = x^3 + ax + b,$$

where a, $b \in \mathbb{F}_q$ and $\Delta = 4a^3 + 27b^2 \neq 0$.

We set $a = -3$ and choose Jacobian coordinates. The point representation using (x, y) determines the affine coordinates introduced by Jacobian. Another point representation with the form $(X : Y : Z)$ defines the projective coordinates which has very efficient point operations that is inversion-free. Jacobian coordinates are a special case of projective coordinates. where the equivalence class of a Jacobian projective point $(X : Y : Z)$ is

$$(X : Y : Z) = \{(\lambda^2 X, \lambda^3 Y, \lambda Z) : \lambda \in \mathbb{F}_p^*\}$$

Table 12 shows the cost, as summarized by P. Longa and C. Gebotys [19], associated with elliptic curve point operations using Jacobian coordinates with no stored values. For the remaindering, doubling (2P), tripling (3P), quintupling (5P), septupling(7P), addition (P+Q) and mixed addition (P+Q) are denoted by D, T, Q, ST, A and mA respectively, where mixed addition means that one of the addends is given in affine coordinates [20]. Explicit formulae for these point operations can be found in [19]. Costs are expressed in terms of field multiplication (M) and field squaring (S). In the prime fields, we make the usual assumption that $1S = 0.8M$ and for purposes of simplification disregard field additions/subtractions and discard multiplications/divisions by small constants.

C.2: Elliptic Curve of Jacobi Quartic Form over Prime Fields

The elliptic curve in the Jacobi quartic form is another form of elliptic curve which is defined by the projective curve

$$Y^2 = X^4 + 2aX^2Z^2 + Z^4,$$

where $a \in \mathbb{F}_q$ and $a^2 \neq 1$. The projective point $(X : Y : Z)$ corresponds to the affine point $(X/Z, Y/Z^2)$.